AF674243

Engineering Physics

Engineering Physics

Ashutosh Asthana

Assistant Professor, Department of Physics,
R.R Institute of Modern Technology, Lucknow

An unit of **BSP Books Pvt., Ltd.**

4-4-309, Giriraj Lane, Sultan Bazar,
Hyderabad - 500 095
Phone : 040 - 23445605, 23445688

Published by :

A unit of **BSP Books Pvt., Ltd.**

4-4-309, Giriraj Lane, Sultan Bazar,
Hyderabad - 500 095
Phone : 040 - 23445605, 23445688
e-mail : info@bspbooks.net

ISBN : 978-93-85433-16-0 (HB)

Dedicated to

Tejashvi, Manisha

My Parents and Family Members

"May we be blessed with the wealth of maximum capabilities.
May we be blessed by the sermon of divine knowledge leading to our progress"

Rig Veda

Preface

During the past three decades Engineering has witnessed dramatic changes. In the days gone by, one would think impossible what is possible now. The modern technological advancement has acquired a unique status and reputation in almost all the segments of engineering and physics the base of all.

There is a great relation between Physics and Engineering and most of the technological developments and their advancements have been the result of joint efforts of physicists and engineers. Therefore, the subject physics is very important and indispensable for engineering students and this textbook would serve the purpose very well.

The present Book "Engineering Physics" deals with

UNIT - I	:	Relativistic Mechanics
UNIT - II	:	Interference and Diffraction
UNIT III	:	Polarisation and Lasers
UNIT IV	:	Radiation
UNIT V	:	Wave Mechanics and X-ray Diffraction
UNIT VI	:	Dielectric and Magnetic Properties of Materials; Ultrasonics
UNIT VII	:	Electromagnetism
UNIT VIII	:	Superconductivity and Science and Technology of Nanomaterials
Appendix	:	Fiber Optics

This book covers the syllabi of almost all engineering colleges of India viz., Uttarkhand Technical University (UTU), GBTU, MMTU, JNTU, Rajasthan Technical University, RGPV, Engineering Colleges in South India, NITs etc.

Main features of this book "Engineering Physics" are:

1. The text of this book is simple, coherent and complete
2. All topics in the book are dealt in simple language
3. Illustrations and diagrams are used in effective way to explain difficult topics
4. Mathematical steps are clear, complete and step by step
5. Large number of questions on each topic with previous year examination questions of different universities
6. Solved numerical examples on every topic
7. Large number of numerical problems after every chapter
8. Numerous objective problems as well as short questions, according to new pattern of examination

Though the author makes no claim of orginality, yet he has used his rich experience, gained in teaching the basic course in physics in the presetation and the development of subject matter in the simplest possible manner keeping in view the difficulties and the liminations of students.

Author sincerely requirest the readers to inform the author about their feedback, if any about the course content and its presentation and also enrich the author through their suggestions for the improvement of the book. Author has tried his level best to avoid any errors but their existence cannot be totally denied and request the readers to make author aware of any errors and omissions.

-Author

Acknowledgement

With extreme joy, I express my sincere gratitude to the following Professors for their kind support extended in bringing out this book.

Prof. (Dr.) V.N.P. Srivastava

Ex. Prof. KNIT, Sultanpur

Prof. (Dr.) D.C. Rupainwar

Ex. Prof. IT BHU

Prof. (Dr.) R.S. Prasad

Ex. Prof. KNIT, Sultanpur

It is my privilege and pleasure to express my genuine appreciation and gratitude to (Dr.) S.H. Abidi, BBD, Lucknow for his excellent cooperation and support in bringing out this book.

With great sense of respect, I thank Prof. (Dr.) R.A. Yadav and Prof. (Dr.) S.K. Srivastava for their long and refreshing encouragement. I am very much grateful to them for the clarification and suggestions given to me at various instances during the preparation of this book and also in my career.

The excellent cooperation and support extended by Dr. Vivek Kumar Pandey , University of Allahabad, Dr. Animesh Ojha, MNNIT, Allahabad, Dr. Manoj Bhusan Pandey, Dr. Abhishek Singh, IET, Sitapur, Dr. Anoop Awasthi, IET, Sitapur, Er. Ajay Narayan Shukla, IET, Sitapur, Dr. Manju Singh, Bansal Institute, Lucknow, Dr. Jitendra Kumar Singh, Mrs. Anjana Singh, Mr. Savyasachi, Mr. Anil Agarwal Chairman RRIMT, Er. Jitesh pati tripathi, Dr. G.N Tripathi, Dr. Pradeep Tripathi, Er.S.S Rai, Er.Sameer Srivastava, Er. Vinay shukla, Er. K.P Singh.

Mr. Dhirendra, Mr. Sandeep is appreciated with gratitude for their constant encouragement and keen interest in my writing and publishing this book.

I thank faculties of Physics, Dr. A.K. Singh, Dr.Dharmendra Pal, KIET, Kanpur, Dr. Ram Lakhan, Dr.P.K.Singh, SMS, Lucknow, Dr.Alok Mishra, Ambalika Institute of Management and Technology, Lucknow, Dr. Dixit, Feroz Gandhi Institute, Rai Bareilly, Dr.D.B. Singh, Sherwood College, Dr. Shukla and Dr. Sharma, IIMT, Greater Noida, and many more faculties for their kind support extended in bringing out this book.

I pay my high regards to my ever loving colleagues and well wishers for their immense help at each and every step in completing this text.

I am indebted to many students; I had over the years for their enthusiasm and constructive criticism.

I also acknowledge all those who have in one way or the other contributed to this work. I am indeed grateful to all the authors and publishers of books which have been referred to during the preparation of this text.

I express my sincere thanks to B.S Publications, Hyderabad and their members Mr. Manoj Jha, Mr. Naresh, Mr. Awadesh and also the editorial members who have contributed their best in bringing out this edition within the time.

-Author

Contents

CHAPTER - 2A Interference

CHAPTER - 2B Diffraction

CHAPTER - 3A Polarisation

CHAPTER - 3B Lasers

CHAPTER - 4 Radiation

CHAPTER - 5A Wave Mechanics

CHAPTER - 5B X-Ray Diffraction

CHAPTER - 6A Dielectric Properties of Materials

CHAPTER - 6B Magnetic Properties of Materials

CHAPTER - 6C Ultrasonics

CHAPTER - 8A Superconductivity

CHAPTER - 8B Science and Technology of Nanomaterials

ANNEXURE Fiber Optics

CHAPTER 1

Relativistic Mechanics

Put your hand on a hot stove for a minute, and it seems like an hour. Sit with a pretty girl for an hour, and it seems like a minute. That's relativity.

1.1 Introduction

The theory of relativity deals with the consequences of the lack of a universal frame of reference. Special relativity, which is what Einstein published in 1905, treats problems that involve inertial frame of reference.

General relativity, published by Einstein a decade later treats problems that involve frames of reference accelerated with respect to one another.

In particular, Einstein demonstrated that space and time are entangled; that is the time between two events depends on how far apart they occur, and vice versa. Also, the entanglement is different for observers who move relative to each other. One result is that time does not pass at a fixed rate, as if it were ticked off with mechanical regularity on some master grandfather clock that controls the universe. Rather, that rate is adjustable. Relative motion can change the rate at which time passes.

1.2 Frame of Reference

A frame of reference is just a system of coordinate axes which clearly specifies the position of particle or an event in two or three dimensional space. The one fundamental requirement of a frame of reference is that it must be rigid.

For complete information about an event, we must not only know its true location or position, but also its exact time of occurrence. We must therefore have yet another co-ordinate axes of x, y and z namely that of time t, such a reference frame with four coordinate axes of x, y, z and t is known as space-time frame, its four axes defining a four-dimensional continuum called space-time.

1.2.1 Inertial Frame

Any rigid body situated in a remote space, far from attracting matter and fixed with respect to the 'fixed stars' may be taken to be an inertial frame of reference and naturally, so also any other reference frame moving with constant velocity relative to it.

It is important to note that a frame stationary relative to earth is not inertial because earth, apart from the orbital motion about the sun, has spin motion about its own axis also due to which each body on the surface of earth experiences a centripetal acceleration $\omega^2 R = \left[\frac{2\pi}{24 \times 60 \times 60}\right]^2 \times 6.4 \times 10^6 = 3.4 \times 10^{-2}$ m/sec^2 towards the centre of earth.

The earth serves well enough as an inertial frame of reference and hence also the interior of a vehicle moving smoothly with constant velocity relative to it.

The first law of Newton defines and the second law identifies an inertial frame of reference to be non-accelerating and hence also non-rotating. All other frames of reference moving with a constant velocity with respect to it are referred to as inertial frame of reference.

The importance of an inertial frame of reference lies in that:

- All our fundamental laws of physics have been formulated with respect to such a frame.
- It is a basic assumption of the special theory of relativity that all fundamental physical laws can be expressed as to assume the same form in all frames of reference.
- Such a frame of reference is supposed to be isotropic with respect to mechanical and optical experiment.

1.2.2 Non-inertial Frame

The law of inertia is valid only in an inertial frame. The laws of inertia are not valid for an accelerated frame of reference. The frames, relative to which a body, not acted upon by an external force, is accelerated is called non-inertial frames.

Examples:

- A reference frame with linear acceleration (Pseudo force)
- A reference frame in uniform circular motion (Coriolis force)
- If the lift is stationary or is moving up or down with a constant velocity
- If the lift is moving up with a uniform acceleration
- If the lift is moving down with a uniform acceleration

1.3 Galilean Transformation

Equations relating to the two sets of coordinates of the event in the two frames of reference are called transformation equations. If the two reference frame be inertial and if mass and time be assumed to be independent of any frame of reference, in accordance with Newtonian mechanics, the transformation is said to be Galilean transformation.

If a quantity is characterised by the same number in the two frames of reference, it is said to be invariant of the transformation.

Position

Let s and s^1 be two inertial frames of reference with s at rest and s^1 be moving with a constant velocity v with respect to it in the positive direction of x, where v is very much less than the velocity of light in free space (c).

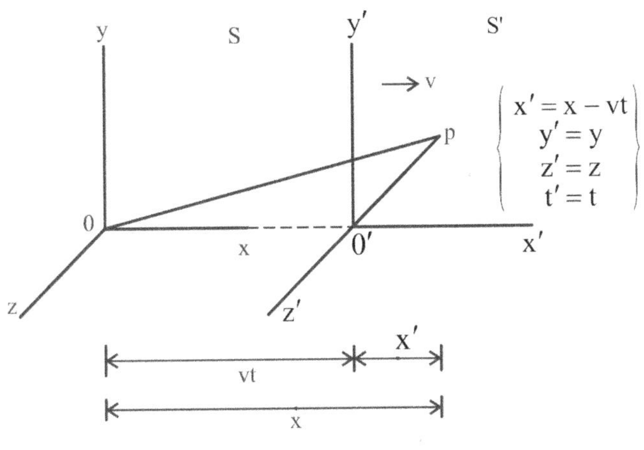

Fig. 1.1

Let the two observers be stationed at the origins of the two frames o and o^1 with their clocks set to zero when o and o^1 coincide. So that time is measured from the instant that the observer just crosses each other.

Let the position of an event at p, as determined by the observer in frame s, after time t be given by the coordinates x, y, z and its position as determined by the observer in frame s^1 by the coordinates x^1, y^1 and z^1 and its time of occurrence by t and t^1.

Since s^1 is moving relative to s with a constant velocity v in the positive direction.

$$\therefore \quad x^1 = x - vt \qquad .....(1.1)$$

Since there is no relative motion between the two frames of reference along the axes of y and z, therefore

$$y^1 = y \qquad .....(1.2)$$

$$z^1 = z \qquad .....(1.3)$$

and since time is supposed to be independent of any frame of reference, therefore

$$t^1 = t \qquad .....(1.4)$$

Thus the four transformation equations for the position of the event as observed in frame s^1 are:

$$\left.\begin{array}{l} x^1 = x - vt \\ y^1 = y \\ z^1 = z \\ t^1 = t \end{array}\right\} \text{Galilean transformation equations}$$

The inverse Galilean transformation equations can be given by replacing v by –v i.e., s^1 is at rest and s is moving with velocity v relative to s^1 in negative x-direction. Therefore

$$\left.\begin{array}{l} x = x' + vt \\ y = y^1 \\ z = z^1 \\ t = t^1 \end{array}\right\} \text{Inverse Galilean transformation equations}$$

Since the position of the event is not the same as measured in the two frame of reference, therefore position is not invariant to Galilean transformation.

Distance or Length

Suppose two events occur at two points p_1 and p_2 some distance apart along the axis of x, such that their coordinated in the stationary frame s are x_1, y_1, z_1 and x_2, y_2, z_2 and in the moving frame s^1 are x_1^1, y_1^1, z_1^1 and x_2^1, y_2^1, z_2^1 respectively.

The distance between the two events is thus

$$(x_2 - x_1) = \Delta x \text{ as measured in frame s}$$

and

$$x_2^1 - x_1^1 = \Delta x^1 \text{ as measured in frame } s^1$$

Now the transformation equation for the event at p_1

$$x_1^1 = x_1 - vt \qquad .....(1.5)$$

$$y_1^1 = y_1 \qquad .....(1.6)$$

$$z_1^1 = z_1 \qquad .....(1.7)$$

Similarly the transformation equation for the event at p_2 are

$$x_2^1 = x_2 - vt \qquad .....(1.8)$$

$$y_2^1 = y_2 \qquad .....(1.9)$$

$$z_2^1 = z_2 \qquad .....(1.10)$$

Clearly, therefore

$$x_2^1 - x_1^1 = \Delta x^1$$

$$\Delta x^1 = (x_2 - vt) - (x_1 - vt)$$

$$\Delta x^1 = (x_2 - x_1)$$

$$\Delta x^1 = \Delta x$$

i.e., the distance between the two events as measured in either frame of reference is the same, irrespective of the values of v and t provided the observations are made simultaneously in the two frames.

Thus the distance or length is invariant to Galilean transformation.

Velocity

The velocity of a particle along the axis of x, as observed in the two frames of reference s and s^1 is given by

$$u = \frac{dx}{dt} \quad .....(1.11)$$

and

$$u^1 = \frac{dx^1}{dt^1} \quad .....(1.12)$$

Now $\quad t^1 = t$ and $x^1 = x - vt$

Therefore

$$u^1 = \frac{dx^1}{dt^1} = \frac{d}{dt}(x - vt)$$

$$u^1 = \frac{dx}{dt} - v$$

$$u^1 = u - v \quad .....(1.13)$$

The inverse transformation is given by

$$u = u^1 + v \quad .....(1.14)$$

Thus the velocities of the particles as observed in the two frames are not the same. Hence, velocity is not invariant to Galilean transformation.

Acceleration

Let the acceleration of a particle along the axis of x, as measured in the two frames of reference s and s^1 be a and a^1 respectively.

Thus,

$$a = \frac{du}{dt} \quad .....(1.15)$$

$$a' = \frac{du'}{dt} \qquad .....(1.16)$$

Since $u' = u - v$

$\therefore$ $a' = \frac{d}{dt}(u - v)$

$$a' = \frac{du}{dt} - 0$$

$\Rightarrow$ $a' = a$

i.e., the acceleration of the particle is the same as measured in either frame of reference thus acceleration is invariant to Galilean transformation.

1.4 Search for the Absolute Frame of Reference: Concept of Ether

In the year 1864, Maxwell had clearly shown the inter-relationship of electricity, magnetism and light. He thus established the existence of electromagnetic waves in space, travelling with the speed of light. Now all the waves known so far e.g. water waves, sound waves etc required a material medium for their propagation. It was thought imperative therefore that there should be a suitable medium for propagation of these new found electromagnetic waves which travelled even through the empty space between the earth and the stars, it come to be called luminiferous ether or ether, and though quite invisible, it was supposed to pervade all space, empty or otherwise.

Then, in order to be able to transmit light a transverse wave motion, it had to be a perfectly transparent rigid solid, and in view of the high velocity of light, it had to have a large shear modulus and yet the earth and the other planets could continue in their regular courses without encountering the slightest resistance from it.

The high sensitivity demand of the apparatus, may be seen from the fact that the velocity of light in free space in 3×10^5 km/sec, the orbital velocity of the earth around the sun and therefore through the ether is only 30 km/sec, which must also be the velocity of the expected ether drift, resulting from it and can be seen that it is a mere.

$$\frac{30 \text{ km/sec}}{3 \times 10^5 \text{ km/sec}} \approx 10^{-4} \text{ of the velocity of light}$$

In view of this Maxwell had actually given up all hope that any terrestrial experiment would be sensitive enough to detect this small ether drift and started looking for any astronomical evidence that might be of help.

A.A. Michelson of the U.S.A, who learnt about the matter in the 1879, accepted the challenge and in collaboration with his colleague E.W. Morley designed and performed the historic experiment "Michelson-Morley experiment" in the year 1887.

1.5 Michelson - Morley Experiment

Michelson Morley experiment is based on interference of light. A parallel beam of monochromatic light from a source S is allowed to fall on an elliptically plane, thin and parallel sided glass plate p_1, thinly silvered on its back face, such that it splits up the incident beam into two separate beams of equal intensity and at right angles to each other at A, a reflected beam upwards and a refracted or transmitted beam in its original direction.

The two beams suffer normal reflections at B and C respectively as they fall on plane mirrors M_1 and M_2 silvered heavily on their front faces to avoid multiple reflections and arranged at right angles to each other at the same distance from A.

Travelling back along their own paths, the two beams again fall on plate P_1, with the former passing through it and later getting reflected at its back surface at A. So as to travel along the same path AE on to the telescope T.

Since the beam AB, suffering reflection at M_1 has to traverse the thickness of plate P_1 thrice and the beam AC, reflected at M_2 only once, their paths through glass are obviously not equal. To make them exactly equal, therefore, an unsilvered compensating plate P_2, identical with plate P_1 in all other respects is arranged parallel to P_2, as shown in Fig. 1.2. So that beam AC has to transverse it twice on its way to and back from, M_2 and is thus made to cover the same total thickness of glass as beam AB.

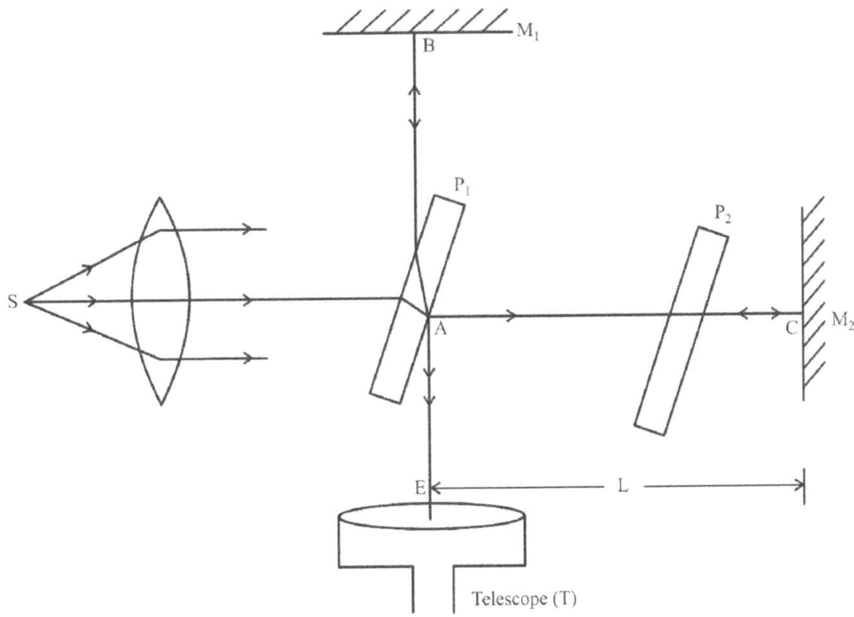

Fig. 1.2

Now if the incident beam falling an plate p_1, be an exactly parallel one and distance AB and AC be exactly equal, the two beams travelling along AB and AC and getting reflected at M_1 and M_2 respectively, arrive at the telescope in the same phase and thus reinforce each other, making the field of view appear bright. But if their paths differ by an odd number of half wavelengths, they arrive at the telescope in apposite phases and thus nullify each other and the field of view appear dark.

Since however the incident beam falling on plate P_1 is almost invariably slightly convergent, some of the rays constituting the two beams arrive at the telescope in phase and some out of phase, so that what we actually see through it is an interference pattern of alternate bright and dark fringes of equal width.

If the mirrors M_1 and M_2 be not exactly perpendicular to each other, take the form of straight and parallel fringes. The apparatus is sensitive enough to be able to detect and measure a shift as small as one hundred of a fringe.

Mathematical Expression

Let the mirrors M_1 and M_2 be at the same distance L from the glass pate P_1. If the apparatus were at rest, the two beams would take the same time to return to P_1.

Let us assume that the earth is moving in space through the ether with a velocity V, therefore the apparatus is moving in space with velocity V.

Suppose C is the velocity of light through the ether. The beam AC moving towards M_2 has a velocity (C – V) relative to the apparatus on the outgoing journey and (C + V) on the return journey.

If T_2 is total time by this beam to go from p_1 to M_2 and back then

$$T_2 = \frac{L}{C-V} + \frac{L}{C+V}$$

$$T_2 = \frac{2LC}{C^2 - V^2}$$

$$T_2 = \frac{2L/C}{1 - V^2/C^2} \quad \text{.....(1.17)}$$

The beam AB that travelled toward mirror M_1 will strike the mirror at M_1^1 but only after P_1 has moved to a new portion P_1^1

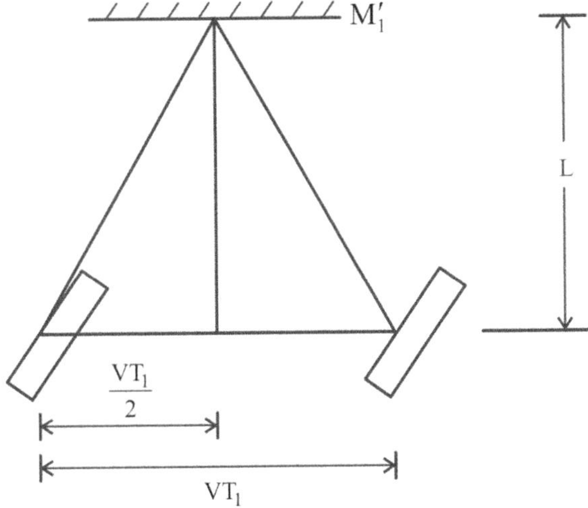

Fig. 1.3

Let the time taken by the beam to move from P_1 be M_1 and back to P_1 be T_1. Then in this time interval the mirror M_1 will move forward through a distance VT_1. The actual distance covered by the light is

$$2\left[L^2+\left(\frac{VT_1}{2}\right)^2\right]^{1/2} = CT_1 \qquad \text{.....(1.8)}$$

On squaring and rearranging we get

$$4L^2 = T_1^2(C^2-V^2) \qquad \text{.....(1.9)}$$

Therefore

$$T_1 = \frac{2L}{\sqrt{C^2-V^2}}$$

$$T_1 = \frac{2L/C}{\sqrt{1-V^2/C^2}} \qquad \text{.....(1.20)}$$

The time difference between two paths due to the motion of the instrument relative to the ether is given by

$$\Delta T = T_2 - T_1 = \frac{2L}{C}\left[\frac{1}{1-V^2/C^2} - \frac{1}{\sqrt{1-V^2/C^2}}\right] \qquad \text{.....(1.21)}$$

Since V is very small compared to C, therefore

$$\frac{1}{1-V^2/C^2} = \left(1-\frac{V^2}{C^2}\right)^{-1} = 1+\frac{V^2}{C^2}$$

Neglecting the higher order terms and

$$\frac{1}{\sqrt{1-V^2/C^2}} = \left(1-\frac{V^2}{C^2}\right)^{-1/2} = 1+\frac{V^2}{2C^2}$$

$$\therefore \qquad \Delta T = \frac{2L}{C}\left[1+\frac{V^2}{C^2}-1-\frac{V^2}{2C^2}\right]$$

$$\Delta T = \frac{LV^2}{C^3} \qquad .....(1.22)$$

If the apparatus is turned through 90°, the role of beam AC and AB is interchanged. Thus

$$T_2{}^1 = \frac{2L/C}{\sqrt{1-V^2/C^2}}$$

and

$$T_1{}^1 = \frac{2L/C}{\left[1-\frac{V^2}{C^2}\right]}$$

$$\Delta T^1 = T_2{}^1 - T_1{}^1$$

$$= \frac{2L}{C}\left[\frac{1}{\sqrt{1-V^2/C^2}} - \frac{1}{1-V^2/C^2}\right]$$

$$\Delta T^1 = -\frac{LV^2}{C^3}$$

Thus

$$\Delta T_{total} = \Delta T - \Delta T^1 = \frac{LV^2}{C^3} - \left(-\frac{LV^2}{C^3}\right)$$

$$= \frac{2LV^2}{C^3} \qquad .....(1.23)$$

The path difference introduced between the components of beam AB and AC will be therefore

$$\delta = C\,(\Delta T_{total}) = C \times \frac{2LV^2}{C^3}$$

$$\delta = \frac{2LV^2}{C^2} \qquad .....(1.24)$$

Since a path difference of one full wavelength λ, results in a shift of the interference pattern through one fringe thus

$$n\lambda = \text{path difference} = \delta$$

$$n\lambda = \frac{2LV^2}{C^2} \quad \text{.....(1.25)}$$

$$n = \frac{2LV^2}{C^2\lambda} \quad \text{.....(1.26)}$$

Michelson and Morley had taken L = 11 m by reflecting the light back and forth several times.

$$\lambda = 6000\ \text{A}^{\circ} \text{ (for visible light)}$$

$$V = 30 \text{ km/s} = 3 \times 10^4 \text{ m/s}$$

$$C = 3 \times 10^8 \text{ m/s}$$

$$n = \frac{2 \times 11 \times (3 \times 10^4 \text{m/sec})^2}{(3 \times 10^8 \text{m/s})^2 \times 6 \times 10^{-7} \text{m}}$$

$$n = 0.37 \approx 0.4$$

n = 2/5 of a fringe and could thus measure even a shift equal to one-hundred of a fringe width. The actual fringe shift observed was negligibly small showing that $(\Delta n \approx 0)$ there was no relative motion between the earth and the ether.

1.5.1 Interpretation of Michelson Morley Experiment

The negative result of the Michelson Morley experiment had two consequences:

1. It rendered untenable the hypothesis of the ether by demonstrating that the ether has no measurable properties.
2. It suggested a new physical principle; the speed of light in free space is the same every where regardless of any motion of source or observer.

1.5.2 Explanation of the Negative Result of Michelson Morley Experiment

1. The result showed that speed of light is same for all observers which is not true for waves which needs a material medium, this is called the principle of constancy of speed of light and is one of the two fundamental postulates of Einstein's special theory of relativity.

2. Lorentz and Fitzerland gave an explanation of the negative result of Michelson-Morley experiment. According to them the size of a moving body is altered due to its motion through stationary ether. They showed that the length of path in the direction of ether flow will be shortened by

$$L\sqrt{1-\frac{V^2}{C^2}} \text{ instead of } L$$

Replacing L by $L\sqrt{1-\frac{V^2}{C^2}}$ in the equation

$$T_2 = \frac{2\frac{L}{C}}{1-\frac{V^2}{C^2}}$$

$$T_2 = \frac{2L\sqrt{1-\frac{V^2}{C^2}}}{C(1-\frac{V^2}{C^2})} = \frac{2L}{C\sqrt{1-\frac{V^2}{C^2}}}$$

$$= T_1$$

Hence $T_2 = T_1$

3. The moving earth drags the ether with it. Hence there is no relative motion between the two so that no shift is observed.

1.6 Einstein's Postulates of Special Theory of Relativity

The following are the two basic postulates of the theory:

1. All the laws of physics have the same identical form in all inertial frames.
2. The velocity of light in free space is the same in all inertial frame i.e. invariant to transformation from one inertial frame to another. It has thus same value 3×10^8 m/sec for all observers irrespective of their state of motion.

1.7 Lorentz Transformation of Space and Time

Galilean transformation is at variance with the constancy of the velocity of light in free space and thus quite independent of the motion of its own source or that of the observer.

Let us consider that if a flash of light is produced at the origin O of a frame S just when the origin O^1 of the frame S^1 crosses it with relative velocity V along the positive x direction, each of the two observes, at O and O' respectively considers himself to be at the centre of the expanding spherical wave front. So that the equation of the wave front for the observer at O in frame S is

$$x^2 + y^2 + z^2 = c^2t^2 \quad .....(1.27)$$

and equation of the wave front for the observer at O^1 in frame S^1 is

$$x^{1^2} + y^{1^2} + z^{1^2} = c^2t^{1^2} \quad .....(1.28)$$

the velocity of light being same in either frame.

In order that both the observers may seems to lie at the centre of the same expanding spherical wave front, we must clearly have equation (1.27) equal to (1.28)

Now in accordance with Galilean transformation

$$\left.\begin{aligned} x^1 &= x - vt \\ y^1 &= y \\ z^1 &= z \end{aligned}\right\} \quad .....(1.29)$$

$$t^1 = t$$

substituting these values in equation (1.28) we have

$$x^2 - 2\,xvt + v^2t^2 + y^2 + z^2 = c^2t^2 \quad .....(1.30)$$

which is clearly not the same as equation (1.27) showing once again the failure of Galilean transformation in the face of constancy of the value of C, the velocity of light in free space.

Derivation of Lorentz Transformation Equation

A transformation that changes all four co-ordinates three space and one time x, y, z, t into x^1, y^1, z^1, t^1 in such a way that the condition, that is the speed of light is constant in all unaccelerated coordinate system is satisfied. Such a transformation set was first obtained by Lorentz and is called the Lorentz transformation equations. These equations are also referred as the relativistic space time transformations. These are based on two postulates of the special theory of relativity.

Let us consider two observers O and O^1 in system s and S^1. System S^1 is moving with a constant velocity $\vec{v}$ relative to system S. Observes O and O^1 are at the origin of the coordinate systems S and S^1 respectively.

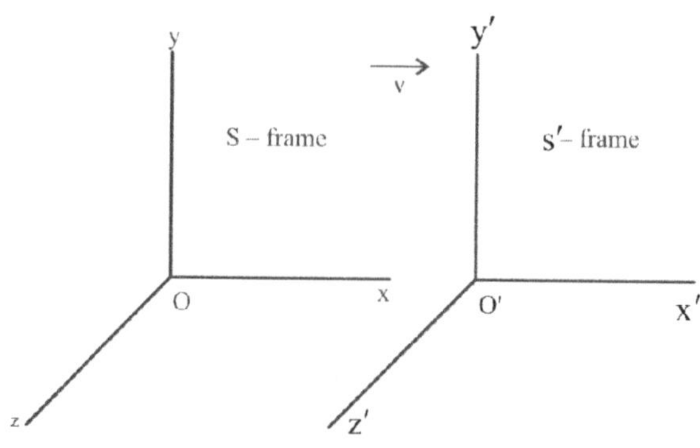

Fig. 1.4

Now we assume that a measurement in positive x direction made in frame S is proportional to that made in frames, that is, the simplest relation suggests itself that

$$x^1 = \gamma (x - vt) \qquad .....(1.31)$$

where γ is a constant of proportionality, independent of both x and t, though possibly dependent on v.

Using First Postulate

As the laws of physics are same in both frames S and S^1 therefore the corresponding equation of x in terms of x^1 and t^1 will be given by

$$x = \gamma (x^1 + vt^1) \qquad .....(1.32)$$

The constant of proportionality (γ) remaining the same, because the first postulate denies the existence of any preferred or absolute frame, and the motion of frame S' being along the positive x direction we have

$$y^1 = y \text{ and } z^1 = z$$

But clearly $t^1 \neq t$ or else therefore (1.31) and (1.32) cannot both be true.

The value of t^1 may be obtained by putting x^1 from (1.31) in (1.32)

$$x = \gamma [\gamma (x - vt) + vt^1]$$

$$x = \gamma^2 [(x - vt)] + \gamma vt^1$$

$$\gamma vt^1 = x - \gamma^2 (x - vt)$$

$$\gamma vt^1 = x - \gamma^2 x + \gamma^2 vt$$

$$\gamma vt^1 = \gamma^2 vt + x(1 - \gamma^2)$$

$$\therefore \quad t^1 = \gamma t + \frac{x(1-\gamma^2)}{\gamma v} \qquad .....(1.33)$$

Using Second Postulate

Let a light signal is given at the origin O at time t = 0, $t^1 = 0$ that is, when O and O^1 coincide. The signal travels with a speed c which is same in both the frames.

x – coordinate of the light pulse in frame S is given by

$$x = ct \qquad .....(1.34)$$

x^1 – coordinate of the light pulse in frame S^1 is given by

$$x^1 = ct^1 \qquad .....(1.35)$$

Putting the value of x^1 and t^1 in (1.35) we get

$$\gamma\,(x - vt) = c\left[\gamma t + \frac{x(1-\gamma^2)}{\gamma v}\right] \qquad .....(1.36)$$

$$\gamma x - \gamma vt = c\gamma t + \frac{cx}{\gamma v}(1-\gamma^2)$$

$$\gamma x - \frac{cx}{\gamma v}(1-\gamma^2) = c\gamma t + \gamma vt$$

$$x\left[\gamma - \frac{c}{\gamma v}(1-\gamma^2)\right] = c\gamma t\left[1+\frac{v}{c}\right]$$

$$x\gamma\left[1-\frac{c}{v}\left(\frac{1}{\gamma^2}-1\right)\right] = c\gamma t\left[1+\frac{v}{c}\right]$$

$$x = \frac{ct\left(1+\frac{v}{c}\right)}{\left[1-\frac{c}{v}\left(\frac{1}{\gamma^2}-1\right)\right]} \qquad .....(1.37)$$

on substituting ct for x from relation (1.34) we get

$$1 = \frac{1+\frac{v}{c}}{1-\frac{c}{v}\left(\frac{1}{\gamma^2}-1\right)}$$

$$1+\frac{v}{c} = 1-\frac{c}{v}\left(\frac{1}{\gamma^2}-1\right)$$

$$\frac{v}{c} = -\frac{c}{v}\left(\frac{1}{\gamma^2}-1\right)$$

$$-\frac{v^2}{c^2} = \frac{1}{\gamma^2}-1 \qquad \text{.....(1.38)}$$

$$1-\frac{v^2}{c^2} = \frac{1}{\gamma^2}$$

$$\therefore \qquad \gamma = \frac{1}{\sqrt{1-\frac{v^2}{c^2}}} \qquad \text{.....(1.39)}$$

Since $\frac{v}{c}$ is usually represented by β

$$\therefore \qquad \gamma = \frac{1}{\sqrt{1-\beta^2}} \qquad \text{.....(1.40)}$$

Now if we substitute this value of γ in (1.31) we get

$$x^1 = \frac{x-vt}{\sqrt{1-\frac{v^2}{c^2}}} = \frac{x-vt}{\sqrt{1-\beta^2}} \qquad \text{.....(1.41)}$$

Since frame S^1 moves along the positive x direction relative to frame S, we have

$$y^1 = y \text{ and } z^1 = z$$

∴ Again from equation 1.33 we have

$$t^1 = \gamma t + \frac{x(1-\gamma^2)}{\gamma v}$$

$$t^1 = \gamma\left[t+\frac{x(1-\gamma^2)}{\gamma^2 v}\right] \qquad \text{.....(1.42)}$$

From (1.38) we have

$$\left(\frac{1}{\gamma^2}-1\right) = -\frac{v^2}{c^2}$$

$$\frac{1-\gamma^2}{\gamma^2} = -\frac{v^2}{c^2}$$

Thus $$t^1 = \gamma\left[t - \frac{x}{v}\left(\frac{v^2}{c^2}\right)\right]$$

$$t^1 = \gamma\left[t - \frac{vx}{c^2}\right]$$

$$t^1 = \frac{t - \frac{vx}{c^2}}{\sqrt{1 - \frac{v^2}{c^2}}} \quad .....(1.43)$$

Thus the Lorentz transformation equations are

$$x^1 = \frac{x - vt}{\sqrt{1 - \frac{v^2}{c^2}}} \quad .....(1.44)$$

$$y^1 = y \quad .....(1.45)$$

$$z^1 = z \quad .....(1.46)$$

$$t^1 = \frac{t - \frac{vx}{c^2}}{\sqrt{1 - \frac{v^2}{c^2}}} \quad .....(1.47)$$

its inverse are $$x = \frac{x^1 + vt}{\sqrt{1 - \frac{v^2}{c^2}}} \quad .....(1.48)$$

$$y = y^1 \quad .....(1.49)$$

$$z = z^1 \quad .....(1.50)$$

and $$t = \frac{t^1 + \frac{vx}{c^2}}{\sqrt{1 - \frac{v^2}{c^2}}} \quad .....(1.51)$$

These transformation equations are referred to as Lorentz transformations after the Dutch Physicist H.A. Lorentz because he was the first to suggest that this was the only possible transformation under which the laws of electricity and Magnetism assume the same forms in all inertial frames.

Note:

1. Since $\frac{v}{c}$ is usually small, only the first two terms of the Binomial expansion of

$$\frac{1}{\sqrt{1-\frac{v^2}{c^2}}} \text{ or } \left(1-\frac{v^2}{c^2}\right)^{-\frac{1}{2}}$$ are retained and γ therefore taken to be equal to $\left(1+\frac{1}{2}\frac{v^2}{c^2}\right)$

this approximation being quite valid up to a value of $\frac{v}{c}$ = 0.5. This discussion will make it ample clear that measurements in space and time are no longer absolute but depend upon the relative motion between the observer and the phenomenon observed.

2. It must be carefully noted, however that if $v << c$ then $\frac{v}{c} \to 0$ we have

$$\gamma = \frac{1}{\sqrt{1-\frac{v^2}{c^2}}} = 1$$

and all Lorentz transformations reduces to corresponding Galilean transformation showing that the classical or Newtonian physics is just a particular case of relativistic physics. Since the orbital velocity of the earth v is only about 10^{-4} of the value of c, therefore $\frac{v}{c} \approx 0$. So that the classical physics is perfectly valid for most of our everyday purposes except of course, when we are dealing with particles like electrons, protons, mesons etc which move with velocities comparable with c.

1.8 Transformation of Length: Length Contraction

Let L_0 be the length of the rod laid along the axis of x in a reference frame s at rest with respect to the observer this length being referred to its proper length. Then if the coordinates of the two end points of the rod be x_1 and x_2 clearly we have

$$\text{Proper length of rod } L_0 = x_2 - x_1$$

If the coordinates of the end-points of the rod in reference frame S', moving with a uniform relative velocity v with respect to frame S, be x_1^1 and x_2^1 as noted simultaneously at the same instant, length of the rod in the moving frame S^1 we have

$$L = x_2^1 - x_1^1$$

Now in accordance with inverse Lorentz transformation

$$x_1 = \frac{x_1^1 + vt^1}{\sqrt{1-\frac{v^2}{c^2}}} \qquad \text{and} \qquad x_2 = \frac{x_2^1 + vt^1}{\sqrt{1-\frac{v^2}{c^2}}}$$

So that

$$L_0 = x_2 - x_1$$

$$L_0 = \frac{x_2^1 + vt^1}{\sqrt{1-\frac{v^2}{c^2}}} - \frac{x_1^1 + vt^1}{\sqrt{1-\frac{v^2}{c^2}}}$$

$$L_0 = \frac{x_2^1 - x_1^1}{\sqrt{1-\frac{v^2}{c^2}}} = \frac{L}{\sqrt{1-\frac{v^2}{c^2}}}$$

$$L_0 = \frac{L}{\sqrt{1-\frac{v^2}{c^2}}}$$

Hence

$$\boxed{L = L_0\sqrt{1-\frac{v^2}{c^2}}}$$

Showing that the length of the rod in the moving frame s^1 is shorter than its proper length L_0.

Note:

1. The faster the motion of the reference frame S^1 carrying the rod the shorter is its length, compared with its proper length L_0 and if the velocity of the rod $v = c$ i.e. if the rod is moving with the velocity of light in free space, its length L would be zero.

   It is thus clear that its proper length when its velocity $v = 0$ is the largest and its length is shortest when it moves with the velocity of light i.e. when $v = c$.

2. There will be no shortening of the rod if it lies along the axis of y and z perpendicular to the axis of x or the direction of motion of the reference frame S^1. If however, the rod is lying inclined to the axis of x, its length will have a component along the axis or the direction of motion of the reference frame and there will be some slight shortening of its length.

3. This shortening or contraction in the length of material object along its direction of motion is known as Lorentz-Fitzgerald contraction.

4. The contraction in length is a reciprocal effect and varies both ways i.e. independent of the direction of velocity but depends only on its magnitude. Thus a length along the axis of x in frame S appears contracted by a fraction $\sqrt{1-\frac{v^2}{c^2}}$ to an observer in frame S^1 and a length along the axis of x^1 in frame S^1 appears equally contacted to an observer in frame S.

   We may therefore say that with relative motion between a material object and an observer, the length of the object as measured by the observer is always shorter than its proper length.
5. A square and a circle in frame S will appear to be a rectangular and an ellipse respectively to an observer in frame S^1.
6. In actual practice we hardly came across speed v comparable with that of light in free space (c) so that $\frac{v}{c} \approx 0$ and $\sqrt{1-\frac{v^2}{c^2}} \approx 1$ and therefore L = L_0 i.e. the Lorentz Fitzgerald contraction in nil and the length of a moving object is the same as its proper length or its length, when at rest with respect to the observer. In other words, the length in all such cases remains an absolute quantity in accordance with classical physics.

1.9 Transformation of Time: Time Dilation

Let us consider that we have reference frame S at rest and a reference frame S' moving with velocity with respect to it along the positive direction of x with velocity v. Suppose that the two reference frames carry identical clocks at their origins O and O^1 such that when frame S^1 just crosses s i.e., when O^1 lies just opposite O, the two clocks show zero time.

Let two events occur at a point x^1 in frame s^1 at time t_1^1 and t_2^1 as noted on the clock carried by it and at times t_1 and t_2 as noted on the clock carried by frame S. Then clearly time-interval between the two events as noted on the clock in S say

$$\Delta t = t_2 - t_1$$

By inverse Lorentz transformation equation we have

$$t_1 = \frac{t_1^1 + \frac{vx^1}{c^2}}{\sqrt{1-\frac{v^2}{c^2}}}$$

$$t_2 = \frac{t_2^1 + \frac{vx^1}{c^2}}{\sqrt{1-\frac{v^2}{c^2}}}$$

$\therefore$
$$\Delta t = t_2 - t_1$$

$$\Delta t = \frac{t_2^1 + \frac{vx^1}{c^2}}{\sqrt{1-\frac{v^2}{c^2}}} - \frac{t_1^1 + \frac{vx^1}{c^2}}{\sqrt{1-\frac{v^2}{c^2}}}$$

$$\Delta t = \frac{t_2^1 - t_1^1}{\sqrt{1-\frac{v^2}{c^2}}}$$

$$\Delta t = \frac{\Delta t^1}{\sqrt{1-\frac{v^2}{c^2}}} = \gamma \Delta t^1$$

Since $\frac{1}{\sqrt{1-\frac{v^2}{c^2}}}$ or γ is greater than 1

Thus the time interval (Δt^1) at a given point in the moving frame (s^1) appears to be longer or dilated by a factor $\frac{1}{\sqrt{1-\frac{v^2}{c^2}}}$ or γ to the observer in the stationary frame.

Thus a moving clock appears to be slowed down to a stationary observer.

1. If v = c, then $\frac{v^2}{c^2} = 1$ this means that $\Delta t = \infty$ i.e. the clock moving with the speed of light will appear to be completely stopped to a stationary observer.

2. If v << c, then

$$\frac{1}{\sqrt{1-\frac{v^2}{c^2}}} = \left(1-\frac{v^2}{c^2}\right)^{-\frac{1}{2}}$$

$$= 1 + \frac{1}{2}\frac{v^2}{c^2}$$

In this $\frac{v^2}{c^2}$ is very small and can be neglected.

$\therefore$ $$\Delta t = \Delta t^1$$

i.e. if the clock is moving with the speed very-very smaller than speed of light, then time interval will remain same for moving observer.

3. Time dilation is real effect
4. Time dilation is a reciprocal effect.

1.9.1 Time Dilation is a real effect

Cosmic ray particles called mesons, μ-mesons are created at high altitudes in the earth atmosphere ≈ 10 km, by the fast cosmic ray photons and are projected towards the earth surface with a very high speed of about 2.994×10^8 m/sec, which is 0.998C. μ-mesons are unstable and decay into electrons or positrons with an average life time of about 2.0×10^{-6} sec. Hence in its life time a μ-meson can travel a distance

$$d = vt = 2.99 \times 10^8 \text{ m/sec} \times 2.0 \times 10^{-6} = 600 \text{ m}$$

But the question is that how μ-mesons travel a distance of 10 km to reach the earth surface. This is possible only because of time dilation.

In its own frame of reference, μ-mesons have an average life time

$$t^1 = 2.0 \times 10^{-6} \text{ sec}$$

In observer's frame of reference

$$t = \frac{t^1}{\sqrt{1 - \frac{v^2}{c^2}}}$$

$$t = \frac{2.0 \times 10^{-6} \text{ sec}}{\sqrt{1 - (0.998)^2}}$$

$$t = \frac{2.0 \times 10^{-6} \text{ sec}}{0.063}$$

$$t = 3.17 \times 10^{-5} \text{ sec}$$

In this dilated life time, a μ-meson can travel a distance

$$d_0 = 2.994 \times 10^8 \text{ m/sec} \times 3.17 \times 10^{-5} \text{ sec}$$

$$d_0 = 10.0 \text{ km}$$

This explains the presence of μ-mesons on the earth surface. Hence we can say that time dilation is a real effect.

1.9.2 Twin Paradox (clock paradox)

It follows as a natural consequence, therefore that the passage of time and hence also all physiological process, that go with it, like pulse and heart-beats, the process of ageing itself gets slowed down in a fast moving reference frame and if the later be moving with the velocity of light, the passage of time and hence also the process of ageing will stop altogether.

This explains what is known as the 'Twin Paradox' namely that if one of the two identical twins stays back on the earth, while the other goes about in space in a high speed rocket, the later on returning to the earth will be found to be younger than the former obviously because his ageing process got slowed down during the time that is spent in the high speed rocket.

Note:

1. $\Delta t_{space} = \Delta t_{earth}\sqrt{1-V^2/C^2}$
2. In our every day life we seldom come across speeds comparable with that of light. So that $\frac{v}{c} \to 0$ and $\frac{1}{\sqrt{1-\frac{v^2}{c^2}}} \approx 1$ and therefore $\Delta t = \Delta t^1$. This clearly implies that for most of our purposes, the time interval between the two events as measured in the stationary and moving frames is the same.

1.10 The simultaneity in Relativity

The simultaneity of two events means their occurrence at exactly the same time. Let us see whether two events occurring simultaneously is a stationary frame S also appear to be so in a reference frame S^1, moving with respect to it with v, velocity v along the positive direction of x.

Let A and B be two points along the axis of x in the stationary frame S, distance x_1 and x_2 from the origin, with P as their mid point. Then a flash of light emitted at P will arrive at A and B at exactly the same time t, because of speed of light being the same in all directions. The two events i.e. the arrival of the light signals thus occur simulataneously at A and B in frame S.

To an observer in the moving frame S^1, the corresponding values of t for the arrival of the signals at A and B will in accordance with Lorentz transformation appear to be

$$t_1^1 = \gamma\left(t - \frac{vx_1}{c^2}\right)$$

$$t_2^1 = \gamma\left(t - \frac{vx_2}{c^2}\right)$$

clearly $x_1 \neq x_2$ and therefore $t_1^1 \neq t_2^1$

The two events will not appear to be simultaneous to the observer in S^1. In fact the time-interval between the two events will appear to be

$$\Delta t' = t_2^1 - t_1^1 = \frac{\gamma v (x_1 - x_2)}{c^2}$$

which clearly depends upon the distance between the two events and the velocity of the moving frame with respect to the stationary frame.

In exactly the same manner, the two events occurring simultaneously in the moving frame S^1 will not appear to be so to an observer in the stationary frame S.

Thus, the simultaneity of event, separated in space, is not absolute in frame of reference in relative motion with respect to each other. This is known as relativity of simultaneity and is clearly a consequence of the invariance of the velocity of light (c) in all frames of reference and in all directions.

1.11 Transformation of Velocity

Suppose a frame of reference S' is moving with velocity v relative to a frame of reference S at rest, only the positive x direction and that a particle is moving with velocity u in frame S, with components.

$$u_x = \frac{dx}{dt}$$

$$u_y = \frac{dy}{dt}$$

$$u_z = \frac{dz}{dt}$$

along the three coordinate axes. If the velocity of the particle as measured in frame S', be u^1 and its components along the three coordinate axes, u_x^1, u_y^1, u_z^1, we have in accordance with Lorentz transformation equation.

$$x^1 = \gamma (x - vt)$$

$$y^1 = y$$

$$z^1 = z$$

and $$t^1 = \gamma\left(t - \frac{vx}{c^2}\right)$$

so that

$$u_x^1 = \frac{dx^1}{dt^1} = \frac{d\,\gamma\,(x - vt)}{d\,\gamma\left(t - \frac{vx}{c^2}\right)}$$

$$= \frac{dx - vdt}{dt - \frac{vdx}{c^2}}$$

$$= \frac{\frac{dx}{dt} - v}{1 - \left(\frac{v}{c^2}\right)\left(\frac{dx}{dt}\right)}$$

$$u_x^1 = \frac{u_x - v}{1 - \left(\frac{v}{c^2}\right)(u_x)}$$

and $$u_y{}^1 = \frac{dy^1}{dt^1} = \frac{dy}{\gamma\left(dt - \frac{v}{c^2}dx\right)}$$

$$u_y^1 = \frac{dy / dt}{\gamma\left[1 - \left(\frac{v}{c^2}\right)\left(\frac{dx}{dt}\right)\right]}$$

$$u_y^1 = \frac{u_y}{\gamma\left[1 - u_x\left(\frac{v}{c^2}\right)\right]}$$

similarly

$$u_z^1 = \frac{u_z}{\gamma\left[1 - u_x\left(\frac{v}{c^2}\right)\right]}$$

If however velocity u of the particle in frame S be itself along the axis of x, the components u_y and u_z in frame S and also components u_y^1 and u_z^1 in frame S^1 are each equal to zero, with

$u_x = u$ and $u_x^1 = u^1$ then we have

velocity of the particle in frame S^1 i.e.

$$u^1 = \frac{u - v}{1 - \frac{uv}{c^2}}$$

Note:

1. It is thus clear that as in the case of Galilean transformation. So also under Lorentz transformation velocity is not invariant in the reference frame.

   The inverse Loretnz transformation for velocities

$$u_x = \frac{u_x^1 + v}{1 + \frac{u_x^1 v}{c^2}} \quad \text{and} \quad u_z = \frac{u_z^1}{\gamma\left(1 + \frac{u_x^1 v}{c^2}\right)}$$

$$u_y = \frac{u_y^1}{\gamma\left(1 + \frac{v u_x^1}{c^2}\right)}$$

and if u^1 be the velocity along the axis of x^1 then

$$u = \frac{u^1 + v}{1 + \frac{u^1 v}{c^2}}$$

the components u_y^1 and u_z^1 in frame S^1 and hence also components u_y and u_z in frame S being zero.

1.11.1 Law for the Addition of Relativistic Velocities

If a particle has a velocity u^1 in a frame of reference S^1, which is moving with velocity v with respect to another frame of reference S, the velocity of the particle in the later frame S is given by

$$u = \frac{u^1 + v}{1 + \frac{u^1 v}{c^2}}$$

It may be pointed out that:

1. If $u_x^1 << c$ then $\frac{u_x^1}{c} \to 0$ therefore all the above expression reduces to that obtained by Galilean transformation.

$$u_x = u_x^1 + v$$

$$u_y = u_y^1$$

$$u_z = u_z^1$$

2. If $u_x^1 = c$ or the velocity of the particle along the axis of x^1 in frame s^1 be the same as the velocity of light i.e. if the particle be a photon, its velocity in frame s is

$$u_x = \frac{c+v}{1+\frac{vc}{c^2}} = \frac{c+v}{\frac{c+v}{c}} = c$$

Thus, only this velocity c is invariant in the two frames of reference.

1.12 Relativity of Mass

We have seen that length and time are not absolute quantities, but depend upon the frame of reference from which they are observed. Relativistic idea of space and time conclude that the mass of a body is not invariable property under all circumstance. In fact, mass is a function of the velocity of the body. It increases with increasing velocity represented by the relation.

$$m = \frac{m_o}{\sqrt{1-\frac{v^2}{c^2}}} \qquad \text{....(1.52)}$$

where m_o is the rest mass of the body, c is the speed of light and v is the velocity of the body.

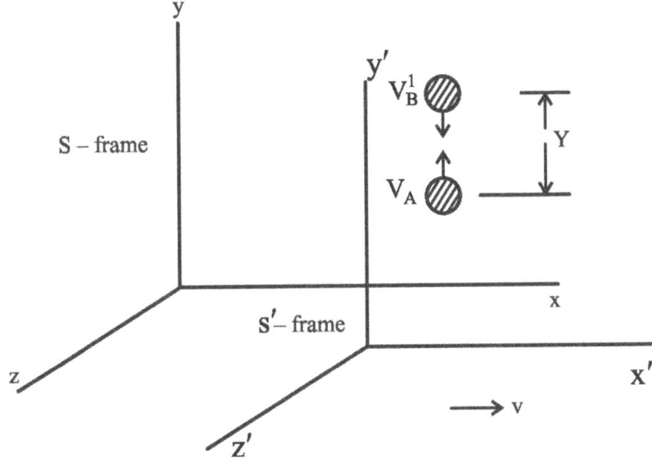

Fig. 1.5

Let us consider an elastic collision between two exactly identical perfectly elastic spherical particles A and B. The frames S and S^1 are oriented as shown in figure, with S^1 moving in the positive x direction with respect to S at the velocity $\vec{V}$.

Before the collision, the particles A had been at rest in frame S and particle B in frame S^1. Then at the same instant, A was thrown in the positive y direction at the speed V_A while B was thrown in the negative y direction at the speed V_B^1 where

$$V_A = V_B^1 \qquad \text{.....(1.53)}$$

Therefore, the behaviour of A as seen from S is exactly same as the behaviour of B as seen from S^1.

When A and B collide, particle A rebounds in negative Y direction at the speed V_A while particle B rebounds in positive Y direction at the speed V_B^1.

Collision occurs at the mid point of the total distance between then i.e. at

$$y = \frac{1}{2} Y$$

for S^1

$$y^1 = y = \frac{1}{2} Y$$

Therefore, round trip T_o for A as measured in frame S, is therefore

$$T_o = \frac{Y}{v_A} \qquad \text{.....(1.54)}$$

If linear momentum is conserved in the S frame it must be true that

$$m_A v_A = m_B v_B \qquad \text{.....(1.55)}$$

where v_A and v_B are velocities as measured in the S-frame

In S the speed v_B is given by

$$v_B = \frac{Y}{T} \qquad \text{.....(1.56)}$$

In S^1, B_s^1 trip requires the time T_o, where

$$T = \frac{T_o}{\sqrt{1 - \frac{v^2}{c^2}}} \qquad \text{.....(1.57)}$$

Equation (1.56) becomes

$$V_B = \frac{V_1}{T_0} \sqrt{1 - \frac{V^2}{C^2}}$$

Equation (1.56) becomes

$$m_A \frac{V_1}{T_0} = m_0 \frac{V_1}{T_0} \sqrt{1-\frac{V^2}{C^2}}$$

$$m_A = m_B \sqrt{1-\frac{V^2}{C^2}} \qquad \text{.....(1.58)}$$

If U_A and $V_B^1 << v$. In this case an observer in S will see B approach A with the velocity v, making a glancing collision

$$m_A = m_o \text{ (rest mass)}$$

$$m_B = m$$

∴ Relativistic mass

$$m = \frac{m_o}{\sqrt{1-\frac{v^2}{c^2}}}$$

This is the formula for the variation of mass with velocity.

Note:

1. As the velocity v of the particle relative to the observer increases, the mass of the particle increases.
2. As $v \to c$, $m \to \infty$, that is no material particle can have velocity equal to or greater than the velocity of light.
3. When $v << c$, then $\frac{v^2}{c^2}$ is very small and may be neglected compared to 1, i.e.

   $m \simeq m_o$ (mass of the particle is equal to its rest mass)
4. The plot of $\frac{m}{m_o}$ versus $\frac{v}{c}$ is shown in figure.

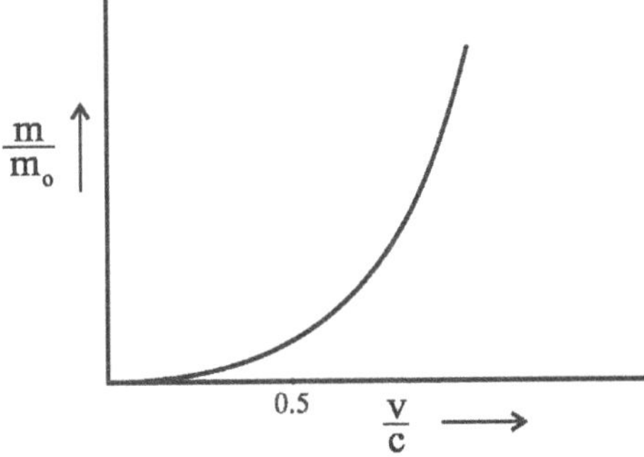

1.13 Mass-Energy Equivalence

The variation of mass with velocity has modified the idea of energy, so that a relationship can be established between mass and energy. The relationship can be derived directly from the definition of the kinetic energy of a moving body i.e. work done in bringing the body from rest to its present state of motion.

Work done $$W = F S$$

$$K.E = F.S$$

$$K.E = \int_{o}^{s} F\, ds$$

In non-relativistic physics, the kinetic energy

$$K.E = \frac{1}{2} m_o v^2$$

In relativistic physics, the kinetic energy

$$dE_k = F.\, ds$$

$$dE_k = \frac{d}{dt}(mv).\, ds$$

$$dE_k = vd\,(mv)$$

Since $$\frac{ds}{dt} = v$$

Where ds is the small change in displacement

$$dE_k = v^2 dm + mvdv$$

Also $$m = \frac{m_o}{\sqrt{1 - \frac{v^2}{c^2}}}$$

$\therefore$ $$m^2\left(1 - \frac{v^2}{c^2}\right) = m_o^2$$

$$m^2 (c^2 - v^2) = m_o^2 c^2$$

$$m^2c^2 - m^2v^2 = m_o^2 c^2$$

$$m^2c^2 = (m^2v^2 + m_o^2 c^2) \qquad .....(1.58)$$

Differentiating this equation we get

$$2m\,c^2\,dm = 2m\,v^2\,dm + 2v\,m^2dv$$

$$c^2dm = v^2dm + mvdv \qquad .....(1.59)$$

comparing equation

$$dE_k = v^2dm + mvdv$$

with equation (1.59) we get

$$dE_k = c^2dm$$

$$E_k = \int_0^{E_k} dE_k \qquad \text{since } E_k = 0 \text{ when } v = 0$$

$$E_k = \int_0^{E_k} dE_k = c^2 \int_{m_0}^{m} dm$$

$$E_k = c^2\,(m - m_o)$$

$$\text{Kinetic energy} = mc^2 - m_oc^2$$

This is the relativistic expression for the kinetic energy of a particle. From this expression it is clear that the increase in kinetic energy is due to the increase in mass of the particle on account of its relative motion and is equal to the product of gain in mass and square of the velocity of light.

The total energy of a moving particle is the sum of kinetic energy of motion and rest energy that is total energy

$$E = \text{rest energy} + \text{relativistic K.E}$$

$$E = m_oc^2 + (m - m_o)\,c^2$$

$$E = mc^2 \qquad .....(1.60)$$

This is well known Einstein mass energy relation

$$E = E_o + K.E \qquad .....(1.61)$$

1.14 Kinetic Energy at Low Speeds

For $v << c$ then the equation

$$K.E = mc^2 - m_oc^2$$

reduces to classical expression $\frac{1}{2}\,m_ov^2$

$$K.E = mc^2 - m_o c^2$$

$$K.E = \frac{m_o c^2}{\sqrt{1-\frac{v^2}{c^2}}} - m_o c^2$$

$$K.E = m_o c^2 \left(1+\frac{v^2}{2c^2}\right) - m_o c^2$$

$$K.E = \frac{1}{2} m_o v^2 \qquad \text{for } v << C$$

1.15 Relation between Total Energy and Relativistic Momentum

We have

$$E = mc^2$$

We can write the above equation as

$$E = \frac{m_o c^2}{\sqrt{1-\frac{v^2}{c^2}}} \qquad \text{.....(1.62)}$$

The momentum P is given by

$$P = mv = \frac{m_o v}{\sqrt{1-\frac{v^2}{c^2}}} \qquad \text{.....(1.63)}$$

Form (1.62)

$$E^2 = \frac{m_o^2 C^4}{\left(1-\frac{v^2}{C^2}\right)} \qquad \text{.....(1.64)}$$

From (1.63)

$$P^2 = \frac{m_o^2 v^2}{\sqrt{1-\frac{v^2}{c^2}}} \qquad \text{.....(1.65)}$$

From (1.64) and (1.65) we have

$$E^2 - P^2c^2 = \frac{m_o^2 c^4 - m_o^2 c^2 v^2}{\left(1-\frac{v^2}{C^2}\right)}$$

$$E^2 - P^2c^2 = \frac{m_o^2c^4 - \left(1 - \frac{v^2}{C^2}\right)}{\left(1 - v^2/C^2\right)}$$

$$E^2 - P^2c^2 = m_o^2 c^4$$

$$\boxed{E^2 = P^2c^2 + m_o^2 c^4} \qquad .....(1.66)$$

Note:

1. $E^2 - P^2c^2$ is invariant under Lorentz transformation because m_o and c both are constant
2. For v < < c

$$E = \frac{P^2}{2m_o}$$

3. For mass less particles like photon

$$m_o = 0$$

then from

$$E^2 = P^2 c^2 + m_0^2 c^4$$

We have

$$E^2 = P^2c^2$$

$$E = Pc$$

Solved Problems

On Michelson Morley Experiment

1. What will be the fringe-shift according to the ether theory in the Michelson-Morley experiment if the effective path length of each path is 7 meter and the wavelength of light is 7000 Å ? The velocity of earth is 3×10^4 m/s.

Sol. The expected fringe shift Δn, according to the ether theory in Michelson-Morley experiment is given by

$$\Delta n = \frac{2 l v^2}{c^2 \lambda}$$

Given,

$$l = 7 \text{ m}$$

$$v = 3 \times 10^4 \text{ m/s}$$

$$c = 3 \times 10^8 \text{ m/s}$$

$$\lambda = 7000 \text{ Å} = 7 \times 10^{-7} \text{ m}$$

$$\therefore \qquad \Delta n = \frac{2 \times (7\text{m}) \times (3 \times 10^4 \text{m/s})^2}{(3 \times 10^8 \text{m/s})^2 \times (7 \times 10^{-7})}$$

$$\Delta n = 0.2$$

2. In Michelson-Moeley experiment, the optical path of each beam was 10 m long and the wavelength of the light used was 550 nm, calculate the fringe shift that would have to obtained if ether existed, given $\frac{v}{c} = 10^{-4}$.

Sol. The fringe shift in Michelson-morley experiment is given by

$$\Delta n = \frac{2 l v^2}{C^2 \lambda}$$

or

$$\Delta n = \frac{2l}{\lambda}\left(\frac{v}{C}\right)^2$$

Given

$$l = 10 \text{ m}$$

$$\lambda = 550 \text{ nm} = 550 \times 10^{-9} \text{ m}$$

$$\therefore \qquad \Delta n = \frac{2 \times (10\text{m})}{(550 \times 10^{-9}\text{m})} \times (10^{-4})^2$$

$$\Delta n = 0.36$$

3. In Michelson-Morley experiment the length of the paths of two beams is 6 meter each. The wavelength of light used is 6000 Å, calculate the phase difference between two interfering beams, velocity of earth relative to ether is 30 Km/sec.

Sol. In Michelson-Morley experiment

The path difference $= \dfrac{l v^2}{C^2}$

$\therefore$ The phase difference $= \dfrac{2\pi}{\lambda} \times$ path difference

$$\text{Phase difference} = \frac{2\pi}{\lambda} \times \frac{l\,v^2}{C^2} \text{ radian}$$

$$= \left(\frac{2\pi}{\lambda}\frac{l\,v^2}{C^2}\right)\left(\frac{180}{\pi}\right) \text{ degree}$$

Given, $l = 6$ m

$v = 30$ Km/sec $= 3 \times 10^4$ m/s

$\lambda = 6 \times 10^{-7}$ m

$C = 3 \times 10^8$ m/s

$$\therefore \quad \text{phase difference} = \frac{360^\circ \times (6\,\text{m})(3\times10^4\ \text{m/s})^2}{(6\times10^{-7})(3\times10^8\ \text{m/s})^2} = 36^\circ$$

4. In Michelson-Morley experiment, the length of the paths of the two beams is 11 metre each. The wavelength of the light used is 6000 Å. If the expected fringe shift is 0.4 fringe, calculate the velocity of earth relative to ether.

Sol. In Michelson-Morley, the fringe shift

$$\Delta n = \frac{2\,l\,v^2}{C^2\lambda}$$

$$v = C\sqrt{\frac{\lambda\,(\Delta n)}{2l}}$$

Given $l = 11$m, $\lambda = 6000$ Å $= 6 \times 10^{-7}$ m

$C = 3 \times 10^8$ m/s

$\Delta n = 0.4$

$$v = (3.0 \times 10^8\ \text{m/s})\sqrt{\frac{(6\times10^{-7}\ \text{m})\times0.4}{2\times(11\text{m})}}$$

$$v = 3.13 \times 10^4\ \text{m/s}$$

On Galilean transformation

5. The position of a point in the frame S' moving relative to S, with a constant velocity of 10 cm/sec along the X-axis is given by (11, 9, 8). Calculate its position with respect to the frame S, if the two frames were in coincidence only 0.5 second before.

Sol. The coordinates of S' are $[x', y', z']$

And that of S are (x y, z)

Then the position of the point in stationary frame S are

$$x = x' + vt$$

$$x = 11 + 10 \times 0.5 = 11 + 5 = 16$$

$$y = y' = 9$$

and $$z = z' = 8$$

Hence in stationary frame S the position of the particle is given by (16, 9, 8)

6. A passenger in a train moving at 30 m/sec passes a man standing on a station platform at $t = t' = 0$ sec. Ten seconds after the train passes him, the man on the platform determines that a bird flying along the tracks in the same direction as the train is 600 m away. What are the coordinates of the bird as determined by the passenger?

Sol. The coordinates assigned to the bird by the man on the station platform are

$$(x, y, z, t) = (600 \text{ m}, 0, 0, 10 \text{ sec})$$

The passenger measures the distance x' to the bird as

$$x' = x - vt$$

$$= 600 - (30 \text{ m/s})(10 \text{ sec})$$

$$= 600 - 300$$

$$= 300 \text{ m}$$

Therefore the bird's coordinates as determined by the passenger are

$$(x', y', z', t') = (300 \text{ m}, 0, 0, 10 \text{ sec})$$

7. A ball has velocity $\left(2\hat{i} - 5\hat{j} + 6\hat{K}\right)$ m/sec relative to a train moving with velocity $\left(3\hat{i} + 4\hat{j}\right)$ m/sec relative to an observer on the ground. Calculate the velocity of the ball relative to the ground.

Sol. Let us consider that the train is equivalent to a moving frame of reference S' and the ground as stationary frame of reference S. According to the question

$$\vec{V} = \left(3\hat{i} + 4\hat{j}\right) \text{ m/s}$$

$$\vec{v}' = \left(2\hat{i} - 5\hat{j} + 6\hat{K}\right) \text{ m/s}$$

Therefore the velocity of the ball relative to the ground is given by

$$\vec{v} = \overline{v}' + \overline{V}$$

$$= \left(2\hat{i} - 5\hat{j} + 6\hat{K}\right) + \left(3\hat{i} + 4\hat{j}\right)$$

$$= \left(5\hat{i} - \hat{j} + 6\hat{K}\right) \text{ m/sec}$$

On the Lorentz Transformation Equations

8. Prove that

$x^2 + y^2 + z^2 = c^2t^2$ is invariant under Lorentz transformations. [UPTU]

Sol. By Lorentz transformation equations we have

$$x' = \frac{x - vt}{\sqrt{1 - \frac{v^2}{c^2}}}$$

$$y' = y$$

$$z' = z$$

and
$$t' = \frac{t - \frac{vx}{c^2}}{\sqrt{1 - \frac{v^2}{c^2}}}$$

Thus inverse Lorentz transformation equations are given by

$$x = \frac{x' + vt'}{\sqrt{1 - \frac{v^2}{c^2}}}$$

$$y = y'$$

$$z = z'$$

$$t = \frac{t' + \frac{vx}{c^2}}{\sqrt{1 - \frac{v^2}{c^2}}}$$

Putting these values in

$$x^2 + y^2 + z^2 = c^2 t^2 \text{ we get}$$

$$\left(\frac{x' + vt'}{\sqrt{1 - \frac{v^2}{c^2}}}\right)^2 + y'^2 + z'^2 = c^2\left(\frac{t' + \frac{vx'}{c^2}}{\sqrt{1 - \frac{v^2}{c^2}}}\right)^2$$

$$y'^2 + z'^2 = \frac{1}{1 - \frac{v^2}{c^2}}\left[c^2t'^2 + \frac{v^2x'^2}{c^2} - x'^2 - v^2t'^2\right]$$

$$= \frac{1}{1 - \frac{v^2}{c^2}}\left[\left(c^2t'^2 - x'^2\right)\left(1 - \frac{v^2}{c^2}\right)\right]$$

$$y'^2 + z'^2 = c^2t'^2 - x'^2$$

Hence $x'^2 + y'^2 + z'^2 = c^2 t'^2$

which is invariant

9. If at time $t = t' = 0$, the origins of systems S and S' just coincide and a spherical pulse of light is produced at the common origin. Show that the speed of propagation of the spherical wave front is the same in both systems – namely c, where systems S' is moving relative to stationary system S.

Sol. The system S is at rest while S' is moving with constant velocity v relative to S.

Let spherical pulse starts from a point where co-ordinates relative to S and S' are (x, y, z, t) and (x', y', z', t') respectively.

We know that

$$x^2 + y^2 + z^2 - c^2 t^2 = 0$$

and $$x'^2 + y'^2 + z'^2 - c'^2 t'^2 = 0 \qquad .....(1)$$

The Lorentz transformation equations are

$$x' = \frac{x - vt}{\sqrt{1 - \frac{v^2}{c^2}}}$$

$$y' = y\,;\; z' = z$$

$$t' = \frac{t - \frac{vx}{c^2}}{\sqrt{1 - \frac{v^2}{c^2}}}$$

By eq. (1)

$$C = \frac{\left(x^2 + y^2 + z^2\right)^{\frac{1}{2}}}{t}$$

and

$$C' = \frac{\left(x'^2 + y'^2 + z'^2\right)^{\frac{1}{2}}}{t}$$

Substituting the values of x', y', z' and t', we get

$$C' = \frac{\left[\left\{\frac{(x - vt)^2}{\left(1 - \frac{v^2}{C^2}\right)}\right\} + y^2 + z^2\right]^{\frac{1}{2}}}{\left[\frac{\frac{(t - vx)}{C^2}}{\sqrt{1 - \frac{v^2}{C^2}}}\right]}$$

or

$$C' = \frac{\left[(x - vt)^2 + \left(y^2 + z^2\right)\left(1 - \frac{v^2}{C^2}\right)\right]^{\frac{1}{2}}}{t - \frac{vx}{C^2}}$$

Adding and subtracting a factor of $x^2\left(1-\frac{v^2}{C^2}\right)$

$$C' = \frac{\left[(x-vt)^2 + (y^2+z^2)\left(1-\frac{v^2}{C^2}\right) + x^2\left(1-\frac{v^2}{C^2}\right) - x^2 + \frac{v^2x^2}{C^2}\right]}{\left(t-\frac{vx}{C^2}\right)}$$

$$C' = \frac{\left[x^2 - 2x\,vt + v^2t^2 + (y^2+z^2)\left(1-\frac{v^2}{C^2}\right) + x^2\left(1-\frac{v^2}{C^2}\right) - x^2 + \frac{v^2x^2}{C^2}\right]}{\left(t-\frac{vx}{C^2}\right)}$$

$$C' = \frac{\left[(x^2+y^2+z^2)\left(1-\frac{v^2}{C^2}\right) + v^2t^2 - 2x\,vt + \frac{x^2v^2}{C^2}\right]^{\frac{1}{2}}}{\left(t-\frac{vx}{C^2}\right)}$$

Using eq. (1), we get

$$C' = \frac{\left[C^2t^2\left(1-\frac{v^2}{C^2}\right) + v^2t^2 - 2x\,vt + \frac{x^2v^2}{C^2}\right]^{\frac{1}{2}}}{\left(t-\frac{vx}{C^2}\right)}$$

$$C' = \frac{\left[C^2t^2 - 2x\,vt + \frac{x^2v^2}{C^2}\right]^{\frac{1}{2}}}{\left(t-\frac{vx}{C^2}\right)}$$

$$C' = \frac{\left[C^2\left(1-\frac{vx}{C^2}\right)^2\right]^{\frac{1}{2}}}{\left(t-\frac{vx}{C^2}\right)}$$

$$C' = C$$

Hence, the speed of propagation of the spherical wavefront is same in both system.

10. As seen by an inertial observer S a certain event takes place at $x_1 = -\frac{L}{2}$ at time $t_1 = \frac{L}{2C}$. Another event takes place at $x_2 = \frac{L}{2}$ at time $t_2 = \frac{L}{2C}$ so that for S the two events are simultaneous. Show that for another inertial observer S' moving along X-axis at velocity v with respect to x the events are not simultaneous, and

$$\Delta t = \frac{-\gamma L v}{C^2}$$

where $\gamma = \left[1 - \frac{v^2}{C^2}\right]^{\frac{-1}{2}}$

Sol. According to Lorentz transformation equations we have

$$x' = \frac{x - vt}{\sqrt{1 - \frac{v^2}{C^2}}} \quad \text{or} \quad x' = \gamma(x - vt)$$

and $$t' = \frac{t - \frac{vx}{C^2}}{\sqrt{1 - \frac{v^2}{C^2}}} \quad \text{or} \quad t' = \gamma\left(t - \frac{vx}{C^2}\right)$$

According to the question,

$$x_1 = -\frac{L}{2}, \qquad t_1 = \frac{L}{2C}$$

$$x_2 = \frac{L}{2}, \qquad t_2 = \frac{L}{2C}$$

In the moving S' frame one event occurs at t_1'

$$t_1' = \gamma\left[\left(\frac{L}{2C}\right) - \left\{v\frac{\left(-\frac{L}{2}\right)}{C^2}\right\}\right]$$

$$= \frac{\gamma L}{2C}\left[1 + \frac{v}{C}\right]$$

and another event occurs at t_2'

$$t_2' = \gamma\left[\left(\frac{L}{2C}\right) - \left\{\frac{v\left(\frac{L}{2}\right)}{C^2}\right\}\right] = \frac{\gamma L}{2C}\left[1 - \frac{v}{C}\right]$$

Thus,

$$\Delta t = t_2' - t_1' = \frac{\gamma L}{2c}\left[\left(1 - \frac{v}{c}\right) - \left(1 + \frac{v}{c}\right)\right]$$

or $$\Delta t = \frac{-\gamma L v}{C^2} \neq 0$$

Hence events are not simultaneous in moving frame S'.

11. A light pulse is emitted at the origin of a frame of reference S' at time $t' = 0$. Its distance x' from the origin after a time t' is given by $x'^2 = C^2 t'^2$. Use the Lorentz transformation to transform this equation to an equation in x and t and show that this is

$x^2 = C^2 t^2$. Discuss the result

Sol. According to the question, the distance of a light pulse x' from the origin after a time t' in the moving frame S' is given as

$$x'^2 = C^2 t'^2 \qquad .....(1)$$

From Lorentz transformation equations, we have

$$x' = \frac{x - vt}{\sqrt{1 - \frac{v^2}{C^2}}}$$

and $$t' = \frac{t - \frac{vx}{C^2}}{\sqrt{1 - \frac{v^2}{C^2}}}$$

Substituting these values of x' and t' in eq. (1), we get

$$\frac{(x-vt)^2}{\left(1-\frac{v^2}{C^2}\right)} = \frac{C^2\left[t-\left(\frac{vx}{C^2}\right)\right]^2}{1-\frac{v^2}{C^2}}$$

or $$(x-vt)^2 = C^2\left(t-\frac{vx}{C^2}\right)^2$$

$$x^2+v^2t^2-2xvt = C^2t^2+\frac{x^2V^2}{C^2}-2txv$$

$$x^2+v^2t^2 = C^2t^2+\frac{x^2v^2}{C^2}$$

or $$x^2-C^2t^2+v^2t^2-\frac{x^2v^2}{C^2}=0$$

$$x^2-C^2t^2-\frac{v^2}{C^2}\left(x^2-C^2t^2\right)=0$$

or $$\left(x^2-C^2t^2\right)\left(1-\frac{v^2}{C^2}\right)=0$$

Since $v \neq C$, therefore

$$\left(1-\frac{v^2}{C^2}\right) \neq 0$$

∴ $$x^2 - C^2t^2 = 0$$

or $$x^2 = C^2t^2$$

This shows that the velocity of light is an absolute constant independent of the frame of reference or the equation is invariant under Lorentz transformation.

12. In an inertial frame S, a red light and a blue light are separated by a distance $\Delta x = 2.45$ Km, with the red light at the larger value of x. The blue light flashes, and 5.35 μs later the red light flashes. Frame S' is moving in a direction of increasing x with speed of u = 0.855 C. What is the distance between the two flashes and the time between them as measured in S'? [UPTU]

Sol. According to Lorentz transformation equations the corresponding positions of two flashes and the time of their occurrence, as measured in moving frame S', are

$$x_1' = \frac{x_1 - vt_1}{\sqrt{1-\frac{v^2}{C^2}}}; \qquad x_2' = \frac{x_2 - vt_2}{\sqrt{1-\frac{v^2}{C^2}}}$$

$$t_1' = \frac{t_1 - \left(\frac{vx_1}{C^2}\right)}{\sqrt{1-\frac{v^2}{C^2}}}; \qquad t_2' = \frac{t_2 - \left(\frac{vx_2}{C^2}\right)}{\sqrt{1-\frac{v^2}{C^2}}}$$

The distance between the red light and blue light flashes with respect to an observer in moving S' is

$$\Delta x' = x_2' - x_1' = \frac{(x_2 - x_1) - v(t_2 - t_1)}{\sqrt{1-\frac{v^2}{C^2}}}$$

$$\Delta x' = \frac{\Delta x - v\Delta t}{\sqrt{1-\frac{v^2}{C^2}}}$$

Here

$$\Delta x = 2.45 \times 10^3 \text{ m}$$
$$\Delta t = 5.35 \times 10^{-6} \text{ sec}$$
$$v = 0.855 \text{ C}$$

and $C = 3 \times 10^8$ m

$$\therefore \quad \Delta x' = \frac{2.45\times10^3 - 0.855C\times5.35\times10^{-6}}{\sqrt{1-\frac{(0.855\,c)^2}{c^2}}}$$

$$\therefore \quad \Delta x' = \frac{2.45\times10^3 - 13.72\times10^2}{\sqrt{1-(0.855)^2}} = \frac{10.78\times10^2}{\sqrt{1-0.731}}$$

$$\Delta x' = \frac{10.78\times10^2}{0.5186}$$

$$\Delta x' = 20.78 \times 10^2$$

$$\Delta x' = 2.08 \text{ Km}$$

The time between the two flashes

$$\Delta t' = t_2' - t_1' = \frac{(t_2 - t_1) - \frac{v(x_2 - x_1)}{C^2}}{\sqrt{1 - \frac{v^2}{C^2}}}$$

$$\Delta t' = \frac{\Delta t - v\frac{\Delta x}{C^2}}{\sqrt{1 - \frac{v^2}{C^2}}}$$

$\therefore$

$$\Delta t' = \frac{5.35 \times 10^{-6} - \frac{(0.855C \times 2.45 \times 10^3)}{C^2}}{\sqrt{1 - (0.855)^2}}$$

$$\Delta t' = \frac{2.094 \times 10^3}{3 \times 10^8 \times 0.5186}$$

$$\Delta t' = -1.346 \times 10^{-5} \text{ sec}$$

$$\Delta t' = -0.135 \ \mu\text{s}$$

–ve sign shows that the red light flash comes before the blue one.

13. Two sources, A of red light and B of blue light, are situated on earth at a distance of 3 Km, with B at larger value of x. The red light flashes, followed after 2 μs by a blue flash. An observer S' moving with a velocity C/3 relative to earth in the direction from A to B observes the events. What is the distance between the two events. What is the distance between the two events and the time interval between them as measured is S'?

Sol. From Lorentz transformation equation

$$x' = \gamma(x - vt) \qquad \text{.....(1)}$$

$$\Delta x' = \gamma(\Delta x - v\,\Delta t)$$

$$t' = \gamma\left(t - \frac{vx}{C^2}\right)$$

$$\Delta t' = \gamma\left(\Delta t - v\frac{\Delta x}{C^2}\right)$$

where $$\gamma = \frac{1}{\sqrt{1-\frac{v^2}{C^2}}} = \frac{1}{\sqrt{1-\frac{1}{9}}} = 1.06$$

Here

$$\Delta x = 3 \times 10^3 \text{ m}$$

$$\Delta t = 2 \times 10^{-6} \text{ sec}$$

$$v = \frac{c}{3} = 10^8 \text{ m/sec}$$

Therefore,

$$\Delta x' = (1.06)\,(3 \times 10^3 - 10^8 \times 2 \times 10^{-6})$$

$$= 2968 \text{ m} = 2.97 \text{ Km}$$

and $$\Delta t' = (1.06)\left[2\times10^{-6} - \frac{10^8 \times 3\times 10^3}{(3\times10^8)^2}\right]$$

$$= -1.41 \times 10^{-6} \text{ sec}$$

$$\Delta t' = -1.41 \text{ µs}$$

The negative sign is $\Delta t'$ tells us that in s'-frame contrary to what is observed in S, the blue flash occurs first. In S'-frame, the time between the flashes is 1.41 µs and the distance between two events is 2.97 km.

14. Show that the circle $x^2 + y^2 = a^2$ in frame S appears to be an ellipse in frame S' which is moving with velocity v relative to S. [UPTU]

Sol. The equation of a circle in a stationary frame is

$$x^2 + y^2 = a^2$$

Due to length contraction

$$x' = x\sqrt{1-\frac{v^2}{C^2}} \quad \text{or} \quad x = \frac{x'}{\sqrt{1-\frac{v^2}{C^2}}}$$

and $$y' = y$$

Therefore, $$\frac{x'^2}{a^2\left(1-\frac{v^2}{C^2}\right)}+\frac{y'^2}{a^2}=1$$

Let, $$b^2 = a^2\left(1-\frac{v^2}{C^2}\right)$$

Therefore, $$\frac{x'^2}{b^2}+\frac{y'^2}{a^2}=1$$

This is equation of an ellipse. Hence in frame S' the shape observed is an ellipse where semi-major axis = a

and semi-minor axis = $a\sqrt{1-\frac{v^2}{C^2}}$

15. What will be the length of a meter rod appear to be for a person travelling parallel to the length of the rod at a speed of 0.8 C relative to the rod? [UPTU]

Sol. According to length contraction

$$\text{Observed length} = \text{Proper length}\left(1-\frac{v^2}{C^2}\right)^{\frac{1}{2}}$$

$$L = L_o\sqrt{1-\frac{v^2}{C^2}}$$

Here $L_o = 1$ m; $V = 0.8$ C

$\therefore$ $$L = \sqrt{1-\left(\frac{0.8C}{C}\right)^2}$$

$$L = 0.6 \text{ m or } 60 \text{ cm}$$

16. At what speed should an object move so that there is 50% contraction in its dimensions in the direction of motion?

Sol. By length contraction formula

$$L = L_o\sqrt{1-\frac{v^2}{C^2}}$$

For 50% contraction we have

$$\frac{L_o}{2} = L_o\sqrt{1-\frac{v^2}{C^2}}$$

$$\frac{v^2}{C^2} = \frac{3}{4}$$

$$V = \frac{\sqrt{3}}{2}C = 0.866\ C$$

17. A square of each side length a is moving with speed v parallel to one of its sides. What is the percentage change in the area if its speed is $C/2$?

Sol. In the rest frame S, the area of the square

$$= a \times a = a^2$$

The contracted length of the side is

$$a' = a\sqrt{1-\frac{v^2}{C^2}}$$

The side perpendicular to the direction of motion will not contract. Thus, when in motion, it will appear a rectangle where area is

$$S' = a \times a' = a^2\sqrt{1-\frac{v^2}{C^2}}$$

The percentage change in area is

$$\frac{S'-S}{S}\times 100 = \left[\left(1-\frac{v^2}{C^2}\right)^{\frac{1}{2}} - 1\right]\times 100$$

$$= \left[\frac{\sqrt{3}}{2} - 1\right]\times 100$$

$$= -\ 13.4\%$$

Therefore, the area a decreases by 13.4%

18. Obtain the volume of a cube, the proper length of each edge of which is L_o when it is moving with a velocity v along one of its edges. [UPTU]

Sol. The contraction in length will occur in the edge of the cube along which it is moving. The lengths of the other edges perpendicular to the direction of motion will remain unaffected.

If L_o is the proper length of each edge is the cube, then the length along which it is moving is given by

$$L_o\sqrt{1-\frac{v^2}{C^2}}$$

The other two edges of the cube remain unaffected and have the same length as before i.e., L_o

Thus the volume of the cube

$$V = l_x\, l_y\, l_z$$

$$V = L_o\sqrt{1-\frac{v^2}{C^2}}.L_o.L_o$$

$$V = L_o^3\sqrt{1-\frac{v^2}{C^2}}$$

19. How much time does a meter stick moving at 0.100 C relative to an observer take to pass the observer? The metre stick is parallel to its direction of motion.

Sol. Since

$$\text{Time} = \frac{\text{Dis tan ce}}{\text{Speed}}$$

$$L = L_o\sqrt{1-\frac{v^2}{C^2}}$$

$$L_o = 1 \text{ m}; \; v = 0.100 \text{ C}$$

$$L = 1\sqrt{1-0.01}$$

$$L = \sqrt{0.99}$$

$$L = 0.922 \text{ m}$$

$$\text{Time} = \frac{0.992 \text{ m}}{0.100 \text{ C}} = \frac{0.992 \text{ m}}{0.1\times 3\times 10^8 \text{ m/sec}}$$

$$\text{Time} = 3.32 \times 10^{-8} \text{ sec}$$

20. How fast should a rocket move relative to an observer in order that its length may appear to be 99% of its proper length?

Sol. We know that

$$L = L_o\sqrt{1-\frac{v^2}{C^2}}$$

$$\therefore \qquad 1-\frac{v^2}{C^2}=\left(\frac{L}{L_o}\right)^2$$

or
$$v = C\sqrt{1-\left(\frac{L}{L_o}\right)^2}$$

$$\therefore \qquad \frac{L}{L_o}=\frac{99}{100}=0.99$$

$$\therefore \qquad v = C\sqrt{1-(0.99)^2} = 0.141C$$

21. Calculate the percentage contraction of a rod moving with a velocity of 0.8C in a direction inclined at 60° to its own length.

Sol. Suppose the proper length of the rod = L_o

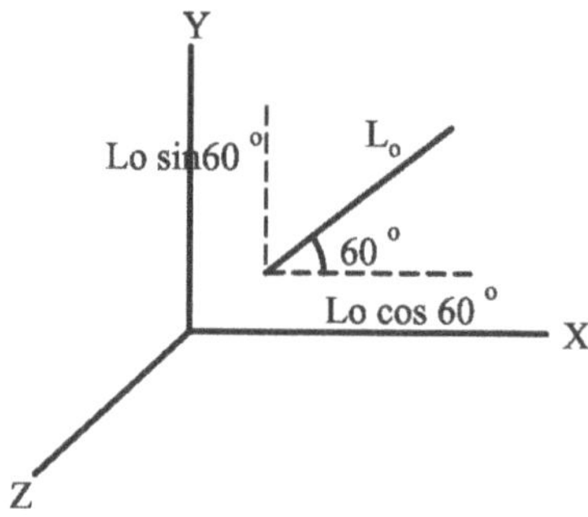

Observed length, along the direction of motion

$$L_A = L_o \cos 60 \sqrt{1-\frac{v^2}{C^2}}$$

$$= L_o \cos 60 \sqrt{1-(0.8)^2}$$

$$L_A = 0.3\ L_o \qquad .....(1)$$

Observed length, perpendicular to the direction of motion

$$L_P = L_o \sin 60^o$$

Since there is no change in perpendicular direction

$$L_p = \frac{\sqrt{3}}{2} L_o \qquad .....(2)$$

$\therefore$ length of moving rod

$$L = \sqrt{L_A^2 + L_P^2}$$

$$L =- \sqrt{(0.3\,L_o)^2 + \left(\frac{\sqrt{3}}{2}L_o\right)^2}$$

$$L = 0.916$$

Percentage contraction in length

$$= \frac{L_o - L}{L_o} \times 100$$

$$= \left(\frac{L_o - 0.916\,L_o}{L_o}\right) \times 100$$

$$= 8.4\%$$

22. A space ship of length 100 m takes 4 μs to an observer on earth. What is its velocity relative to earth?

Sol. In t second observed length L moving with velocity v passes through observer.

$$\therefore \qquad t = \frac{L}{v} = \frac{L_0\sqrt{1-\frac{v^2}{C^2}}}{v} = L_0\left[\frac{1}{v^2} - \frac{1}{C^2}\right]^{\frac{1}{2}}$$

Substituting the values of t and L_o

$$t = 4\ \mu s = 4 \times 10^{-6}\ \text{sec}$$

$$L_o = 100\ \text{m}$$

$$4 \times 10^{-6} = 100\left[\frac{1}{v^2} - \frac{1}{C^2}\right]^{\frac{1}{2}}$$

$$v = 0.083\ C$$

23. A circular lamina move with its plane parallel to the x-y plane of a reference frame S at rest. Assuming its motion to be along the axis of x (or y), calculate the velocity at which its surface area would appear to be reduced to half to an observer in frame S. [UPTU]

Sol. Let the circular lamina be considered moving along the axis x of the reference S with velocity v. Now its radius R along this axis will contract and let it be R' whereas the radius perpendicular to it will remain unaltered.

Therefore, the circular lamina will appear to have assumed an elliptical shape with its major and minor axes equal to R and R' along the axes of y and x respectively.

Therefore the surface area will appear to be $\pi R R'$ while the original surface area of the circular lamina was πR^2

We have $$\pi RR' = \frac{1}{2}\pi R^2$$

where $$R' = \frac{R}{2}$$

We know that $$R' = R\sqrt{1-\frac{v^2}{C^2}}$$

where v is the velocity of the circular lamina with respect to the reference frame S

$\therefore$ $$\frac{R}{2} = R\sqrt{1-\frac{v^2}{C^2}}$$

or $$\frac{1}{4} = 1-\frac{v^2}{C^2}$$

$$\frac{v^2}{C^2} = \frac{3}{4}$$

or $$\frac{v}{C} = \left(\frac{\sqrt{3}}{2}\right)$$

$$v = 0.866\ C$$

Thus, the velocity of the circular lamina relative to frame S must be 0.866 C.

24. An observer S is standing on a platform of length D_o = 65 m on a space station. A rocket passes at a relative speed of 0.80 C moving parallel to the edge of the platform. The observer S notes that the front and back of the rocket simultaneously line up with the ends of the platform at a particular instant.

(a) What is the time necessary according to S for the rocket to pass a particular point on the platform?

(b) What is the rest length of the rocket?

(c) According to an observer S' on the rocket, what is the length D of the platform?

(d) According to S' how along does it take for observer S to pass the entire length of the rocket.

(e) According to S, the ends of the rocket simultaneously line up with the ends of the platform. Are these events simultaneous to S'.

Sol.

(a) According to S, the length of the rocket L matches with the length of the platform D_o so the required time is the proper time

$$T_o = \frac{L}{v} = \frac{65}{0.8 \times 3 \times 10^8} = 0.278 \ \mu s$$

(b) Shall measure the contracted length of the rocket, hence the proper length L_o will be

$$L_o = \frac{L}{\sqrt{1-\frac{v^2}{C^2}}} = \frac{65}{\sqrt{1-(0.8)^2}} = 108\,\text{m}$$

(c) To S' the length of the platform will appear contracted i.e.,

$$D = D_o\sqrt{1-\frac{v^2}{C^2}}$$

$$D = 65\ \sqrt{1-(0.8)^2}$$

$$D = 39\ \text{m}$$

(d) According to S', the time required to pass entire length of the rocket is equal to move a distance equal to the proper length

$$T' = \frac{108}{0.8 \times 3 \times 10^8} = 0.45\ \mu s$$

(e) According to S', the rocket ha a rest length L_o = 108 m, whereas the platform has contracted to length D = 39 m. Therefore, for S' the two ends shall not align with the platform simultaneously. In fact, these events will be separated by a time interval taken to pass the difference in the length of rocket and platform

$$\Delta T' = \frac{108-39}{0.80 \times 3 \times 10^8}$$

$$\Delta T' = 0.29\ \mu s$$

On Time Dilation

25. At what speed should a clock be moved that it may appear to lose 1 minute in each hour? [UPTU]

Sol. If the clock is to lose 1 min in every 60 min; then the clock records (60 – 1) = 59 minutes for each 60 minutes,

Thus,

$$\text{Proper time } = T_o = 59 \text{ min}$$

$$\text{Observed time } T = 60 \text{ min}$$

By time dilation formula

$$T = \frac{T_o}{\sqrt{1-\frac{v^2}{C^2}}}$$

We have, $\text{Proper time } (T_o) = \text{Observed time } (T) \times \sqrt{1-\frac{v^2}{C^2}}$

$$T_o = T\sqrt{1-\frac{v^2}{C^2}}$$

$$59 = 60\left(1-\frac{v^2}{C^2}\right)^{\frac{1}{2}}$$

$$\frac{59}{60} = \left(1-\frac{v^2}{C^2}\right)^{\frac{1}{2}}$$

$$\frac{v^2}{C^2} = 1-\left(\frac{59}{60}\right)^2 \Rightarrow v^2 = \left[1-\left(\frac{59}{60}\right)^2\right]C^2$$

$$v = 5.45 \times 10^7 \text{ m/sec}$$

26. Nakul goes to the pole star and comes back to the earth on a rocket. Calculate the age difference between him and his twin brother, Sahdev who preferred to stay on the earth. The rocket velocity $v = \left(\frac{4}{5}\right)C$ and the distance between the earth and the pole star is 40 light years.

(The light year is used by astronomers as a unit of distance;

light years = $3 \times 10^8 \times 60 \times 60 \times 24 \times 365$ m)

Sol. For the person who stays back on the earth i.e., for Sahdev, the journey takes

$$= 2 \times 40 \times \frac{C}{v} = 2 \times 40 \times \frac{5}{4}$$

$$= 100 \text{ years}$$

For Nakul who makes the to and fro journey, the time interval appears to be

$$T_{space} = T_{earth}\sqrt{1-\frac{v^2}{C^2}}$$

$$T_{space} = 100\sqrt{1-\frac{v^2}{C^2}}$$

$$= 100\ \sqrt{1-(0.8)^2}$$

$$= 100\ \sqrt{1-0.64}$$

$$= 60 \text{ years}$$

Thus Nakul is 40 years younger than his twin brother, Sahdev.

27. An airplane is moving with respect to the earth with a speed of 600 m/sec. As determined by a earth clocks, how long will it take for the airplane's clock to fall behind by two microseconds.

Sol. We know that

$$\Delta t_{space} = \Delta t_{earth}\sqrt{1-\frac{v^2}{C^2}}$$

$$\therefore \qquad \Delta t_{earth} = \frac{\Delta t_{plane}}{\sqrt{1-\frac{v^2}{C^2}}}$$

$$\Delta t_{earth} = \frac{\Delta t_{plane}}{\sqrt{1-\left(\frac{6\times10^2\,m/s}{3\times10^8\,m/s}\right)}}$$

$$\approx \frac{\Delta t_{plane}}{1-2\times10^{-12}}$$

$$(1 - 2 \times 10^{-12})\ \Delta t_{earth} = \Delta t_{plane}$$

$$\Delta t_{earth} - 2 \times 10^{-12}\ \Delta t_{earth} = \Delta t_{plane}$$

$$\Delta t_{earth} - \Delta t_{plane} = 2 \times 10^{-12}\ \Delta t_{earth}$$

$$2 \times 10^{-6} = 2 \times 10^{-12}\ \Delta t_{earth}$$

$$\Delta t_{earth} = 10^6 \text{ seconds}$$

$$\Delta t_{earth} = 11.6 \text{ days}$$

28. A clock keeps correct time with what speed should it be moved relative to an observer so that it may appear to lose 1 minute per day?

Sol. Given

Proper time $T_o = 1$ day $= 24 \times 60$ minutes

Observed time $T = (24 \times 60 + 1)$ minutes

Since clock keeps correct time

$$\text{Proper time} = \text{Observed time} \times \sqrt{1 - \frac{v^2}{C^2}}$$

$$T_o = T\sqrt{1 - \frac{v^2}{C^2}}$$

$$\frac{24 \times 60}{(24 \times 60 + 1)} = \left(1 - \frac{v^2}{C^2}\right)^{\frac{1}{2}}$$

$$1 - \frac{v^2}{C^2} = 0.9986$$

$$v = 0.0374\ C$$

29. A beam of particle cross 10 m length of a laboratory in 4.0×10^{-8} sec. and half of these particles decay during this time. What would be the length of the laboratory observed by an observer moving along with particles?

Sol. The proper time is $T_o = 4 \times 10^{-8}$ second

$\therefore$ Velocity $v = \dfrac{10}{4 \times 10^{-8}} = 2.5 \times 10^8$ m/s

The laboratory length observed in particle's frame is

$$L = L_o\sqrt{1 - \frac{v^2}{C^2}} = 10\sqrt{1 - \left(\frac{2.5}{3}\right)^2} = 10 \times 0.553$$

$$= 5.53 \text{ m}$$

30. The period of a pendulum is measured to be 3 sec in the reference frame of the pendulum. What is the period that is measured by an observer moving at a speed of 0.95 C relative to the pendulum?

Sol. The proper period of pendulum = 3 sec

With respect to the observer, pendulum is moving with speed 0.95 second. The pendulum clock is delayed.

$$\therefore \qquad T = \frac{T_o}{\sqrt{1-\frac{v^2}{C^2}}}$$

$$T = \frac{3}{\sqrt{1-(0.95)^2}}$$

$$T = 9.60 \text{ seconds}$$

31. A μ meson has a life time of 2×10^{-6} seconds [UPTU]

(a) What is the mean life time when the particle is moving with a speed 0.98 C?

(b) How for does it go during one mean life?

(c) What distance will be traversed by the meson without relativistic effect?

Sol. (a) According to time dilation

$$\text{Proper time} = \text{Observed time } \sqrt{1-\frac{v^2}{C^2}}$$

$$T_o = T\sqrt{1-\frac{v^2}{C^2}}$$

$$\therefore \qquad 2 \times 10^{-6} = T\sqrt{1-\left(\frac{0.98\,C}{C}\right)^2}$$

$$\therefore \qquad T = 31.71 \times 10^{-6} \text{ seconds}$$

(b) The distance travelled by the particle during one mean life

$$d = vT$$

$$= 0.98 \times 3 \times 10^8 \times 31.71 \times 10^{-6}$$

$$= 9510 \text{ m}$$

(c) Distance transversed without relativistic effect

$$= vT_o$$

$$= 0.98 \times 3 \times 10^8 \times 2 \times 10^{-6}$$

$$= 588 \text{ m}$$

32. Ram is 25 years old and his father Shyam is 45 years old, travel in a fast spaceship. On returning from his space journey, Shyam finds himself only 55 years old whereas his son has become 75 years old on the earth. Calculate the velocity of the spaceship.

Sol. Shyam's clock is the proper clock. The proper time interval is

$$T_o = 55 - 45 = 10 \text{ years}$$

Dilated time interval

$$T = 75 - 25 = 50 \text{ years}$$

By time dilation formula

$$T = \frac{T_o}{\sqrt{1 - \frac{v^2}{C^2}}}$$

$$\sqrt{1 - \frac{v^2}{C^2}} = \frac{T_o}{T}$$

$$1 - \frac{v^2}{C^2} = \left(\frac{T_o}{T}\right)^2$$

$$\frac{v^2}{C^2} = 1 - \left(\frac{T_o}{T}\right)^2$$

$$v = C\sqrt{1 - \left(\frac{T_o}{T}\right)^2}$$

$$v = C\sqrt{1 - \left(\frac{10}{50}\right)^2}$$

$$v = \frac{2C}{\sqrt{5}} = \frac{2 \times 3 \times 10^8}{\sqrt{5}} = 2.68 \times 10^8 \text{ m/sec.}$$

33. Twins A and B are 20 years each. Twin A travels towards a star 30 light years at speed 0.8 C. He than returns home. Twin B remains at the home. What are ages of A and B as observed by A and as observed by B?

Sol. For the twin B who stays at home, the journey takes

$$t = \frac{2d}{v} = 2 \times 30 \text{ light year} = 2 \times 30 \times C$$

$$\therefore \qquad t = \frac{2 \times 30 \times C}{0.8C} = 75 \text{ years}$$

For the twin A who makes the to and fro journey. The observed time is T = 75 years (observed by A)

$$\therefore \qquad \text{Proper time } T_o = T\sqrt{1 - \frac{v^2}{C^2}}$$

$$T_o = 75\sqrt{1 - \left(\frac{0.8C}{C}\right)^2}$$

$$T_o = 45 \text{ years}$$

Ages as observed by

(a) A — Age of A = 20 + 45 = 65 years
Age of B = 20 + 45 = 65 years

(b) B — Age of B = 20 + 75 = 95 years
Age of A = 20 + 75 = 95 years

34. At 6.00 pm IST, the street light in Delhi and Mumbai are switched on simultaneously in the reference frame of the earth. Given that the nautical distance between Delhi to Mumbai is 1400 km. What will be the time difference as reckoned in the reference frame of the travelling from Delhi to Mumbai at a speed of 0.9 C?

Sol. Let us consider the x-axis from Delhi to Mumbai

We have

$$\Delta x = 1400 \times 10^3 \text{ m}$$

$$v = 0.9 \text{ C or } \frac{v}{C} = 0.9$$

From

$$t' = \frac{t - \frac{vx}{C^2}}{\sqrt{1 - \frac{v^2}{C^2}}}$$

We have

$$\Delta t' = \frac{-v\frac{\Delta x}{C^2}}{\sqrt{1 - \frac{v^2}{C^2}}}$$

$$\Delta t' = \frac{-0.9 \times 1400 \times \dfrac{10^3}{3 \times 10^8}}{\sqrt{1-(0.9)^2}}$$

$$\Delta t' = -9.6 \text{ ms}$$

The minus sign indicates that lights in Delhi are switched on later.

On Relativistic Velocity Addition

35. What is the velocity of photon as determined by an observer in another photon?

Sol. By relativistic velocity addition formula

$$V_x = \frac{V_x' + v}{1 + \dfrac{V_x' v}{C^2}} = \frac{C + C}{1 + \dfrac{CC}{C^2}} = C$$

36. Show that no signal can travel faster than light

Sol. By velocity addition theorem

$$V_x = \frac{V_x' + v}{1 + \dfrac{v V_x'}{c^2}}$$

If $V_x' = C$ and $v = C$

$$V_x = \frac{C + C}{1 + \dfrac{C.C}{C^2}} = \frac{2C}{1+1} = C$$

Hence, the addition of any velocity to the velocity of light simply reproduces the velocity of light. Hence we can conclude that no signal can travel faster than light.

37. Two particles come toward each other, each with speed 0.9 C with respect to the laboratory. What is their relative speed? [UPTU]

Sol. By velocity addition theorem

$$V_x = \frac{V_x' + v}{1 + \dfrac{v V_x'}{C^2}}$$

$$V_x' = 0.9\,C$$

$$v = 0.9\,C$$

$$V_x = \frac{0.9\,C + 0.9\,C}{1 + \frac{0.9\,C \times 0.9\,C}{C^2}}$$

$$V_x = 0.995\ C$$

38. Space craft Alpha is moving at 0.9 C relative to the earth. If space craft Beta is to pass Alpha at a relative speed of 0.5 c in the same direction, what speed must Beta have with respect to the earth?

Sol. By relativistic velocity addition theorem

$$V_x^1 = 0.50\,c$$

$$\nu = 0.90\,c$$

$$V_x = \frac{V^1{}_x + \nu}{1 + \frac{\nu V^1{}_x}{c^2}} = \frac{0.50\,c + 0.90\,c}{1 + \frac{0.90c \times 0.50\,c}{c^2}}$$

$$V_x = 0.97\ c$$

39. A spacecraft A has a velocity 0.9 c relative to the earth. Another spacecraft B overtakes A with relative velocity of 0.50 c. Find the velocity of spacecraft B relative to the ground.

Sol. By relativistic velocity addition theorem

$$V_x = \frac{V^1{}_x + \nu}{1 + \frac{\nu\, V^1{}_x}{c^2}} = \frac{0.5\,c + 0.9\,c}{1 + \frac{0.9\,c \times 0.5\,c}{c^2}}$$

$$V_x = 0.9655\ c$$

Classically $V_x = V_x' + \nu = 0.9\ c + 0.5\ c = 1.4\ c$

which is relativistically impossible

40. A radioactive atoms moving with respect to laboratory at a speed 0.3 C in x-direction. If it emits an electron having a speed of 0.8 C in the rest frame of the atom, find the velocity of the electron with respect to laboratory observer when

(a) Electron is ejected in the x-direction

(b) It is ejected in the y-direction

Sol. (a) We have

$$V_x = \frac{V_x' + v}{1 + \frac{v V_x'}{C^2}}$$

$$= \frac{0.8\,C + 0.3\,C}{1 + \frac{0.8\,C \times 0.3\,C}{C^2}}$$

$$= \frac{1.1\,C}{1.24} = 0.89\,C$$

(b)

$$V_y = \frac{V_y' \sqrt{1 - \frac{v^2}{c^2}}}{1 + v\frac{V_x'}{c^2}}$$

$$= \frac{0.8\,C\sqrt{1-(0.3)^2}}{1+0}$$

$$= 0.8 \times C\sqrt{1-(0.3)^2}$$

$$= 0.76\ C$$

$$V_x = \frac{V_x' + V}{1 + \frac{v V_x'}{c^2}} = \frac{0+0.3\,C}{1+0} = 0.3\,C$$

Therefore

$$V = \sqrt{V_x^{\,2} + V_y^{\,2}}$$

$$V = \sqrt{(0.3\,C)^2 + (0.76\,C)^2} = 0.82\,C$$

The speed of electron is 0.82 C. It makes angle θ with the x-axis, where

$$\theta = \tan^{-1}\left(\frac{V_y}{V_x}\right) = \tan^{-1}\left(\frac{0.76}{0.30}\right)$$

$$= \tan^{-1}\left(\frac{0.76}{0.30}\right)$$

$$= 68.5^\circ$$

41. While moving with velocity of 0.2 C in the lab-frame, a radioactive nucleus emits a β-particle which moves with a speed 0.6 C relative to the nucleus. If it is emitted perpendicular to the direction of motion of the nucleus, what is the velocity of the β-particle in lab frame?

Sol. Let the nucleus is moving along the x-direction; frame S' is attached to the nucleus. The nucleus emits β-particle along the y-direction in the S'-frame. We determine the x- and y-components of velocity in lab s-frame.

The component V_x in the lab-frame is

$$V_x = \frac{V_x' + v}{1 + \frac{vV_x'}{C^2}}$$

Here $v = 0.2\text{ C}$; $V_x' = 0$ and $V_y' = 0.6\text{ C}$

$$V_x = \frac{0 + 0.2\text{ C}}{1 + 0} = 0.2\text{ C}$$

and
$$V_y = \frac{V_y'\sqrt{1 - \frac{v^2}{C^2}}}{1 + \frac{v\,V_x'}{C^2}} = \frac{0.6\text{ C}\sqrt{1 - (0.2)^2}}{1 + 0}$$

$$\simeq 0.9\text{ C}$$

Therefore the magnitude of the resultant velocity in lab frame is

$$V = \sqrt{V_x{}^2 + V_y{}^2} = \sqrt{0.04\text{ C}^2 + 0.35\text{ C}^2} = 0.62\text{ C}$$

The β-particle makes an angle θ with the direction of motion of the nucleus.

$$\theta = \tan^{-1}\left(\frac{V_y}{V_x}\right)$$

$$\theta = \tan^{-1}\left(\frac{0.59}{0.2}\right)$$

$$\theta = \tan^{-1}(2.95)$$

$$\theta = 71.3^\circ$$

42. An air force rocket is chasing enemy's spaceship. From earth, it is found that the speed of air force rocket is 2.55×10^8 m/sec while that of enemy's ship is 2.25×10^8 m/sec. What is the relative velocity of

(a) Enemy's ship as seen by air force rocket

(b) Air force rocket as seen by enemy's ship

(c) Air force rocket with respect to enemy's ship as seen from earth

(d) Enemy's ship with respect to air force rocket as seen from earth.

Sol. According to the relativistic addition formula

$$V_x = \frac{V_x' + v}{1 + \frac{V_x' v}{C^2}}$$

Here $V_x' = 2.55 \times 10^8$ m/sec

$v = 2.25 \times 10^8$ m/sec

$C = 3.0 \times 10^8$ m/sec

$$\therefore \qquad V_x = \frac{2.55\times10^8 - 2.25\times10^8}{\left[1 - \frac{2.55\times10^8 \times 2.25\times10^8}{3.0\times10^8}\right]}$$

$$V_x = \frac{0.3\times10^8}{1-0.6375}$$

$$V_x = 0.83 \times 10^8 \text{ m/sec}$$

(a) The relative velocity of enemy's ship as seen by air force rocket will be 0.83×10^8 m/sec.

(b) Similarly, the relative velocity of air force rocket as seen by enemy's ship towards him, is also 0.83×10^8 m/sec

(c) The relative velocity of air force rocket with respect to enemy's ship as seen from the earth is

$$= V_x' - v$$

$$= 2.55 \times 10^8 - 2.25 \times 10^8$$

$$= 0.3 \times 10^8 \text{ m/sec}$$

Therefore, the observer situated at earth assuming enemy's ship at rest will see the rocket is rushing towards the enemy's ship with a velocity 0.3×10^8 m/sec

(d) Considering air force rocket to be at rest, the observer will see that the enemy's ship is approaching towards the air force rocket with a velocity 0.3×10^8 m/sec.

43. A huge missile explodes and divides into the equal parts, which take off with velocities $c/2$, one to the right and the other to the left . The piece that moves to the right again gets divided in to two, by an explosion such that with respect to its own rest frame the resulting two pieces take off with velocities $C/2$ to the right and left. Calculate the velocities of the last two pieces with respect to the earth.

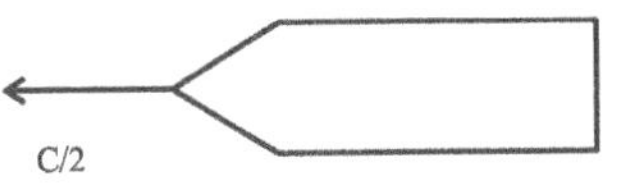

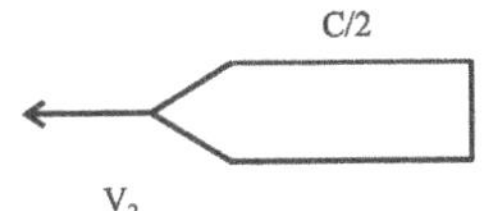

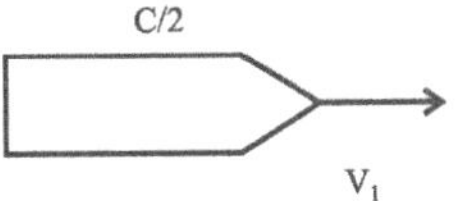

Sol. Let us consider that v_1 and v_2 denote the velocities of the last two pieces with respect to the earth

By velocity addition theorem

$$v_1 = \frac{\frac{C}{2} + \frac{C}{2}}{1 + \frac{C^2}{4C^2}} = \frac{4}{5}C$$

$$v_2 = \frac{\frac{C}{2} + \frac{C}{2}}{1 - \frac{C^2}{4C^2}} = 0$$

On Relativity of Mass

44. Show that a particle which travels with the speed of light must have a zero rest mass.

Sol. The relativistic mass of a particle with speed v is

$$m = \frac{m_o}{\sqrt{1 - \frac{v^2}{C^2}}}$$

Its rest mass is

$$m_o = m\sqrt{1 - \frac{v^2}{C^2}}$$

when $v = C$; then $m_o = 0$

45. A stationary bomb explodes into two fragments of rest mass 2.0 Kg each moving apart with a speed of 0.6 C. Find the rest mass of the bomb.

Sol. From the relativistic conservation of energy

$$m_oC^2 = \frac{m_{o1}C^2}{\sqrt{1-\left(\frac{v_1}{C}\right)^2}} + \frac{m_{o2}C^2}{\sqrt{1-\left(\frac{v^2}{C}\right)^2}}$$

Here $m_{o1} = m_{o2} = 2Kg$

and $v_1 = v_2 = 0.6\ C$

$$m_oC^2 = \frac{2\times C^2 \times 2}{\sqrt{1-(0.6)^2}} = 5.0\ Kg$$

or $m_oC^2 = 5.0\ Kg$

46. Calculate the velocity of a particle at which its mass will be 5 times the mass at rest.

Sol. Here

$$m = \frac{m_o}{\sqrt{1-\frac{v^2}{C^2}}}$$

$$M = 5m_o \text{ (given)}$$

$$5 = \frac{1}{\sqrt{1-\frac{v^2}{C^2}}}$$

or $$1-\frac{v^2}{C^2} = \frac{1}{25}$$

$$v^2 = \frac{24}{25}C^2$$

∴ $$v = 0.98\ C$$

47. A particle of rest mass m_o moves with velocity $c/\sqrt{2}$. Calculate its mass, momentum, total energy and kinetic energy. [UPTU]

Sol. The relativistic mass of the particle is given by

$$m = \frac{m_o}{\sqrt{1-\frac{v^2}{C^2}}} = \sqrt{2}\ m_o$$

Momentum of the particle

$$P = mv = m_o\sqrt{2} \times \frac{C}{\sqrt{2}} = m_o C$$

$$\text{Total energy } = mC^2 = \sqrt{2}\ m_o C^2$$

$$\text{Kinetic energy} = mC^2 - m_o C^2$$

$$= \sqrt{2} m_o C^2 - m_o C^2$$

$$= 0.41\ m_o C^2$$

48. A particle is moving with 20% of the velocity of light. Compare its relativistic mass with its rest mass.

Sol.
$$m = \frac{m_o}{\sqrt{1 - \frac{v^2}{C^2}}}$$

Here
$$\frac{v}{C} = \frac{90}{100} = 0.9$$

$\therefore$
$$\frac{m}{m_o} = \frac{1}{\sqrt{1-(0.9)^2}} = 2.3$$

Hence the relativistic mass is 2.3 times the rest mass of the particle.

49. Show that the relativistic form of Newton's second law, when $\vec{F}$ is parallel to $\vec{v}$ is [UPTU]

$$F = m_o \frac{d\vec{v}}{dt}\left(1 - \frac{v^2}{C^2}\right)^{\frac{-3}{2}}$$

Sol. As we know that according to Newton's second law of motion the force acting on the body is equal to the rate of change of momentum i.e.,

$$F = \frac{dp}{dt} = \frac{d}{dt}(mv)$$

We know that

$$m = \frac{m_o}{\sqrt{1 - \frac{v^2}{C^2}}}$$

$$\therefore \qquad F = \frac{d}{dt}\left[\frac{m_o v}{\sqrt{1-\frac{v^2}{C^2}}}\right]$$

$$F = m_o\left[\frac{1}{\sqrt{1-\frac{v^2}{C^2}}}\frac{dv}{dt} + \frac{v.\left(-\frac{1}{2}\right)\left(1-\frac{v^2}{C^2}\right)^{-\frac{1}{2}}\left(-\frac{2v}{C^2}\right)}{\left(1-\frac{v^2}{C^2}\right)}\frac{dv}{dt}\right]$$

$$F = m_o\left[\frac{1}{\sqrt{1-\frac{v^2}{C^2}}}\frac{dv}{dt} + \frac{\frac{v^2}{C^2}}{\left(1-\frac{v^2}{C^2}\right)^{\frac{3}{2}}}\frac{dv}{dt}\right]$$

$$F = m_o\frac{dv}{ft}\left(1-\frac{v^2}{C^2}\right)^{-\frac{3}{2}}\left[\left(1-\frac{v^2}{C^2}\right)+\frac{v^2}{C^2}\right]$$

$$F = m_o\frac{dv}{dt}\left(1-\frac{v^2}{C^2}\right)^{\frac{-3}{2}}$$

Hence proved

On-Mass Energy Relation

50. A stationary Bomb explodes into two fragments of rest mass 2.0 Kg each moving apart with a speed of 0.6 C. Find the rest mass of the bomb.

From the relativistic conservation of energy

$$m_o C^2 = \frac{m_{o1}C^2}{\sqrt{1-\frac{v_1^2}{C^2}}} + \frac{m_{o2}C^2}{\sqrt{1-\frac{v_2^2}{C^2}}}$$

Here $m_{o1} = m_{o2} = 2$ Kg

and $v_1 = v_2 = 0.6$ C

$$\therefore \qquad m_o C^2 = \frac{2\times C^2}{\sqrt{1-(0.6)^2}}\times 2$$

$$m_oC^2 = \frac{4C^2}{\sqrt{1-(0.6)^2}}$$

$$m_o = 5.0 \text{ Kg}$$

51. A stationary bomb explodes into two fragments of rest mass 1.5 Kg each that move apart with velocity 0.8 C. What is the rest mass of the original bomb?

Sol. We know that

$$m_oC^2 = \frac{m_{o1}C^2}{\sqrt{1-\frac{v^2}{C^2}}} + \frac{m_{o2}C^2}{\sqrt{1-\frac{v^2}{C^2}}}$$

Here $m_{o1} = m_{o2} = 1.5$ Kg

and $v = 0.8$ C

$\therefore$

$$m_oC^2 = \frac{1.5 \times C^2}{\sqrt{1-(0.8)^2}} \times 2$$

$\Rightarrow$

$$m_o = 5.0 \text{ Kg}$$

52. Show that the momentum of a particle of a rest mass m_o and kinetic energy KE is given by the expression. [UPTU]

$$P = \sqrt{\frac{KE^2}{C^2} + 2m_oKE}$$

Sol. We know that

$$E^2 = mo^2C^4 + P^2\,C^2$$

$$E = \left(m_o^2C^4 + P^2C^2\right)^{\frac{1}{2}}$$

Also we know that the total energy E is given by

$$E = \text{rest energy} + \text{kinetic energy}$$

$$E = m_oC^2 + K.E$$

Equating the two values of E, therefore, we have

$$\left(m_o^2C^4 + P^2C^2\right)^{\frac{1}{2}} = m_oC^2 + K.E$$

Squaring both sides of the equation, we have

$$m_o^2C^4 + P^2C^2 = m_o^2C^4 + 2m_oC^2\,K.E. + K.E^2$$

$$P^2C^2 = KE^2 + 2m_oC^2\,KE$$

$$P^2 = \frac{KE^2}{C^2} + 2m_o\,KE$$

$$\therefore \qquad P = \sqrt{\frac{KE^2}{C^2} + 2m_oKE}$$

53. The binding energy (BE) per nucleon for C^{12} is 7.68 MeV and that for C^{13} = 7.47 MeV. Calculate the energy required to remove one neutron from C^{13}.

Sol.

$$\text{Required energy} = \text{BE of } C^{13} - \text{BE of } C^{12}$$
$$= (7.47 \times 13 - 7.68 \times 12)\text{ MeV}$$
$$= 4.95\text{ MeV}$$

54. Prove that a particle having rest mass zero is moving always with the velocity of light.

Sol. Given that rest mass = 0

$$\therefore \qquad E = \sqrt{P^2C^2 + m_o^2C^4} = PC$$

Since $\quad E = mC^2$

$\therefore \quad mC^2 = PC$

and $\quad P = mv$

$\therefore \quad mC^2 = m\,v\,C$

$\Rightarrow \quad C = v$

55. A nucleus of mass m emits a gamma ray photon of frequency ν_o. Show that the loss of internal energy by the nucleus is

$$h\nu_o = \left(1 + \frac{h\nu_o}{2mC^2}\right)$$

Sol. The momentum of the emitted photon is given by

$$\frac{h\nu_o}{C}$$

The recoil momentum of the nucleus $= \dfrac{h\nu_o}{C}$

$$\therefore \qquad \text{Recoil energy} = \frac{\left(\dfrac{h\,\nu_o}{c}\right)^2}{2m}$$

Therefore

Total loss in internal energy

$$= h\nu_o + \frac{(h\nu_o)^2}{2mC^2}$$

$$= h\nu_o\left(1+\frac{h\nu_o}{2mC^2}\right)$$

56. Calculate the mass of an electron accelerated to a kinetic energy of 2MeV. Given that rest mass of the electron is 9.1×10^{-31} Kg.

Sol. We know that

$$K.E = (m - m_o)C^2$$

$$\therefore \quad m = m_o + \frac{KE}{C^2}$$

$$= 9.1 \times 10^{-31} + \frac{2\times10^6 \times 1.6\times10^{-19}}{3\times10^8 \times 3\times10^8}$$

$$= 9.1 \times 10^{-31} + 35.55 \times 10^{-31}$$

$$= 44.65 \times 10^{-31} \text{ Kg}$$

57. In a T.V picture tube an electron moves with a speed of 0.25 C. Find its total energy and kinetic energy.

Sol.

$$\text{Total energy} = E = \frac{m_oC^2}{\sqrt{1-\frac{v^2}{C^2}}}$$

$$E = \frac{0.511\,\text{MeV}}{\sqrt{1-(0.25)^2}}$$

$$E = 0.528 \text{ MeV}$$

$$\therefore \quad K.E = mC^2 - m_oC^2$$

$$= 0.528 - 0.511$$

$$= 0.017 \text{ MeV}$$

58. Calculate the amount of work to be done to increase the speed of an electron from 0.6 C to 0.8 C. Given that rest energy of electron = 0.5 MeV.

Sol. We know that

$$K = (m - m_o)C^2$$

$$= C^2\left[\frac{m_o}{\sqrt{1-\frac{v^2}{C^2}}} - m_o\right]$$

$$= m_oC^2\left[\frac{1}{\sqrt{1-\frac{v^2}{C^2}}} - 1\right]$$

$\therefore$

$$K_1 = m_oC^2\left[\left(1-\frac{(0.6C)^2}{C^2}\right)^{-\frac{1}{2}} - 1\right]$$

$$K_1 = m_oC^2\,(1.25 - 1)$$

$$K_1 = 0.25 \times 0.5 \times 10^6\ \text{eV} = 1.25 \times 10^5\ \text{eV}$$

Similarly

$$K_2 = m_oC^2\,(1.67 - 1) = m_oC^2 \times 0.67$$

$$= 0.67 \times 0.5 \times 10^6\ \text{eV}$$

$$= 3.35 \times 10^5\ \text{eV}$$

The amount of work, to be done

$$= K_2 - K_1$$

$$= 2.1 \times 10^5\ \text{eV} = 2.1 \times 10^5 \times 1.6 \times 10^{-19}$$

$$= 3.36 \times 10^{-14}\ \text{J}$$

59. The mass of a moving electron its 11 times its rest mass. Find its kinetic energy and momentum. [UPTU]

Sol. The kinetic energy $K = mC^2 - m_oC^2$

Here $m = 11\ m_o$

$\therefore$ $m_o = 9.11 \times 10^{-31}$ Kg and $C = 3 \times 10^8$ m/sec

$\therefore$ $K = 11\ m_oC^2 - m_oC^2 = 10\ m_oC^2$

$$K = \frac{10 \times 9.11 \times 10^{-31} \times (3 \times 10^8)^2\ \text{eV}}{1.6 \times 10^{-19}}$$

$$K = 5.12 \times 10^6\ \text{eV}$$

The momentum

$$P = mv$$

$$v = C\sqrt{1-\left(\frac{m_o}{m}\right)^2}$$

$$v = 3.0 \times 10^8 \sqrt{1-\left(\frac{m_o}{11m_o}\right)^2}$$

$$v = 3 \times 10^8 \times 0.996$$

$$v = 2.99 \times 10^8 \text{ m/sec}$$

$P = 11\ m_o v$

$P = 11\ \times 9.11 \times 10^{-31} \times 2.99 \times 10^8$

$P = 2.99 \times 10^{-21}$ Kg m/sec

Conceptual Questions and Answers

1. Distinguish between inertial and non-inertial frames of reference

Ans. Inertial frame of reference is that frame in which the bodies obey Newton's law of inertia and other laws of Newtonian mechanics. On the other hand, in non-inertial frame of reference a body not acted by an external force, is accelerated and Newton's laws are not applicable.

2. What were the consequences of the negative result of Michelson-Morley experiment?

Ans. The negative result of the Michelson-Morley experiment had two consequences:

(a) It rendered untenable the hypothesis of the ether by demonstrating that the ether has no measurable properties.

(b) It suggested a new physical principle; the speed of light in free space is the same everywhere, regardless of any motion of source or observer.

3. Why did Michelson and Morley repeat the experiment during day and night and during all seasons of the year

Ans. This is done to find out if there was any effect due to the change in the direction of orbital velocity of the earth as it was different at different times of the year. The experiment was also repeated at different locations of the earth for the same reason.

The orbital as well as the rotational velocity of earth is non-uniform and varies in magnitude as well as direction. So to avoid the effect of variation in direction and magnitude of the velocity of the earth, the experiment was repeated during day and night and all seasons of the year.

4. Had the Michelson-Morley experiment given a positive result, what would it have implied?

Ans. Had the Michelson-Morley experiment given a positive result, it would have implied not only the existence of the ether, but also a special frame of reference. The ether frame would have had the unique property that the speed of light in this frame is equal to C.

5. Define Proper length and proper time.

Ans. Proper frame is that frame in which the observed body is at rest. The length of rod as measured in this frame is called the proper length and the time recorded by a clock fixed with respect to the observed event is called proper time.

6. A woman travelling at high speed in a space ship holds a meter stick in her hand. What does she notice about the length of the stick as she rotates it from a position parallel to the line of motion to a position perpendicular?

Ans. She notices no change in the stick's length. The length contraction effect concern objects moving at high speed relative to the observer. The metre-stick is at rest relative to the woman.

7. What is the rest mass of a photon?

Ans. We know that photon travels with the velocity of light and its momentum is given by

$$P = mv = \frac{m_o v}{\sqrt{1 - \frac{v^2}{c^2}}}$$

But according to de-Broglie concept, photon of wavelength λ has momentum P given by

$$P = \frac{h}{\lambda}$$

$$\therefore \qquad \frac{h}{\lambda} = \frac{m_o v}{\sqrt{1 - \frac{v^2}{c^2}}}$$

or $$m_o = \frac{h\sqrt{1-\frac{v^2}{c^2}}}{v\lambda}$$

Now for a photon v = c, therefore we have

$$m_o = 0$$

which implies that the rest mass of photon is zero.

8. If photons have a speed C in one reference frame, can they be found at rest in any other frame?

Ans. No, when a body moves with a speed less than C in one reference frame, we can find another reference frame in which it is at rest. But when the body moves with speed C in one reference frame, it will move with the same speed C in all reference frames.

9. At what speed does the mass of a particle become infinite?

Ans. At the speed equal to the velocity of light.

10. If a particle could move with the velocity of light, how much kinetic energy would it possess?

Ans.
$$K.E = (m - m_o)C^2$$
$$= \left(\frac{m_o}{\sqrt{1-\frac{v^2}{c^2}}} - m_o\right)C^2$$

Of v = C then K.E. = ∞

11. Sow that the velocity of a body can never be grater than the velocity of light in vacuum.

Ans. According to Einstein's theory of relativity, the mass of a body moving with velocity v is given by

$$m = \frac{m_o}{\sqrt{1 - v^2/C^2}}$$

when v > C, m becomes imaginary which is not possible.

12. Does light or photons have mass? If no, then how photons have momentum?

Ans. We know that the light is composed of photons and the photon is a massless particle. Also the photon has energy and momentum but no mass. The relativistic formula for momentum, which works for photons as well as marine objects is

$$P^2 = \left(\frac{E}{C}\right)^2 - (m_o C^2)^2$$

where E is the total energy due to motion and possibly due to rest mass and m_o, is the rest mass. Since photons have no rest mass so the equation reduces to

$$P = E/C \text{ and hence momentum.}$$

13. A spring is compressed by tying its ends together and then completely dissolved in acid. Where did its stored potential energy go?

Ans. When we compress a spring, we give it elastic potential energy ΔU. Its mass increases by an amount $\Delta U/C^2$. Thus the mass of a compressed spring is greater than that of an uncompressed one. If the compressed spring is dissolved in acid, the reaction products are of slightly greater mass than if the spring were uncompressed.

14. In relativity, quantities like space, time and mass are not absolute but depend on the reference frame of the observer. But the speed of light C remains the same for all observers. Are there other such relativistic invariants?

Ans. Yes, one invariant is electric charge. Although an electron's mass increases with speed, its charge does not; all observers measure the same value for the charge.

In relativity, distance depends on the observer so does time. But the quantity space time interval remains invariant in relativity. The space time interval ΔS is given by the expression.

$$(\Delta S)^2 = c^2(\Delta t)^2 - \left[(\Delta x)^2 + (\Delta y)^2 + (\Delta z)^2\right]$$

15. Tabulate the following as variants and invariants under the Galilean transformation.

Space, Time, Mass, Velocity, Acceleration, Force, Momentum, Kinctic energy, Impulse, Work, Power, Torque and Angular momentum.

Ans. *Variants:* Velocity, Momentum, Kinetic energy and Angular momentum.

Invariants: Space, Time, Mass, Acceleration, Force Impulse, Work Power, Torque.

Objective Questions

1. Einstein proposed the general theory of relativity in

   (a) 1916 (b) 1905 (c) 1904 (d) 1906

2. Michelson and Morley experiment showed that:

   (a) Newtonian Mechanics is correct for all low and high velocities

   (b) there is an absolute ether frame

   (c) there is no absolute frame, but all frames are relative

   (d) velocities of light is relative in all cases.

3. A cube, when moved along one of its faces at very high speed will look like a

   (a) rectangle (b) cube (c) sphere (d) rectangular Parallelepiped

4. If x and x^1 are the co-ordinates of a particle in two frames of reference s and s^1 moving w.r.t each other with a velocity v along the x-axis and having the coordinate axes parallel to each other, then which of the following is correct relation?

   (a) $x = x^1$ (b) $\frac{dx}{dt} = \frac{dx^1}{dt^1}$ (c) $\frac{d^2x}{dt^2} = \frac{d^2x^1}{dt'^2}$ (d) none of these

5. An inertial S' frame is moving along the Y axis with a velocity v with respect to another inertial frame S. If the co-ordinate axes of the two frames are parallel to each other then which of the following equations is correct?

   (a) $x^1 = x - vt$ (b) $z^1 = z - vt$ (c) $y^1 = y - vt$ (d) none of these

6. Which of the following is invariant under Galilean transformation?

   (a) space coordinate (b) velocity (c) force (d) momentum

7. One of the postulates of special theory of relativity is

   (a) mass is relative (b) speed of light is same in all reference frame

   (c) speed of light is relative (d) time is relative

8. A proton has a charge q when at rest. When it acquires a velocity $\frac{c}{2}$, its charge becomes

   (a) infinity (b) $q\sqrt{1-\left(\frac{1}{2}\right)^2}$ (c) q (d) $\frac{q}{\sqrt{1-\left(\frac{1}{2}\right)^2}}$

9. "All the inertial frames are equivalent". This statement is called the principle of
(a) equivalence (b) correspondence (c) relative motion (d) inertia

10. Given the relativistic mass of a particle

$$m = \frac{m_o}{\sqrt{1-\frac{v^2}{c^2}}}$$

where m_o = rest mass; v = velocity, c = velocity of light, which of the following is true?
(a) increase in mass is due to its increase potential energy
(b) mass increases only when v = 0
(c) increase in mass is equal to increase in kinetic energy divided by c^2
(d) there is no increase in mass

11. Einstein mass-energy relations $E = mc^2$ shows that:
(a) mass disappears to reappear as energy
(b) energy disappears to reappear as mass
(c) mass and energy are two different forms of the same entity
(d) all the above statements are correct

12. The length of a meter rod moving parallel to its length at 80% of the velocity of a light will be
(a) 0.5 m (b) 0.6 m (c) 0.7 m (d) 0.8 m

13. Rest volume L_o is connected to relativistic volume as, where $\beta = \frac{v}{c}$
(a) $L_0^3(1-\beta^2)^{\frac{3}{2}}$ (b) $\frac{L_o^3}{(1-\beta^2)}$ (c) $L_0^3\sqrt{1-\beta^2}$ (d) $\frac{L_0^3}{\sqrt{1-\beta^2}}$

14. According to relativity, the length of a rod in motion
(a) is same as its rest length
(b) is more than its rest length
(c) is less than its rest length
(d) may be less or more than or equal to rest length depending as the speed of rod

15. The postulates of special theory of relativity are applicable to:
(a) stationary frame (b) inertial frame
(c) accelerated frame (d) none of these

16. Frame s^1 moves along the positive x-direction of s frame. A rod placed along y^1-axis if observed from s frame appears
(a) contracted (b) elongated (c) uncharged (d) none of these

17. A stone is raised to the top of Mt. Everest. Its rest mass has

(a) increased (b) decreased (c) remains same (d) none of these

18. Of the two twin brothers, one goes on a relativistic tour and comes back. The brother on tour will

(a) become younger (b) become older
(c) will be of the same age (d) none of these

19. Which of the following is not assumed to be absolute in Newtonian mechanics?

(a) space (b) time (c) mass (d) state of rest or motion

20. The kinetic energy of a particle moving with relativistic speed v is given by (m_o = rest mass)

(a) $\frac{1}{2}m_o v^2$ (b) $\frac{1}{2}\frac{m_o v^2}{\sqrt{1-\frac{v^2}{c^2}}}$ (c) $\left[\frac{m_o}{\sqrt{1-\frac{v^2}{c^2}}}-m_o\right]c^2$ (d) $\left[\frac{m_o c^2}{\sqrt{1-\frac{v^2}{c^2}}}\right]$

21. If the speed of light were $\frac{2}{3}$ of its present value, the energy released in a given atomic explosion will

(a) decrease by a factor of $\frac{2}{3}$ (b) decrease by a factor of $\frac{4}{9}$

(c) decrease by a factor of $\frac{5}{9}$ (d) remains unchanged

22. An observer moves with a speed of $\frac{c}{2}$ towards a stationary source of light, the speed of light appears to the observer to be:

(a) c (b) 2c (c) $\frac{c}{2}$ (d) $\frac{3c}{2}$

23. The rest energy of an electron is 0.511 Mev. This electron is accelerated from rest to a velocity 0.5 c. The change in its energy will be

(a) 0.026 Mev (b) 0.051 Mev (c) 0.079 Mev (d) 0.105 Mev

24. At what speed would the mass of an electron be double its rest mass?

(a) 1.3×10^8 m/s (b) 2.6×10^8 m/s (c) 1.9×10^8 m/s (d) 2.25×10^8 m/.s

25. Rest mass of photon is zero, its mass in motion is

    (a) zero (b) $\frac{E}{C^2}$ (c) $\frac{E}{C}$ (d) $\frac{P}{C}$

26. A body moves with 0.2C velocity. The ratio of the moving mass to rest mass is

    (a) 1.02 (b) 1.2 (c) 0.2 (d) 1.0

27. The mass of a body moving with velocity 0.6C is 2.5 Kg. The rest mass of the body is

    (a) 1.8 Kg (b) 1.60 Kg (c) 1.2 Kg (d) 2.0 Kg

28. Two photons recede from each other. Their relative velocity will be

    (a) Zero (b) $\frac{C}{2}$ (c) 2C (d) C

29. Decay of μ meson supports

    (a) Length construction (b) Time dilation
    (c) Relativity of mass (d) Relativity of energy

Theoretical Questions

1. What do you understand by frame of reference? Is earth an inertial frame of reference? If not why?
2. Differentiate the inertial and non-inertial frames of reference and show that inertial frames move with constant velocity relative to each other.
3. Explain the meaning of fictitious force in reference to a non-inertial frame.
4. What do you mean by Galilean transformations? Write down the Galilean transformations for the frames of reference in relative motion with a uniform velocity and show that they are inertial and the length is invariant.
5. Derive the Galilean transformation equations.
6. Discuss the Galilean transformation equations with respect to position, velocity and acceleration.
7. Show that the motion of a body is identical in all the inertial frames.
8. What was the objective of conducting the Michelson-Morley experiment.
9. Describe Michelson-Morley experiment with the help of neat diagram and discuss the negative results. What is the final conclusion?
10. State and explain the postulates of special theory of relativity.

11. State the fundamental postulates of the special theory of relativity. Deduce the Lorentz transformation equations.
12. Derive Lorentz transformation equations for space and time co-ordinates. Show that when $v < < C$, then these transformations convert into Galilean transformations.
13. Derive the inverse Lorentz transformation equations
14. Derive the Lorentz fitzerland contraction formula. Explain its significance.
15. Deduce an expression for the time dilation on the basis of Lorentz transformation.
16. Prove that time dilation is a real effect.
17. Define proper time interval. Explain why a moving clock appears to go slow to a stationary observer.
18. Define proper length. Explain negative results of Michelson-Morley experiment based on Lorentz contraction formula.
19. Deduce the relativistic velocity addition theorem. Show that it is consistent with Einstein's second postulate.
20. Show that no signal can travel faster than light.
21. Deduce an expression for the variation of mass with velocity.
22. Obtain Einstein's mass energy relation $E = mc^2$. Give some evidence showing its validity.
23. Show from Lorentz transformation that two events simultaneous ($t_1 = t_2$) at different positions ($x_1 \neq x_2$) in a reference frame S are not in general simultaneous in another reference frame.
24. Show that
    (a) time dilation and length contraction are reciprocal
    (b) the particle having rest mass zero is moving always with velocity of light
25. What should have happened to classical mechanism if the speed of light were not so high?
26. Momentum is conserved in a collision of two objects as measured by an observer on a uniformly moving spacecraft. Show that momentum is also conserved for the ground observer.
27. Consider a body B at rest in frame S. Two pulses of radiations each of energy $E/2$ are incident on B moving in + x and – x direction respectively. These pulses are absorbed by B and its energy increases by E. Observe this collision from

frame S' moving with velocity v relative to S in the y-direction. By conserving, momentum, arrive at $E = \Delta mc^2$ (Einstein's derivation).

28. Derive the relation

$$E^2 = P^2c^2 + m_o^2c^4 \qquad \text{where P is the momentum}$$

29. Show that is the variation of mass with velocity is taken into account, the kinetic energy of a particle of rest mass m_o and moving with a velocity v is given by

$$K = m_oC^2\left[\left(1-\frac{V^2}{C^2}\right)^{-1/2} - 1\right]$$

where C is the speed of light

30. A nucleus of mass m emits a gamma photon of frequency v_o. Show that the loss of internal energy by the nucleus is not $hv_o\left[1+\frac{hv_o}{2m_oc^2}\right]$.

31. Prove that $\frac{1}{2}mv^2$ where $m = \frac{m_o}{\sqrt{1-v^2/C^2}}$ does not equal the kinetic energy of a particle moving at relativistic speed.

32. Show that the relativistic form of Newton's second law is $F = m_o\frac{dv}{dt}\left(1+\frac{v^2}{C^2}\right)^{-3/2}$

33. Obtain relativistic acceleration transformations

$$a_x^1 = a_o\frac{dv}{dt}\frac{\left(1-\frac{v^2}{C^2}\right)^{-3/2}}{\left(1-v_n\frac{v}{C^2}\right)}$$

Numerical Problems

1. The mean life of meson is 2×10^{-8} sec., calculate the mean life of a meson moving with a velocity 0.8 C.

2. The mass of a moving electron is 11 times its rest mass. Calculate its kinetic energy and momentum.

3. If the kinetic energy of a body its twice its rest mass energy, find its velocity.

4. Find the speed of an electron of kinetic energy 0.1 MeV according to classical and relativistic mechanics.

5. Show that relativistic kinetic energy of a particle is given by $K = (m - m_o) C^2 = m_o C^2 \left[\left(1 - \frac{V^2}{C^2}\right)^{\frac{1}{2}} - 1\right]$ and hence, calculate the work done to increase the speed of electron of rest energy 0.5 MeV from 0.6 C to 0.8C.

6. A train whose length is 150 m when at rest has to pass through a tunnel of length 125 m. The train is moving with uniform speed of 2.4×10^8 m/s, towards the tunnel. Find the length of train and that of tunnel as observed by an observer at the tunnel.

7. A rod has length 100 cm. When the rod is in a satellite moving with a velocity one half velocity of light relative to laboratory. What is the length of the rod as determined by an observer?

   (a) in the satellite (b) in the laboratory

8. What will be the length of a meter rod appear to be for a person travelling parallel to the length of the rod at a speed of 0.8 C relative to the rod [0.6 m]

9. A man leaves the earth in a rocket ship that makes a round trip to the nearest star which is 4 light years away at a speed of 0.8C. How much younger will he be on his return than his twin brother who preferred to stay behind [4 yrs]

10. A man weighs 50 Kg on the earth. When he is in a rocket ship in flight, his mass is 50.5 kg as measured by an observer on earth. What is the speed of the rocket?

    [0.42×10^8 m/s]

11. A wrist watch keeps correct time on the surface of the earth. A pilot wears this watch and leaves the earth in a spaceship which moves at a constant speed of 10^7 m/s. How many seconds will the watch lose per day with respect to an observer on the earth? [50.4 sec]

12. How fast would a rocket have to go relative to an observer for its length to be contracted to 99% of its length at rest? [0.42×10^8 m/s]

13. The period of a pendulum is measured to be 3 seconds in the inertial frame of reference of the pendulum. What is the period T of the pendulum when measured by an observer moving at a speed of 0.95 C w.r.t to the pendulum?

    [3.3 sec]

14. Calculate the % contraction of a rod moving with a velocity of 0.6 C in a direction inclined at 30° to its own length [0.854]

15. Show that by the application of Lorentz transformation that $x^2 + y^2 + z^2 + w^2 = 0$ is invariant, where w = ict where $i = \sqrt{-1}$

16. A rod has a length of 1m. It is moving in a spaceship moving with a velocity of 0.4 C relative to the earth. Calculate its length as measured by an observer

    (a) on spaceship (b) on earth [(a) 1 m (b) .92 m]

17. Consider two twins A and B, aged 20 yrs. twin B takes a round trip space voyage to a star at velocity v = 0.99C. According to those of us on earth, the star is 40 light years away. What will be ages of A and B, when B finishes his trip.

    [A = 100.8 yrs B = 31.4 yrs]

18. Two photons are moving in opposite direction. Calculate the relative speed of one photon with respect to the other? [3×10^8 m/s]

19. At what speed will the mass of a body is 2.25 times its rest mass

    [2.687×10^8 m/s]

20. Find the mass and speed of 2 MeV electrons? [3.56×10^{-30} kg, 2.9×10^8 m/s]

21. How much does a proton gain in mass when accelerated to a kinetic energy of 500 MeV? [8.89×10^{-28} Kg]

22. The rest mass of a proton is 1.67×10^{-27} Kg. At what speed will its mass be double to its rest mass? [2.598×10^8 m/s]

23. How fast must an electron move in order that its mass equal to the rest mass of the proton? [2.99×10^8 m/s]

24. Find the speed of a 0.1 MeV electron according to classical and relativistic mechanics? [1.87×10^8 m/s, 1.64×10^8 m/s]

25. An electron which has a mass of 9.31×10^{-31} Kg moves with a speed of 0.75C. Find its relativistic momentum and compare this with the momentum calculated from the classical expression. [3.1×10^{-22} Kg m/s; 2.05×10^{-22} Kg m/s]

26. Calculate the gain in mass of an electron whose kinetic energy is 1 MeV?

    [1.28×10^{-38} Kg]

27. The binding energy of an electron to the proton (in hydrogen atom) is 13.6 eV. Find the loss of mass in the formation of 1 atom of hydrogen [2.4×10^{-35} Kg]

28. A certain particle has a life time of 10^{-7} sec when measured at rest. How far does it go before decaying if speed is 0.99 C when it is created? [213 m]

29. 1 Kg of water is heated from 20 °C to 100 °C. Calculate the increase in the mass of water. Given that the specific heat of water is 4.2×10^3 J Kg °C

[4.27×10^{-12} Kg]

30. Ice at 0 °C melts into water at the same temperature and gains a mass of 1 kg in this process. Calculate the initial mass of ice. Given that the latent heat of fusion of ice is equal to 3.36×10^5 J/kg. [2.68×10^{11} kg]

31. A woman sitting in a train pours two cup of coffee, one 10 minutes after the other. The train is moving in a straight line with a velocity of 20 m/sec. What is the distance of separation between the two pouring as measured by a person on the ground?

[12,000 m]

Answers to Objective Questions

1. (a)	2. (c)	3. (d)	4. (c)	5. (c)	6. (c)
7. (b)	8. (c)	9. (c)	10. (c)	11. (d)	12. (b)
13. (c)	14. (c)	15. (b)	16. (c)	17. (a)	18. (a)
19. (d)	20. (c)	21. (b)	22. (a)	23. (c)	24. (b)
25. (a)	26. (a)	27. (d)	28. (d)	29. (b)	

CHAPTER 2A

Interference

'The wave nature of light was demonstrated convincingly for the first time in 1801 by Thomas Young by a wonderfully simple experiment… He let a ray of sunlight in to a dark room, placed a dark screen in front of it, pierced with two small pinholes, and beyond this, at some distance a white screen. He then saw two darkish lines at both sides of a bright line, which gave him sufficient encouragement to repeat the experiment, this time with spirit flame as light source, with a little salt in it, to produce the bright sodium light. This time he saw a number of dark lines, regularly spaced; the first clear proof that light added to light can produce darkness. This phenomenon is called interference. Thomas young has expected it because he believed in the wave theory of light.

–"Dennis Gabor"

2A.1 Introduction

The interference in light is as real effect as interference in water or sound waves, and there is one example of it familiar to everybody, the bright colours of a thin film of oil spread out on a water surface. There are two reasons why the interference of light is observed in some cases and not in others? Firstly, light waves have very short wavelengths – the visible part of the spectrum extends only from 400 nm for violet light to 700 nm for red light. Secondly, every natural source of light emits light waves only as short trains of random pulses, so that any interferences that occurs is averaged out during the period of observation by the eye, unless special procedures are used.

Is it possible that light added to light produces darkness? Yes when the two light waves are out of phase.

Light + Light = Brighter light

If the two waves are in phase, this phenomenon is called interference, which is based on the principle of superposition.

"When two waves of same frequency and having a constant phase difference traverse in the same region of medium simultaneously and cross each other, then the disturbance at each point of the region of superposition is the resultant of the disturbance due to individual waves". But beyond this region the waves come through completely un-influenced by each other.

Hence there is a modification in the intensity of the light in the medium in the region of superposition, which is in general difference from the sum of intensities due to waves at the point. The resultant intensity at any point depends upon the amplitudes and the phase relationships of the component waves. This modification in the intensity resulting from the superposition of two or more waves of light is called interference.

2A.2 Constructive Interference

At certain points the waves superimpose in such a way that the resultant intensity is greater than the sum of the intensities due to individual waves. The interference produced at these points is called constructive interference or reinforcement.

2A.3 Destructive Interference

At certain points the resultant intensity is less than the sum of intensities due to individual waves. The interference produced at these points is called destructive interference.

2A.4 Principle of Superposition of Waves

When a wave passes through a medium, the particles of the medium vibrate to and fro, about their equilibrium position. Suppose two or more waves pass through a medium at the same time, each wave makes the particles of the medium vibrate independently. The resultant displacements of a particle at any instant, is equal to the algebraic sum of the displacements due to each wave and depends on the phase-difference. This is called the principle of superposition

$$\vec{y} = \vec{y_1} + \vec{y_2} + \vec{y_3} + \ldots..$$

where y is the resultant displacement and $\vec{y_1}$, $\vec{y_2}$, $\vec{y_3}$ are the displacements of each individual particles.

Superposition of two troughs, two crests and a crest and trough of two waves are as shown in Fig. 2A.1 (a), (b) and (c).

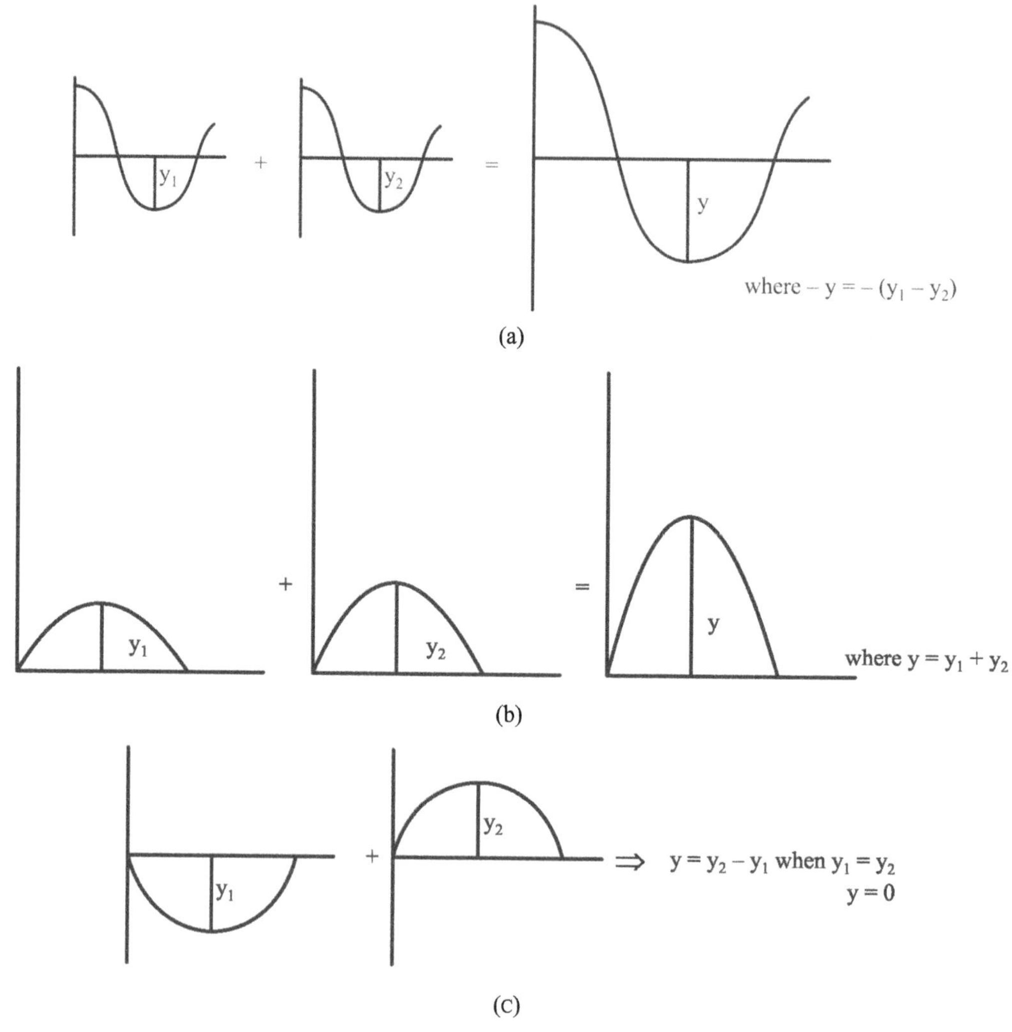

Fig. 2A.1

2A.5 Superposition of two Waves of same Frequency but having Constant Phase Difference

Let us consider the superposition of two wave trains of same frequency $\frac{\omega}{2\pi}$ but different in phase, in a homogeneous medium, if a_1 and a_2 are the amplitudes of the two waves, the displacement due to one wave at any instant t is represented as

$$y_1 = a_1 \sin \omega t \qquad .....(2A.1)$$

The displacement due to another wave at the same instant t is

$$y_2 = a_2 \sin(\omega t + \delta) \quad \text{.....(2A.2)}$$

where δ is the phase difference between the two waves at instant t.

Now according to Young's principle of superposition when a particle is distributed simultaneously by two or more waves, the resultant displacement at any instant will be given by

$$y = y_1 + y_2 \quad \text{.....(2A.3)}$$

$$\therefore \quad y = a_1 \sin \omega t + a_2 \sin(\omega t + \delta)$$

$$y = a_1 \sin \omega t + a_2 \sin \omega t \cos \delta + a_2 \cos \omega t \sin \delta$$

$$y = (a_1 + a_2 \cos \delta) \sin \omega t + (a_2 \sin \delta) \cos \omega t \quad \text{.....(2A.4)}$$

Let

$$a_1 + a_2 \cos \delta = A \cos \phi \quad \text{.....(2A.5)}$$

$$a_2 \sin \delta = A \sin \phi \quad \text{.....(2A.6)}$$

where A and ϕ are new constants

Squaring eq. 2A.5 and 2A.6 and then adding we get

$$A^2 = a_1^2 + a_2^2 + 2a_1a_2 \cos \phi \quad \text{.....(2A.7)}$$

Dividing eq. 2A.6 by 2A.5 we get

$$\tan \phi = \frac{a_2 \sin \delta}{a_1 + a_2 \cos \delta} \quad \text{.....(2A.8)}$$

Using eqs 2A.5 and 2A.6, eq. 2A.4 can be written as

$$y = A \cos \phi \sin \omega t + A \sin \phi \cos \omega t$$

$$y = A \sin(\omega t + \phi) \quad \text{.....(2A.9)}$$

Obviously the resultant disturbance has amplitude A and phase ϕ given by 2A.7 and 2A.8 respectively

The intensity at any point is given by

$$I = A^2$$

$$I = a_1^2 + a_2^2 + 2a_1a_2 \cos \delta \quad \text{.....(2A.10)}$$

Eq, 2A.10 can be written in terms of intensity of the individual waves

$$I = I_1 + I_2 + 2\sqrt{I_1 I_2} \cos \delta \quad \text{.....(2A.11)}$$

Eq. 2A.10 can also be written as

$$I = a_1^2 + a_2^2 + 2a_1a_2 \cos\delta$$

$$I = a_1^2 + a_2^2 + 2a_1a_2 \left(2\cos^2 \delta/2 - 1\right)$$

$$I = a_1^2 + a_2^2 - 2a_1a_2 + 4a_1a_2 \cos \delta/2 \quad \text{.....(2A.12)}$$

Thus we can say that the intensity variation is of the cosine square form.

Condition for maximum intensity

For maximum intensity $\cos\delta = +1$ where phase difference $\delta = 2n\pi$; $n = 0, 1, 2, 3, \ldots$

or the intensity is maximum when phase difference is even multiple of π or path difference is even multiple of half wavelength $(\lambda/2)$

$\therefore$

$$I_{max} = a_1^2 + a_2^2 + 2a_1a_2$$

$$I_{max} = (a_1 + a_2)^2$$

$$I_{max} > a_1^2 + a_2^2$$

$$I_{max} > I_1 + I_2$$

Thus the resultant intensity at a certain point is greater than the sum of the individual intensities of the waves.

For minimum intensity

For minimum intensity $\cos\delta = -1$, when phase difference

$$\delta = (2n - 1)\pi$$

where $n = 1, 2, 3, 4, \ldots\ldots$

i.e., the intensity is minimum when phase difference is odd multiple of π or the path differences is odd multiple of $(\lambda/2)$

$$I_{min} = a_1^2 + a_2^2 - 2a_1a_2$$

$$I_{min} = (a_1 - a_2)^2$$

Thus

$$I_{min} < a_1^2 + a_2^2$$

$$I_{min} < I_1 + I_2$$

Thus the resultant intensity at certain points is less than the sum of the individual intensities of the waves.

Energy distribution and conservation of energy

The average energy at any point in the region of superposition is equal to the sum of the intensity of the individual waves. This may be shown analytically by averaging the energy corresponding to one wave as follows

$$I_{av} = \frac{\int_0^{2\pi} I d\delta}{\int_0^{2\pi} d\delta}$$

$$I_{av} = \frac{\int_0^{2\pi} \left(a_1^2 + a_2^2 + 2a_1a_2 \cos\delta\right) d\delta}{\int_0^{2\pi} d\delta}$$

$$= \frac{\left[a_1^2\delta + a_2^2\delta + 2a_1a_2 \sin\delta\right]_0^{2\pi}}{\left[\delta\right]_0^{2\pi}}$$

$$= \frac{\left(a_1^2 + a_2^2\right)2\pi}{2\pi} = a_1^2 + a_2^2 = I_1 + I_2$$

This means that the energy of the maxima is at the expense of the energy of the minima. The energy $2a_1a_2$ is transferred from minima to maxima. Thus in the phenomenon of interference, the energy is conserved.

The curve showing the variation of intensity with phase is shown in Fig. 2A.2.

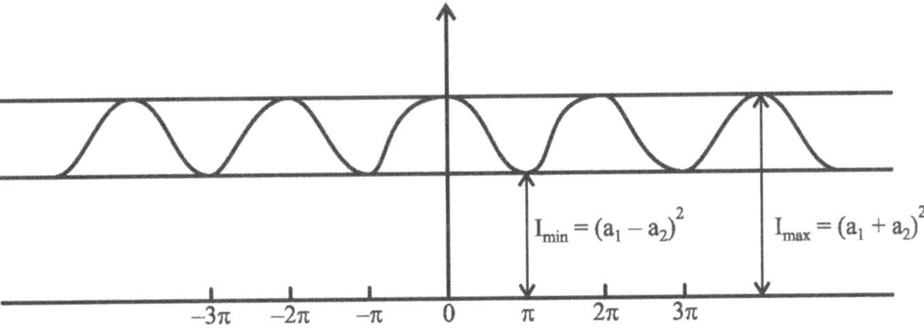

Fig. 2A.2

2A.6 Coherence: Coherent Sources

The word coherence refers to phase, in optics the phase relationship between the light emitted by two sources is expressed by the term coherence.

Coherent sources are a pair of sources which emit light of same wavelength, same frequency and equal intensity or amplitude (nearly equal amplitude) in the same direction. The two light waves must be in the same phase or with a constant initial phase difference. The light emitted by the two sources must be in the same state of polarisation i.e., the light energy emitted by the two sources must be confined to the same plane. If the light energy emitted by one source is in the vertical plane, then the light energy emitted by the second source should also lie in the vertical plane.

Consider a source of light emitting light radiation. The electric field of the light emitted by the source must be sinusoidal. The field should remain sinusoidal for all values of time. But this is possible only for ideal source of light (Fig. 2A.3).

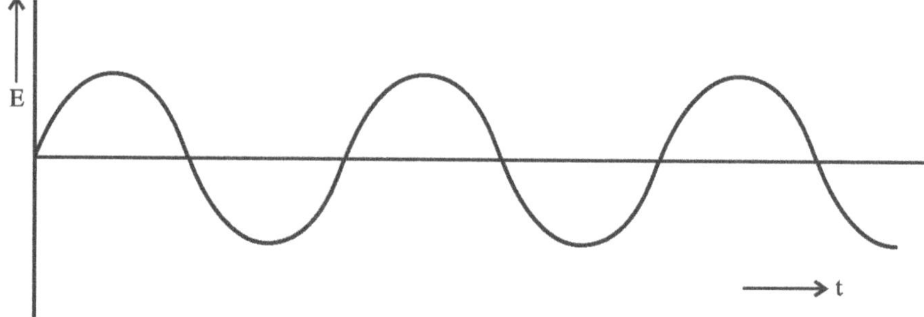

Fig. 2A.3

But for actual sources, the field remains sinusoidal only for about 10^{-10} sec or so. After this time the phase changes abruptly. So the field due to a real source of light would be of the form as shown in Fig. 2A.4.

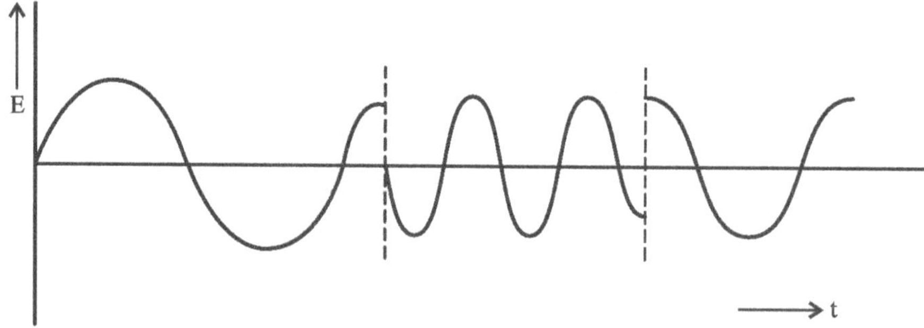

Fig. 2A.4

"The average time for which the field remains sinusoidal with a definite phase through out is called coherence time or temporal coherence." The distance for which the field remains sinusoidal is called coherence length L.

where

$$L = \text{Coherence time} \times \text{Velocity of light}$$

For sodium yellow light, the coherence length is about 3 cm so that the coherence time is

$$\frac{L}{C} = \frac{3}{3 \times 10^{10}} = 10^{-10} \text{ sec}$$

Coherence sources are two narrow sources, separated by a very small distance, emitting light waves having the same wavelength or frequency, equal intensity or amplitude, in the same state of polarization, with a constant phase difference or in the same phase. The phase difference between any pair of points in the two sources should always remain constant, so that the interference pattern remains stationary.

2A.7 Condition for Sustained interference of Light Waves

To obtain sustained interference pattern, the intensity must be maximum or zero at points corresponding to constructive and destructive interference respectively. For this purpose following conditions must be fulfilled.

1. The two interfering sources must be coherent.
2. The two interfering sources should emit light of the same wavelength or frequency. If this condition is not satisfied the phase difference between the interfering waves will continuously vary. As a result the resultant intensity at any point will vary with time, being alternatively maximum/minimum.
3. The interfering waves must have equal amplitudes otherwise the minimum intensity will not be zero and there will be general illumination.
4. The two interfering waves must be propagated along the same line otherwise their vibration will not be along a common line.
5. The separation between the two sources must be as small as possible. If this condition is not satisfied the fringes of maximum and minimum intensity will lie so close together that the fringes will not be separately visible. This is because the distance between consecutive dark or bright fringes (i.e., fringe width β) is directly proportional to the wavelength of light used and is inversely proportional to the distance between the two sources.

6. The two sources must be narrow; this is due to the fact that a broad source is equivalent to a large number of fine sources lying side by side. Each pair of fine sources will give its own interference pattern. The fringes of different pattern will overlap causing a general illumination.
7. If the interfering waves are polarised, they must be in the same state of polarisation. This is the condition for interference in polarised light.

Note : In practice, a sustained interference pattern cannot be obtained by using two independent sources of light. It is because of the following reasons.

1. Two independent sources of light cannot emit waves continuously.
2. The waves emitted by two independent sources of light do not have same phase or constant phase difference.

2A.8 Production of Coherent Sources

If two sources are derived from a single source by same device, then any phase-change in one is simultaneously accompanied by the same phase-change in the other. Thus the phase difference between the sources remains constant. The devices for creating coherent sources of light are as follows.

1. ***Young's double slits:*** For experimental purpose two virtual sources are obtained from a single parent source. In such a case all the random phase changes occurring in the parent source are repeated in the virtual or secondary sources, thus maintaining a constant phase difference between them these virtual sources will be coherent. In this device, two narrow slits S_1 and S_2 receive light from the source narrow slit S. Hence S_1 and S_2 act as a coherent source (Fig. 2A.5).

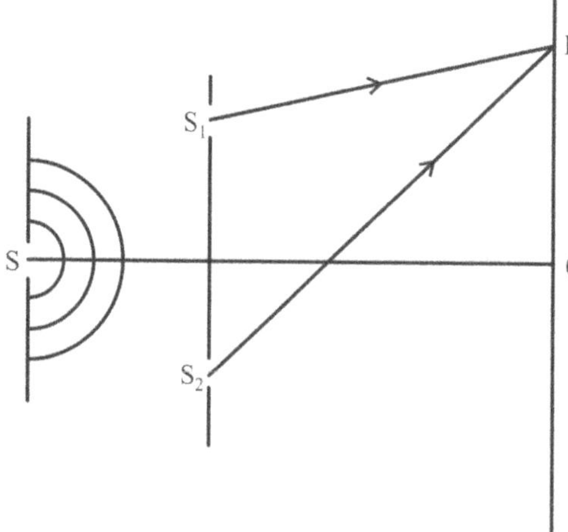

Fig. 2A.5

2. ***Lloyd's single mirror method:*** In this method Lloyd used a plane mirror of size 30 cm length and 6 to 8 cm in breadth. The mirror is a perfectly polished metal or glass plate such that no reflection could take place from back surface of the mirror. S is a source of light and S' is its image formed by reflection. Hence S and S′ are the two coherent sources where S is the original source and S′ is a virtual source formed by reflection (Fig. 2A.6).

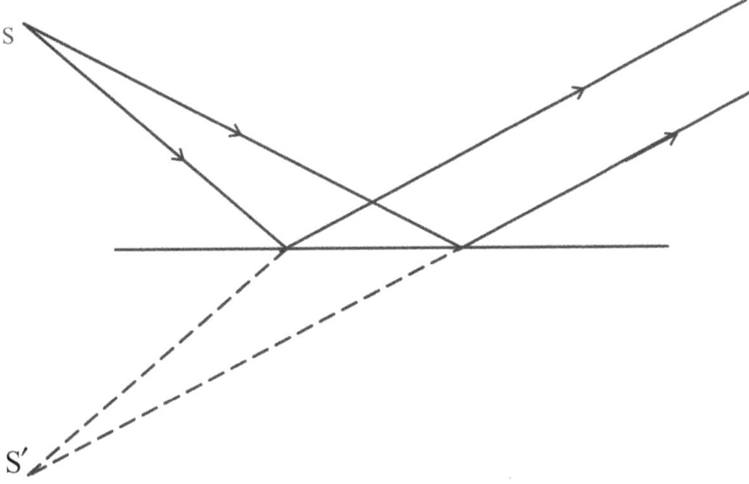

Fig. 2A.6

3. ***Fresnel's Biprism method:*** The Biprism abc consists of a single prism equivalent to two acute angle prism placed base to base of obtuse angle of 179°. The acute angle on both sides is 30′. The prism is placed with its replacing edge parallel to the source of light S such that Sa is normal to the face bc of the prism. When light falls from S on the upper position of the prism, it bends down and appears to come from virtual source S_1, similarly the light falling from S on the lower position of the prism is bent upwards and appears to be coming from virtual source S_2. This is the virtual source S_1 and S_2 derived from the light source S are the coherent sources produced by the biprism (Fig. 2A.7)

2A.9 Phase Difference and Path Difference

The difference between optical paths of two rays, which are in constant phase difference with each other reuniting at a particular point, is known as path difference. For example, let the two coherent sources traversed different path and meet at a particular point P (Fig. 2A.8). Then the path difference is given as

$$\text{Path difference} = S_2P - S_1P$$

Suppose for a path difference x, the phase difference is δ

for a path difference λ the phase difference = 2π

$\therefore$ for a path difference x

The phase difference = $\dfrac{2\pi}{\lambda}x$

Phase difference,

$$\delta = \frac{2\pi}{\lambda} \times (\text{path difference})$$

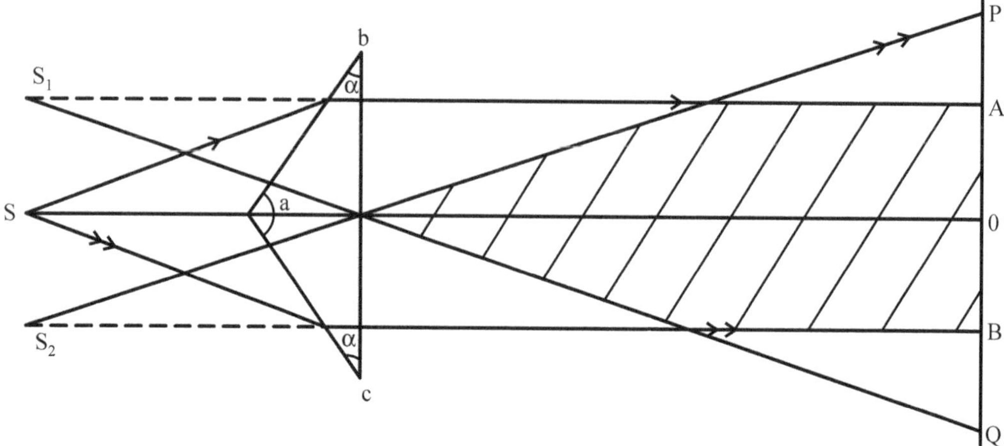

Fig. 2A.7

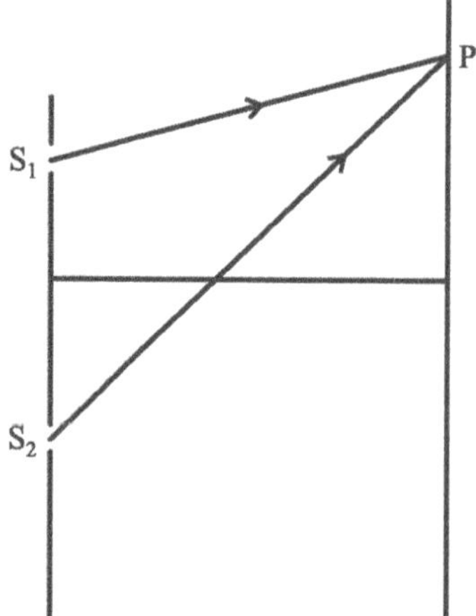

Fig. 2A.8

2A.10 Analytical Theory of Interference Fringes in a Two Slit Arrangement

Let S_1 and S_2 be two coherent sources separated by a distance 2d, let the distance of the screen from the coherent sources be D. Let O be the foot of the perpendicular drawn from C, the mid point of S_1 and S_2 on the screen. Obviously point O is equidistant from S_1 and S_2. Therefore the path difference between the two waves at point O is zero. Thus the point O has maximum intensity. Draw S_1M and S_2N perpendicular from S_1 and S_2 on the screen (Fig. 2A.9). Now consider a point P on the screen at a distance y from O.

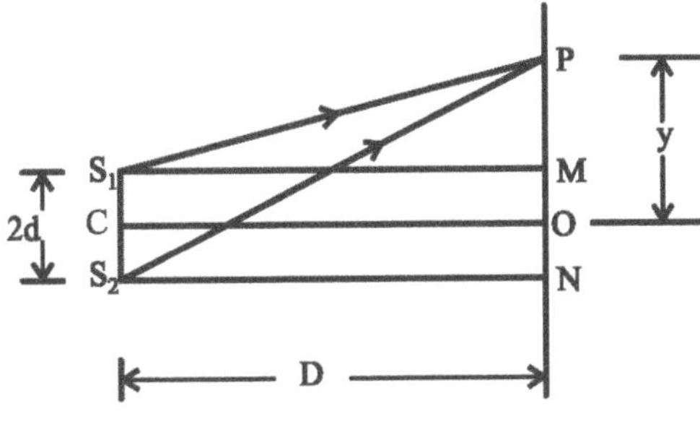

Fig. 2A.9

then

$$PM = y - d$$

$$PN = y + d$$

In right angled triangle $\Delta\, S_2\, P\, N$

$$S_2\,P = [(S_2N)^2 + (PN)^2]^{1/2}$$

$$S_2\,P = [D^2 + (y+d)^2]^{1/2}$$

$$S_2\,P = D\left[1 + \left(\frac{y+d}{D}\right)^2\right]^{1/2}$$

Using Binomial theorem and neglecting higher terms, Since $D >> y$ and we get

$$S_2\,P = D\left[1 + \frac{(y+d)^2}{2D^2}\right]$$

$$S_2\,P = D + \frac{(y+d)^2}{2D} \qquad \text{.....(2A.13)}$$

Similarly in right angled triangle $\Delta\ S_1MP$, we get

$$S_1P = \left[(S_1M)^2 + (MP)^2\right]^{1/2}$$

$$S_1P = \left[D^2 + (y-d)^2\right]^{1/2}$$

$$S_1P = D\left[1 + \frac{(y-d)^2}{D^2}\right]^{1/2}$$

$$S_1P = D\left[1 + \frac{(y-d)^2}{2D^2}\right]$$

$$S_1P = D + \frac{(y-d)^2}{2D} \qquad .....(2A.14)$$

∴ Path difference

$$S_2P - S_1P = \left\{D + \frac{(y+d)^2}{2D}\right\} - \left\{D + \frac{(y+d)^2}{2D}\right\}$$

$$S_2P - S_1P = \left\{D + \frac{(y+d)^2}{2D} - D - \frac{(y+d)^2}{2D}\right\}$$

$$S_2P - S_1P = \frac{1}{2D}\left\{(y+d)^2 - (y-d)^2\right\}$$

$$S_2P - S_1P = \frac{4yd}{2D} = \frac{2yd}{D}$$

∴ Phase difference

$$\delta = \frac{2\pi}{\lambda}.\Delta = \frac{2\pi}{\lambda}\left(\frac{2yd}{D}\right) \qquad .....(2A.15)$$

2A.10.1 Position of Bright Fringes

For bright fringes or maximum intensity at P, the path difference must be an even multiple of the half wavelength of the light used

$$\frac{2yd}{D} = 2n.\frac{\lambda}{2} = n\lambda$$

where n = 0, 1, 2, 3, ...

$$y = \frac{n\,D\,\lambda}{2d}$$

This equation gives the distance of the nth bright fringe from the point O. Therefore

$$y_n = \frac{n\,\lambda\,D}{2d} \qquad \text{.....(2A.16)}$$

2A.10.2 Position of Dark Fringes

For dark fringes or minimum intensity at P, the path difference must be an odd multiple of half wavelength of light used.

$$\frac{2y\,d}{D} = (2n+1)\lambda/2 \qquad \text{where } n = 0, 1, 2, 3, \ldots.$$

$$\frac{2y\,d}{D} = (2n+1)\lambda/2$$

$$y = \frac{(2n+1)\lambda D}{4d}$$

This equation gives the distance of nth dark fringes from O.

Thus

$$y_n = \frac{(2n+1)D\,\lambda}{4d}$$

2A.11 Fringe Width (β)

The distance between any two consecutive bright fringes or any two consecutive dark fringes is called the fringe width (β).

(a) ***For bright fringes:*** If y_{n+1} and y_n denote the distance of two consecutive bright fringes from O, then we have

$$y_{n+1} = \frac{(n+1)D\,\lambda}{2d}$$

$$y_n = \frac{n\,D\,\lambda}{2d}$$

$$\therefore \text{ fringe width } \beta = y_{n+1} - y_n$$

$$\beta = \frac{(n+1)D\,\lambda}{2d} - \frac{n\,D\,\lambda}{2d} = \frac{D\lambda}{2d}$$

$$\beta = \frac{D\,\lambda}{2d}$$

(b) ***For dark fringes:*** If y_{n+1} and y_n are the distances of the two consecutive dark fringes from O, then we have

$$y_{n+1} = (2n+3)\frac{D\lambda}{4d}$$

$$y_n = (2n+1)\frac{D\lambda}{4d}$$

$$\therefore \quad \beta = y_{n+1} - y_n = (2n+3)\frac{D\lambda}{4d} - (2n+1)\frac{D\lambda}{4d}$$

$$= \frac{2D\lambda}{4d} = \frac{D\lambda}{2d}$$

$$\therefore \quad \beta = \frac{D\lambda}{2d}$$

Results

1. As β is independent of n, all the fringes are of same width.
2. As β is directly proportional to λ, the fringes produced by light of shorter wave will be narrower as compared to those produced by light of longer wavelength.
3. For appreciable fringe width D should be large and d is small.
4. On immersing the apparatus in a liquid (e.g., water) the wavelength of light decreases $\lambda' = \frac{\lambda}{\mu}$, thus when the apparatus is immersed in a liquid, the fringe width will decrease.
5. Since $\beta = \frac{\lambda D}{2d}$

Thus the fringe width increases as

(i) Wavelength increases

(ii) Distance D of the screen from the source increases

(iii) The distance 2d between the source decreases

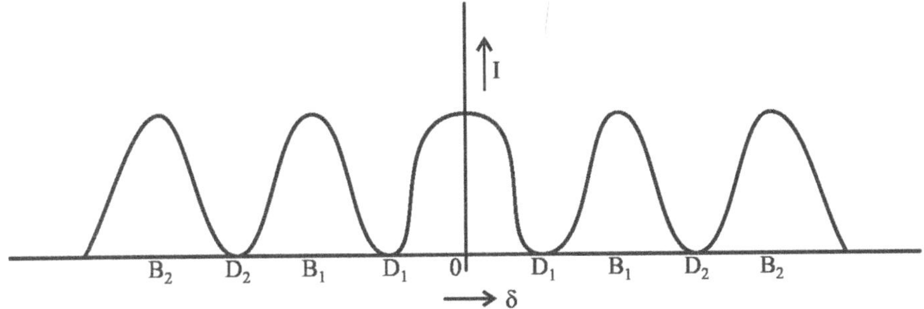

Fig. 2A.10

2A.12 Angular Fringe Width

The angular fringe width is defined as the angular separation between consecutive bright or dark fringes and is denoted by β_θ.

$$\beta_\theta = \frac{\beta}{D}$$

$$= \frac{\frac{\lambda D}{2d}}{D} = \frac{\lambda}{2d}$$

Thus $$\beta_\theta = \frac{\lambda}{2d}$$

2A.13 Shape of the Interference Fringes

At any point P on the screen, there is maximum or minimum intensity according to

$$S_2 P - S_1 P = n\lambda$$

$$S_2 P - S_1 P = (2n-1)\,\lambda/2$$

So for a given valve of n, the locus of points of maximum or minimum intensity is given by

$$S_2 P - S_1 P = \text{Constant}$$

This is the equation of a hyperbola with foci S_1 and S_2. Since the wavelength of light is extremely small of the order of 10^{-5} cm, the value of $S_2 P - S_1 P$ is also of that order. Hence these hyperbolae appear, more or less as straight lines.

2A.14 Fresnel's Biprism

Fresnel used a biprism to show interference phenomenon. The biprism abc consists of two acute angled prisms placed base to base. Actually it is constructed as a single prism of abtuse angle of about 179°. The acute angle α on both sides is about $30'$.

2A14.1 Experimental Arrangement of Biprism

The Biprism is placed with its refracting edge parallel to the line source S (slit) such that Sa is normal to the face bc of the prism. When light falls from S on the lower portion of the biprism, it is bent upwards and appears to come from the virtual source B. Similarly light falling from S on the upper portion of the prism, is bent downwards and appears to come from the virtual source A (Fig. 2A.11). Therefore A and B act as two coherent sources. Suppose the distance between A and B = 2d. If a screen is placed at C, interference fringes of equal width are produced between E and F but beyond E and F fringes of large width are produced which arc due to diffraction.

MN is a stop to limits the rays. To observe the fringes, the screen can be replaced by an eyepiece or low power microscope and fringes are seen in the field of view. If the point C is at the principle focus of the eye piece, the fringes are observed in the field of view.

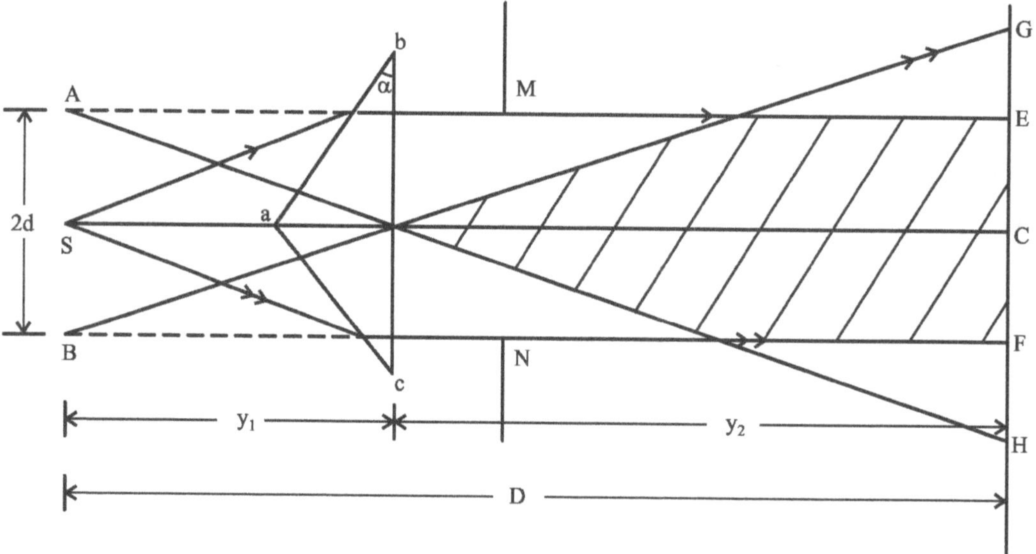

Fig. 2A.11

2A.14.2 Theory of Fresnel's Biprism

The point C is equidistant from A and B. Therefore it has maximum intensity on both sides of C, alternatively bright and dark fringes are produced. The width of the bright fringe or dark fringe $\beta = \frac{\lambda D}{2d}$, move over any point on the screen will be at the centre of a bright fringe if the distance from C is

$$\beta = \frac{n\lambda D}{2d} \qquad \text{where } n = 0, 1, 2, 3, \ldots$$

The point will be at the centre of a dark fringe if its distance from C is

$$\frac{(2n+1)\lambda D}{4d} \qquad \text{where } n = 0, 1, 2, 3, \ldots \text{ etc.}$$

2A.14.3 Determination of Wavelength of Light

Fresnel's biprism can be used to determine the wavelength of a given source of monochromatic light.

A fine vertical slit S is adjusted close to a source of light and the refracting edge is also set parallel to the slit S such that bc is horizontal. They are adjusted on an optical bench. A micrometer eyepiece is placed on the optical bench at some distance from the prism to view the fringe in its focal plane (at its cross wires).

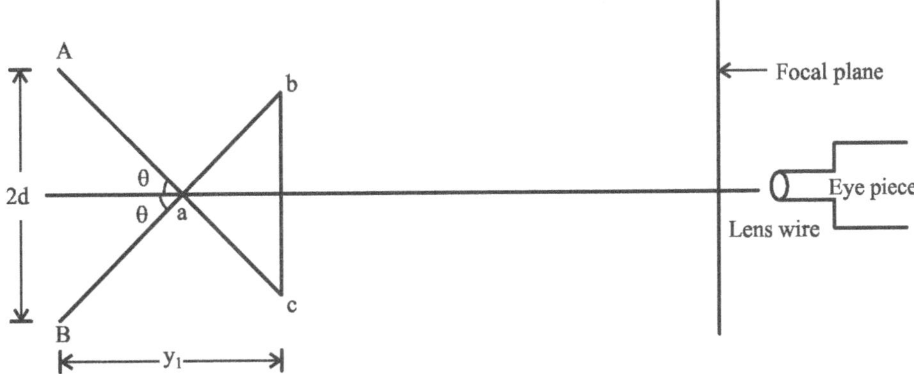

Fig. 2A.12

Suppose the distance between the source and the eyepiece = D and the distance between the two virtual sources A and B = d, eyepiece is moved horizontally (perpendicular to the length of the bench) to determine the fringe width, suppose for crossing 20 bright fringes from the field of view, the eyepiece has moved through a distance l.

Then the fringe width

$$\beta = \frac{l}{20}$$

But the fringe width

$$\beta = \frac{\lambda D}{2d}$$

$$\therefore \quad \lambda = \frac{\beta(2d)}{D} \qquad \text{.....(2A.17)}$$

2A.14.4 Determination of the Distance Between the Two Virtual Source (2d)

For this purpose, we apply displacement method. A convex lens is placed between the biprism and the eyepiece in such a position, that the images of the virtual sources A and B are seen in the field of view of the eyepiece. Suppose the lens is in the position L_1, measure the distance between the images of A and B as seen in the eyepiece. Let it be d_1 (Fig. 2A.13).

In this case

$$\frac{d_1}{2d} = \frac{v}{u} = \frac{n}{m} \qquad \text{.....(2A.18)}$$

Now move the lens towards the eyepiece and bring it to some other position L_2, so that again the images of A and B are seen, clearly in the field of view of the eyepiece. Measure the distance between the two images in this case also. Let it be equal to d_2 (Fig. 2A.14)

Here $\quad v = m$ and $u = n$

$$\frac{d_2}{2d} = \frac{v}{u} = \frac{m}{n} \qquad \text{.....(2A.19)}$$

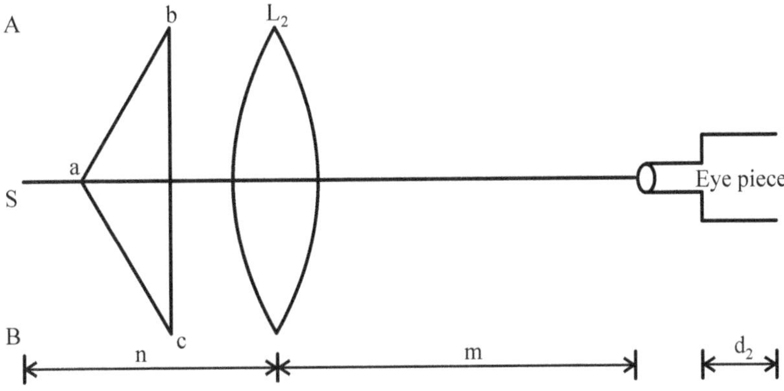

Fig. 2A.13

From eq. 2A.18 and 2A.19, we have

$$\frac{d_1\, d_2}{(2d)^2} = 1$$

$$d_1\, d_2 = (2d)^2$$

$$\therefore \quad \boxed{2d = \sqrt{d_1\, d_2}}$$

Here d_1 will be greater than d_2 and 2d is the geometrical mean of d_1 and d_2. Therefore d can be calculated. Substituting the values of 2d, β and D in eq. 2A.17, the wavelength of the given monochromatic light can be calculated.

The second method to find 2d is to measure accurately the refracting angle α. As the angle is small, the deviation produced

$$\theta = (\mu - 1)\,\alpha$$

Therefore the total angle between Aa and Ba is 2θ

$$= 2(\mu - 1)\,\alpha$$

If the distance between the prism and the slit S is y_1, then

$$\boxed{2d = 2(\mu - 1)\,\alpha\, y_1}$$

Therefore the distance 2d can be calculated

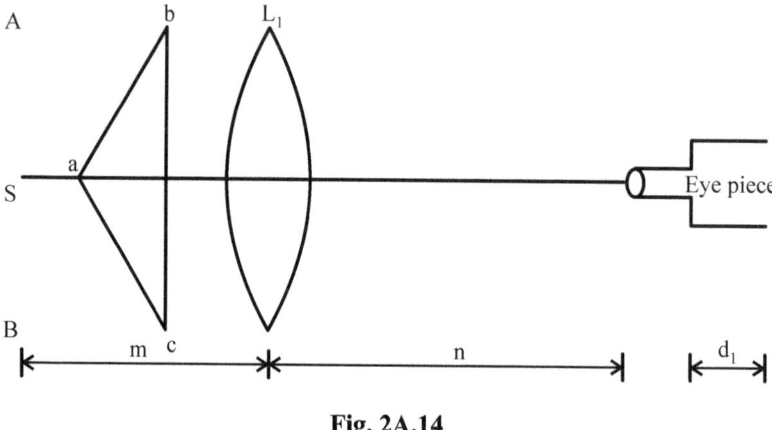

Fig. 2A.14

2A.15 Displacement of the Fringes
Determination of Thickness of Thin Transparent Sheet

Let us consider the effect of introducing a thin transparent plate of any material e.g., glass or mica in the path of the two interfering waves of light.

Let S_1 and S_2 be two coherent monochromatic sources giving light of wavelength λ (Fig. 2A.15). Let a thin plate of thickness t be introduced in the path of the light from S_1. Let μ be the refractive index of the plate for monochromatic light used.

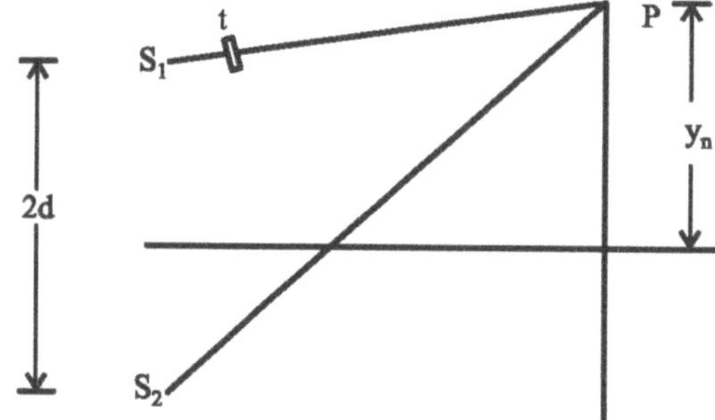

Fig. 2A.15

Here the light from S_1 travels partly in air and partly in the plate. The distance travelled from S_1 to P in air is (S_1 P – t) and that in the plate is t. If C and v are the velocities of light in air and the plates respectively, then time taken by light from S_1 to P

$$= \frac{S_1 P - t}{C} + \frac{t}{C} \qquad \text{since } \mu = \frac{C}{v}$$

$$= \frac{S_1 P - t}{C} + \frac{\mu t}{C}$$

$$= \frac{S_1 P + (\mu - 1)t}{C}$$

This is equivalent to supposing that the effective path in air from S_1 to P is $S_1P + (\mu - 1)\,t$ i.e., the air path S_1P has been increased by an amount $(\mu - 1)\,t$ by the introduction of the plate.

Let O be a point such that in the absence of the plate the optical paths S_1O and S_2O are equal and hence O is the position of the central maximum. But on introducing the plate, the optical paths S_1O and S_2O become unequal and therefore the two waves from S_1 and S_2 do not arrive at O simultaneously. Hence the waves from S_1 arrives $\frac{(\mu - 1)\,t}{C}$ seconds later than the wave from S_2. Therefore the central fringe is shifted to a point O′ situated on the side of the plate such that at point O′ the optical paths from S_1 and S_2 become equal. Such a shift results for all the fringes.

Now at any point P, the effective path differences

$$= S_2 P - [S_1 P + (\mu - 1)\,t]$$

$$= S_2 P - S_1 P - (\mu - 1)\,t] \qquad \text{.....(2A.20)}$$

If 2d is the distance between S_1 and S_2, D is the distance of the screen from the sources and $OP = y_n$, then we have

$$S_2 P - S_1 P = \frac{2d}{D}\, y_n \qquad \text{.....(2A.21)}$$

Substituting this value in eq. 2A.20, we have effective path difference at P

$$= \frac{2d}{D}\, y_n - (\mu - 1)t$$

If the point P is now the centre of nth bright fringe, we have

$$\frac{2d}{D}\, y_n - (\mu - 1)t = n\,\lambda$$

i.e.,

$$y_n = \frac{D}{2d}\,[n\lambda + (\mu - 1)t] \qquad \text{.....(2A.22)}$$

In the absence of the plate t = 0, the distance of the nth maximum from O

$$= \frac{D}{2d}\, n\,\lambda \qquad \text{.....(2A.23)}$$

∴ the displacement of nth bright fringe y_o is given by

$$y_o = \frac{D}{2d}\,[n\,\lambda + (\mu - 1)\,t] - \frac{D}{2d}\,n\,\lambda$$

$$y_o = \frac{D}{2d}\,[(\mu - 1)\,t] \qquad .....(2A.24)$$

This expression is independent of n. This indicates that all the bright fringes (including the central or zero order fringe) are displaced through the same amount equal to $\frac{D}{2d}$ $(\mu - 1)\,t$. Similarly it can be shown that the displacement of any dark fringe is also

$$= \frac{D}{2d}\,(\mu - 1)t$$

The fringe width β is given by

$$\beta = y_n - y_{n-1}$$

$$\beta = \frac{D}{2d}\,[n\,\lambda + (\mu - 1)\,t] - \frac{D}{2d}\,[(n - 1)\,\lambda + (\mu - 1)\,t]$$

$$\beta = \frac{D\lambda}{2d} \qquad .....(2A.25)$$

Which is identical with the width in the absence of the plate

Substituting value of $\frac{D}{2d}$ from (2A.25) and (2A.24) we get the shift

$$y_o = \frac{\beta}{\lambda}\,(\mu - 1)\,t \qquad .....(2A.26)$$

Thus we see that the introduction of the plate in the path of one of the interfering waves simply displaces the entire fringe system through a distance $\frac{D}{2d}$ $(\mu - 1)$ t or $\frac{\beta}{\lambda}$ $(\mu - 1)$ t towards the side on which the plate is placed and there is no other change in the appearance of fringes.

If the central fringe moves through a distance formerly occupied by n^{th} bright fringe, we have shift.

$$y_o = n\,\beta$$

$$\frac{\beta}{\lambda}\,(\mu - 1)\,t = n\,\beta$$

or $$n\lambda = (\mu - 1)\,t$$

$\therefore$ $$t = \frac{n\lambda}{\mu - 1} \qquad .....(2A.27)$$

Results

1. This equation gives us the number of order or bands through which the fringe system has shifted.
2. If λ and μ are known, then by measuring n and substituting the values in eq. (2A.27), the thickness of the plate t can be calculated.
3. If we use monochromatic light, central and all other fringes are exactly similar. Thus the shift of any fringes will not be observable and hence with the monochromatic light, the shifts of fringes cannot be measured.
4. ***With white light:*** When white light is used a few coloured fringes with a central white fringe are obtained on the screen in the absence of the plate. When plate is introduced in any one of the interfering beams, the central and other fringes are displaced by an amount $\frac{D}{2d}(\mu - 1)$, which depends on λ. Since μ changes with λ, for smaller wavelength μ is larger and the violet central fringe in shifted further than the red one.
5. In order to measure the shift in the fringes we must use white light and a very thin transparent plate (e.g., mica sheet).

2A.16 Change of Phase on Reflection: Stoke's Treatment Stoke's Statement

"When a light wave is reflected from denser medium, it always suffers a phase change of π or a path difference of $\frac{\lambda}{2}$".

Derivation

Let a ray AO of amplitude a be incident at O on the boundary M_1M_2 of two media 1 and 2 (Fig. 2A.16). Let the medium 2 be optically denser than medium 1. The ray AO is partly reflected along OB and partly refracted along OC. If r is the reflection coefficient i.e., the fraction of the amplitude reflected and t is the transmission coefficient (i.e.,) the fraction of amplitude refracted or transmitted then, we must have

$$r^2 + t^2 = 1 \qquad .....(2A.28)$$

This equation represents conservation of energy. The amplitude of the reflected wave OB will be ar and the amplitude of the transmitted wave OC be at.

Now if the paths of the reflected wave OB and the refracted wave OC are reversed, then by the law of reversibility they should give rise to the wave of original amplitude a, provided we assume that there is no dissipation of energy by absorption.

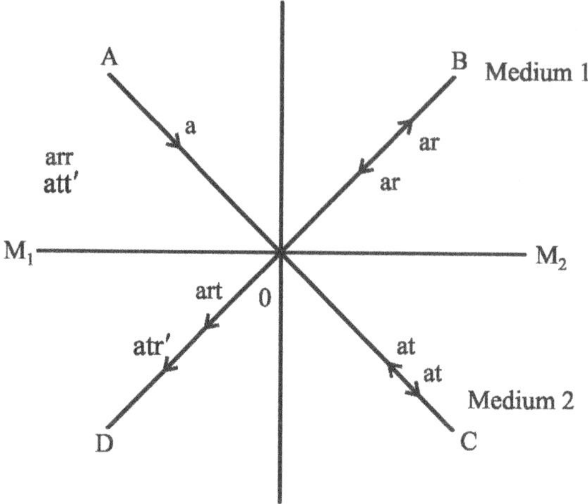

Fig. 2A.16

If we reverse the transmitted wave OC, we get one wave repacted along OA of amplitude att′ where t′ is the transmission coefficient from medium 2 to medium 1 and the other wave reflected along OD of the amplitude at r′ where r′ is the reflection coefficient for medium 2. Similarly if we reverse the reflected wave OB, we get a reflected wave along OA of amplitude ($arr = ar^2$) and a transmitted wave along OD of amplitude art.

According to principle of reversibility, we have

$$ar^2 + att' = a \qquad \text{.....(2A.29)}$$

As there was no wave originally along OD, if the light is retracting its path, there should be no wave along OD i.e.,

$$art + atr' = 0 \qquad \text{.....(2A.30)}$$

This gives

$$r = -\,r' \qquad \text{.....(2A.31)}$$

Equation (2A.29) gives

$$r^2 = 1 - tt' \qquad \text{.....(2A.32)}$$

From (2A.28) and (2A.32) we get

$$t = t' \qquad \text{.....(2A.33)}$$

eq.2A.33 represents that the coefficient of reflection from the surface of denser medium 1 is equal in magnitude to the coefficient of reflection from the surface of a rarer medium 2 but opposite in sign. The negative sign in amplitude corresponds to a phase change of π radians or a path difference of $\frac{\lambda}{2}$, where λ is the wavelength of incident wave.

Results

The reflection from the surface of the denser medium involves a phase change of π-radians, while reflection from the surface of rarer medium does not involves any phase change.

2A.17 Thin Films

2A.17.1 Interference Due to reflected Light (Thin Films)

Consider a transparent film of thickness t and respective index μ. A ray SA incident on the upper surface of the film is partly reflected along AT and partly refracted along AB. At B part of it is reflected along BC and finally emerges out along CQ. The difference in path between the two rays AT and CQ can be calculated. Draw CN normal to AT and AM normal to BC. The angle of incidence is *i* and the angle of refraction is r. Also produce CB to meet AE produce at P.

Here $\angle APC = r$

The optical path difference

$$x = \mu\,(AB + BC) - AN$$

Since $\mu = \frac{\sin i}{\sin r} = \frac{AN}{CM}$

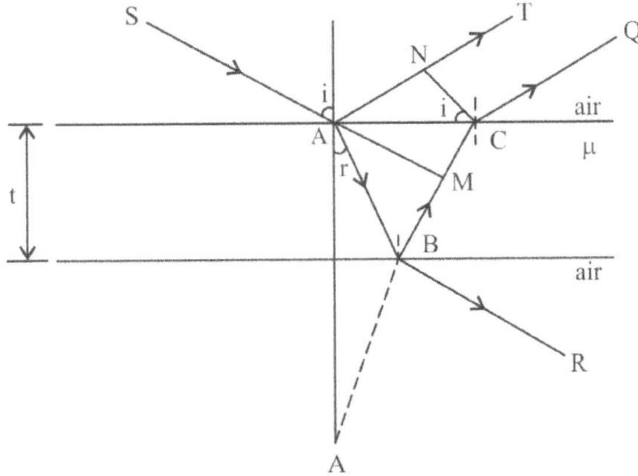

Fig. 2A.17

$$\therefore \qquad AN = \mu\, CM$$

$$x = \mu\,(AB + BC) - \mu\,(CM)$$

$$x = \mu\,(AB + BC - CM) = \mu\,(PC - CM)$$

$$x = \mu\, PM$$

In the Δ APM

$$\cos r = \frac{PM}{AP}$$

$$PM = AP \cos r = (AE + EP) \cos r$$

$$= 2t \cos r$$

$$\therefore \qquad AE = EP = t$$

$$\therefore \qquad x = \mu PM = 2\mu t \cos r \qquad .....(2A.34)$$

This equation in the case of reflected light does not represent the correct path difference but only the apparent. It has been established on the basis of electromagnetic theory that when light is reflected from the surface of an optically denser medium (air-medium interference) a phase change of π equivalent to a path difference of $\frac{\lambda}{2}$ occurs.

Therefore the correct path difference in this case is

$$x = 2\mu t \cos r - \frac{\lambda}{2} \qquad .....(2A.35)$$

Results

1. If the path difference $x = n\lambda$ where $n = 0, 1, 2, 3,$ constructive interference takes place and the film appear bright.

2. If the path difference $x = (2n+1)\frac{\lambda}{2}$ where $n = 0, 1, 2, 3, ...$ etc., destructive interference takes place and the film appears dark

$$\therefore \qquad 2\mu t \cos r - \frac{\lambda}{2} = (2n + 1)\frac{\lambda}{2} \qquad .....(2A.36)$$

or

$$2\mu t \cos r = (n + 1)\,\lambda$$

$$\boxed{2\mu t \cos r = n\,\lambda}$$

Hence n is an integer only, therefore (n + 1) can also be taken as n.

2A.17.2 Interference Due to Transmitted Light (Thin Films)

Consider a thin transparent film of thickness t and refractive index μ. A ray SA after reflection goes along AB. At B it is partly reflected along BC and partly refracted along BR. The ray BC after reflection at C, finally emerges along DQ. Here at B and C reflection takes place at the rarer medium. Therefore, no phase change occurs. Draw BM normal to CD and DN normal to BR. The optical path difference between DQ and BR is given by

The path difference is given by

$$x = \mu\,(BC + CD) - BN$$

Also $$\mu = \frac{\sin i}{\sin r} = \frac{BN}{MD}$$

or $$BN = \mu MD$$

In the above figure

$$\angle BPC = r \text{ and } CP = BC = CD$$

∴ $$BC + CD = PD$$

∴ $$x = \mu\,(PD) - \mu\,(MD)$$

$$x = \mu\,(PD - MD)$$

$$x = \mu PM$$

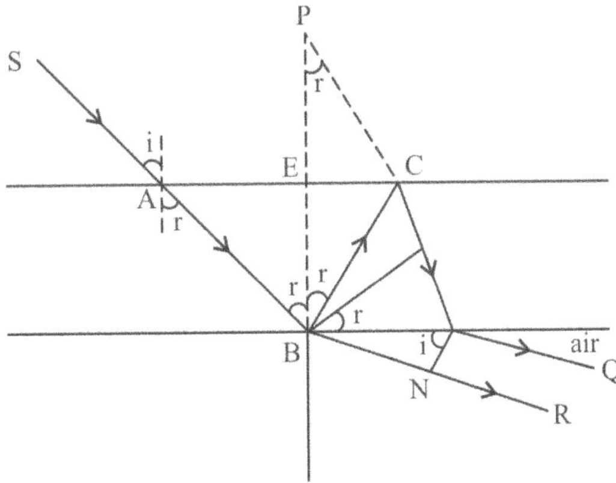

Fig. 2A.18

In the ΔBPM, $\cos r = \frac{PM}{BP}$ or

$$PM = BP \cos r$$

But $$BP = 2t$$

$$PM = 2t \cos r$$

$\therefore$ $$x = \mu PM$$

$$x = 2\mu t \cos r \qquad \text{.....(2A.37)}$$

1. For a bright fringe, the path difference

$$x = n\lambda$$

$$2\mu t \cos r = n\lambda$$

Where $n = 0, 1, 2, 3, \ldots$etc

2. For dark fringes, the path differences

$$x = (2n + 1)\frac{\lambda}{2}$$

$\therefore$ $$2\mu t \cos r = \frac{(2n+1)\lambda}{2}$$

where $n = 0, 1, 2, 3, \ldots$ etc

In the case of transmitted light, the interference fringes obtained are less distinct because the difference in amplitude between BR and DQ is very large. However when the angle of incidence is nearly 45° the fringes are more distinct.

2A.17.3 Colours of Thin Films

When white light is incident on a thin film, the light which comes from any point from it will not include the colour where wavelength satisfies the equation

$$2\mu t \cos r = n\lambda$$

in the reflected system.

Therefore the film will appear coloured and the colour will depend upon the thickness and the angle of inclination. If r and t are constant, the colour will be uniform. In the case of oil on water, different colours are seen because r and t vary.

2A.18 Interference due to Wedge-shaped thin Film i.e., Thin Film of Increasing Thickness

Let us consider a wedge-shaped film of refractive index μ bounded by two plane surfaces OA and OB at an angle θ. The thickness of the film is increasing from O to A.

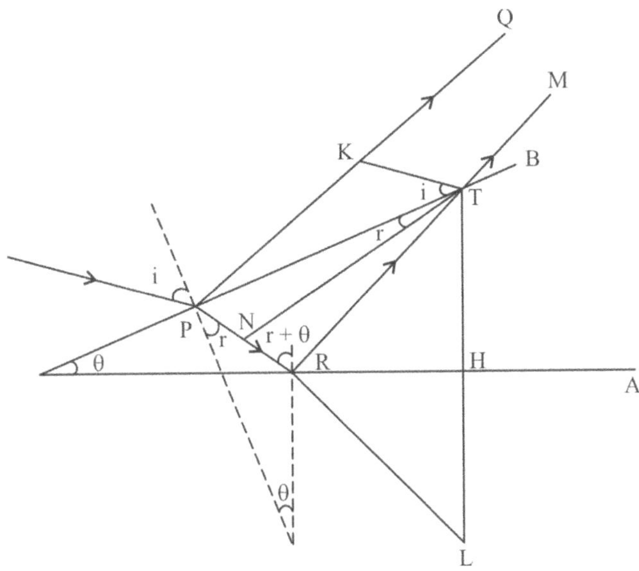

Fig. 2A.19

Let SP is the incident ray of monochromatic light on the upper surface of the film. PQ is the reflected ray and PR is the refracted ray (Fig. 2A.19). After reflection from the lower surface of the film at R it emerges out from the film in the direction TM. Let us draw perpendicular TN on PR and TK on PQ. The optical path difference between PQ and TM is therefore, gives as

Path difference = Path (PR + RT) in film – PK in air

$$= \mu\,(PR + RT) - PK$$

$$= \mu\,(PN + NR + RT) - PK \qquad .....(2A.38)$$

From simple geometry it may be shown that

$$\angle PTK = i \text{ and } \angle PTN = r$$

From ΔPKT

$$\sin i = \frac{PK}{PT} \qquad \text{and}$$

From ΔPNT

$$\sin r = \frac{PN}{PT}$$

$$\therefore \qquad \mu = \frac{\sin i}{\sin r} = \frac{PK}{PN}$$

$$\therefore \qquad PK = \mu PN \qquad .....(2A.39)$$

$$\text{Path difference} = \mu\,(PN + NR + RT) - \mu PN$$

$$= \mu\,(NR + RT) \qquad .....(2A.40)$$

Now triangle RTH and RHL are congruent and

$$\angle RTH = \angle RLH = r + \theta$$

$$\angle RHT = \angle RHL = 90^\circ \qquad \text{and} \quad \text{RH is common}$$

Hence $\qquad RT = RL \qquad .....(2A.41)$

Putting eq. 2A.41 in 2A.40 we get

$$\text{Path Difference} = \mu\,(NR + RL)$$

$$= \mu\,(NL) \qquad .....(2A.43)$$

From right angled ΔTNL

$$\frac{NL}{TL} = \cos(r + \theta)$$

$$\frac{NL}{TL} = \cos(r + \theta)$$

$$NL = TL \cos(r + \theta)$$

$$NL = (TH + HL) \cos(r + \theta)$$

$$NL = 2t \cos(r + \theta) \qquad .....(2A.44)$$

Because

$$TH = HL = t$$

Substituting the value of NL from (2A.44) in (2A.45) we get

$$\text{Path difference} = 2\mu t \cos(r + \theta) \qquad \text{.....(2A.45)}$$

Since the ray along PQ is the reflected ray from a denser medium hence there occurs an additional phase change of π radian or path difference of $\frac{\lambda}{2}$.

Therefore the effective path difference between the interfering rays PQ and TH is given as

$$\text{Path difference} = (2\mu t)\cos(r + \theta) - \frac{\lambda}{2}$$

Condition for maxima and minima

(a) ***For maxima or constructive interference***

$$2\mu t \cos(r + \theta) - \frac{\lambda}{2} = 2n\frac{\lambda}{2}$$

$$2\mu t \cos(r + \theta) = (2n + 1)\frac{\lambda}{2} \qquad \text{where } n = 0, 1, 2, 3, \ldots$$

$$2\mu t \cos(r + \theta) = (2n - 1)\frac{\lambda}{2} \qquad \text{where } n = 1, 2, 3, 4, \ldots$$

(b) ***For destructive interference or minima***

$$2\mu t \cos(r + \theta) - \frac{\lambda}{2} = (2n - 1)\frac{\lambda}{2}$$

$$2\mu t \cos(r + \theta) = n\lambda \qquad \text{where } n = 0, 1, 2, 3, \ldots$$

2A.19 Fringe Width

The condition for bright or maxima is given by

$$2\mu t \cos(r + \theta) = (2n - 1)\frac{\lambda}{2} \qquad \text{where } n = 1, 2, 3, \ldots \qquad \text{.....(2A.46)}$$

Let the distance of the nth bright fringe from the edge of the film is X_n.

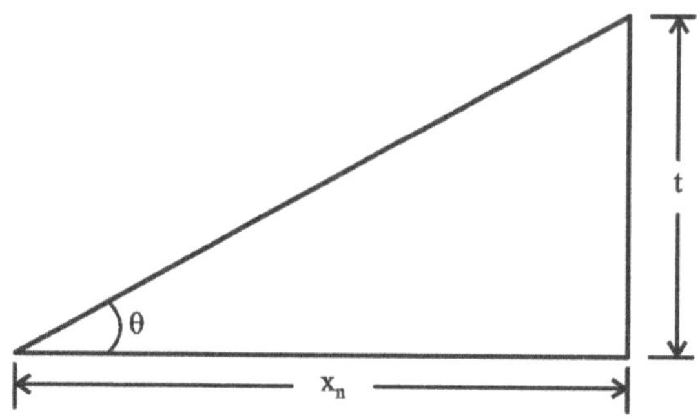

Fig. 2A.20

Thus from the Fig. 2A.20

$$\frac{t}{X_n} = \tan\theta$$

$$t = X_n \tan\theta \quad \text{.....(2A.47)}$$

Substituting the value of t from 2A.47 in 2A.46, we get

$$2\mu X_n \tan\theta \cos(r+\theta) = (2n-1)\frac{\lambda}{2} \quad \text{.....(2A.48)}$$

Similarly if X_{n+1} is the distance of the $(n+1)^{th}$ bright fringe, then

$$2\mu X_{n+1} \tan\theta \cos(r+\theta) = (2n+1)\frac{\lambda}{2} \quad \text{.....(2A.49)}$$

Subtracting eq. 2A.48 from 2A.49, we get

$$2\mu (X_{n+1} - X_n) \tan\theta \cos(r+\theta) = \lambda$$

Since $X_{n+1} - X_n$ is equal to the fringe width β

then $$2\mu\beta \tan\theta \cos(r+\theta) = \lambda$$

$$\beta = \frac{\lambda}{2\mu \tan\theta \cos(r+\theta)}$$

This is the same for two consecutive dark fringes also

For normal incidence

$$i = r = 0 \qquad \cos(r+\theta) = \cos\theta$$

$$\beta = \frac{\lambda}{2\mu \tan\theta \cos\theta} = \frac{\lambda}{2\mu \sin\theta}$$

If θ is very small then $\sin\theta \approx \theta$

$$\beta = \frac{\lambda}{2\mu\theta}$$

2A.20 Newton's Rings

When a Planoconvex lens of large focal length is placed in contact with a plane glass plate, we get a thin air film between the lower surface of the lens and the upper surface of the plate. The thickness of this film is very small at the joint of contact and gradually increases as we proceed outwards from the centre or from the point of contact. When monochromatic light falls normally on such a film we get an inner dark spot surrounded by alternatively bright and dark circular rings when seen by reflected light and vice versa by transmitted light.

If the incident light is white, a series of concentric coloured rings is seen around about the point of contact in both the reflected and transmitted light. The air film is wedge shaped and the loci of the points of equal thickness are circles concentric with the point of contact. The interference fringes or rings of equal thickness so formed were first investigated by Newton and hence they are called Newton's rings.

2A.20.1 Newton's Rings by Reflected Light

(a) ***Experimental arrangement:*** S is an extended source of monochromatic light placed at the focus of lens L_1. The light rays emitted from S are made parallel by a lens L_1. These horizontal parallel rays falls on a glass plate G at 45° and are partly reflected from it. These reflected rays falls normally on the lens L placed on the glass plate E. Interference occurs between the rays reflected from the upper and lower surfaces of the film. The interference rings are viewed with a low power microscope focused on the air film where the rings are formed (Fig. 2A.21).

(b) ***Theory of Newton's Ring:*** The phenomenon of Newton's Rings was satisfactorily explained by Young. These rings are produced as a result of interference between the light waves reflected from the upper and the lower surfaces of the air film formed between the plano convex lens and the glass plate. In Fig. 2A.22 '1' and '2' are the reflected interfering rays corresponding to an incident ray SP.

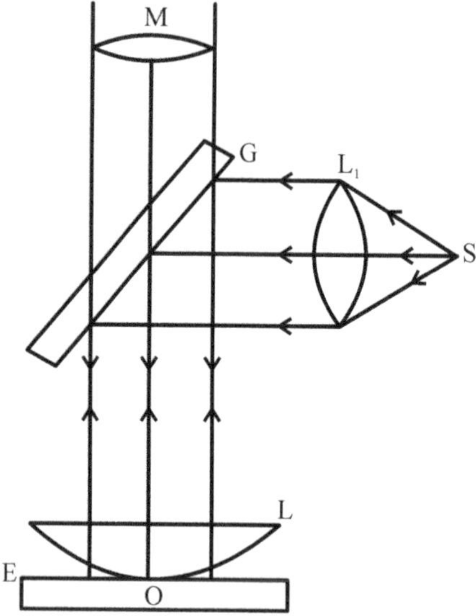

Fig. 2A.21

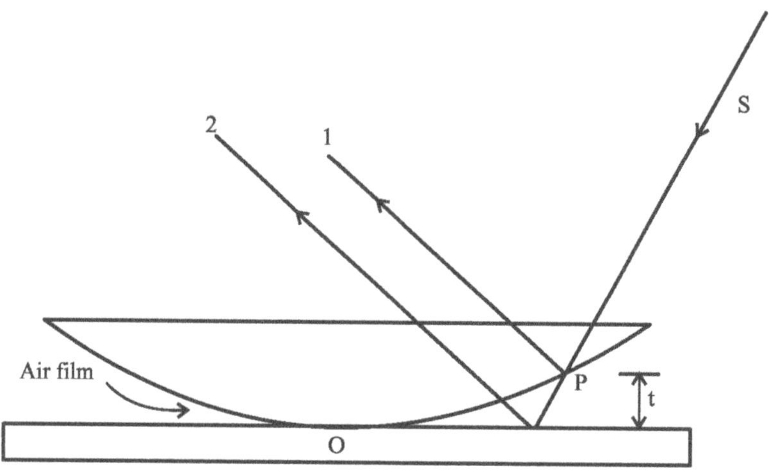

Fig. 2A.22

The effective path difference between the interfering rays in reflected light

$$2\mu t \cos(r + \theta) - \frac{\lambda}{2}$$

where μ is the refractive index of the film (air), t is the thickness of the film at the point P, r is the angle of refraction inside the film and θ is the angle of the wedge.

At the point of contact O, t = o; therefore the path difference is equal of $\frac{\lambda}{2}$. This is the condition of minimum intensity. Hence the central spot of the ring is dark.

The condition for maximum intensity or for a bright ring

$$2\mu t - \frac{\lambda}{2} = \frac{2n\lambda}{2} \qquad \text{where } n = 0, 1, 2, 3$$

$$2\mu t = (2n + 1)\frac{\lambda}{2} \qquad .....(2A.50)$$

or

$$2\mu t = (2n - 1)\frac{\lambda}{2} \qquad \text{where } n = 1, 2, 3, ...$$

The condition for minimum intensity i.e., for dark rings

$$2\mu t - \frac{\lambda}{2} = (2n - 1)\frac{\lambda}{2} \qquad \text{where } n = 0, 1, 2, 3......$$

$$2\mu t = n\lambda \qquad .....(2A.51)$$

It is clear from 2A.50 and 2A.51 that for a bright or dark ring of any particular order, t should be constant. But here in the air film the locus of points having the same thickness is a circle with its centre at the point of contact. Hence the fringes are circular rings having the common centre at the point of contact.

(c) ***Diameters of darks and bright rings:*** In Fig. 2A.23 L is a plano convex lens placed on the glass plate E with O as the point of contact. Let R be the radius of the curved surface of the lens and t the thickness of the film at the certain point.

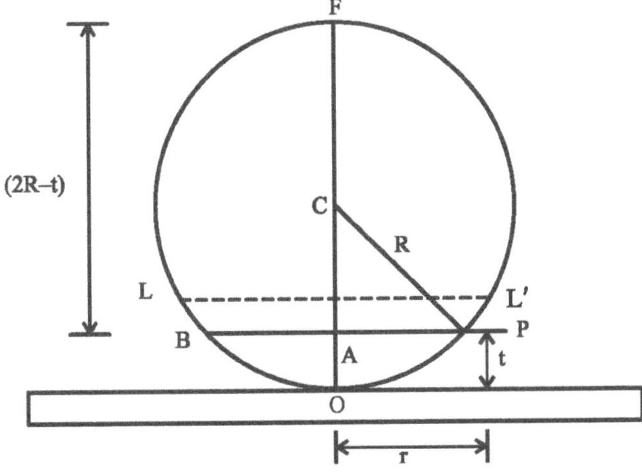

Fig. 2A.23

From the property of the circle

$$AP^2 = OA \times OF$$

But AP = AB = r, the radius of the ring passing through P.

$$OA = PH = t$$

We have

$$r^2 = t\,(2R - t) = 2Rt - t^2$$

In actual practice R is quite large and t is very small. Therefore t^2 may be neglected in comparison with 2Rt.

$$r^2 = 2Rt$$

$$t = \frac{r^2}{2R} \quad \text{.....(2A.52)}$$

(a) ***For bright rings:*** Substituting this value of t from equation 2A.52 in 2A.50, we get

$$2\mu t = (2n - 1)\,\frac{\lambda}{2}$$

$$2\mu.\ \frac{r^2}{2R} = (2n - 1)\,\frac{\lambda}{2}$$

$$r^2 = \frac{(2n-1)\lambda R}{2\mu}$$

This denotes the radius of nth bright ring, thus we have

$$r_n^2 = \frac{(2n-1)\lambda R}{2\mu} \quad \text{.....(2A.53)}$$

If D_n is the diameter of nth bright ring, we have

$r_n = \dfrac{D_n}{2}$ therefore equation 2A.53 becomes

$$\left(\frac{D_n}{2}\right)^2 = \frac{(2n-1)\lambda R}{2\mu}$$

$$D_n^2 = \frac{4(2n-1)\lambda R}{2\mu}$$

$$D_n^2 = \frac{2(2n-1)\lambda R}{\mu} \qquad \text{.....(2A.54)}$$

For air film $\mu = 1$

$$D_n^2 = 2(2n-1)\lambda R \qquad \text{.....(2A.54(a))}$$

$$D_n = \sqrt{2(2n-1)\lambda R}$$

$$D_n = \sqrt{2\lambda R}\ \sqrt{2n-1} \qquad \text{.....(2A.55)}$$

$$D_n = K\sqrt{2n-1} \qquad \text{where } K = \sqrt{2\lambda R}$$

As n = 1, 2, 3 ...... (2n – 1) is an odd number. Thus the diameters of successive bright rings are proportional to the square root of the odd natural numbers.

(b) *For dark rings*

Substituting the value of t from 2A.52 in 2A.51, we get

$$2\mu\frac{r_n^2}{2R} = n\lambda$$

or

$$r_n^2 = \frac{n\lambda R}{\mu} \qquad \text{.....(2A.56)}$$

If D_n is the diameter of the nth dark ring, we have

$$r_n = \frac{D_n}{2}$$

Therefore equation 2A.56 gives

$$D_n^2 = \frac{4n\lambda R}{\mu} \qquad \text{.....(2A.57)}$$

For air film $\quad \mu = 1$

$$D_n^2 = 4n\lambda R$$

$$D_n = \sqrt{4n\lambda R}$$

$$D_n = \sqrt{4\lambda R}\,.\sqrt{n} \qquad \text{.....(2A.58)}$$

or $$D_n = K'\sqrt{n} \qquad \text{where } K' = \sqrt{4\lambda R}$$

Thus the diameter of successive dark rings is proportional to the square root of the natural numbers.

2A.20.2 Determination of Refractive Index of Liquid by Newton's Rings

The liquid whose refractive index μ is to be determined is placed between the lens and the glass plate.

The diameter of the nth dark ring is given by

$$D_n^2 = \frac{4n\lambda R}{\mu} \qquad \text{.....(2A.57)}$$

Similarly for $(n + p)^{th}$ ring

$$D_{n+p}^2 = \frac{4(n+p)\lambda R}{\mu} \qquad \text{.....(2A.59)}$$

Subtracting (2A.57) from 2A.59, we get

$$\left[D_{n+p}^2 - D_n^2\right]_{liquid} = \frac{4p\lambda R}{\mu} \qquad \text{.....(2A.60)}$$

For air film $\mu = 1$, therefore

$$\left[D_{n+p}^2 - D_n^2\right]_{air} = 4p\lambda R \qquad \text{.....(2A.61)}$$

Dividing 2A.61 by 2A.60 we get

$$\mu = \frac{\left[D_{n+p}^2 - D_n^2\right]_{air}}{\left[D_{n+p}^2 - D_n^2\right]_{liquid}} \qquad \text{.....(2A.62)}$$

Thus μ can be determined by measuring the diameter of Newton's rings with liquid film and air film and using the same value of p in calculation.

Note :

With liquid film the diameter of nth dark ring is

$$\left(D_n^2\right)_{liquid} = \frac{4n\lambda R}{\mu} \qquad \text{.....(2A.63)}$$

and with air film the diameter of nth dark ring is

$$\left(D_n^2\right)_{air} = 4n\lambda R \qquad .....(2A.64)$$

$$\therefore \qquad \left(D_n\right)_{liquid} = \frac{\sqrt{4n\lambda R}}{\sqrt{\mu}} \qquad .....(2A.65)$$

$$\text{and} \qquad \left(D_n\right)_{air} = \sqrt{4n\lambda R} \qquad .....(2A.66)$$

dividing eq. 2A.65 by 2A.66

$$\frac{\left(D_n\right)_{liquid}}{\left(D_n\right)_{air}} = \frac{1}{\sqrt{\mu}}$$

$$\text{i.e.,} \qquad \frac{\text{Diameter of a ring in liquid film}}{\text{Diameter of the same ring in air film}} = \frac{1}{\sqrt{\mu}} \qquad .....(2A.67)$$

Thus when a little liquid is introduced between the lens and the plate, the rings are contracted according to the relation (2A.67).

Since $\mu > 1$

Therefore $\left(D_n\right)_{liquid} < \left(D_n\right)_{air}$

Newtons' rings with both surfaces curved

In Newton's ring arrangement, if the plane glass plate is replaced by a curved surface, the rings are formed due to reflection from the upper and lower surfaces of the air film. The following two experimental arrangements are possible.

Case I

When the lower surface is convex as shown in Fig. 2A.24

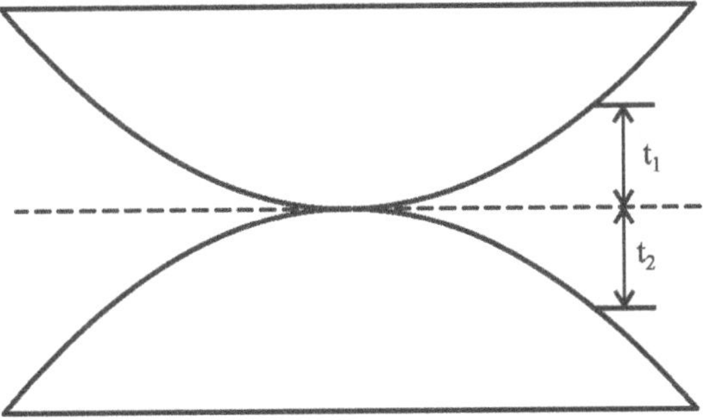

Fig. 2A.24

The total thickness of the air film at any point as shown in the Fig. 2A.24 is given by

$$t = t_1 + t_2 \quad \text{......(2A.68)}$$

If R_1 and R_2 be the radii of curvatures of the upper and lower curved surfaces respectively, then

$$t = \frac{r_n^2}{2R_1} + \frac{r_n^2}{2R_2}$$

or

$$2t = \frac{r_n^2}{R_1} + \frac{r_n^2}{R_2} \quad \text{.....(2A.69)}$$

where r_n is the radius of curvature of nth ring corresponding to thickness t.

The condition for dark rings at normal incidence and air film is given by

$$2t = n\lambda \quad \text{.....(2A.70)}$$

From (2A.69) and (2A.70)

$$r_n^2 \left[\frac{1}{R_1} + \frac{1}{R_2}\right] = n\lambda \quad \text{.....(2A.71)}$$

If D_n is the diameter of the nth dark ring, then

$$r_n = \frac{D_n}{2} \quad \text{.....(2A.72)}$$

∴ from 2A.71 and 2A.72 we have

$$\frac{D_n^2}{4}\left[\frac{1}{R_1} + \frac{1}{R_2}\right] = n\lambda$$

$$D_n^2 = \frac{4n\lambda}{\left[\frac{1}{R_1} + \frac{1}{R_2}\right]} \quad \text{.....(2A.73)}$$

Case 2

When the lower surface is concave as shown in Fig. 2A.25.

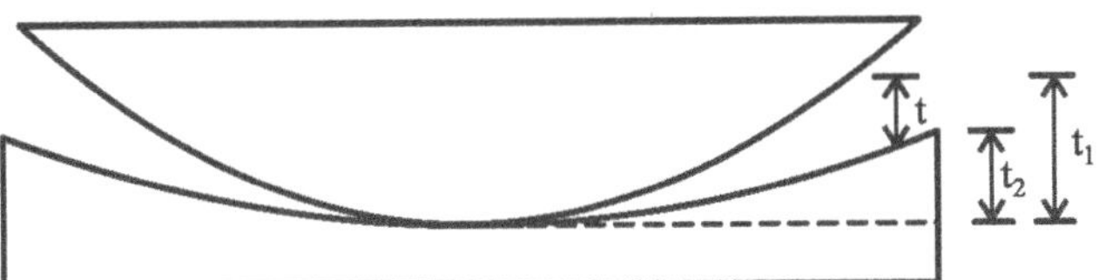

Fig. 2A.25

It is clear from the Fig. that

$$t = t_1 - t_2 \qquad .....(2A.74)$$

Substituting $t_1 = \frac{r_n^2}{2R_1}$ and $t_2 = \frac{r_n^2}{2R_2}$ in Eq. 2A.74, we get

$$t = \frac{r_n^2}{2R_1} - \frac{r_n^2}{2R_2}$$

$$2t = r_n^2\left[\frac{1}{R_1} - \frac{1}{R_2}\right] \qquad .....(2A.75)$$

The condition for dark rings at normal incidence and air film is

$$2t = n\lambda \qquad .....(2A.76)$$

$$\therefore \qquad 2t = r_n^2\left[\frac{1}{R_1} - \frac{1}{R_2}\right]$$

Let D_n be the diameter of the nth dark ring then

$$r_n = \frac{D_n}{2}$$

$$2t = \frac{D_n^2}{4}\left[\frac{1}{R_1} - \frac{1}{R_2}\right]$$

$$n\lambda = \frac{D_n^2}{4}\left[\frac{1}{R_1} - \frac{1}{R_2}\right]$$

$$\frac{D_n^2}{4} = \frac{4n\lambda}{\left[\frac{1}{R_1} - \frac{1}{R_2}\right]} \qquad .....(2A.77)$$

Solved Numerical Problems

1. Two coherent sources whose intensity ratio is 100 : 1 produce interference fringes. Deduce the ratio of maximum intensity to minimum intensity in fringe system.

Sol. We know that

$$I_{max} = (a_1 + a_2)^2$$

and $$I_{min} = (a_1 - a_2)^2$$

$$\therefore \quad \frac{I_{max}}{I_{min}} = \frac{(a_1 + a_2)^2}{(a_1 - a_2)^2}$$

Given,

$$\frac{I_1}{I_2} = \frac{a_1^2}{a_2^2} = \frac{100}{1}$$

$$\therefore \quad \frac{a_1}{a_2} = \frac{10}{1}$$

$$\therefore \quad \frac{I_{max}}{I_{min}} = \frac{(10a_2 + a_2)^2}{(10a_2 - a_2)^2} = \left(\frac{11}{9}\right)^2 = \frac{121}{81}$$

$$\therefore \quad \frac{I_{max}}{I_{min}} = \frac{121}{81}$$

2. In an interference pattern with two coherent sources, the amplitude of intensity variation is found to be 5% of the average intensity. Calculate the relative intensities of the interfering sources. [UPTU]

Sol. Let the amplitude ratio be a : 1

where a > 1

$$\therefore \quad \frac{I_{max}}{I_{min}} = \frac{(a + 1)^2}{(a - 1)^2}$$

Here, if average intensity is 100 units, then the maximum is 105 and minimum is 95. Therefore, the ratio is

$$\frac{I_{max}}{I_{min}} = \frac{105}{95} = \left(\frac{a + 1}{a - 1}\right)^2$$

or $$\frac{a+1}{a-1} = 1.051$$

or $$a + 1 = 1.051\ (a - 1)$$

or $$a = \frac{2.051}{0.051}$$

$$a \simeq 40$$

Therefore, the intensity ratio of the two interfering sources is

$$I_1 : I_2 = 1600 : 1$$

3. Two coherent sources of intensity ratio α interfere. Prove that in the interference pattern. [UPTU]

$$\frac{I_{max} - I_{min}}{I_{max} + I_{min}} = \frac{2\sqrt{\alpha}}{1+\alpha}$$

Sol. Let I_1 and I_2 be the intensities, and a_1 and a_2 the amplitudes of the disturbances from the two coherent sources.

Since, the intensity of wave is proportional to the square of the amplitude, we can write

$$\frac{a_1^2}{a_2^2} = \left(\frac{I_1}{I_s}\right)$$

or $$\sqrt{\left(\frac{I_1}{I_2}\right)} = \frac{a_1}{a_2}$$

since $$\frac{I_1}{I_2} = \frac{a_1^2}{a_2^2} = \alpha$$

$\therefore$ $$\frac{a_1}{a_2} = \sqrt{\alpha}$$

$$\frac{I_{max} - I_{min}}{I_{max} + I_{min}} = \frac{(a_1 + a_2)^2 - (a_1 - a_2)^2}{(a_1 + a_2)^2 + (a_1 - a_2)^2}$$

$$\frac{I_{max} - I_{min}}{I_{max} + I_{min}} = \frac{2a_1a_2}{a_1^2 + a_2^2}$$

or
$$\frac{I_{max} - I_{min}}{I_{max} + I_{min}} = \frac{2\left(\frac{a_1}{a_2}\right)}{\left[\left(\frac{a_1}{a_2}\right)^2 + 1\right]}$$

$$\frac{I_{max} - I_{min}}{I_{max} + I_{min}} = \frac{2\sqrt{\alpha}}{\alpha + 1}$$

4. Calculate the separation between two consecutive bright and dark fringes. [UPTU]

Sol. The portion of nth bright fringe is given by

$$y_n = \frac{n\lambda D}{2d} \qquad \text{..... (1)}$$

where λ is the wavelength of light, 2d the distance between two sources and D their distance from the screen

The position of nth order dark fringe is given by

$$y_n = (2n - 1)\frac{\lambda D}{4d} \qquad \text{..... (2)}$$

Therefore, the separation between two consecutive bright and dark fringes is

$$y_n - y_n^1 = \frac{n\lambda D}{2d} - \left(n - \frac{1}{2}\right)\frac{\lambda D}{2d}$$

$$y_n - y_n^1 = \frac{\lambda D}{4d} = \frac{1}{2}\left(\frac{\lambda D}{2d}\right)$$

5. In a two slit interference pattern at a point we observe 10th order maxima for $\lambda = 7000\ \mathring{A}$. What order will be visible here if the source of light is replaced by light of wavelength $5000\ \mathring{A}$. [UPTU]

Sol. For maximum intensity or maxima at point on the screen in two slit interference pattern the path difference between the rays originating from two coherent sources must be an even multiple of half wavelength of light used, that is

Path difference

$$\Delta = 2n\left(\lambda/2\right)$$

$$\Delta = n\lambda \quad ; \qquad \text{where } n = 0, 1, 2, 3, \ldots..$$

Suppose at a point, n^{th} order maximum for wavelength λ_1 and n_2^{th} order maximum for wavelength λ_2 will be visible, then

$$\Delta = n_1 \lambda_1 = n_2 \lambda_2$$

Given $\lambda_1 = 7000\ \mathring{A}$

$n_1 = 10$

$\lambda_2 = 5000\ \mathring{A}$

$n_2 = ?$

$$n_2 = \frac{n_1\lambda_1}{\lambda_2} = \frac{10 \times 7000}{5000} = 14$$

That is 14th order maximum will be visible.

6. Two coherent sources of monochromatic light of wavelength 6000 Å produce an interference pattern on a screen kept at a distance of 1 metre from them. The distance between two consecutive bright fringes on the screen is 0.5 mm. Find the distance between the two coherent sources. [UPTU]

Sol. The fringe width β is given by

$$\beta = \frac{D\lambda}{2d}$$

or the distance between sources

$$2d = \frac{D\lambda}{\beta}$$

Given $D = 1m = 100\ cm$

$\lambda = 6000\ \mathring{A} = 6000 \times 10^{-8}\ cm$

and $\beta = 0.5\ mm = 0.05\ cm$

$$\therefore \quad 2d = \frac{100 \times 6000 \times 10^{-8}}{0.05}$$

$2d = 0.12\ cm$

$2d = 1.2\ mm$

7. In front of a lecture hall, a coherent beam of monochromatic light from a helium-neon laser (λ = 632.8 nm) illuminates a double slit. From there it travels a distance of 20.0 m to a mirror at the back of the hall and returns the same distance to a screen. [UPTU]

1. In order that the distance between interference maxima be 10.0 cm, what should be the distance between the slits?
2. State what will you see if the lecturer slips a thin sheet of cellophone over one of the slits. The path through the cellophane contains 2.5 more waves than the path through air of same geometrical thickness.

Sol. 1. If 2d is the distance between the slits, then fringe width β is given by

$$\beta = \frac{\lambda D}{2d}$$

or $$2d = \frac{\lambda D}{\beta}$$

Here $D = 20 \text{ m} + 20 \text{ m} = 40 \text{ m}$

$$\lambda = 632.8 \text{ nm} = 6.328 \times 10^{-7} \text{ m}$$

and $\beta = 10 \text{ cm} = 0.1 \text{ m}$

∴ $$2d = \frac{6.328 \times 10^{-7} \times 40}{0.1}$$

$$2d = 0.253 \times 10^{-3} \text{ m}$$

2. If the lecturer slips a thin sheet cellophane over one of the slits, which accommodates 2.5 waves more than in air, then the entire fringe pattern will shift towards the cellophane sheet, keeping its nature unchanged.

Due to an additional path difference of $2\lambda + \lambda/2$ which is effectively $\lambda/2$, the dark and bright fringes will be mutually interchanged.

8. In Young's double slit experiment the slits are 0.5 mm apart and interference is observed on a screen placed at a distance of 100 cm from the slits. It is found that the 9th bright fringe is at a distance of 8.835 mm from the second dark fringe from the centre of the pattern. Find the wavelength of light used.

Sol. The position of nth bright fringe in the interference pattern is

$$y_n = \frac{n\lambda D}{2d}$$

Here n = 9, therefore

$$y_9 = \frac{9\lambda D}{2d} \qquad .....(1)$$

Similarly, the position of nth dark fringe is

$$y_n = \frac{(2n-1)\lambda D}{4d}$$

Hence n = 2, therefore

$$y_2 = \frac{(2n-1)\lambda D}{4d}$$

$$y_2 = \frac{3\lambda D}{4d} \qquad ..... (2)$$

But $y_9 - y_2 = 8.835$ mm

$y_9 - y_2 = 8.835 \times 10^{-3}$ m

$\therefore \quad \frac{9\lambda D}{2d} - \frac{3\lambda D}{4d} = \frac{15\lambda D}{2(2d)} = 8.835 \times 10^{-3}$

or $\quad \lambda = \frac{8.835 \times 10^{-3} \times 2 \times 2d}{15D}$

Given $2d = 0.5$ mm

$2d = 0.5 \times 10^{-3}$ m

and $D = 100$ cm $= 1$m

$\therefore \quad \lambda = \frac{8.835 \times 10^{-3} \times 2 \times 0.5 \times 10^{-3}}{15 \times 1}$

$\therefore \quad \lambda = 0.589 \times 10^{-6}$ m

$\lambda = 5890$ Å

9. Two coherent sources are 2.0 mm apart and are illuminated with a monochromatic light of wavelength 5896 Å. Fringes are observed at a distance 60 cm from the sources. Find the fringe width. Comment on the shape of the fringes.

Sol. As the fringe width is given by

$$\beta = \frac{\lambda D}{2d}$$

Where $2d = 2.0 \text{ mm} = 2.0 \times 10^{-3} \text{ m}$

$$\lambda = 5896 \text{ Å} = 5896 \times 10^{-10} \text{ m}$$

$$D = 60 \text{ cm} = 0.60 \text{ m}$$

$$\beta = \frac{(0.60\text{m}) \times (5896 \times 10^{-10}\text{m})}{(2.0 \times 10^{-3}\text{m})}$$

$$\beta = 1.77 \times 10^{-4} \text{ m}$$

Actual shape of fringes is hyperboidal but due to extremely small wavelength of light, fringes appear almost straight.

10. In a Young's double slit experiment the angular width of a fringe formed on a distant screen is 0.1°. The wavelength of light used is 6000 Å. What is the spacing between the slits.

Sol. Angular fringe width is given

$$\theta = \frac{\lambda}{2d}$$

$$\theta = 0.1^\circ = \frac{0.1^\circ \times 3.14}{180^\circ}$$

$$\theta = 1.74 \times 10^{-3} \text{ rad}$$

$$\lambda = 6000 \text{ Å} = 6000 \times 10^{-10} \text{ m}$$

$$2d = \frac{6000 \times 10^{-10}\text{m}}{1.74 \times 10^{-3}\text{rad}}$$

$$2d = 3.4 \times 10^{-4} \text{ m}$$

$$2d = 0.34 \text{ mm}$$

11. A beam of light consisting of two wavelengths 6500 Å and 5200 Å is used to obtain interference fringes in a Young's double slit experiment.

(i) Find the distance of the third bright fringe on the screen from the central maximum for $\lambda = 6500$ Å.

(ii) What is the least distance from the central maximum when the bright fringe due to both the wavelengths coincides. The distance between the slits is 2 mm and the distance between the plane of the slits and the screen is 120 cm.

Sol. (i) The distance of nth maxima from central maxima is given by

$$y_n = \frac{n\lambda D}{2d}$$

$$n = 3$$

$$D = 1.20 \text{ m}$$

$$2d = 2 \times 10^{-3} \text{ m}$$

$$\lambda = 6500 \text{ A}^{\circ} = 6.5 \times 10^{-7} \text{ m}$$

$$y_n = \frac{3 \times 1.20 \times 6.5 \times 10^{-7}}{2 \times 10^{-3}} = 1.17 \times 10^{-3} \text{ m} = 1.17 \text{ mm}$$

(ii) Let n^{th} bright fringe of $\lambda_1 (6500\ \mathring{A})$ and $(n + 1)^{th}$ bright fringe of $\lambda_2\ (5200\ \mathring{A})$ coincide, then

$$\frac{nD\lambda_1}{2d} = (n+1)\frac{D\lambda_2}{2d}$$

$$\therefore \quad n\lambda_1 = (n + 1)\lambda_2$$

$$\frac{\lambda_1}{\lambda_2} = \frac{n+1}{n}$$

$$\frac{6500\ \mathring{A}}{5200\ \mathring{A}} = \frac{n+1}{n}$$

$$\frac{5}{4} = \frac{n+1}{n}$$

$$5n = 4(n + 1)$$

$$n = 4$$

$\therefore$ Required least distance from the central maximum

$$y = 4 \times \frac{D\lambda_1}{2d}$$

$$y = 4 \times \frac{1.20 \times 6500 \times 10^{-10}}{2 \times 10^{-3}} \text{ m}$$

$$y = 1.56 \text{ mm}$$

12. In a two slit interference pattern width $\lambda = 6000 \overset{o}{A}$, the zero order and tenth order maxima fall at 12.34 mm and 14.73 mm respectively. If λ is changed to $5000 \overset{o}{A}$, deduce the positions of the zero – order and twentieth fringes, other arrangement remaining the same.

Sol. Given that $\lambda = 6000 \overset{o}{A}$

The distance between zero-order and tenth order fringe is

$$= 14.73 - 12.34$$

$$= 2.39 \text{ mm}$$

∴ fringe width

$$\beta = \frac{\lambda D}{2d}$$

∴ $$\beta = \frac{2.39\text{mm}}{10 - 0} = 0.239 \text{ mm}$$

$$\frac{(\beta)_{6000}}{(\beta)_{5000}} = \frac{6000 \overset{o}{A}}{5000 \overset{o}{A}} = \frac{6}{5}$$

$$(\beta)_{5000} = \frac{5}{6} (\beta)_{6000}$$

$$= \frac{5}{6} \times 0.239 = 0.199 \text{ mm}$$

Thus width $\lambda = 5000 \overset{o}{A}$, the zero – order fringe will still be at 12.34 mm while the twentieth order fringe will be at

$$= 12.34 \text{ mm} + (0.199 \text{ mm} \times 20)$$

$$= 16.32 \text{ mm}$$

13. In a biprism experiment, the distance between the slits and the screen is 160.0 cm. The biprism is 40 cm away from the slit and its refractive index is 1.52. When a source of wavelength 5893 Å is used, the fringe width is found to be 0.01 cm. Find the angle of prism.

Sol. The fringe width, $\beta = \frac{\lambda D}{2d}$

or $$2d = \frac{\lambda D}{\beta} \qquad .....(1)$$

But $$2d = 2y_1(\mu - 1)\alpha \qquad .....(2)$$

Where y_1 is the distance of biprism from the slit

from relations (1) and (2) we have

$$2y_1(\mu - 1)\alpha = \frac{\lambda D}{\beta}$$

∴ angle of prism,

$$\alpha = \frac{\lambda D}{\beta . 2y_1(\mu - 1)}$$

Given, $\lambda = 5893\ A^o = 5893 \times 10^{-8}$ cm

$D = 160.0$ cm

$\beta = 0.01$ cm

$y_1 = 40$ cm

and $\mu = 1.52$

∴ $$\alpha = \frac{5893 \times 10^{-8} \times 160}{0.01 \times 2 \times 40 \times (1.52 - 1)}$$

$$\alpha = 0.0226 \text{ rad}$$

or $$\alpha = 0.0226 \times \frac{180^o}{\pi}$$

$$\alpha = \frac{0.0226 \times 180^o}{3.14}$$

$$\alpha = 1.3$$

14. In an experiment with Fresnel's biprism sodium light is used and bands of 0.0196 cm in width are observed at a distance of 100 cm from the slit. A convex lens is then put between the observer and the prism so as to give an image of the source at a distance of 100 cm from the slit. The distance apart of the images is found to be 0.70 cm, the lens being 30 cm from the slit. Calculate the wavelength of sodium light.

Sol. We know that the wavelength of light λ is given by

$$\lambda = \frac{2d}{D}\beta$$

Given $\beta = 0.0196$ cm

$D = 100$ cm

When the lens is at distance of 30 cm from the slit, the separation of two coherent images becomes 0.70 cm, that is

$$d_1 = 0.70 \text{ cm}$$

From the magnification formula

$$M = \frac{\text{size of image}}{\text{size of object}} = \frac{\text{Distance of image}}{\text{Distance of object}}$$

$$M = \frac{I}{O} = \frac{v}{u}$$

Given $I = 0.70$ cm

$u = 30$ cm

$O = 2d$, the distance of the image from the lens,

$v = 100 - 30 = 70$ cm

$$\therefore \quad \frac{0.70}{2d} = \frac{70}{30}$$

or
$$2d = \frac{0.70 \times 30}{70} = 0.30 = 0.30 \text{ cm}$$

Thus
$$\lambda = \frac{2d}{D}\beta$$

$$\lambda = \frac{0.30}{100} \times 0.0196$$

$$\lambda = 5880 \times 10^{-8} \text{ cm}$$

15. A fresnel biprism with apex angle 1°30' is used to form interference fringes. Its refractive index is 1.52. Calculate the fringe width for λ = 6563 A°, when the distance between the source and the biprism is 20 cm and that between the biprism and screen is 80 cm.

Sol. The separation 2d between the two virtual sources in biprism experiment is given by

$$2d = 2y_1(\mu - 1)\alpha$$

where y_1 is the distance between the source and the biprism, α is the refracting angle

$$y_1 = 20 \text{ cm}$$

$$\mu = 1.52$$

$$\alpha = 1^\circ 30' = \frac{1.5 \times \pi}{180} \text{ radian} = \frac{\pi}{120} \text{ radian}$$

$$2d = 2 \times 20(1.52 - 1) \times \frac{3.14}{120} \text{ cm}$$

$$2d = 0.544 \text{ cm}$$

∴ fringe width

$$\beta = \frac{\lambda D}{2d} = \frac{(20 + 80)\text{cm} \times 6563 \times 10^{-8} \text{cm}}{0.544 \text{ cm}}$$

$$\beta = 0.0121 \text{ cm}$$

16. The distance between the slits and the biprism and between the biprism and the screen are 50 cm each. The obtuse angle of biprism is 179° and its refractive index is 1.5. If the width of the fringes is 0.0135 cm, calculate the wavelength of light. [UPTU]

Sol. As fringe width

$$\beta = \frac{\lambda D}{2d}$$

$$\lambda = \frac{\beta(2d)}{D}$$

Given $\beta = 0.0135$ cm and $D = (50 + 50) = 100$ cm

∴ $2d = 2y_1 (\mu - 1)\alpha$

Given $y_1 = 50$ cm

$$\mu = 1.5$$

$$\therefore \quad \alpha = \frac{180^\circ - 179^0}{2} = \left(\tfrac{1}{2}\right)^0 = \frac{\pi}{360} \text{ rad}$$

$$\therefore \quad 2d = 2 \times 50 \times (1.5 - 1) \times \frac{\pi}{360} = 0.436 \text{ cm}$$

$$\therefore \quad \lambda = \frac{\beta(2d)}{D} = \frac{0.0135 \text{ cm} \times 0.436 \text{ cm}}{100 \text{ cm}}$$

$$\lambda = 5.89 \times 10^{-5} \text{ cm}$$

$$\lambda = 5890 \text{ Å}$$

17. On introducing a thin sheet of mica (thickness = 12×10^{-5} cm) in the path of one of the interfering beans in a biprism arrangement, the central fringe is shifted through a distance equal to the spacing between successive bright fringes. Calculate the refractive index of mica ($\lambda = 6 \times 10^{-5}$ cm) [UPTU]

Sol. The shift y_o of the central fringe is

$$y_o = \frac{\beta}{\lambda}(\mu - 1)t$$

so that $$(\mu - 1)t = \frac{y_o \lambda}{\beta}$$

Here $$y_o = \beta$$

$$\lambda = 6 \times 10^{-5} \text{ cm}$$

$$t = 12 \times 10^{-5} \text{ cm}$$

$$\therefore \quad (\mu - 1)t = \lambda$$

$$\mu - 1 = \frac{\lambda}{t}$$

$$\mu = \frac{\lambda}{t} + 1$$

$$\mu = \frac{6 \times 10^{-5} \text{ cm}}{12 \times 10^{-5} \text{ cm}} + 1$$

$$\mu = 0.5 + 1$$

$$\mu = 1.5$$

18. A thin mica sheet ($\mu = 1.6$) of 7 microns thickness introduced in the path of one of the interfering beams in a biprism arrangement shifts the central fringe to a position normally occupied by the 7th bright fringe from the centre. Find the wavelength of light used. ($1\mu = 10^{-4}$ cm).

Sol. The shift is given by

$$y_o = \frac{\beta}{\lambda}(\mu - 1)t$$

so that

$$\lambda = \frac{\beta}{y_o}(\mu - 1)t$$

Given $y_o = 7\beta$

$$\lambda = \frac{\beta}{7\beta}(\mu - 1)t$$

$$\lambda = \frac{1}{7}(\mu - 1)t$$

$$\lambda = \frac{1}{7}(1.6 - 1) \times 7 \times 10^{-4} \text{ cm}$$

$$\lambda = \frac{0.6 \times 7 \times 10^{-4}}{7} \text{ cm}$$

$$\lambda = 6 \times 10^{-5} \text{ cm}$$

$$\lambda = 6000 \text{ Å}$$

19. On placing a thin sheet of mica of thickness 1.2×10^{-4} cm in the path of one of the interfering beams in biprism experiment using light of wavelength 5890 Å, the central fringe shifts to a position originally occupied by 12th bright fringe. Calculate the refractive index of the sheet. [UPTU]

Sol. On placing a thin sheet of mica in the path of one of the interfering beam, the displacement of the central bright fringe is given by

$$y_o = n\beta = \beta/\lambda \ (\mu - 1)\ t$$

or $$n\beta = \beta/\lambda \ (\mu - 1)\ t$$

∴ $$(\mu - 1)\ t = n\lambda$$

Given $\lambda = 5890\ \overset{o}{A} = 5890 \times 10^{-8}$ cm

$n = 12$

and $t = 1.2 \times 10^{-4}$ cm

$$\mu = 1 + \frac{12 \times 5890 \times 10^{-8}}{1.2 \times 10^{-4}}$$

$$\mu = 1 + 5.89$$

$$\mu = 6.89$$

20. In Fresnel's biprism experiment with a source of light of wavelength 5890 A°, a thin mica sheet of refractive index 1.6 is placed normally in the path of one of the interfering beams and the central bright fringe is shifted to a position of third bright fringe from the centre. Calculate the thickness of mica sheet. [UPTU]

Sol. For the thickness t of the transparent mica sheet,

We have

$$(\mu - 1)t = n\lambda$$

$$t = \frac{n\lambda}{\mu - 1}$$

Given $\mu = 1.6$; $n = 3$ and $\lambda = 5890$ A°

$$t = \frac{3 \times 5890}{(1.6 - 1)} = \frac{3 \times 5890}{0.6} = 29450\ \overset{o}{A}$$

21. A monochromatic light of λ = 5000 A° is incident on two slit separated by a distance of 5×10^{-4} m. The interference pattern is seen on a screen placed at a distance of 1 m from the slits. A thin glass plate of thickness 1.5×10^{-6} m and refractive index $\mu = 1.5$ is placed between one of the slits and the screen. Find the intensity at the centre of the screen. Also find the lateral shift of the central maximum. [UPTU]

Sol. The lateral shift of the central Maximum is given by

$$y_o = \frac{D}{2d}(\mu - 1)t$$

Given

$D = 1$m

$2d = 5 \times 10^{-4}$ m

$\mu = 1.5$

and $t = 1.5 \times 10^{-6}$ m

$$y_o = \frac{1}{5 \times 10^{-4}} \times (1.5 - 1) \times (1.5 \times 10^{-6}) = 1.5 \text{ mm}$$

Path difference created due to introduction of thin transparent glass plate.

$$\Delta = (\mu - 1)t$$

$$\Delta = (1.5 - 1) \times 1.5 \times 10^{-6}$$

$$\Delta = 0.75 \times 10^{-6} \text{ m}$$

$\therefore$ phase difference

$$\delta = \frac{2\pi}{\lambda} \times \Delta$$

$$\delta = \frac{2\pi}{(5 \times 10^{-7})} \times 0.75 \times 10^{-6} = 3\pi$$

Intensity, I, at the centre of the screen

$$= a_1^2 + a_2^2 + 2a_1a_2 \cos\delta$$

Here, $\delta = 3\pi$; $\cos\delta = -1$ and $a_1 = a_2 = a$

$\therefore$ $I = a^2 + a^2 - 2a^2 = 0$ i.e., centre of screen is dark

22. Calculate the thickness of the thinnest film (μ = 1.4) in which interference of violet component (l = 4000 Å) of incident light can take place by reflection.

Sol. The condition for constructive interference of light reflected from a film of thickness t is

$$2\,\mu t \cos r = (2n - 1)\, \lambda/2$$

where $n = 0, 1, 2, \ldots$

For thinnest film, n = 0 and cos r = 1 (normal incidence)

So that

$$2\,\mu t = \frac{\lambda}{2}$$

$$t = \frac{\lambda}{2}$$

$$t = \frac{4000 \times 10^{-10} \text{ m}}{4 \times 1.4}$$

$$t = 0.7142 \times 10^{-7} \text{ m}$$

23. A man whose eyes are 150 m above the oil film on water surface observes greenish colour at a distance of 100 cm from his feet. Calculate the probable thickness of the film ($\lambda_{green} = 5000$ Å, $\mu_{oil} = 1.4$; $\mu_{water} = 1.33$). [UPTU]

Sol. If t is the thickness and r the angle of refraction, then the condition of maxima in the reflected light is given by

$$2\,\mu t \cos r = (2n - 1)\,\frac{\lambda}{2}$$

where n = 1, 2, 3….

$$t = \frac{(2n-1)\lambda}{4\mu \cos r} \qquad \text{.....(1)}$$

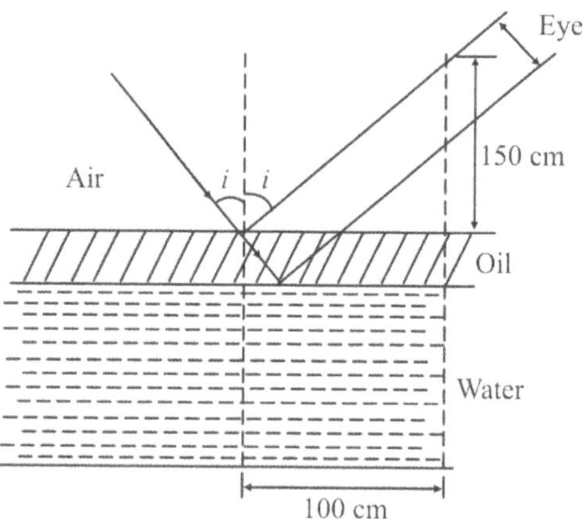

From the figure

$$\tan i = \frac{100}{150} = \frac{2}{3}$$

$$\sin i = \frac{2}{\sqrt{13}}$$

But $$\mu = \frac{\sin i}{\sin r}$$

$$\therefore \qquad \sin r = \frac{\sin i}{\mu} = \frac{2/\sqrt{13}}{1.4}$$

$$\sin r = \frac{2}{5.0477} = 0.3962$$

$$\therefore \qquad \cos r = \sqrt{1 - \sin^2 r}$$

$$\cos r = \sqrt{1 - (0.3962)^2}$$

$$\cos r = \sqrt{1 - 0.1569}$$

$$\text{or} \qquad \cos r = \sqrt{0.8431}$$

$$\cos r = 0.9182$$

Putting this in equation

$$t = \frac{(2n-1)\lambda}{2\mu \cos r}$$

we get

$$t = \frac{(2n-1) \times 5 \times 10^{-5}}{4 \times 1.4 \times 0.9182}$$

$$t = 9.725 \times 10^{-6} (2n - 1) \text{ cm}$$

where n = 1, 2, 3....

24. A soap film of refractive index 1.43 is illuminated by white light incident at an angle of 30°. The refracted light is examined by a spectroscope in which dark band corresponding to the wavelength 6×10^{-7} m is observed. Calculate the thickness of the film. [UPTU]

Sol. The condition of destructive interference or dark bands in reflected light is

$$2 \mu t \cos r = 2n \frac{\lambda}{2}$$

$$\text{or} \qquad t = \frac{n\lambda}{2 \mu \cos r}$$

Given i = 30°; μ = 1.43, n = 1 and $\lambda = 6 \times 10^{-7}$ m

Since from snell's law

$$\mu = \frac{\sin i}{\sin r} \quad \Rightarrow$$

$$\sin r = \frac{\sin i}{\mu} = \frac{\sin 30^\circ}{1.43} = 0.38$$

$$\therefore \quad \cos r = \sqrt{1-\sin^2 r} = \sqrt{1-(0.38)^2}$$

$$\cos r = 0.92$$

$$\therefore \quad t = \frac{n\lambda}{2\,\mu \cos r} = \frac{1\times 6\times 10^{-7}}{2\times 1.43\times 0.92}\ \text{m}$$

$$t = 2.28 \times 10^{-7}\ \text{m} \qquad \text{or}$$

$$t = 2.28 \times 10^{-5}\ \text{cm}$$

25. A soap film of refractive index μ is illuminated with white light incident at an angle i. The light refracted by it is examined and two bright bands focussed corresponding to wavelengths λ_1 and λ_2. Show that the thickness of the film is

$$t = \frac{\lambda_1 \lambda_2}{\lambda_1 - \lambda_2}\left[\frac{1}{2\sqrt{\mu^2 - \sin^2 i}}\right]$$ [UPTU]

Sol. The condition for bright bands in refracted light is

$$2\,\mu t \cos r = n\,\lambda$$

If n and (n + 1) are the orders of the bright bands due to wavelengths λ_1 and λ_2 respectively, then

$$2\,\mu t \cos r = n\lambda_1 = (n+1)\,\lambda_2$$

$$n = \frac{\lambda_2}{\lambda_1 - \lambda_2}$$

$$\therefore \quad 2\,\mu t \cos r = n\lambda_1 = \frac{\lambda_1 \lambda_2}{\lambda_1 - \lambda_2}$$

$$\therefore \quad t = \frac{\lambda_1 \lambda_2}{\lambda_1 - \lambda_2}\left(\frac{1}{2\,\mu \cos r}\right)$$

But $$\cos r = \sqrt{1-\sin^2 r}$$

$$\sin r = \frac{\sin i}{\mu}$$

$$\therefore \qquad \cos r = \sqrt{\left\{1 - \left(\frac{\sin i}{\mu}\right)^2\right\}}$$

$$\cos r = \sqrt{\frac{\mu^2 - \sin^2 i}{\mu^2}}$$

$$t = \frac{\lambda_1 \lambda_2}{\lambda_1 - \lambda_2}\left[\frac{1}{2\sqrt{\mu^2 - \sin^2 i}}\right]$$

26. Calculate the thickness of a soap bubble film (refractive index = 1.46) that will result in constructive interference in reflected light, if the film is illuminated with light whose wavelength in free space is 6000 $\overset{\circ}{A}$. [UPTU]

Sol. The condition for the constructive interference in the reflected light for the brightness of soap bubble film is

$$2\,\mu t \cos\; r = (2n - 1)\,\frac{\lambda}{2},$$

where $\qquad n = 1, 2, 3....$

If the light is incident normally on the film, then

$$r = 0 \text{ and } \cos r = 1$$

$$\therefore \qquad 2\,\mu t = (2n - 1)\,\frac{\lambda}{2}$$

$$t = \frac{(2n-1)\lambda}{4\,\mu}$$

For the least thickness of the film, n = 1

∴ least thickness of the film,

$$t = \frac{\lambda}{4\,\mu}$$

$$t = \frac{6000}{4 \times 1.46} = 1027.4\ \overset{\circ}{A}$$

27. Light of wavelength 5893 Å is reflected at nearly normal incidence from a soap film of refractive index (μ) = 1.42. What is the least thickness of the film that will appear (i) dark (ii) bright. [UPTU]

Sol. (i) The condition for the darkness of the film is given by

$$2\ \mu t \cos r = n\ \lambda$$

For normal incidence, r = 0 and cos r = 1

$$2\ \mu t = n\ \lambda$$

or $$t = \frac{n\lambda}{2\mu}$$

For the least thickness of the film; n = 1

∴ the least thickness of the film

$$t = \frac{\lambda}{2\mu} = \frac{5893\times10^{-8}}{2\times1.42} = 2.075 \times 10^{-5}\ \text{cm}$$

$$= 2075\ \text{Å}$$

(ii) The condition for the brightness of the film in reflected system is given by

$$2\ \mu t \cos r = (2n-1)\ \frac{\lambda}{2}$$

For normal incidence, r = 0 and cos r = 1

∴ $$2\ \mu t = (2n-1)\ \frac{\lambda}{2}$$

or $$t = \frac{(2n-1)\lambda}{4\mu}$$

For least thickness, n = 1

∴ the least thickness of the film

$$t = \frac{\lambda}{2\mu} = \frac{5893\times10^{-8}}{4\times1.42}$$

$$t = 1037.5 \times 10^{-5}\ \text{cm}$$

$$t = 1037.5\ \text{Å}$$

28. White light is incident on a soap film at an angle $\sin^{-1} 4/5$ and the reflected light is observed with a spectroscope. It is found that two consecutive dark bands correspond to wavelengths 6.1×10^{-5} and 6.0×10^{-5} cm. If the refractive index of the film be $4/3$, calculate the thickness. [UPTU]

Sol. We know that the condition for dark band or fringe in the reflected light is

$$2\,\mu t \cos r = n\,\lambda$$

If n and (n + 1) are the orders of consecutive dark bands for wavelengths λ_1 and λ_2 respectively, then

$$2\,\mu t \cos r = n\,\lambda_1 \quad \text{.....(1)}$$

and $$2\,\mu t \cos r = (n+1)\,\lambda_2$$

$\therefore$ $$2\,\mu t \cos r = n\,\lambda_1 = (n+1)\,\lambda_2$$

$\therefore$ $$n\,\lambda_1 = (n+1)\,\lambda_2$$

$\therefore$ $$n\,\lambda_1 = n\,\lambda_2 + \lambda_2$$

or $$n(\lambda_1 - \lambda_2) = \lambda_2$$

or $$n = \frac{\lambda_2}{\lambda_1 - \lambda_2}$$

Substituting this value of n in eq. (1), we get

$$2\,\mu t \cos r = \frac{\lambda_1 \lambda_2}{(\lambda_1 - \lambda_2)}$$

or $$t = \frac{\lambda_1 \lambda_2}{(\lambda_1 - \lambda_2)} \frac{1}{2\,\mu \cos r} \quad \text{.....(2)}$$

But

$$\cos r = \sqrt{1 - \sin^2 r} = \sqrt{\left[1 - \left(\frac{\sin i}{\mu}\right)^2\right]}$$

From snell's law; $\mu = \dfrac{\sin i}{\sin r}$

Given

$$\mu = 4/3 \text{ and } \sin i = 4/5$$

$$\therefore \qquad \cos r = \sqrt{1 - \frac{\left(\frac{4}{5}\right)^2}{\left(\frac{4}{3}\right)^2}}$$

$$\cos r = \sqrt{1 - \left(\frac{3}{5}\right)^2}$$

$$\cos r = \sqrt{\frac{16}{25}} = \frac{4}{5}$$

Given $\lambda_1 = 6.1 \times 10^{-5}$ cm

$\lambda_2 = 6.0 \times 10^{-5}$ cm

$\mu = \frac{4}{3}$

Putting the value of cos r in eq. (2) we get

$$\therefore \qquad t = \frac{6.1 \times 10^{-5} \times 6.0 \times 10^{-5}}{\left(6.1 \times 10^{-5} - 6.0 \times 10^{-5}\right) \times 2 \times \left(\frac{4}{3}\right) \times \left(\frac{4}{5}\right)}$$

$$t = \frac{6.1 \times 10^{-5} \times 6.0 \times 10^{-5} \times 3 \times 5}{0.1 \times 10^{-5} \times 2 \times 4 \times 4}$$

$$t = 0.0017 \text{ cm}$$

29. White light is reflected from an oil film of thickness 0.01 mm and refractive index 1.4 at an angle of 45° to the vertical. If the reflected light falls on the slit of a spectrometer, calculate the number of dark bands seen between wavelengths 4000 and 5000 Å.

Sol. The condition of dark bands in reflected light is

$$2\,\mu t\, \cos r = n\,\lambda$$

For $\lambda_1 = 4000$ Å, we have

$$2\,\mu t \cos r = n_1\,\lambda_1$$

$$n_1 = \frac{2\,\mu t \cos r}{\lambda_1}$$

For $\lambda_2 = 5000\ \mathring{A}$

$$2\ \mu t \cos r = n_2\ \lambda_2$$

$$n_2 = \frac{2\ \mu t \cos r}{\lambda_2}$$

Here $\mu = 1.4$; $t = 0.01$ mm $= 0.001$ cm

Angle of reflection = angle of incidence

$$i = 45^o$$

We have
$$\mu = \frac{\sin i}{\sin r}$$

i.e.,
$$\sin r = \frac{\sin i}{\mu}$$

$$\cos r = \sqrt{1 - \sin^2 r}$$

$$\cos r = \sqrt{1 - \frac{\sin^2 i}{\mu^2}}$$

$$\cos r = \sqrt{1 - \frac{1}{2\mu^2}}$$

$$\cos r = \sqrt{1 - \frac{1}{2(1.4)^2}} = 0.86$$

$$n_1 = \frac{2\ \mu t \cos r}{\lambda_1}$$

$$n_1 = \frac{2 \times 1.4 \times 0.001 \times 0.86}{4000 \times 10^{-8}} = 60$$

and

$$n_2 = \frac{2\ \mu t \cos r}{\lambda_2} = \frac{2 \times 1.4 \times 0.001 \times 0.86}{5000 \times 10^{-8}} = 48$$

The number of dark bands seen within wavelengths $4000\ \mathring{A}$ and $5000\ \mathring{A}$ is

$$n_1 - n_2 = 60 - 48 = 12$$

30. A drop of oil of volume 0.2 cm^3 is dropped on the surface of a tank of water of area 1 m^2. The film spreads uniformly over the whole surface and white light reflected normally is observed through a spectrometer. The spectrum is seen to contain first dark band whose centre has wavelength of 5.5×10^{-7} m (in air). Find the refractive index of oil.

Sol. We know that

Volume = Area × Thickness

$$\therefore \quad \text{Thickness } t = \frac{\text{Volume}}{\text{Area}}$$

$$t = \frac{0.2 \times 10^{-6}\ \text{m}^3}{1\,\text{m}^2}$$

The film will appear dark by reflected light for wavelength λ, if

$$2\,\mu t \cos r = n\,\lambda$$

For normal incidence

$$r = 0 \quad \text{so that} \quad \cos r = 1$$

$$n = 1$$

and $$\lambda = 5.5 \times 10^{-7}\ \text{m}$$

$$\therefore \quad 2\,\mu t = n\,\lambda$$

$$\mu = \frac{n\,\lambda}{2t}$$

$$\mu = \frac{1 \times 5.5 \times 10^{-7}}{2 \times 2 \times 10^{-7}}$$

$$\mu = 1.375$$

31. Light of wavelength 6000 Å falls normally on thin wedge shaped film of refractive index 1.35, forming fringes that are 2.0 mm apart. Find the angle of the wedge. [UPTU]

Sol. For normal incidence, the fringe width is given by

$$\beta = \frac{\lambda}{2\,\mu\,\theta}$$

$$\theta = \frac{\lambda}{2\,\mu\,\beta}$$

Given $\lambda = 6000\,\mathring{A} = 6000 \times 10^{-8}$ cm

$\mu = 1.35$

$\beta = 2.0$ mm $= 0.20$ cm

$\therefore$ $$\theta = \frac{6000 \times 10^{-8}\text{ cm}}{2 \times 1.35 \times 0.20\text{ cm}}$$

$$\theta = 1.1 \times 10^{-4}\text{ radian}$$

$$\theta = (1.1 \times 10^{-4}) \times \frac{180^\circ}{3.14}$$

$$\theta = 0.0063^\circ$$

as $$1^\circ = 3600''$$

$\therefore$ $$\theta = 22.68''$$

32. Using light of $\lambda = 5.9 \times 10^{-7}$ m, it is found that in a thin film of air 7.4 fringes occur between two points. Deduce the difference of film thickness between these points.

Sol. Given that the number of fringes between two points where thickness are t_2 and $t_1 = 7.4$

$$\lambda = 5.9 \times 10^{-7}\text{ m}$$

$$\mu = 1 \qquad \text{(for air film)}$$

Now

$$t_2 - t_1 = \frac{\lambda \times \text{No. of fringes}}{2\mu}$$

$$t_2 - t_1 = \frac{5.9 \times 10^{-7}\text{ m} \times 7.4}{2 \times 1}$$

$\therefore$ $$t_2 - t_1 = 0.2183 \times 10^{-5}\text{ m}$$

33. Using sodium light ($\lambda = 5893\,\mathring{A}$) interference fringes are formed by reflection from a thin air wedge. When viewed perpendicularly, 10 fringes are observed in a distance of 1 cm, calculate the angle of the wedge.

Sol. Fringe width for air wedge ($\mu = 1$) is

$$\beta = \frac{\lambda}{2\theta}$$

$$\theta = \frac{\lambda}{2\beta}$$

Given that

$$\lambda = 5893\ \mathring{A} = 5893 \times 10^{-8}\ \text{cm}$$

$$\beta = \frac{1\,\text{cm}}{10}$$

$$\theta = \frac{5893 \times 10^{-8}\ \text{cm}}{2 \times \frac{1}{10}}$$

$$\theta = 2946.5 \times 10^{-7}$$

$$\theta = 2.95 \times 10^{-4}\ \text{radian}$$

34. Two plane glass surfaces in contact along one edge are separated at the opposite edge by a thin wire. If 20 fringes are observed between these edges in sodium light for normal incidence, what is the thickness of the wire.

Sol. The fringe width is given by

$$\beta = \frac{\lambda}{2\theta} \qquad \text{.....(1)}$$

Let t be the thickness of the wire and x be the length of the glass surface

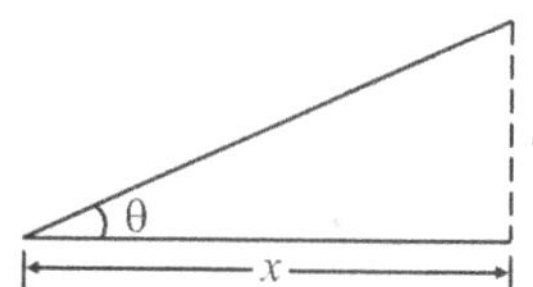

Then $\qquad \theta = \frac{t}{x}$

Then $\qquad \beta = \frac{\lambda x}{2t}$

If n fringes are seen in the entire film, then

$$x = n\,\beta$$

Therefore,

$$\beta = \frac{\lambda n \beta}{2t}$$

or $$t = \frac{n\lambda}{2}$$

Given n = 20 and $\lambda = 5890 \overset{o}{A}$ (for sodium light) $= 5890 \times 10^{-8}$ cm

$$t = \frac{20 \times 5890 \times 10^{-8}}{2}$$

$$t = 5.89 \times 10^{-4} \text{ cm}$$

35. Two plane rectangular piece of glass are in contact at one edge and separated by a hair at opposite edge so that a wedge is formed. When light of wavelength $6000 \overset{o}{A}$ falls normally on the wedge, nine interference fringes are observed. What is the thickness of hair. [UPTU]

Sol. The fringes width in air-wedge for normal incidence is given by

$$\beta = \frac{\lambda}{2\theta}$$

$\therefore$ $$\theta = \frac{t}{x}$$

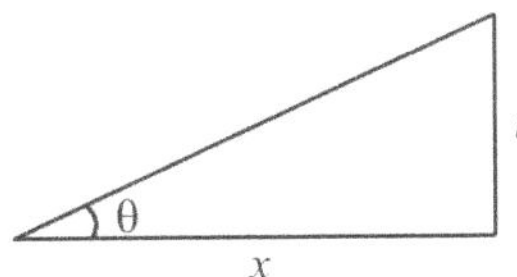

where t is the thickness of the hair and x be the length of the glass surface from the point of contact.

If n fringes are seen in the entire film then

$$x = n\beta$$

$\therefore$ $$\beta = \frac{\lambda}{2\theta} = \frac{\lambda x}{2t}$$

$$\beta = \frac{\lambda(n\beta)}{2t} \Rightarrow t = \frac{n\lambda}{2}$$

Given n = 9 and $\lambda = 6000 \overset{o}{A}$

$$\therefore \qquad t = \frac{n\lambda}{2}$$

$$t = \frac{9 \times 6000 \text{ Å}}{2}$$

$$t = 27000 \text{ Å}$$

36. A square piece of cellophane film with index of refraction 1.5 has a wedge shaped section, so that its thickness at two opposite sides it t_1, and t_2. If the number of fringes appearing with wavelength $\lambda = 6000$ Å is 10, calculate the difference $(t_1 - t_2)$ [UPTU]

Sol. If the order of the fringe appearing at one end of the film be n, then the order of the fringe appearing at other end will be (n + 10). For n^{th} and $(n + 10)^{th}$ dark fringes

$$\therefore \qquad 2\,\mu t_1, \cos(r + \theta) = n\,\lambda \qquad .....(1)$$

$$2\,\mu t_2, \cos(r + \theta) = (n + 10)\,\lambda \qquad .....(2)$$

Subtracting (1) and (2) we get

$$2\,\mu\,(t_2 - t_1)\cos(r + \theta) = 10\,\lambda \qquad .....(3)$$

If the fringe is seen normally and the angle of wedge is very small, then r = 0, so that

$$\cos(r + \theta) = \cos\theta = 1 \qquad .....(4)$$

Substituting the value of cos (r + θ) from equation (4) to eq. (3), we get

$$2\,\mu\,(t_2 - t_1) = 10\,\lambda$$

$$t_2 - t_1 = \frac{10\,\lambda}{2\mu} = \frac{5\lambda}{\mu}$$

Given $\mu = 1.5$ and $\lambda = 6000 \text{ Å} = 6000 \times 10^{-7}$ cm

$$t_2 - t_1 = \frac{5 \times 6000 \times 10^{-7}}{1.5}$$

$$t_2 - t_1 = 2 \times 10^{-4} \text{ cm}$$

37. Light containing two wavelengths λ_1 and λ_2 fall normally on a plano-convex lens plate of radius of curvature R resting on a glass place. If the nth dark ring due to λ_1

coincides with $(n + 1)^{th}$ dark ring due to λ_2, prove that the radius of the nth dark ring of λ_1 is $\sqrt{\frac{\lambda_1\lambda_2 R}{(\lambda_1 - \lambda_2)}}$ [UPTU]

Sol. Radius of the nth dark ring in the Newton's ring pattern is given by

$$r_n = \sqrt{n\lambda R} \quad .....(1)$$

n^{th} dark ring due to λ, coincides with the $(n + 1)^{th}$ dark ring due to λ_2. That is

$$D_n^2 = 4n\lambda_1 R = 4(n+1)\lambda_2 R$$

or $$n\lambda_1 = (n + 1)\lambda_2$$

or $$n = \frac{\lambda_2}{\lambda_1 - \lambda_2}$$

Substituting this value of n in eq. (1) we get

$$r_n = \sqrt{n\,\lambda_1\,R}$$

$$r_n = \sqrt{\frac{\lambda_1\lambda_2 R}{\lambda_1 - \lambda_2}}$$

38. Newton's rings are observed normally is reflected light of wavelength 6000 Å. The diameter of the 10th dark ring is 0.50 cm. Find the radius of curvature of the lens and the thickness of the film. [UPTU]

Sol. The diameter of nth dark ring is given by

$$D_n^2 = 4\,n\,\lambda\,R$$

or $$R = \frac{D_n^2}{4n\lambda}$$

Given that

$$D_n = 0.50 \text{ cm} ; \lambda = 6000 \text{ Å} = 6.0 \times 10^{-5} \text{ cm};$$

and n = 10

$$R = \frac{0.50 \times 0.50}{4 \times 10 \times 6.0 \times 10^{-5}} = \frac{0.25 \times 10^5}{240} = \frac{25000}{240} = 104.166$$

$$R = 104.166 \text{ cm}$$

If t is the thickness of the film corresponding to a ring of D diameter, then

$$2t = \frac{D^2}{4R}$$

$$t = \frac{D^2}{8R} = \frac{0.50 \times 0.50}{8 \times 104.166} = \frac{0.25}{8 \times 104.166}$$

$$t = 3 \times 10^{-4} \text{ cm}$$

39. Newton's rings are observed by keeping a spherical surface of 100 cm radius on a plane glass plate. If the diameter of the 15th bright ring is 0.590 cm and the diameter of the 5th ring is 0.336 cm, what is the wavelength of light used. [UPTU]

Sol. If D_{n+p} and D_n be the diameter of $(n + p)^{th}$ and n^{th} bright ring, then

$$\lambda = \frac{D_{n+p}^2 - D_n^2}{4PR}$$

Given $D_{15} = 0.590$ cm

$D_5 = 0.336$ cm

$P = 10$

and $R = 100$ cm

$$\therefore \quad \lambda = \frac{(0.590)^2 - (0.336)^2}{4 \times 10 \times 100}$$

$$\lambda = 5.88 \times 10^{-5} \text{ cm}$$

$$\lambda = 5880 \text{ Å}$$

40. In Newton's ring experiment the diameter of 4th and 12th dark rings are 0.400 cm and 0.700 cm respectively. Deduce the diameter of 20th dark ring. [UPTU]

Sol. If D_{n+p} and D_n be the diameter of (n + p)th and n^{th} dark rings respectively, then

$$D_{n+p}^2 - D_n^2 = 4p\lambda R$$

Given $n = 4$, $n + p = 12$; $D_4 = 0.400$ cm; $D_{12} = 0.700$ cm

$$\therefore \quad D_{12}^2 - D_4^2 = 4 \times 8 \times \lambda \times R$$

and

$$D_{20}^2 - D_4^2 = 4 \times 16 \times \lambda \times R$$

$$\therefore \quad \frac{D_{12}^2 - D_4^2}{D_{20}^2 - D_4^2} = \frac{4 \times 8}{4 \times 16} = \frac{1}{2}$$

$$2\left(D_{12}^2 - D_4^2\right) = \left(D_{20}^2 - D_4^2\right)$$

or $$D_{20}^2 = 2D_{12}^2 - D_4^2$$

$$D_{20}^2 = 2(0.700) \times (0.7.00) - (0.400)(0.400)$$

$\therefore$ $$D_{20}^2 = 0.98 - 0.16 = 0.82$$

So the diameter of 20th dark ring

$$= \sqrt{0.82} = 0.906 \text{ cm}$$

41. Newton's rings are formed in reflected light of wavelength 6000 $\overset{o}{A}$ with a liquid between the plane and curved surfaces. If the diameter of the 6th bright ring is 3.1 mm and the radius of curvature of the curved surface is 100 cm, calculate the refractive index of the liquid. [UPTU]

Sol. The diameter of nth bight ring is given by

$$D_n^2 = \frac{2(2n-1)\lambda R}{\mu}$$

or $$\mu = \frac{2(2n-1)\lambda R}{D_n^2}$$

In the given question

$$n = 6, \lambda = 6000 \overset{o}{A} = 6000 \times 10^{-8} \text{ cm}; R = 100 \text{ cm}$$

and $D_n = 3.1 \text{ mm} = 0.31 \text{ cm}$

$\therefore$ $$\mu = \frac{2(2n-1)\lambda R}{D_n^2} = \frac{2(12-1) \times 6000 \times 10^{-8} \times 100}{(0.31)(0.31)}$$

$$\mu = \frac{2 \times 11 \times 6000 \times 10^{-8} \times 100}{0.31 \times 0.31} = 1.373$$

42. In a Newton's ring arrangement, of a drop of water ($\mu = 4/3$) be placed between the lens and the plate, the diameter of the 10th dark ring is found to be 0.6 cm. Obtain the radius of curvature of the face of the lens in contact with the plate. The wavelength of light used is 6000 $\overset{o}{A}$. [UPTU]

Sol. We know that

$$D_n^2 = \frac{4n\lambda R}{\mu}$$

$$D_n^2 = \frac{4\times10\times6000\times10^{-8}\ \text{cm}\times R}{\frac{4}{3}} = 0.0018\ R$$

Since $D_n = 6$ cm

$\therefore$
$$R = \frac{0.6\times0.6\ \text{cm}^2}{3\times10\times6000\times10^{-8}\ \text{cm}} = 200\ \text{cm}$$

43. In Newton's rings arrangement the diameter of 10th dark ring changes from 1.5 cm to 1.27 cm when a liquid is introduced between the lens and the glass plate. Calculate refractive index of the liquid. [UPTU]

Sol. Refractive index

$$\mu = \frac{(D_n)^2_{air}}{(D_n^2)_{liquid}}$$

$$\mu = \left[\frac{1.50\ \text{cm}}{1.27\ \text{cm}}\right]^2$$

$$\mu = 1.395$$

44. In an arrangement for observing Newton's rings with two different media between the glass surfaces, then nth rings have diameters as 10:7. Find the ratio of the refractive indices of the two media.

Sol. As refractive index

$$\mu = \frac{(D_n^2)_{air}}{(D_n^2)_{liquid}}$$

$\therefore$
$$\frac{\mu_1}{\mu_2} = \frac{(D_n^2)_{air}}{(D_n^2)_{liquid(1)}} \times \frac{(D_n^2)_{liquid(2)}}{(D_n^2)_{air}}$$

$$\frac{\mu_1}{\mu_2} = \frac{(7)^2}{(10)^2} = 49:100$$

45. Two plano-convex lenses, each of radius of curvature 100 cm, are placed with their curved surfaces in contact with each other. Newton's rings are formed by using a light of wavelength 6×10^{-5} cm. Find the distance between 10th and 20th rings.

Sol. In this case thickness of air – film

$$t = t_1 + t_2$$

which gives

$$\frac{1}{R_1} + \frac{1}{R_2} = \frac{4n\lambda}{D_n^2}$$

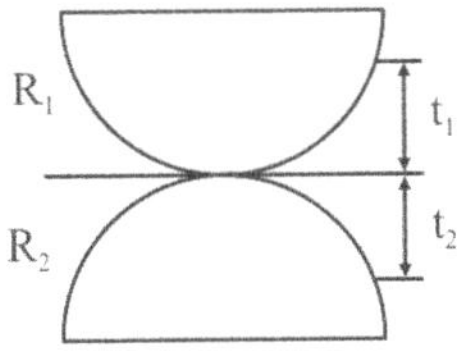

Given $R_1 = R_2 = 100$ cm; $\lambda = 6 \times 10^{-5}$ cm

$$r_n^2 = \frac{n\lambda R_1 R_2}{R_1 + R_2}$$

$$r_n = \sqrt{\frac{n\lambda R_1 R_2}{R_1 + R_2}}$$

$$\therefore \quad r_{20} - r_{10} = \sqrt{\frac{n\lambda R_1 R_2}{R_1 + R_2}}\left[\sqrt{20} - \sqrt{10}\right]$$

$$= \sqrt{\frac{6 \times 10^{-5} \times 10^4}{200}}\left[\sqrt{20} - \sqrt{10}\right]$$

$$= 0.0717 \text{ cm}$$

Conceptual Questions and Answers

1. **What is interference of light?**

Ans. The phenomenon of non-uniform distribution of energy in the medium due to super position of two light waves is called interference of light.

2. **Can two independent sources of light be coherent**

Ans. No

3. State two conditions to obtain sustained interference

Ans. (a) The two light waves should be of same wavelength

(b) The two light waves should either be in phase or should have a constant phase different

4. What happens to the interference pattern, if the phase difference between the two sources continuously varies?

Ans. If the phase difference between the two sources continuously varies, the position of minima and maxima will also vary. Such as interference pattern will not be sustained one and it will not be observed.

5. Two silts in Young's double slit experiment are illuminated by two different sodium lamp emitting light of same wavelength. Do you observe any interference pattern on the screen?

Ans. No it is because the phase difference between the light waves from the two independent sources keeps on changing continuously.

6. What is the principle of superposition of light waves?

Ans. It states that when two or more wave trains of light-travelling in a medium superpose upon each other, then the resultant displacement at any instant is equal to the vector sum of the displacement due to individual waves.

7. When a low flying aircraft passes overhead, we sometimes notice a slight shaking of the picture on our TV screen, suggest a possible explanation.

Ans. A low flying aircraft reflects the TV signal. Due to interference between the direct signal and the reflected signal the picture on TV screen shows a slight shaking.

8. Can white light produce interference? What is its nature.

Ans. Yes, white light can produce interference. However only first few coloured fringes can be detected and beyond that the fringes of different colours overlap to produce general illumination, white in colour.

9. Why do the oil films on the surface of water appear to be coloured?

Ans. The oil films on the surface of water appear to be coloured due to the interference of white light reflected from the two surfaces of the film.

10. Why does an excessively thin film appear black in reflected light?

Ans. In case of an excessively thin film (t = 0), the path difference is just $\lambda/2$. The path difference of $\lambda/2$ occurs due to the reason that one of the two rays suffers reflection at the denser medium. As the path difference between the two rays is $\lambda/2$, the film appears dark.

11. Bubbles of colourless soap solution appear coloured in sunlight. Why?

Ans. It is due to interference of white light from the thin film of the soap bubbles.

12. Can two electric bubs, point-like and having filaments of the same material placed closed together, produce, interference? Explain.

Ans. The two independent sources (such as two electric bulbs) even point-like, cannot act as coherent sources. So the sustained interference cannot be produced. The emission of light in the bulbs is due to millions of atoms in which electrons jump from some higher orbit to the lower orbit. This process takes place in a time of about 10^{-8} sec. Thus, the phase difference can remain constant for about 10^{-8} sec only. In other words, in one second, the phase will change 10^8 times approximately. Such a fast variation in the interference pattern cannot be detected even with the most superior camera available so far. Therefore, the interference will be produced, best it will go undetected.

13. Why is no interference pattern observed, when two coherent sources are (i) infinitely close to each other (ii) far apart from each other.

Ans. Use know that

$$\beta = \frac{\lambda D}{d}$$

(i) When the two sources are close to each other, (d quite small) sufficiently wide fringes are formed. In case the two sources are infinitely close to each other, general illumination take place

(ii) When the sources are moved far apart (d very large) the fringe width will be very small and they will not be separately visible.

14. Why does the colour of the oil film on the surface of water continuously change?

Ans. The location of the bright and dark fringes produced by the thin oil film depends upon the thickness of the film. The thickness of the oil film on the surface of water continuously varies and as a result, the portion of the coloured fringes also changes. This appears as a variation in the colour of the oil films.

15. Why we don't see interference pattern if we look at a glass window?

Ans. The thickness of the glass window is very large in comparison to the wavelength of light. Also, the window is not perfectly flat, so slight variation in thickness are usually more than a wavelength. These two problems together make it very difficult for the path length difference to be uniform over a large enough regions to observe interference patterns.

16. Why does a compact disc show a rainbow of colours?

Ans. The grooves in a compact disc are very fine and close to each other to distance comparable to typical wavelength in the visible range. Different light rays reflected on different grooves subsequently interfere and reinforce or cancel different wavelengths depending upon the angle of incidence, which results in rainbow of colours.

17. In relation to Newton's rings pattern what will happen if:

(a) we use plano concave lans and place it with concave side on glass plate

(b) we use mirror in place of plane glass plate below the plano convex lens

(c) we use plano convex lens of smaller redius of curvature

(d) the plano convex lens is raised by height Δh from the surface of plane glass plate.

Ans. (a) In this case fringes are still circular. The lens touches the glass plate at the circumference hence dark rings shall be observed there.

The thickness of air film increases as we move from circumstance to centre and therefore order of rings is reversed i.e. highest order ring at the centre and zero order ring at the periphery of the lens. If t is the thickness of air film and t_n be the thickness of film at position of nth ring, then radius of the nth ring from centre is

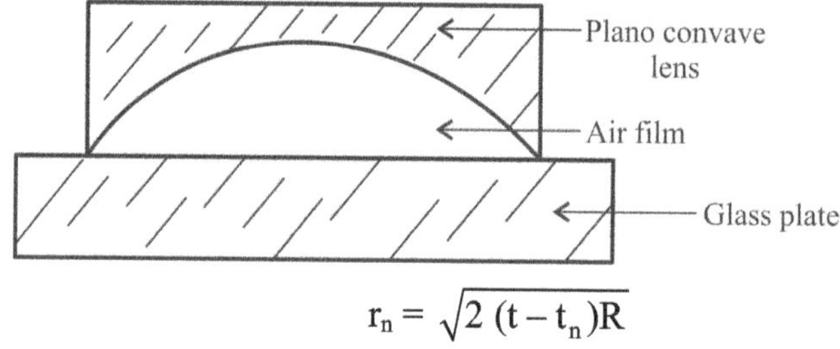

$$r_n = \sqrt{2\,(t - t_n)R}$$

For n^{th} dark ring $2t_n = n\lambda$

$$\therefore \qquad r_n = \sqrt{2t - n\lambda)R}$$

where R is the radius of curvature. The spacing between two consecutive rings decreases as we move towards the centre of the lens.

(b) When we use mirror, the interference pattern due to reflected and transmitted light shall superimpose. Since these two patterns are complimentary their super position gives uniform illumination.

(c) The radius of given Newton's ring is proportional to square root of radius of curvature of plano convex lens. Therefore decrease in the radius of curvature of plano convex lens shall reduce the radii of Newton's rings.

(d) Film thickness at the position of nth ring is

$$t_n = \Delta t + \frac{r_n^2}{2R}$$

whence R is the radius of curvature of the plano-convex lens. For n^{th} dark ring $2t_n = n\lambda$ which

$$r_n = \sqrt{n\lambda R - 2\Delta t R} \qquad \text{(for dark)} \qquad .....(1)$$

Similarly for the nth bright ring

$$2t_n = \left(n - \frac{1}{2}\right)\lambda$$

$$\therefore \qquad r_n = \sqrt{\left(n - \frac{1}{2}\right)\lambda R - 2\Delta t R} \qquad \text{(for bright)} \qquad .....(2)$$

If $\Delta t = \frac{\lambda}{4}$, the centre becomes bright as $r_n = 0$ from eq. (2) for $n = 1$.

We also observe that as t increases, r decreases, the radius of fringes decreases while raising the lens.

Objective Questions

1. Young's experiment is an

   (a) Interference of radio waves (b) Diffraction of radio waves

   (c) Interference of light (d) Diffraction of light

2. In interference

   (a) There is loss of energy in great amount

   (b) There is no loss of energy

   (c) There is loss of energy in less amount

   (d) None of these

3. The order of wavelength of light waves is

   (a) 10^{-2} cm (b) 10^{-3} cm (c) 10^{-4} cm (d) 10^{-5} cm

4. Intensity (energy) is

   (a) Directly proportional to the square root of amplitude

   (b) Directly proportional to the square of amplitude

   (c) Directly proportional to the cube root of amplitude

   (d) None of these

5. Path difference for wave emanating from two coherent sources separated by distance d and screen be placed at a distance D from coherent sources is

   (a) $\frac{2\pi}{x}\frac{\lambda d}{D}$ (b) $\frac{2\pi}{\lambda}\frac{x d}{D}$

   (c) $\frac{2\pi}{\lambda}\frac{x D}{d}$ (d) None of these

   Where x is the distance of point on screen from C

6. Fringe width is defined as

   (a) Distance between any two consecutive bright fringes

   (b) Distance between any two consecutive dark bright fringes

   (c) Distance between any two consecutive coloured fringes

   (d) (a) or (b)

7. The fringe width β is

   (a) $\frac{\lambda d}{D}$ (b) $\frac{\lambda D}{d}$ (c) λDd (d) λd

8. In Young's double slit experiment the separation of the slits is 1.9 mm and fringe spacing is 0.31 mm at distance of 1 m from slits. The wavelength of light is

   (a) 6890 Å (b) 5890 Å (c) 4890 Å (d) 3890 Å

9. Green light of wavelength 5100 Å from a narrow slit is incident on a double slit. If the overall separation of 10 fringes on a screen 200 cm away is 2 cm, then the slit separation is

   (a) 0.071 cm (b) 0.061 cm (c) 0.051 cm (d) 0.041 cm

10. Interference in thin film takes place due to

   (a) Reflected light (b) Refracted light

   (c) Transmitted light (d) both (a) and (c)

11. The path difference for thin films is given as $x = 2\mu t \cos r - \frac{\lambda}{2}$. In this case interference is by

(a) Reflected light (b) Transmitted light

(c) Refracted light (d) All are wrong

12. The condition for obtaining bright fringes in thin films by interference due to transmitted light is

(a) $2\mu t \cos r = n\lambda$ (b) $2\mu t \cos r = (2n + 1) \lambda/2$

(c) $2\mu \cos r = (2n + 1) \lambda/2$ (d) $2\mu t \cos r = (2n - 1) \lambda/2$

13. The fringe width produced by wedge shaped thin film is

(a) $\beta = \frac{\lambda}{\theta}$ (b) $\beta = \frac{\lambda}{2}$

(c) $\beta = \frac{\lambda}{2\theta}$ (d) $\beta = \frac{\lambda}{3\theta}$

14. The fringe width produced by wedge shaped thin film having a refractive index μ is

(a) $\beta = \frac{\lambda\mu}{\theta}$ (b) $\beta = \frac{\lambda}{2\mu}$

(c) $\beta = \frac{\lambda\mu}{2\theta}$ (d) $\frac{\lambda}{2\mu\theta} = \beta$

15. When a plano-convex lens of long focal length is placed on a glass plate, a thin film of air is enclosed between the lens and upper surface of the plate. The fringes of the thin film are concentric circles. There rings are known as

(a) Eye rings (b) Circular rings

(c) Newtons rings (d) None of these

16. The quality $\Delta = 2\sqrt{n\lambda R} = 0$ corresponds to

(a) First Newton's ring (b) Second Newton's ring

(c) Third Newton's ring (d) Centre of the Newton's ring

17. The radius of dark Newton's by reflected light is given by

(a) $r = \sqrt{n\lambda}$ (b) $r = \sqrt{n\lambda R}$

(c) $r = \sqrt{\frac{(2n-1)}{2}\lambda R}$ (d) $r = \sqrt{nR}$

18. The radius of dark Newton's ring by transmitted light is given by

(a) $r = \sqrt{n\lambda}$ (b) $r = \sqrt{n\lambda R}$

(c) $r = \sqrt{\frac{(2n-1)}{2}\lambda R}$ (d) $r = \sqrt{nR}$

19. Which is not correct about Newton's rings

(a) It helps in the determination of the wavelength of sodium light

(b) It helps in the determination of refractive index of a liquid

(c) Newton's rings are easily obtainable from refracted light

(d) (a) and (b)

20. The wavelength of the sodium light using Newton's rings is given by

(a) $\lambda = \frac{(D_n)^2 - (D_{n+p})^2}{4R}$ (b) $\lambda = \frac{(D_{n+p})^2 - (D_n)^2}{4P}$

(c) $\lambda = \frac{(D_{n+p}) - (D_n)}{4PR}$ (d) $\lambda = \frac{(D_{n+p})^2 - (D_n)^2}{4PR}$

21. With white light, the Newton's rings are

(a) Bright (b) Dark

(c) Alternately dark and bright (d) Coloured

22. Alternatively dark and bright Newton's rings are obtained by

(a) Monochromatic source (b) Sodium light

(c) White light (d) Both (a) and (b)

23. The Michelson's interferometer is used to observe

(a) Diffraction pattern (b) Interference pattern

(c) Newton's rings (d) None of these

24. The circular fringes obtained in the Michelson interferometer are known as

(a) Newton's rings (b) Haidinger's fringes

(c) Fresnel's rings (d) None of these

25. Fringes of equal inclination are known as

(a) Newton's fringes (b) Localised fringes

(c) Fresnel's fringes (d) Haidinger's fringes

26. The light waves from two coherent sources of intensity I, interfere. At the minima, intensity of light is zero. The intensity of light at the maxima is

    (a) 4I (b) I (c) 2I (d) I^2

27. In double slit experiment (Young's) the separation between the slits is halved and distance between the slits and the screen is doubled. The fringe width is

    (a) Unchanged (b) Halved

    (c) Doubled (d) Quadrupled

28. The maximum number of possible interference maxima for slit separation equal to twice the wavelength in Young's double slit experiment is

    (a) Infinite (b) Five (c) Three (d) Zero

29. Colours appear on a thin soap film and soap bubbles is due to the phenomenon of

    (a) Interference (b) Scattering

    (c) Diffraction (d) Dispersion

30. The interference phenomenon can take place

    (a) In all waves (b) In transverse waves only

    (c) In longitudinal waves only (d) In standing waves only

31. In Young's double slit experiment, intensity at a point is one-fourth of the maximum intensity. Angular position of this point is

    (a) $\sin^{-1}\left(\lambda/d\right)$ (b) $\sin^{-1}\left(\lambda/2d\right)$

    (c) $\sin^{-1}\left(\lambda/3d\right)$ (d) $\sin^{-1}\left(\lambda/4d\right)$

Theoretical Questions

1. What is coherent sources? How are they obtained in practice?
2. Discuss why two independent sources of light of same wavelength cannot show interference?
3. What do you understand by coherent sources? How are they realized in practice?
4. State the basic condition for observing the phenomenon of interference of light.

5. Explain the phenomenon of interference of light. Sketch the diagram showing the formation of maxima and minima due to over lapping of monochromatic light from two narrow slits of double slit experiment.

6. Discuss briefly the phenomenon of interference with relation to law of conservation of energy.

7. Find an expression for fringe width in case of young's double slit experiment

8. Describe Fresnel's biprism. Describe how the wavelength of monochromatic source of light can be calculated by using it?

9. What is a Fresnel's Biprism? Give the production and theory of fringes produced by it.

10. In the Fresnel's biprism arrangement show that $d = 2y_1(\mu - 1)\propto$, where y_1, is the distance of the source from the base of the biprism, α is the refracting angle of the biprism and μ is the refractive index of the material of the biprism.

11. Explain the interference pattern produced by a biprism with white light?

12. Discuss the effect of introducing a thin mica sheet in the path of one of the interfering beams. Deduce an expressions for the displacement of fringes?

13. When a thin mica sheet is introduced in the path of one of the interfering beams, show that fringe width remains same. Describe how this method is used for finding the thickness of a thin glass plate?

14. Explain the colours when a thin film illuminated by white light, is observed in reflected light.

15. Why a broad source is necessary for observing colours in thin films? Why are colours not observed in the case of a thick film?

16. Discuss the formation of interference fringes in a thin wedge-shaped film. Find the expression for fringe width?

17. Describe and explain the formation of Newton's rings in reflected monochromatic light. Explain briefly why Newton's rings are circular?

18. What are Newton's rings? Describe and explain the formation of Newton's rings in reflected mono chromatic light. Prove that in reflected light:

    (i) diameter of the dark rings are proportional to the square root of natural numbers,

    (ii) diameters of bright rings are proportional to the square root of odd natural numbers.

19. Describe Newton's rings method for measuring the wavelength of monochromatic light and the refractive index of a liquid. Give the necessary theory.

20. Why the centre of Newton's rings is dark? What will happen if:

    (i) a little water is introduced between the lens and plane,

    (ii) a plane mirror is used instead of a glass plate and

    (iii) white light is used.

21. How can Newton's rings be used to determine the refractive index of a liquid? Derive the necessary formula?

22. Explain the formation of Newton's rings. Why do we prefer a convex lens of large radius of curvature for producing Newton's rings?

23. Show that the diameters of Newton's rings when two surfaces of radii R_1 and R_2 are placed in contact, are related by the equation

$$\frac{1}{R_1} - \frac{1}{R_2} = \frac{y_n \lambda}{d_n^2}$$

Numerical Problems

1. If in an interference pattern, the ratio between maximum and minimum intensities is 36:1, find the ratio between the amplitude and intensities of the two interfering waves?

   [49 : 25]

2. Two waves of amplitudes 3 mm and 4 mm respectively travel in the same direction of the phase difference between these waves at a point in their path is $\frac{\pi}{2}$, what will be the resultant amplitude at that point due to these waves

   [$\pm$ 5mm]

3. Two sources of intensities I and 4I are used in an interference experiment. Obtain intensities at points where the waves from two sources superimpose with a phase difference of

   (i) 0 (ii) $\pi/2$ (iii) π

   [9I, 5I, I]

4. Two coherent beams of wavelength 5000 A° reaching a point would individually produce intensities 1.44 and 4.00 units if they reach there together, the intensity is 0.90 units. Calculate the lowest phase difference with which the beams reach that point?

[161°]

5. In an interference pattern, the amplitude of intensity variation is found to be 5% of the average intensity. Calculate the relative intensities of the interfering sources.

[1600 : 1]

6. In a Young's double slit experiment the angular width of a fringe termed on a distant screen is 0.1°. The wavelength of light used is 6000 A°. What is the spacing between the slits

[0.34 mm]

7. In a two slit interference pattern with λ = 6000 A°, the zero order and tenth order maxima fall at 12.34 mm and 14.73 mm respectively. If λ is changed to 5000 A°, deduce the position of the zero-order and twentieth fringes, other arrangement remaining the same?

[16.32 mm]

8. In Young's experiment, the distance of screen from the two slits is 1.0 m. When light of wavelength 6000 A° is made incident, fringes of width 2.0 mm are obtained on the screen. Calculate

   (i) the distance between the slits, [0.3 mm]

   (ii) the fringe width if the incident light is 4800 A° [1.6 mm]

9. Two coherent sources are placed 0.18 mm apart and the fringes are observed on a screen 80 cm away. It is found that with a certain monochromatic source of light the fourth bright fringe is situated at a distance of 10.8 mm from the central fringe. Find the wavelength of light?

[6075 A°]

10. Given light of wavelength 5100 A° from a narrow slit is incident on a double slit. If the separation of 10 fringes on a screen 200 cm away is 2.0 cm, find the double slit separation?

[0.51 mm]

11. In a particular two slit interference pattern with $\lambda = 6000$ A°, the zero order and 10th oder maxima fall at micrometer reading 12.34 mm and 14.73 mm. If λ is changed to 5000 A°, deduce the position of the zero order and 20th order fringes, other arrangement remaining the same.

[16.32 mm, 8.36 mm]

12. In Young's double slit experiment the slits are 0.5 mm apart and interference is observed on a screen placed at a distance of 100 cm from the slits. It is found that the 9th bright fringes is at a distance of 8.835 mm from the second dark fringe from the centre of the pattern. Find the wavelength of light used?

[5890 A°]

13. In Fresnel Biprism experiment the angle of glass prism is 3° and $\mu = 1.5$. Interference fringes are formed with a source of wavelength 6000 A° located 10 cm from the prism and source screen distance is 100 cm. Find the maximum number of fringes that can be observed?

[45]

14. In a Fresnel biprism experiment the fringe width observed is 0.087 mm. What will it become if the slit to biprism is reduced to 3/4 the original distance (all else remaining the same)?

[0.0116 cm]

15. A Fresnel biprism arrangement is set with sodium light ($\lambda = 5893$A°) and in the field of view of the eye-piece we get 62 fringes. How many fringes shall we get if we replace the source by mercury lamp using

(a) green filter ($\lambda = 5461$ A°] [67]

(b) violet filter ($\lambda = 4358$ A°] [83]

16. In a double slit interference arrangement one of the slit is covered by a thin mica sheet whose refractive index is 1.58. The slits are 0.1 cm apart and the fringes are observed on a screen placed 50 cm away from the slit. Due to the introduction of mica sheet the central fringe get shifted by 0.2 cm. Determine the thickness of mica sheet?

[6.7×10^{-4} cm]

17. A thin sheet of transparent material ($\mu_D = 1.60$) is placed in the path of one of the interfering beams in a biprism experiment using sodium light ($\lambda = 5890 \times 10^{-8}$ cm). The central fringe shifts to a position normally occupied by the 12th bright fringe. Calculate the thickness of the sheet?

[1.178×10^{-3} cm]

18. Calculate the thickness of the thinnest film ($\mu = 1.4$) in which interference of violet component ($\lambda = 4000$ A°) of incident light can take place by reflection.

[714.3 Å]

19. Two glass plates enclose a wedge shaped air film, touching at one edge and are separated by a wire of 0.05 mm diameter at a distance of 15cm from the edge.

Calculate the fringe width. Monochromate light of $\lambda = 6000$ Å from a broad source falls normally on the film.

[0.09 cm]

20. If the angle of wedge is 0.25° of arc and the wavelength of sodium D lines are 5890×10^{-8} and 5896×10^{-8} cm, find the distance from the apex of the wedge at which the maximum due to each wavelength first coincide?

[6.63 cm]

21. Newton's rings are made with light $\lambda = 6400$ Å and a thin layer of oil $\mu = 1.60$ formed between the curved surface of a plano convex lens (radius of curvature 80 cm, $\mu = 1.65$) and a plane glass plate $\mu = 1.55$. Calculate the radius of the smallest dark ring.

[0.04 cm]

22. In an arrangement for observing Newton's rings with two different media between the glass surfaces, the nth rings have diameters as 10 : 7. Find the ratio of the refractive indices of the two media.

[49 : 100]

Answers to Objective Questions

1. (c)	2. (b)	3. (d)	4. (b)	5. (b)	6. (d)
7. (b)	8. (b)	9. (c)	10. (d)	11. (a)	12. (a)
13. (c)	14. (d)	15. (c)	16. (d)	17. (b)	18. (d)
19. (c)	20. (d)	21. (d)	22. (d)	23. (b)	24. (b)
25. (d)	26. (a)	27. (d)	28. (d)	29. (a)	30. (a)
31. (c)					

CHAPTER 2B

Diffraction

"No one has ever been able to define difference between interference and Diffraction satisfactorily. It is just a question of usage and there is no specific, important physical difference between them". The best we can do is, roughly speaking, is to say that where there are only a few sources say two, interfering them the result is usually called interference, but if there is a large number of them, it seems that the word diffraction is more often used.

"Richard Feynman".

2B.1 Introduction

The phenomenon of bending of light around the corner of an obstacle and their spreading into the geometrical shadow of an object is called diffraction. To illustrate the phenomenon of diffraction, let us consider a point source s and a screen X, Y placed in front of it. In Fig. 2B.1(a) AB is a narrow opaque obstacle placed in front of light source S. if the rectilinear propagation of light is assumed, and then AB will cast its shadow on screen as CD. However, it is observed that dark and bright fringes near the edges are seen. Similarly, when the light from a source S is allowed to pass through an aperture and fall on a screen X Y (Fig. 2B.1(b)) then it is observed that the intensity does not fall to zero sharply outside the area CD but shows some fluctuations in the form of dark and bright fringes. It is evident, therefore that the light passing through a small aperture or by edge of small obstacle, bends to some extent into the region of geometrical shadow.

This phenomenon cannot be explained by the principle of rectilinear propagation of light. It was satisfactorily explained by A.J. Fresnel in 1815 by wave theory of light. He considered the diffraction phenomenon as a result of interference between several secondary wavelets produced by the un obstructed portion of the same wave front.

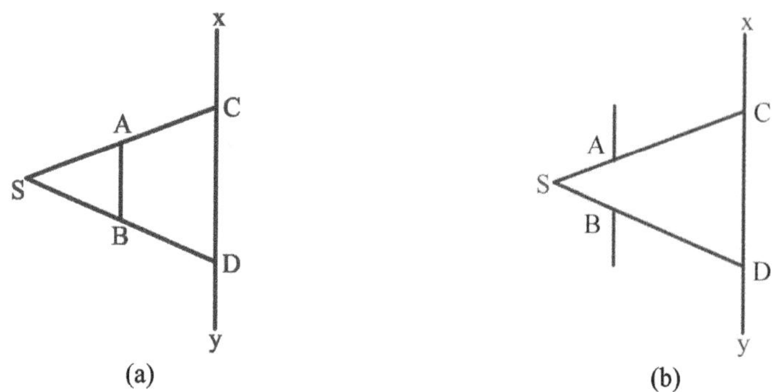

Fig. 2B.1

2B.2 Fresnel and Fraunhofer Diffraction

The diffraction phenomenon are usually divided into two following classes:

1. Fresnel diffraction, and
2. Fraunhofer diffraction

In the Fresnel class of diffraction the source of the light and the screen are generally at a finite distance from the diffracting aperture. In the Fraunhofer class of diffraction, the source of light and screen are at infinite distance from the diffracting aperture (Fig. 2B.2(a) and 2B.2(b)).

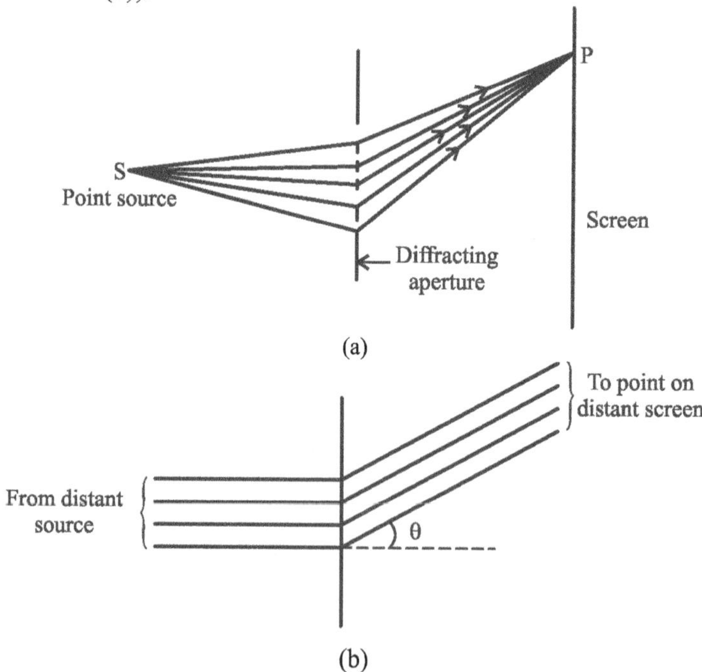

Fig. 2B.2

Fresnel diffraction is more general and it includes Fraunhofer diffraction as a special case.

The difference between Fresnel's diffraction and Fraunhofer diffraction are given in Table 2B.1.

2B.2.1 Difference between Fresnel and Fraunhofer Diffraction

Table 2B.1

Fresnel Diffraction	Fraunhofer Diffraction
1. The source and the screen are at finite distance from the diffracting aperture.	The source and the screen are at infinite distance from the diffracting aperture.
2. The wave fronts are divergent either spherical or cylindrical.	The wave fronts are plane which is realized by using convex lens.
3. No mirrors or lenses are used for observation.	Diffracted light is collected by a lens as in a telescope.
4. For obtaining Fresnel diffraction zone plates are used.	For this, single, double slits or plane diffraction gratings are used.
5. The centre of diffraction pattern may be bright or dark depending up on the number of Fresnel zones.	The centre of the diffraction pattern is always bright for all paths parallel to the axis of the lens.

2B.3 Difference between Interference and Diffraction

The difference between Interference and Diffraction is given below:

Interference	Diffraction
1. This phenomenon is the result of interaction taking place between two separate wave fronts originating from two coherent sources.	1. This phenomenon is the result of interaction of light between the secondary wavelets originating from different points of the same wave front.
2. The regions of minimum intensity are usually almost perfectly dark.	2. The regions of minimum intensity are not perfectly black.
3. Interference fringes may or may not be of the same width.	3. Diffraction fringes are not the same width.
4. All maxima are of same intensity.	4. Maxima are of varying intensity

2B.4 Resultants of n Harmonic Waves

Let us consider n simple Harmonic vibrations having equal amplitude a and common phase difference δ between successive vibrations. To find the resultant amplitude, a polygon is constructed as shown in Fig. 2B.3.

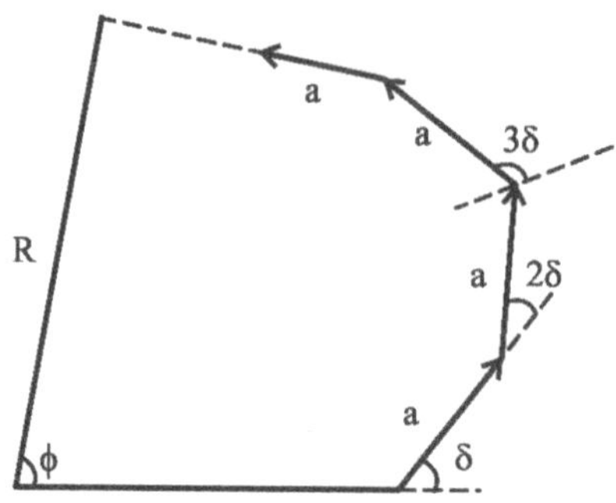

Fig. 2B.3

The closing side of the polygon will give the resultant amplitude R and a resultant phase ϕ. Now resolving the amplitude parallel and perpendicular to the first vibration, we get

$$R\cos\phi = a + a\cos\delta + a\cos 2\delta + \dots\dots\dots + a\cos(n-1)\,\delta \qquad \dots\dots(2B.1)$$

and

$$R\sin\phi = a\sin\delta + a\sin 2\,\delta + \dots\dots\dots + a\sin(n-1)\,\delta \qquad \dots\dots(2B.2)$$

Multiplying eq. 2B.1 by $2\sin\frac{\delta}{2}$, we get

$$2\,R\,\cos\phi\,\sin\frac{\delta}{2} = a\left[2\sin\frac{\delta}{2} + 2\sin\frac{\delta}{2}\cos\delta + 2\sin\frac{\delta}{2}\cos 2\delta + \dots + 2\sin\frac{\delta}{2}\cos(n-1)\delta\right]$$

With the help of Trigonometric formula 2 sin A cos B = sin (A + B) – sin (A – B), the above equation becomes,

$$2R\cos\phi\sin\frac{\delta}{2} = a\left[\sin\frac{\delta}{2} + \sin\left(\frac{3\delta}{2} - \frac{\sin\delta}{2}\right) + \left(\frac{\sin 5\delta}{2} - \frac{\sin 3\delta}{2}\right) + \dots\dots\left\{\sin\left(n-\frac{1}{2}\right)\delta - \sin\left(n-\frac{3}{2}\right)\delta\right\}\right]$$

$$2R\cos\phi\sin\frac{\delta}{2} = a\left[\sin\frac{\delta}{2} + \sin\left(n-\frac{1}{2}\right)\delta\right]$$

$$2R\cos\phi\sin\frac{\delta}{2} = 2\,a\sin\frac{n\delta}{2}\times\cos\frac{(n-1)\delta}{2}$$

$$R\cos\phi = \frac{a\sin\left(\frac{n\delta}{2}\right)\cos\frac{(n-1)\delta}{2}}{\sin\frac{\delta}{2}}$$

Thus,

$$R\cos\phi = \frac{a\sin\left(\frac{n\delta}{2}\right)\cos\frac{(n-1)\delta}{2}}{\sin\frac{\delta}{2}} \qquad \text{.....(2B.3)}$$

Similarly, multiplying eq. (2B.2) by 2 sin $\frac{\delta}{2}$ and simplifying, we get

$$R\sin\phi = \frac{a\sin\left(\frac{n\delta}{2}\right)\sin\frac{(n-1)\delta}{2}}{\sin\frac{\delta}{2}} \qquad \text{.....(2B.4)}$$

Squaring equation (2B.3) and (2B.4) and adding, we get

$$R^2 = \frac{a^2\sin^2\left(\frac{n\delta}{2}\right)}{\sin^2\left(\frac{\delta}{2}\right)} \qquad \text{..... (2B.5)}$$

$$R = a\sin\frac{\left(\frac{n\delta}{2}\right)}{\sin\left(\frac{\delta}{2}\right)} \qquad \text{.....(2B.6)}$$

Dividing eq. (2B.4) by eq. (2B.3), we get

$$\tan\phi = \tan\frac{(n-1)\delta}{2}$$

$$\phi = \frac{(n-1)\delta}{2} \qquad \text{.....(2B.7)}$$

When a and δ are very small and n is infinitely large, such that na = A, again taking $n\delta = 2\alpha$ then eq. (2B.6) takes the form

$$R = \frac{a \sin \alpha}{\sin\left(\frac{\alpha}{n}\right)} = na \frac{\sin \alpha}{\alpha} \qquad \text{.....(2B.8)}$$

Since $\frac{\sin \alpha}{n} \cong \frac{\alpha}{n}$ as n is very large, hence $\frac{\alpha}{n}$ is very small

or
$$R = \frac{A \sin \alpha}{\alpha} \qquad \text{.....(2B.9)}$$

and
$$\phi = \frac{(n-1)\delta}{2} \cong \frac{n\delta}{2} = \alpha \qquad \text{.....(2B.10)}$$

2B.5 Fraunhofer Diffraction due to a single Slit

Let S is point monochromatic source of light of wavelength, placed at the focus of a collimating lens L_1. Light beam is incident normally from S on a narrow slit A B of width a and is diffracted from it. The diffracted beam is focussed at the screen XY by another converging lens L_2. The diffraction pattern having a central bright band followed by alternative dark and bright bands of decreasing intensity on both sides is obtained (Fig. 2B.4).

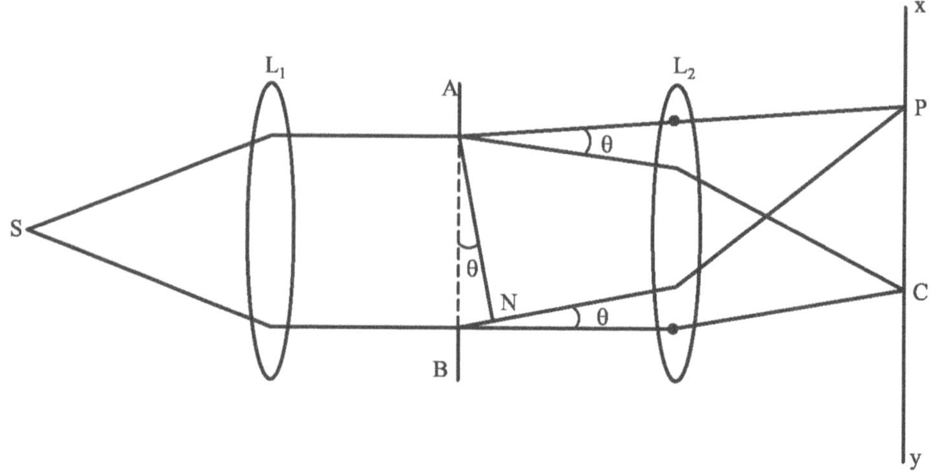

Fig. 2B.4

Analytical Explanation

Light from the point source S is incident as a plane wave front on the slit AB. According to the Huygen's wave theory, every point in AB sends out secondary waves in all directions. The undeviated rays from AB are focussed at C on the screen by the lens L_2.

While the rays diffracted through an angle θ are focussed at point P on the screen. The rays from the ends A and B reach at C in same phase and hence the intensity is maximum.

To find the intensity at P, let us draw normal AN on BN. Since the path beyond AN is the same, therefore, the path difference between the extreme rays is

$$BN = AB \sin\theta = e \sin\theta$$

where AB = e which is the width of the slit.

or path difference $= \dfrac{2\pi}{\lambda} \times e \sin\theta$

Let AB is divided into a large number n of equal parts then there may be infinitely large number of point sources of secondary wavelets between A and B. The phase difference between any two consecutive parts is, therefore.

$$\frac{1}{n} \cdot \frac{2\pi}{\lambda} \; e \sin\theta = \delta \text{ (say)}$$

Thus, we have to find the resultant amplitude of n such vibrations each of amplitude a and phase difference δ between successive vibrations.

According to the theory of composition of n vibrations each of amplitude a and common phase difference δ between successive vibrations, the resultant amplitude at P is given by

$$R = \frac{a \sin\left(\dfrac{n\delta}{2}\right)}{\sin\left(\dfrac{\delta}{2}\right)}$$

$$R = \frac{a \sin\left\{\left(\dfrac{\pi e \sin\theta}{\lambda}\right)\right\}}{\sin\left\{\left(\dfrac{\pi e \sin\theta}{n}\right)\right\}}$$

Let us substitute

$$\frac{\pi e \sin\theta}{\lambda} = \alpha \text{, then}$$

$$R = \frac{a \sin \alpha}{\sin\left(\frac{\alpha}{n}\right)}$$

Since for large value of n, $\frac{\alpha}{n}$ is very small, therefore $\sin\left(\frac{\alpha}{n}\right)$ may be replaced by $\left(\frac{\alpha}{n}\right)$. The above equation then takes the form

$$R = \frac{a \sin \alpha}{\frac{\alpha}{n}} = \frac{na \sin \alpha}{\alpha}$$

$$R = \frac{A \sin \alpha}{\alpha} \qquad \text{where } A = na$$

The resultant intensity at P is proportional to the square of the amplitude. Taking the proportionality constant as unity, the resultant intensity at P is given by

$$I = A^2 \left(\frac{\sin \alpha}{\alpha}\right)^2 \qquad \text{.....(2B.11)}$$

2B.5.1 Condition for Maxima and Minima

The resultant amplitude as given by equation $R = \frac{A \sin \alpha}{\alpha}$ can be expanded as

$$R = \frac{A \sin \alpha}{\alpha} = \frac{A}{\alpha}\left[\alpha - \frac{\alpha^3}{3!} + \frac{\alpha^5}{5!} - \frac{\alpha^7}{7!} + \ldots\ldots\right]$$

$$R = A\left[1 - \frac{\alpha^2}{3!} + \frac{\alpha^4}{5!} - \frac{\alpha^6}{7!} + \ldots\right]$$

R will be maximum if the negative terms in the above equation vanish. This is possible only when $\alpha = 0$ or

$$\frac{\pi}{\lambda} e \sin \theta = 0$$

or $\qquad \theta = 0$

The resultant intensity at P will be maximum for $\theta = 0$ and called the principal maxima. Hence the intensity of the principal maxima = A^2.

Position of Minima

It is clear from equation

$$I = A^2\left(\frac{\sin\alpha}{\alpha}\right)^2$$

That the intensity will be minimum when $\sin\alpha = 0$ but $\alpha \neq 0$ i.e.,

$$\alpha = \pm\, m\pi \qquad \text{where } m = 1, 2, 3, \ldots..$$

or

$$\frac{\pi e \sin\theta}{\lambda} = \pm\, m\pi$$

$\therefore$

$$e \sin\theta = \pm\, m\lambda \qquad \ldots..(2B.12)$$

The values of m = 1, 2, 3 …… gives the direction of first, second, third minima.

Secondary Maxima

In diffraction pattern, there are secondary maxima in addition to principal maxima. The condition of secondary maxima may be obtained by differentiating eq. 2B.11 with respect to α and equating it to zero. Hence

$$\frac{dI}{d\alpha} = \frac{d}{d\alpha}\left[A^2\left(\frac{\sin\alpha}{\alpha}\right)^2\right]$$

$$= A^2\,\frac{2\sin\alpha}{\alpha}\left(\frac{\alpha\cos\alpha - \sin\alpha}{\alpha^2}\right) = 0$$

$\therefore$ either

$$\frac{\sin\alpha}{\alpha} = 0$$

or

$$\alpha\cos\alpha - \sin\alpha = 0$$

or

$$\alpha = \tan\alpha$$

The condition $\frac{\sin\alpha}{\alpha} = 0$ gives the position of principal maxima except $\alpha = 0$.

The position of secondary maxima is given by

$$\alpha = \tan\alpha \qquad \ldots..(2B.13)$$

The eq. 2B.13 can be solved graphically by plotting the curve

$$y = \alpha \quad \text{and} \quad y = \tan \alpha \qquad \text{.....(2B.14)}$$

The curve is shown in Fig. 2B.5.1. The equation y = α gives a straight line passing through the origin and making an angle of 45° with x-axis. The equation y = tan α gives a discontinuous curve.

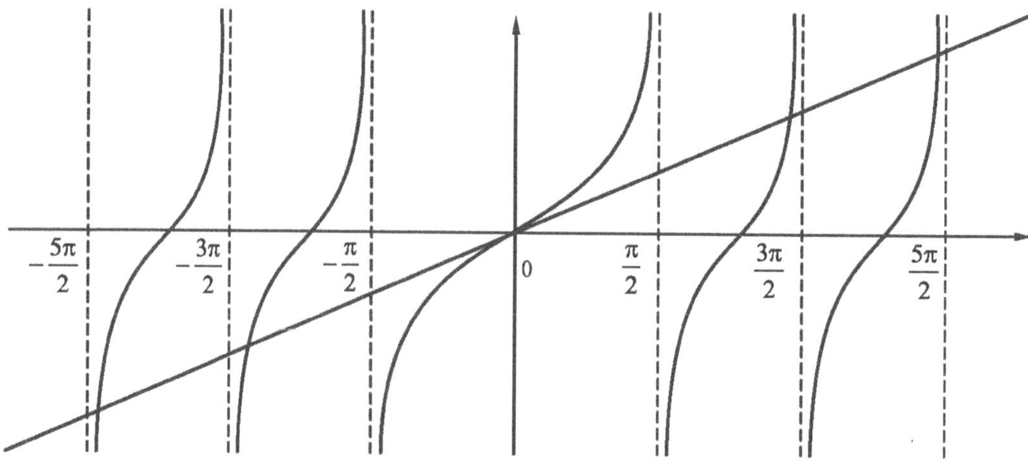

Fig. 2B.5.1

The point of intersection of these two curves gives the value of α satisfying the equation.

$$\alpha = \tan \alpha$$

These points corresponds to the value of

$$\alpha = 0, \ \pm\frac{3\pi}{2}, \ \pm\frac{5\pi}{2}, \ \pm\frac{7\pi}{2}, \ldots\ldots..$$

The first value α = 0 gives the position of principal maxima, while the value of

$\alpha = \pm\frac{3\pi}{2}, \pm\frac{5\pi}{2}, \pm\frac{7\pi}{2}, \ldots\ldots$ gives the position of first secondary maxima. Second, secondary maxima and so on

$$I_1 = A^2\left(\frac{\sin \alpha}{\alpha}\right)^2$$

$$I_1 = I_0 \left(\frac{\sin \frac{3\pi}{2}}{\frac{3\pi}{2}} \right)^2$$

$$= I_0 \left[\frac{-1}{\frac{3\pi}{2}} \right]^2$$

$$= \frac{4}{9\pi^2} I_o = \frac{I_0}{22}$$

Where I_o is the intensity of principal maxima. Similarly the intensity of secondary maxima is given by

$$I_2 = A^2 \left(\frac{\sin \frac{5\pi}{2}}{\frac{5\pi}{2}} \right)^2$$

$$I_2 = I_0 \left(\frac{2}{5\pi} \right)^2 = \frac{4}{25\pi^2} I_0 = \frac{I_0}{61}$$

Thus it is obvious from the values of I_0, I_1, I_2, ... that diffraction pattern consists of a bright central followed by minima of zero intensity and then secondary maxima of decreasing intensity on either side of it.

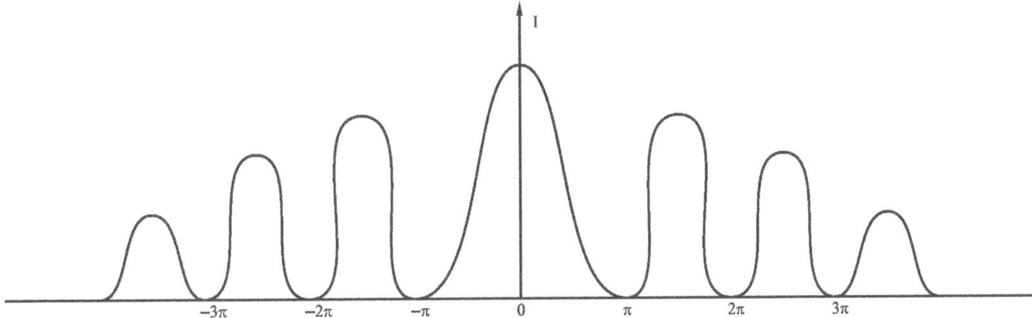

Fig. 2B.5.2

The relative intensities of successive maxima are nearly

$$1:\frac{4}{9\pi^2}:\frac{4}{25\pi^2}:\frac{4}{49\pi^2}\cdots$$

or

$$1:\frac{1}{22}:\frac{1}{61}:\frac{1}{121}\cdots$$

2B.5.2 Width of Central Maxima

Let the distance of the first secondary minimum from the centre of the principal maximum be x.

$\therefore$ width of central maximum = 2x

If the lens L lies very close to the slit or screen is very far from the lens, then the distance between the slit and the screen is very large. Let this distance be D.

$$\therefore \tan\theta = \sin\theta = \frac{x}{D}$$

for the first minimum

$$\sin\theta = \pm\frac{\lambda}{e}$$

$$\pm\frac{\lambda}{e} = \frac{x}{D}$$

$$\therefore \qquad x = \pm\frac{\lambda D}{e}$$

If f is the focal length of the lens and the lens is very near to the slit, then D = f

$$\therefore \qquad x = \pm\frac{\lambda f}{e}$$

$\pm$ sign indicates, on either side to central maxima

$\therefore$ width of central maximum

$$= x - (x) = 2x$$

$$= \frac{2\lambda f}{e} \qquad \text{.....(2B.15)}$$

This shows that the width of the central maximum is directly proportional to the wavelength of light (λ) and inversely proportional to the slit width (e).

2B.5.3 Effect of Slit Width

As $\sin\theta = \pm\frac{\lambda}{e}$

If slit width e is large, then for a given wavelength of light, $\sin\theta$ is small and hence θ is small. This means that the maxima and minima lie very close to the central maximum.

If slit width e is narrow, θ is large and then diffraction maxima and minima are quite distinct and clear.

2B.6 Fraunhofer Diffraction at a Double Slit

Let a plane wave front be incident normally on two slits AB and CD of equal width e and separated by an opaque distance d. The diffracted light is focussed on the screen XY, as shown in Fig. 2B.6

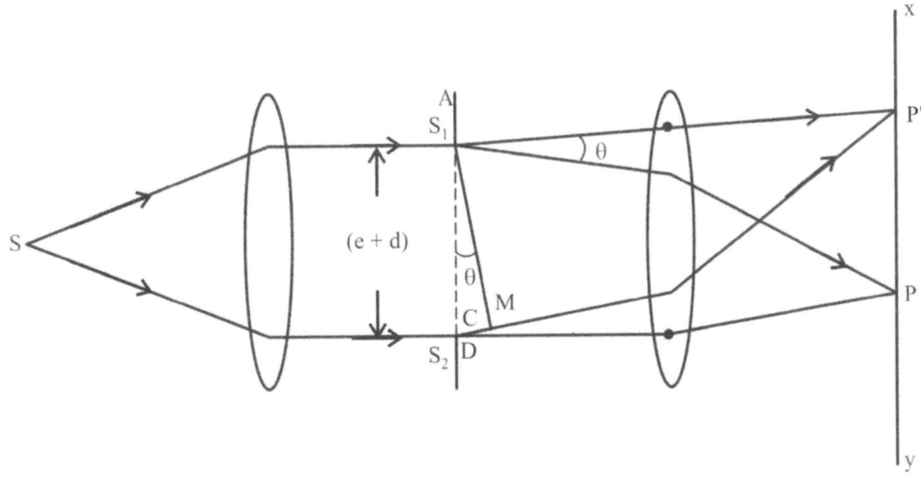

Fig. 2B.6

The diffraction pattern on the screen consists of equally spaced bright and dark fringes due to interference of light from both the slits and modulated by diffraction pattern from individual slits.

The diffraction pattern due to double slit can be explained considering the following points.

1. All points in slits S_1 and S_2 will send secondary waves in all directions.

2. All the secondary waves moving along the incident wave will be focussed at P and the diffracted waves will be focussed at P'.

3. The amplitude at p' is the resultant from two slits, each of amplitude $R = \frac{A \sin \alpha}{\alpha}$

4. The two waves from two slits S_1 and S_2 will interfere at P'.

Let us draw a perpendicular S_1M from S_1 on S_2M.

The path difference between the waves from S_1 and S_2

$$= S_2 M = (e + d) \sin \theta \qquad .....(2B.16)$$

And the corresponding phase difference $= \frac{2\pi}{\lambda}(e + d) \sin \theta = 2\beta$ ......(2B.17)

The resultant amplitude at P' can be obtained by the Fig. 2B.6.1.

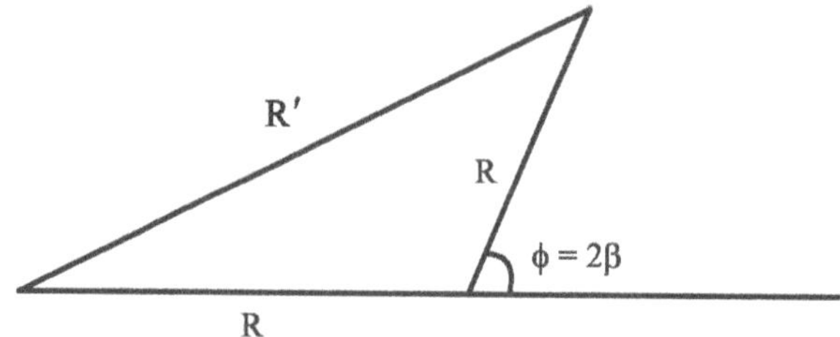

Fig. 2B.6.1

Thus the resultant amplitude at P' is then given by

$$R'^2 = R^2 + R^2 + 2RR \cos(2\beta)$$

$$= 2R^2(1 + \cos 2\beta)$$

$$= 2R^2 . 2\cos^2 \frac{2\beta}{2} = 4R^2 \cos^2 \beta \qquad .....(2B.18)$$

where $$R = \frac{A \sin \alpha}{\alpha}$$

Therefore

$$R'^2 = 4 A^2 \frac{\sin^2 \alpha}{\alpha^2} \cos^2 \beta \qquad .....(2B.19)$$

where $$\beta = \frac{\pi}{\lambda}(e + d) \sin \theta \qquad ..... (2B.20)$$

Therefore the resultant intensity at P is

$$I = R'^2 = 4A^2 \frac{\sin^2 \alpha}{\alpha^2} \cos^2 \beta$$

Hence, the resultant intensity depends upon two factors

1. $\frac{\sin^2 \alpha}{\alpha^2}$ and 2. $\cos^2 \beta$

$\frac{\sin^2 \alpha}{\alpha^2}$ gives diffraction pattern due to each individual slit and $\cos^2 \beta$ gives interference pattern due to diffracted light waves from the two slits.

$\frac{\text{Sin}^2 \alpha}{\alpha^2}$ gives a central maximum in the direction $\theta = 0$, having alternatively minima and subsidiary maxima of decreasing intensity on either side, as shown in Fig. 2B.6.2(a), the minima are obtained in the direction given by

$$\sin \alpha = 0$$

$$\alpha = \pm m \pi$$

$$\alpha = \frac{\pi e \sin \theta}{\lambda}$$

$$\therefore \quad \pi e \sin \theta = \pm m \pi \lambda$$

or $$e \sin \theta = \pm m \lambda \qquad \text{.....(2B.21)}$$

where m = 1, 2, 3, … (except zero)

The term $\cos^2 \beta$ in the intensity pattern gives a set of equidistant dark and bright fringes as shown in Fig. 2B.6.2(b). The bright fringes are obtained in the directions given by

$$\cos^2 \beta = 1$$

$$\beta = \pm n \pi$$

or $$\frac{\pi}{\lambda} (e + d) \sin \theta = \pm n \pi$$

$$(e + d) \sin \theta = \pm n \pi \qquad \text{.....(2B.22)}$$

where n = 0, 1, 2, …. corresponds to zero-order, first-order, second-order maximas.

The resultant intensity distribution pattern is shown in Fig. 2B.6.2(c).

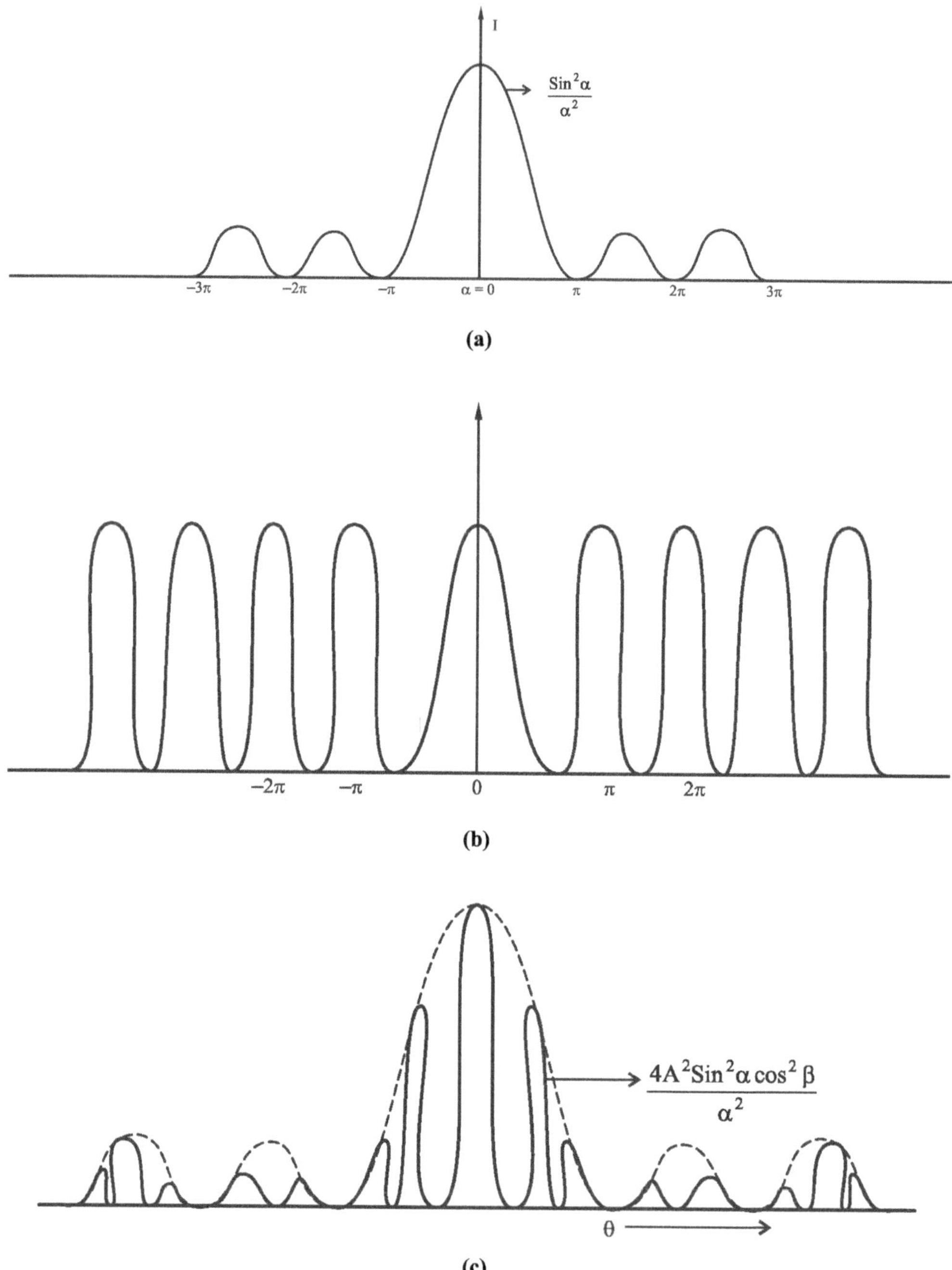
I
$\frac{\text{Sin}^2\alpha}{\alpha^2}$
−3π
−2π
−π
α = 0
π
2π
3π
(a)
−2π
−π
0
π
2π
(b)
$\frac{4A^2\text{Sin}^2\alpha\cos^2\beta}{\alpha^2}$
θ
(c)

Fig. 2B.6.2

2B.6.1 Dependence on the Slit Width and Distance between the Slits

(a) ***Effect of increasing the slit width:*** On increasing the slit width, e, the central peak will become sharper, but the fringe spacing remains unchanged. Hence less interference maxima fall within the central diffraction maximum.

(b) ***Effect of increasing the distance between slits:*** On increasing the separation between silt d keeping the slit width constant, the fringes become closer together but the envelope of the pattern remains unchanged. Hence more interference maxima fall within the central envelope.

2B.7 Absent Order

For certain values of d, certain interference maxima become absent from the pattern

As, the directions of interference maxima are given by

$$(e + d) \sin\theta = n\lambda \qquad \text{.....(2B.23)}$$

The directions of diffraction minima are given by

$$e \sin\theta = m\lambda \qquad \text{.....(2B.24)}$$

If the values of e and d are such that both the equations are satisfied for the same value of θ, then a certain interference maximum will overlap the diffraction minimum and hence the spectrum order will be missing (absent).

Dividing eq. 2B.23 by eq. 2B24, we get

$$\frac{e+d}{e} = \frac{n}{m}$$

If $e = d$; $\frac{n}{m} = 2$ or $n = 2m$

If $m = 1, 2, 3,$ etc; then $n = 2, 4, 6$ etc

This means that the 2, 4, 6, etc. orders of interference maxima will be missing in the diffraction pattern. Thus the central diffraction maxima will have three interference maxima (the zero order and two first-order)

If $d = 2e$; $\frac{e+2e}{e} = \frac{n}{m}$ or $n = 3m$

If $m = 1, 2, 3, \ldots$; then $n = 3, 6, 9, \ldots$

This means that 3rd, 6th, 9th etc. orders of interference maxima will be missing in the diffraction pattern. On both sides of the central maximum, the number of interference maximum is 2 and hence there will be five interference maxima in the central diffraction maximum.

2B.8 Distinction between Single Slit and Double Slit Diffraction Patterns

Single Slit Diffraction Pattern	Double slit Diffraction Pattern
1. In this, diffraction pattern consists of a central bright maximum with secondary maxima and minima of gradually decreasing intensity	1. Diffraction pattern consists of equally spaced interference maxima and minima with the central maximum
2. Intensity $I = I_o \left(\frac{\sin\alpha}{\alpha}\right)^2$ where $\alpha = \frac{\pi e \sin\theta}{\lambda}$, e = slit width; $\therefore$ for central maxima $I = I_o$	2. Intensity $I = 4\ I_o \left[\frac{\sin\alpha}{\alpha}\right]^2 \cos\beta$ where $\beta = \frac{\pi(e+d)\sin\theta}{\lambda}$; $\therefore$ for central maxima $I = 4\ I_o$
3. On increasing the slit width, e, for a given wavelength of light, the maxima and minima will come closer to the central maxima	3. On increasing the slit width e, the central peak will become sharper, but the fringe spacing remains unchanged.
4. The single slit diffraction pattern is shown: I; -3π -2π $-\pi$ 0 π 2π 3π	4. The double slit diffraction pattern is shown: I; 0

2B.9 Fraunhofer Diffraction due to N parallel Slits (Diffraction Grating)

A diffraction grating consists of a large number of parallel slits of equal width and separated from one another by equal opaque spaces. It may be constructed by ruling a large number of parallel and equidistant lines on a plane glass plate with the help of a diamond point. The replicas of the original grating are prepared by passing a thin layer of colloidal solution over it and then allowed to harden. This layer is then removed from the original grating and fixed between two glass plates which serve as a plane transmission grating. Generally there are 10,000 to 15,000 lines per inch in a plane transmission grating.

Theory

Since a plane diffraction grating in a N-slit arrangement, the diffraction due to it will be combined diffraction effect of all such slits. Let a plane wave front of monochromatic light be incident normally on N-parallel slits of the grating. Each point in the slits then sends out secondary wavelets in all directions as shown in Fig. 2B.7. Let e be the width of each slit and d be the separation between any two consecutive slits, then (e + d) is known as grating element. The diffracted rays from each slit are focused at a point P on the screen XY with the help of a convex lens L. Let S_1, S_2, S_3.... be the middle point of each slit and S_1M_1, S_2M_2,....S_{n-1} M_{n-1}, let the perpendicular drawn as shown in Fig. 2B.7. The waves diffracted from each slit is equivalent to a single wave of amplitude

$$R = \frac{A \sin \lambda}{\alpha}$$

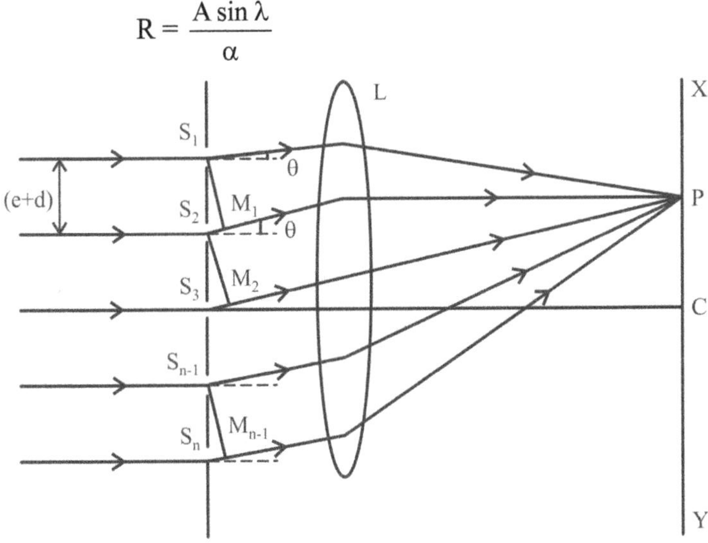

Fig. 2B.7

The path difference between the waves from S_1 and S_2 is

$$S_2 \ M_1 = (e + d) \sin \theta$$

The path difference between the waves from S_2 and S_3 is

$$S_3 \ M_2 = (e + d) \text{ Sin } \theta$$

The path difference between the waves from S_{n-1} and S_n is

$$S_n M_{n-1} = (e + d) \sin \theta$$

Thus it is obvious that the path difference between all the consecutive waves is the same and equal to $(e + d) \sin \theta$.

The corresponding phase difference

$$= \frac{2\pi}{\lambda}(e + d) \sin \theta = 2\beta \qquad \text{(say)}$$

Thus the resultant amplitude at P is the resultant amplitude of N waves, each of amplitude R and common phase difference (2β). Hence the resultant amplitude at P is given by

$$R' = \frac{R \sin \frac{2N\beta}{2}}{\sin \frac{2\beta}{2}}$$

or

$$R' = R \frac{\sin N\beta}{\sin \beta}$$

$$R' = \frac{A \sin \alpha}{\alpha} \frac{\sin N\beta}{\sin \beta}$$

The resultant intensity at P is given by

$$I = R'^2 = \frac{A^2 \sin^2 \alpha}{\alpha^2} \cdot \frac{\sin^2 N\beta}{\sin^2 \beta} \qquad \text{.....(2B.25)}$$

The factor $\frac{A^2 \sin^2 \alpha}{\alpha^2}$ gives the intensity pattern due to a single slit, while the factor $\frac{\sin^2 N\beta}{\sin^2 \beta}$ gives the distribution of intensity due to interference from all the N-slits

2B.9.1 Principal Maxima

From eq. 2B.25, it is clear that the intensity will be maximum when

$\sin \beta = 0$ or $\beta = \pm n\pi$.

where n = 0, 1, 2, 3,

But under this condition sin Nβ is also equal to zero. Hence the term $\frac{\sin N\beta}{\sin\beta} = \frac{0}{0}$ which is indeterminate. To find the value, the numerator and denominator should be differentiated. Hence,

$$\lim_{\beta\to\pm n\pi}\frac{\sin N\beta}{\sin\beta} = \lim_{\beta\to\pm n\pi}\frac{\frac{d}{d\beta}(\sin N\beta)}{\frac{d}{d\beta}(\sin\beta)}$$

$$= \lim_{\beta\to\pm n\pi}\frac{N\cos N\beta}{\cos\beta} = N \qquad .....(2B.26)$$

Substituting this value of $\frac{\sin N\beta}{\sin\beta}$ in eq. (2B.25), we get

$$I = \frac{A^2\sin^2\alpha}{\alpha^2}N^2 \qquad .....(2B.27)$$

Thus the condition for principle maxima is

$$\sin\beta = 0 \qquad \text{or} \qquad \beta = \pm n\pi$$

or $$\frac{\pi}{\lambda}(e+d)\sin\theta = +n\pi$$

$$(e+d)\sin\theta = \pm n\pi \qquad .....(2B.28)$$

For n = 0, we get θ = 0 and this gives the zero order principal maximum. The values of n = 1, 2, 3, … gives the direction of first, second, third, …. order principle maxima.

2B.9.2 Minima

The intensity is minimum, when

$$\sin N\beta = 0, \text{ but } \sin\beta \neq 0$$

$$N\beta = \pm m\pi$$

or $$N.\frac{\pi}{\lambda}(e+d)\sin\theta = \pm m\pi$$

$$N.(e+d)\sin\theta = \pm m\lambda$$

Where m can take all integral values except 0, N, 2N, 3N, ... because for these values of m, $\sin\beta = 0$, which gives the position of principle maxima.

2B.9.3 Secondary Maxima

It is clear from the above condition of minima, that there are (N-1) minima between two successive principle maxima. Hence there is (N-2) other maxima with alternate minima between two successive principal maxima. These (N-2) maxima are called secondary maxima. To find the condition of secondary maxima eq. 2B.25 is differentiated with respect to β and equated to zero.

$$\frac{dI}{d\beta} = \frac{A^2 \sin^2\alpha}{\alpha^2} \times 2\left(\frac{\sin N\beta}{\sin\beta}\right)\left[\frac{N\cos N\beta \sin\beta - \sin N\beta \cos\beta}{\sin^2\beta}\right]$$

or $$N\cos N\beta \sin\beta - \sin N\beta \cos\beta = 0$$

$$\tan N\beta = N\tan\beta \qquad .....(2B.29)$$

Let us construct a triangle as shown in Fig. 2B.8

From this triangle, $\sin N\beta = \dfrac{N\tan\beta}{\sqrt{1+N^2\tan^2\beta}}$

Substituting this value of sin Nβ in eq. 2B.25, we get

$$I' = \frac{A^2\sin^2\alpha}{\alpha^2}\frac{N^2\tan^2\beta}{\left(1+N^2\tan^2\beta\right)\sin^2\beta}$$

$$I' = \frac{A^2\sin^2\alpha}{\alpha^2}\frac{N^2}{\left(1+N^2\tan^2\beta\right)\cos^2\beta}$$

$$I' = \frac{A^2\sin^2\alpha}{\alpha^2}\frac{N^2}{\cos^2\beta + N^2\sin^2\beta}$$

$$I' = \frac{A^2\sin^2\alpha}{\alpha^2}\frac{N^2}{\left(1-\sin^2\beta + N^2\sin^2\beta\right)}$$

$$I' = \frac{A^2\sin^2\alpha}{\alpha^2}\frac{N^2}{1+\left(N^2-1\right)\sin^2\beta} \qquad .....(2B.30)$$

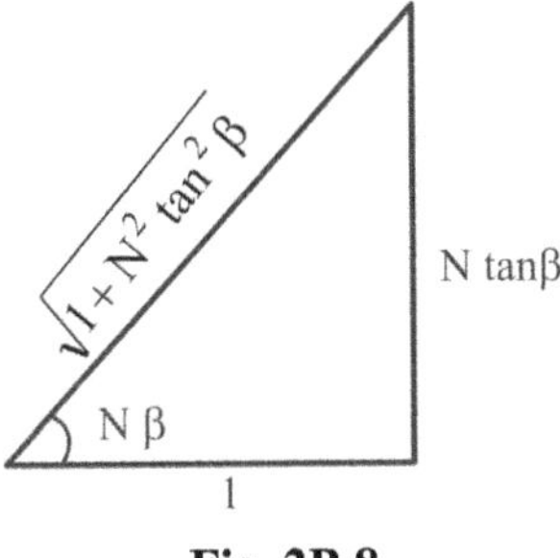

Fig. 2B.8

Dividing eq.2B.30 by eq.2B.27, we have

$$\frac{\text{Intensity of secondary maxima}}{\text{Intensity of principal maxima}} = \frac{1}{1+\left(N^2-1\right)\sin^2\beta} \qquad \text{.....(2B.31)}$$

It is obvious from eq. 2B.31 that when N increases, the intensity of secondary maxima decreases. The intensity distribution in diffraction pattern due to a plane transmission grating is shown in Fig. 2B.9.

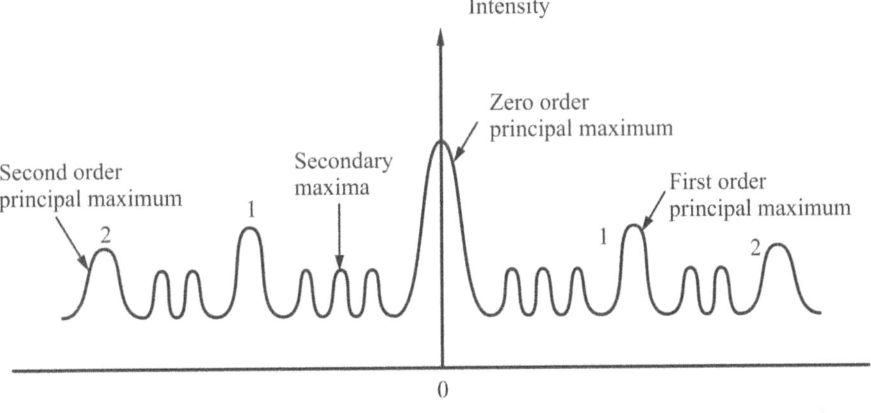

Fig. 2B.9

2B.10 Formation of Spectra with a Grating

From eq. 2B.28, it is clear that the direction of principal maxima are given by

$$(e + d) \sin \theta = \pm n\lambda$$

The following conclusions may be drawn from the above equation:

1. For fixed values (e + d) and n; θ is different for different wavelength.

   As $\lambda_r > \lambda_v$, then $\theta_r > \theta_v$

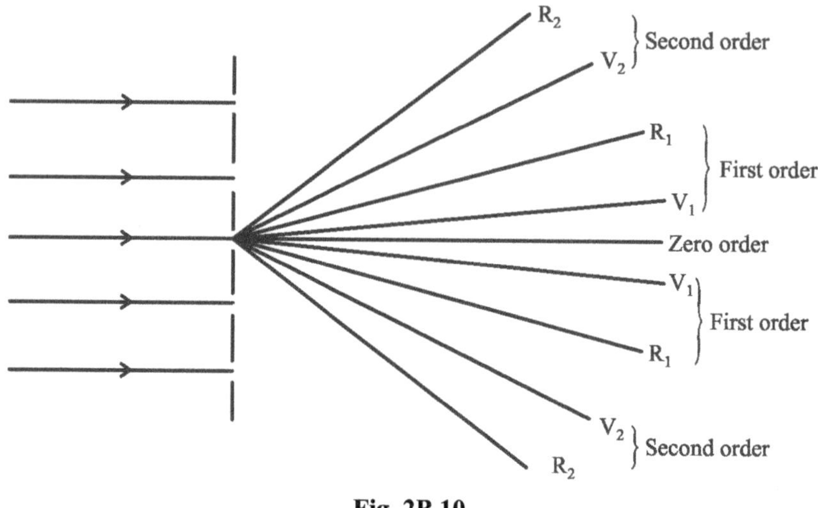

Fig. 2B.10

2. For fixed value of (e + d) and λ, θ is different for different order n.
3. For n = 0, θ = 0, for all values of λ. Hence, the zeroth order principal maxima is always white.
4. Most of the intensity of light is concentrated in zeroth order principal maxima and it decreases with increasing order. The pattern of grating spectra is shown in Fig. 2B.10.

2B.11 Condition for Absent Spectra

As the resultant intensity due to N-parallel slits (plane diffraction grating) is given by

$$I = A^2 \frac{\sin^2 \alpha}{\alpha^2} \frac{\sin^2 N\beta}{\sin^2 \beta} \quad \text{.....(2B.32)}$$

where $\alpha = \dfrac{\pi e \sin \theta}{\lambda}$ and $\beta = \dfrac{\pi (e+d) \sin \theta}{\lambda}$

Now the direction of principle maxima in grating spectrum is given by

$$(e + d) \sin \theta = n\lambda \quad \text{.....(2B.33)}$$

where n is the order of the maximum

The direction of minima in a single-slit pattern is given by

$$e \sin \theta = m \lambda \quad \text{.....(2B.34)}$$

where m = 1, 2, 3,

If both conditions are simultaneously satisfied, a particular maximum or order n will be absent in the grating spectrum, these are known as absent spectra (or missing order spectrum).

Now, dividing eq. (2B.33) by eq. (2B.34) becomes, we get

$$\frac{e+d}{e} = \frac{n}{m} \quad \text{.....(2B.35)}$$

This is the condition for the spectrum of the order n to be absent.

If d = e, then equation (2B.35) becomes

$$n = 2m \qquad \text{where } m = 1, 2, 3, ...$$

$$= 2, 4, 6, ...$$

Hence, when the width of transparencies and opacities of the grating are equal (e = d), then 2nd, 4th, 6th... order spectra are missing.

$$d = 2e,$$

then $$n = 3m$$

$$= 3, 6, 9, ...$$

That is 3rd, 6th, 9th.... order spectra will be absent.

Maximum number of orders

As the grating equation is

$$(e + d) \sin\theta = n\lambda$$

where (e + d) is grating element, θ is the angle of diffraction and n the order of the maximum.

We can write

$$n = \frac{(e+d)\sin\theta}{\lambda} \quad \text{.....(2B.36)}$$

Maximum possible value of $\theta = 90^\circ$

Maximum possible order is given by

$$n_{max} = \frac{(e+d)\sin 90^\circ}{\lambda}$$

$$= \frac{(e+d)}{\lambda} \quad \text{.....(2B.37)}$$

This shows that if (e + d) < 2λ, then

$$n_{max} < \frac{2\lambda}{\lambda} < 2 \qquad .....(2B.38)$$

Hence for normal incidence only first order will be obtained.

2B.11.1 Determination of Wavelength of Light with a Plane Diffraction Grating

The nth order principal maxima for normal incidence is given by

$$(e + d) \sin \theta = n \lambda$$

$$\lambda = \frac{(e+d)\sin\theta}{n} \qquad .....(2B.39)$$

Thus the determination of λ, involves the measurement of (e + d), the grating element, θ the angle of diffraction and n, the order of spectrum.

2B.11.2 Determination of (e + d)

The number of lines per inch is written on the grating. If N is the number of lines per inch, then the grating element

$$(e+d) = \frac{2.54}{N} \text{ cm}$$

2B.12 Grating Under Oblique Incidence

Let us consider a monochromatic beam of light of wavelength λ is incident obliquely on the grating at an angle of incidence i (Fig. 2B.11). Let secondary wavelets starts from the mid point of the slits and diffracted at an angle θ. Let S_1L be the perpendicular on R_2S_2 and S_1K be the perpendicular on $S_2 D_2$.

Hence the path difference between diffracted rays D_1 and D_2 is

$$S_2L + S_2K = S_1S_2 \sin i + S_1 S_2 \sin \theta$$
$$= (e + d) \sin i + (e + d) \sin \theta$$
$$= (e + d) [\sin i + \sin \theta] \qquad .....(2B.40)$$

where $S_1S_2 = (e + d)$ = grating element.

The grating equation would be

$$(e + d) [\sin i + \sin \theta] = n\lambda$$

$$\text{or} \qquad 2(e+d)\sin\left(\frac{i+\theta}{2}\right)\cos\left(\frac{i-\theta}{2}\right) = n\lambda$$

Now $(i+\theta)$ is the deviation δ suffered by the rays.

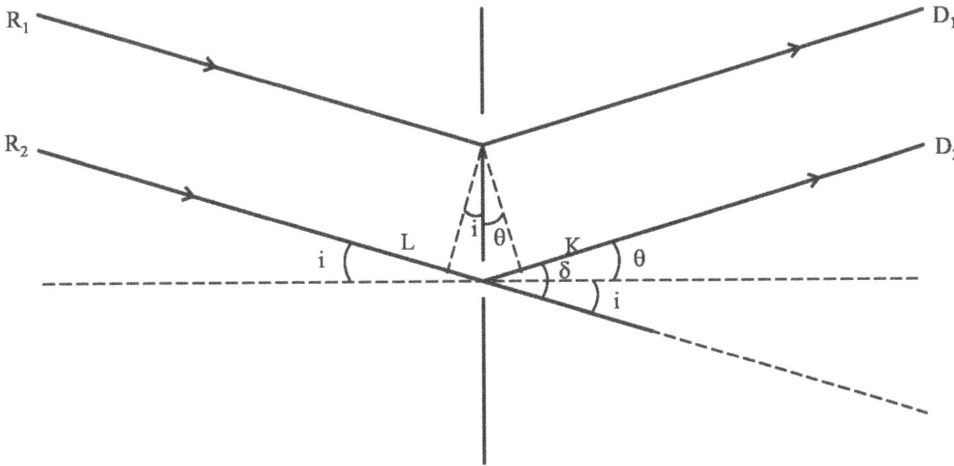

Fig. 2B.11

$$\therefore \qquad 2(e+d)\sin\frac{\delta}{2}\cos\left(\frac{i-\theta}{2}\right) = n\,\lambda \qquad .....(2B.41)$$

Now $\sin\frac{\delta}{2}$ and hence δ would be minimum when $\cos\left(\frac{i-\theta}{2}\right)$ is maximum

$$\text{or} \qquad i-\theta = 0$$

$$i = \theta \qquad .....(2B.42)$$

Thus the rays would suffer minimum deviation, δ_{min} when the angle of incidence equals the angle of diffraction.

$\therefore$ The grating equation would then becomes

$$2(e+d)\sin\frac{\delta_{min}}{2} = n\lambda \qquad .....(2B.43)$$

2B.13 Dispersive Power of a Grating

The dispersive power of a diffraction grating is defined as the rate of change of the angle of diffraction with the wavelength of light.

It is expressed as $\frac{d\theta}{d\lambda}$

As the grating equation is

$$(e + d) \sin \theta = n\lambda$$

Differentiating it with respect to λ, we get

$$(e + d) \cos \theta \frac{d\theta}{d\lambda} = n$$

or $$\boxed{\frac{d\theta}{d\lambda} = \frac{n}{(e+d)\cos\theta}} \qquad \text{....(2B.44)}$$

This the dispersive power of a grating.

According to this expression, the dispersive power is:

1. Directly proportional to the order n
2. Inversely proportional to the grating element (e + d)
3. Inversely proportional to cos θ

If the value of θ is large, then cos θ will be smaller and dispersive power $\frac{d\theta}{d\lambda}$ is high.

As θ (red) > θ (violet) in a given spectrum, therefore the dispersion in the red region is greater than that in the violet region.

If θ is small, cos θ = 1, hence $d\theta \propto d\lambda$. Such a spectrum is known as normal spectrum.

2B.14 Angular Half Width

The nth order principal maximum is obtained in the direction θ_n given by

$$(e + d) \sin \theta_n = n\lambda \qquad \text{.....(2B.45)}$$

Let the first minimum adjacent to the nth maximum be obtained in the direction $(\theta_n + d\theta_n)$

where $d\theta_n$ is called the 'angular half-width' of the nth maximum.

The minima are obtained in the directions given by

$$N (e + d) \text{ Sin } \theta = m \lambda$$

where N = total number of parallel slits

m = integer; except 0, N, 2N, …, nN

As the first minimum adjacent to n^{th} order principal maximum is obtained in the direction $\theta_n + d\theta_n$.

Hence this gives

$$m = nN + 1$$

Hence eq.(2B.45) becomes

$$N\,(e + d)\sin(\theta_n + d\theta_n) = (nN + 1)\lambda$$

or $$N\,(e + d)(\sin\theta_n \cos\theta_n + \cos\theta_n \sin d\theta_n) = (nN + 1)\lambda$$

As $d\theta_n$ is small

$$\cos d\theta_n = 1 \qquad \text{and} \qquad \sin d\theta_n = d\theta_n$$

$\therefore$ $$N(e + d)\sin\theta_n + N\,(e + d)\cos\theta_n\, d\theta_n = nN\,\lambda + \lambda \qquad \text{.....(2B.46)}$$

But from eq. 2B.45,

$$(e + d)\ \mathrm{Sin}\ \theta_n = n\,\lambda$$

Eq.2B.46 becomes,

$$Nn\lambda + N\,(e + d)\cos\theta_n\, d\theta_n = Nn\,\lambda + \lambda$$

or $$N\,(e + d)\cos\theta_n\, d\theta_n = \lambda$$

$\therefore$ $$d\theta_n = \frac{\lambda}{N(e + d)\cos\theta_n} \qquad \text{.....(2B.47)}$$

Using eq. 2B.45

$$(e + d)\sin\theta_n = n\,\lambda$$

$$(e + d) = \frac{n\,\lambda}{\sin\theta_n}$$

$\therefore$ eq.2B.47 can also be written as

$$d\theta_n = \frac{\lambda}{N.\dfrac{n\lambda}{\sin\theta_n}.\cos\theta_n}$$

or $$d\theta_n = \frac{1}{Nn\cot\theta_n} \qquad \text{.....(2B.48)}$$

This is the expression for angular half width

2B.15 Difference between Prismatic and Grating Spectra

Prism spectrum	Grating spectrum
1. The prism spectrum is formed by dispersion	1. The grating spectrum is formed by diffraction
2. The prism spectrum is only of one order	2. The grating spectrum forms spectra of different order
3. The prism spectrum depends on the material of the prisms	3. The grating spectrum is independent of the material of the grating
4. The prism spectrum does not show the fine structure of spectral lines	4. The grating spectrum shows the fine structure of spectral lines
5. The prism spectral lines are curved	5. The grating spectral lines are almost straight
6. In prism spectrum there is greater deviation in the violet region as compared to red region	6. In grating spectrum there is greater deviation in the red region as compared to violet region

2B.16 Rayleigh Criterion of Resolution

The diffraction of light limits resolution. Rayleigh has proposed a criterion for resolution. Two images from fine slits or point source are said to be just resolved if the position of the central maximum of intensity curve due to one source coincides with the first minimum of the diffraction intensity curve of the second and vice-versa. Fig. 2B.12 shows three situations – unresolved, just resolved and well resolved under Rayleigh criterion.

When the intensities of the two sources happen to be different or spectral widths are different then these do not appear distinct even if above criterion is satisfied. In fact resultant intensity shows a dip or hump as shown in Fig. 2B.13(a) and 23B.13(b). The Rayleigh criterion has to be defined numerically.

The single slit intensity distribution is given by

$$I = I_0 \left(\frac{\sin \alpha}{\alpha}\right)^2$$

where
$$\alpha = \frac{\pi}{\lambda} e \sin \theta$$

The first minimum appears at $\alpha = \pm \pi$. According to Rayleigh criterion, central maximum due to one coincide with first minimum due to another. If this happens, the intensity at the mid point between central maximum and first minimum is twice I at

$$\alpha = \pm \pi/2$$

$$I_{mid} = 2\, I_o \left[\frac{\sin \pi/2}{\pi/2} \right]^2$$

$$I_{mid} \simeq \frac{8}{\pi^2}\, I_o = 0.81\, I_o$$

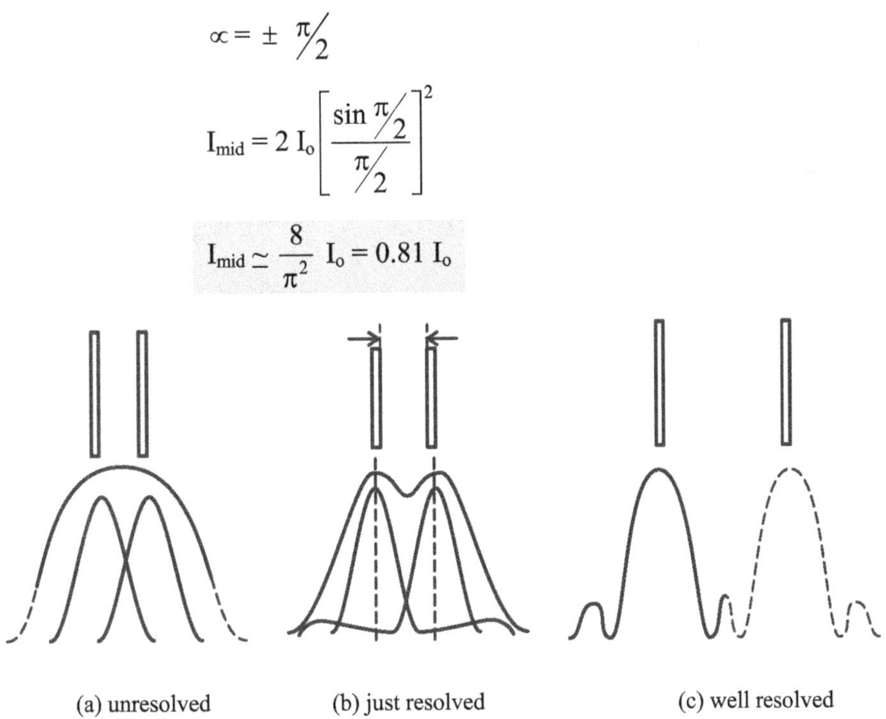

Fig. 2B.12 [Rayleigh criterion of resolution limit of resolution]

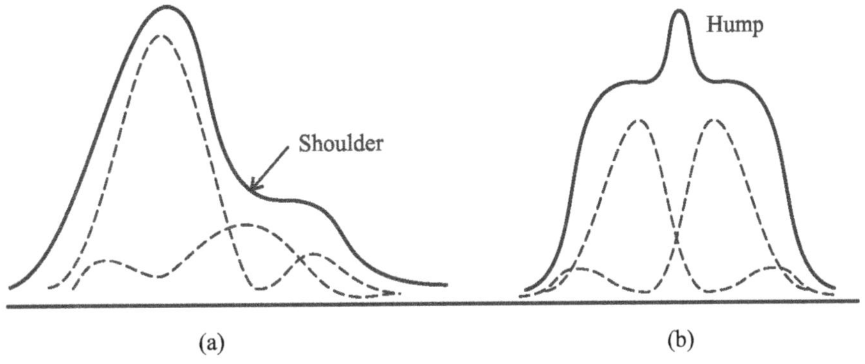

Fig. 2B.13

Thus according to Rayliegh Criterion, two images or spectral lines will be just resolved when intensity at dip in the middle falls to 81% of maximum intensity or intensity reduces by 19%.

2B.17 Resolving Power of Telescope and Microscopes

(a) ***Telescope:*** As per Rayleigh criterion, the two far away objects seen from telescope shall be resolved if the separation between two maxima formed by light diffracted from lens is equal to half width of maxima as shown in Fig. 2B.14.

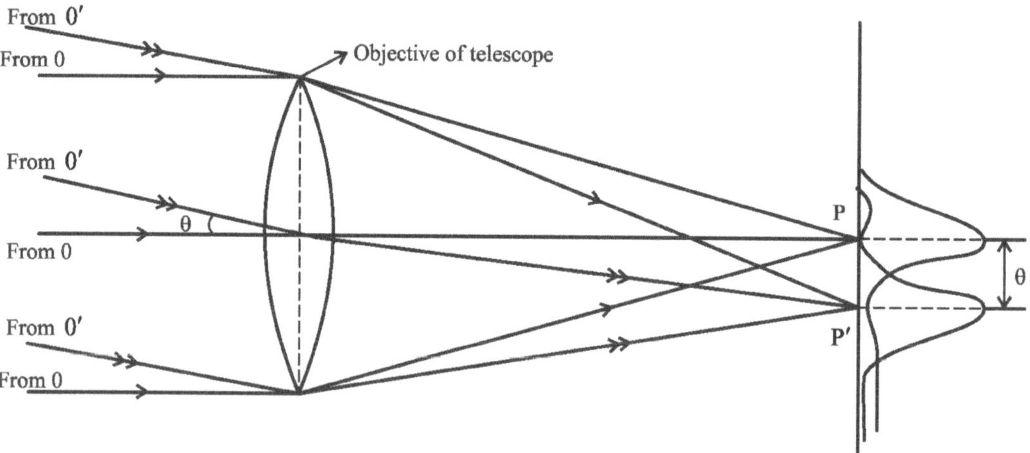

Fig. 2B.14 light rays from two nearly but distinct objects, O and O′ from diffraction maxima or minima. (the images will be just resolved when maximum of one coincoides with minimum of the other. Angular separation equals the ratio of Airy's disk)θ.

Hence

$$\text{Resolving limit for telescope} = \theta_{min} = \frac{1.22\,\lambda}{D}$$

where D is the diameter of the objective of telescope

$$\therefore \qquad RP = \frac{1}{\theta_{min}} = \frac{D}{1.22\,\lambda}$$

If r be the radius of the central spot called 'Airy's disk' and f focal length of lens, then

$$\theta_{min} = \frac{r}{f} = \frac{1.22\,\lambda}{D}$$

$$r = \frac{1.22\,\lambda\, f}{D}$$

(b) ***Microscope:*** The smallest separation between two nearby objects seen distinct by the microscope is the resolving limit of microscope. The object will be just resolved when diffraction maximum due to one object source coincides with the diffraction minimum due to other object source as required by Rayleigh criterion. The position of first diffraction minimum falls at position where path difference happens to be 1.22 λ, the factor 1.22 is due to circular geometry. From the Fig. 2B15.

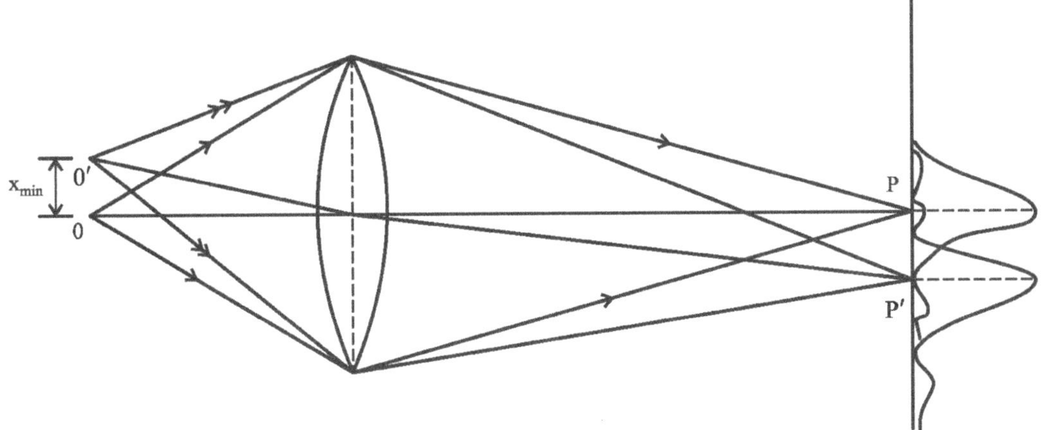

Fig. 2B.15

we have

$$2\,x_{min} \sin\alpha = 1.22\,\lambda$$

where α is the semi-vertical angle made by the point source on to the lens

Hence $$x_{min} = \frac{1.22\,\lambda}{2 \sin\alpha}$$

$\therefore$ resolving power

$$R.P = \frac{2 \sin\alpha}{1.22\,\lambda}$$

If the object is dipped in a refractive index μ

then $$R.P = \frac{2 \sin\alpha.\mu}{1.22\,\lambda}$$

2B.18 Resolving Power of Plane Diffraction Grating

The resolving power of grating is defined as its ability to form separate diffraction principal maxima of two wavelengths which are very close to each other.

Let a light consisting of two wavelengths λ_1 and λ_2 is incident normally on a grating of grating element (e + d). The spectral lines corresponding to wavelength λ_1, and λ_2 will form its diffraction pattern which can be seen through the telescope.

The spectral lines may be just resolved, if they satisfy the Rayleigh criterion, i.e., if the central maxima due to one spectral line falls over the first minima of the other.

The direction of the nth principal maxima for wavelength λ_1 is given by

$$(e + d) \sin \theta_n = n \lambda_1$$

$$N (e + d) \sin \theta_n = Nn\lambda_1 \quad \text{.....(2B.49)}$$

where N is the total number of lines on the grating.

The first minima in the direction $(\theta_n + d\theta_n)$ is given by

$$N (e + d) \sin (\theta_n + d\theta_n) = n \lambda_1 \quad \text{.....(2B.50)}$$

Therefore, the first minima adjacent to the nth maxima in the direction $(\theta_n + d\theta_n)$ is possible only when

$$m = (nN + 1)$$

Hence eq. 2B.50 may be written as

$$N (e + d) \sin (\theta_n + d\theta_n) = (nN + 1) \lambda_1 \quad \text{.....(2B.51)}$$

The direction of principal maxima of the diffraction pattern due to wavelength λ_2 in the direction $(\theta_n + d\theta_n)$ is given by

$$(e + d) \sin (\theta_n + d\theta_n) = n\lambda_2$$

$$N (e + d) \sin (\theta_n + d\theta_n) = Nn\lambda_2 \quad \text{......(2B.52)}$$

For Rayleigh's Criterion to be satisfied, the principal maxima of wavelength λ_2 should coincide with the first minima of wavelength λ_1 in the same direction $(\theta_n + d\theta_n)$.

Hence comparing eq. 2B.51 and 2B.52, we get

$$(nN + 1) \lambda_1 = Nn\lambda_2 \quad \text{.....(2B.53)}$$

Let $\lambda_1 = \lambda$; $\lambda_2 = \lambda + d\lambda$ and $\lambda_2 - \lambda_1 = d\lambda$. Then from eq. 2B.53, we have

$$(nN + 1)\lambda = Nn (\lambda + d\lambda)$$

$$\lambda = Nnd\lambda \quad \text{.....(2B.54)}$$

The resolving power of grating is equal to $\frac{\lambda}{d\lambda}$

$\therefore$ from eq. 2B.54

$$R.P = \frac{\lambda}{d\lambda} = nN \quad \text{.....(2B.55)}$$

Hence R.P = order of spectrum × total number of lines on grating

Further $(e + d) \sin \theta = n\lambda$

$$n = \frac{(e+d)\sin\theta}{\lambda} \quad \text{.....(2B.56)}$$

Substituting the value of n from eq.2B.56 in 2B.55, we get

$$R.P = \frac{\lambda}{d\lambda} = \frac{N(e+d)\sin\theta}{\lambda} \quad \text{.....(2B.57)}$$

It may be noted from eq. 2B.55 that

1. R.P is proportional to number of lines N on the grating within the limit $(e + d) > \lambda$
2. R.P is proportional to order of spectrum. Spectral lines in second order are more resolved as compared to first order. However, at higher orders, intensity limits their use to $\left(I \propto \frac{1}{n^2}\right)$
3. RP is independent of grating element $(e + d)$ of the grating.

2B.19 Fresnel's Half Period Zones

Let us consider in Fig. 2B.16 plane wave front ABCD (perpendicular to the plane of paper) of monochromatic light of wavelength λ moving in the direction from O to P where P is an external point at which the effect of the wave front is to be found. Fresnel subdivided the wavefront into a number of half period zones. To get these zones let us drop a perpendicular PO from the point P on the wave front and let OP = b. The point O on the wave front is the pole with respect to P. Taking the point P as center, let us draw the series of spheres of radii.

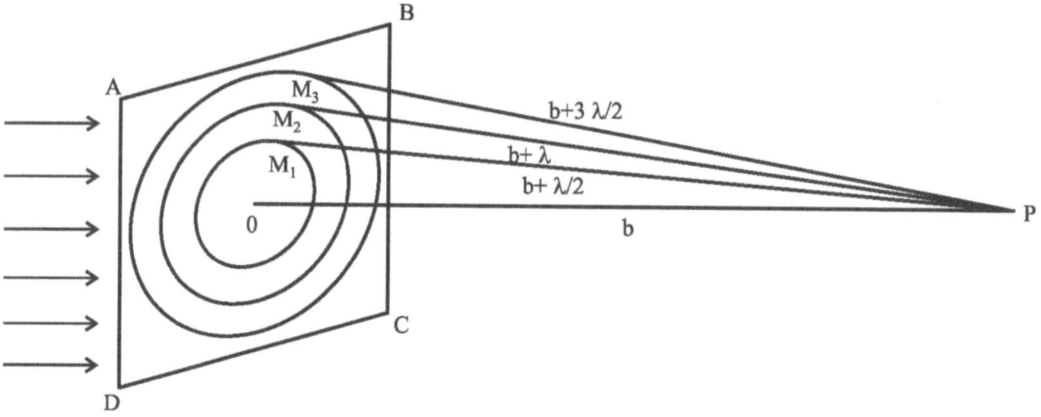

Fig. 2B.16

$$PM_1 = b + \frac{\lambda}{2}, PM_2 = b + \frac{2\lambda}{2}, PM_3 = P + \frac{3\lambda}{2} \quad \ldots\ldots\ldots\ldots$$

which cross the wave front ABCD in concentric circles of radii.

$$OM_1 = \sqrt{\left(b + \frac{\lambda}{2}\right)^2 - b^2} = \sqrt{b\lambda}$$

$$OM_2 = \sqrt{\left(b + \frac{2\lambda}{2}\right)^2 - b^2} = \sqrt{2b\lambda} \quad \ldots\ldots(2B.58)$$

Thus

$$OM_n = \sqrt{\left(b + \frac{n\lambda}{2}\right)^2 - b^2}$$

$$OM_n = \sqrt{nb\lambda} \quad \ldots\ldots(2B.59)$$

where the term λ^2 have been ignored. Thus the area of first half period zone (i.e., area enclosed in the inner most circle of radium OM_1) is

$$\pi\, OM_1^2 = \pi\, b\, \lambda \quad \ldots\ldots(2B.60)$$

The area of the second half period zone (i.e., the annular area between the first circle and the second circle) is

$$\pi\left[OM_2^2 - OM_1^2\right] = \pi\, b\, \lambda \quad \ldots\ldots(2B.61)$$

and so on

The path difference $\frac{\lambda}{2}$ between the secondary wavelets originally from O to M_1 and reaching the point P is equivalent to time difference of half period. It is the reason as calling these zones as half period zones. Eq. 2B.58 shows that the radii of half period zones are proportional to the square root of natural numbers. On the other hand eq. 2B.59 and 2B.60 shows that the area of each zone is $\pi b \lambda$, irrespective of its order. However these radii and areas depend upon λ and the distance b of the point P from the wave front.

For considering the resultant amplitude at P due to all the zone, we should keep in mind that the amplitude of disturbance due to a zone varies directly as area of the zone, inversely varies as the distance of P from the zone and directly varies as obliquity factor $(1 + \cos\theta)$. Further more, since the average distance of P from two consecutive zones differ by $\frac{\lambda}{2}$, the secondary waves from two consecutive zones reach P in opposite phase. Hence if $R_1, R_2, R_3, \ldots\ldots.., R_n$ are the amplitudes of the disturbances at P due to first, second, third,... n^{th} half period zones, then the resultant amplitude at P due to whole wave front is

$$R = R_1 - R_2 + R_3 \ldots \pm R_n, \qquad \text{.....(2B.62)}$$

where the sign of R_n is positive for odd n and negative for even n.

The successive amplitudes $R_1, R_2, R_3\ldots.R_n$ are of gradually decreasing magnitude in view of obliquity factor and the distance from P i.e.,

$$R_3 < R_2 < R_1; \qquad R_5 < R_4 < R_3; \text{ and so an}$$

We may have the approximation

$$R_2 = \frac{R_1 + R_3}{2}, \qquad R_4 = \frac{R_3 + R_5}{2} \qquad \text{.....(2B.63)}$$

These eq. (2B.61) may be written as

$$R = \frac{R_1}{2} + \left(\frac{R_1}{2} - R_2 + \frac{R_3}{2}\right) + \left(\frac{R_3}{2} - R_4 + \frac{R_5}{2}\right) + \ldots.$$

or

$$R = \frac{R_1}{2} + \frac{R_n}{2} \qquad \text{if n is odd} \qquad \text{.....(2B.64)}$$

and

$$R = \frac{R_1}{2} + \frac{R_{n-1}}{2} \qquad \text{if n is even}$$

where we have used the relation (2B.62)

For large n,

$$R_1 >> R_n \qquad \text{and} \qquad R_1 >> R_{n-1}$$

On account of obliquity factor and the distance from the point P. Hence eq. (2B.64) simply reduces to

$$R = \frac{R_1}{2} \qquad \text{.....(2B.65)}$$

which shows that the resultant amplitude at any point due to entire wave front is only one half of that which would be produced by the first half period zone alone. Hence the intensity at any point due to whole wave front is only one fourth of that due to the first half period zone alone.

2B.20 Zone Plate

Zone plate is specially constructed diffraction screen such that light from every alternate zone is cut off. It is a simple device for to using light rays by diffraction. Of even half period zones are obstructed by any device and only odd ones are exposed then from equation

$$R = R_1 - R_2 + R_3 \ldots + R_n$$

the resultant amplitude at the point becomes

$$R = R_1 + R_3 + R_5 + \ldots\ldots\ldots + R_n \approx \left(\frac{n+1}{2}\right) R_1$$

or

$$R = R_1 + R_3 + R_5 + \ldots\ldots\ldots + R_{n-1} \approx \frac{n}{2} R_1$$

accordingly as n is odd or even respectively. Thus the intensity at P will be increased $\left(\frac{n+1}{2}\right)^2$ times or $\left(\frac{n}{2}\right)^2$ times by obstructing even half period zones. Similar result follows if odd half period zones are obstructed and even zones are exposed. Hence the intensity at the point P will be many times greater than that due to all the zones.

2B.20.1 Theory of Zone Plate

Let us consider a transparent plate AB at right angle to the paper containing line SP where S is the location of point source of monochromatic light of wavelength λ giving out the spherical waves and P is the position of screen for the bright image. With O as the centre and radii equal to OM_1, OM_2, OM_3,.....OM_4, we divide the plate AB into half period zones such that there is an increase of path difference $\frac{\lambda}{2}$ from one zone to the next i.e.,

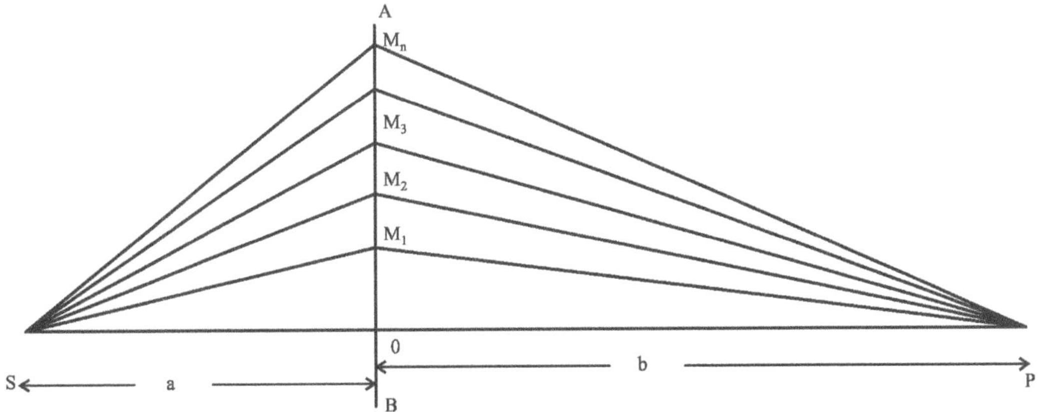

Fig. 2B.17

$$SM_1 + M_1P - (SO + OP) = \frac{\lambda}{2}$$

$$SM_2 + M_2P - (SM_1 + M_1P) = \frac{\lambda}{2}$$

....

....

$$SM_n + M_nP - (SM_{n-1} + M_{n-1}P) = \frac{\lambda}{2}$$

which may also be written as

$$SM_1 + M_1P = SO + OP + \frac{\lambda}{2}$$

$$SM_2 + M_2P = SO + OP + \frac{2\lambda}{2} \qquad \text{.....(2B.66)}$$

....

....

$$SM_n + M_nP = SO + OP + \frac{n\lambda}{2}$$

If $r_n = OM_n$ = radius of nth zone, then we have

$$SM_n^2 = SO^2 + r_n^2$$

$$SM_n^2 = a^2 + r_n^2$$

$$SM_n = a\left[1 + \frac{r_n^2}{a^2}\right]^{1/2}$$

$$SM_n \approx a + \frac{r_n^2}{2a}$$

Since r_n is much smaller than a (the distance of source from O)

Similarly, we have

$$M_nP = b + \frac{r_n^2}{2b}$$

Substituting these relations in last of eq. (2B.66) , we get

$$a + b + \frac{r_n^2}{2}\left[\frac{1}{a} + \frac{1}{b}\right] = a + b + \frac{n\lambda}{2}$$

which gives

$$r_n^2 = n\lambda\left(\frac{ab}{a+b}\right) \qquad \text{.....(2B.67)}$$

$$r_n = \left(\frac{ab\,\lambda}{a+b}\right)^{1/2}\sqrt{n} \qquad \text{.....(2B.68)}$$

Showing that the radii of the zones are proportional to the square roots of the natural numbers.

The area of nth zones is

$$\pi\left[r_n^2 - r_{n-1}^2\right] = \frac{\pi\, ab\,\lambda}{a+b} \qquad \text{.....(2B.69)}$$

which is independent of n and hence the area of all the zones is the same. But the average distance of the zone from P increases and the obliquity factor $(1 + \cos\theta)$ decreases with the increase in the order of zone and hence the amplitude at P due to the zone decreases with the order of the zone.

When the successive transparent zones are alternate (as in the case of zone plate), then the wave from them will reach P in the same phase and hence the resultant amplitudes

$$R = R_1 + R_3 + R_5 + \ldots.$$

is many times larger than $\frac{R_1}{2}$ which is the resultant of the entire zone (before obstructing alternative zones). Thus the point P is extremely bright and it may be treated as image of the source S. It shows the focussing action of the zone plate.

From eq. 2B.67 we have

$$\frac{1}{a} + \frac{1}{b} = \frac{n\lambda}{r_n^2} \qquad \text{.....(2B.70)}$$

where a, the distance between source and zone plate, may be taken as u (object to lens distance) and b, the distance between zone plate and image, may be taken as v (image to lens distance). Then eq. (2B.70) may be written as the famous lens formula

$$\frac{1}{u} + \frac{1}{v} = \frac{1}{f} \qquad \text{.....(2B.71)}$$

where $f = \frac{r_n^2}{n\lambda}$ is the focal length of zone plate which behaves lines a convergent lens.

2B.20.2 Difference between a Zone Plate and a Convex Lens

Similarities

1. Both form real images of an object on the other side of the object and the distances of object and images are connected by similar relations.
2. In both cases focal length depends on λ and so both shows chromatic abbreviations.

Differences

1. In the case of lens the rays are brought to focus by refraction while in the zone plate, the image is formed by diffraction.
2. In the case of convex lens, all the rays reaching the point where the image is formed travel the same optical path while in zone plate; the waves reaching the image through two alternate zones have a path difference of λ.
3. Image formed by a convex lens is more intense than that due to a zone plate.

4. For convex lens $f_v < f_r$ while for zone plate $f_r < f_v$.

5. The convex lens has only one focus on one side of it, while a zone plate has multiple foci distances

$$\frac{r_n^2}{n\lambda}, \frac{r_n^2}{3n\lambda}, \frac{r_n^2}{5n\lambda}, \ldots\ldots \frac{r_n^2}{(2m+1)n\lambda}$$

The intensity reduces as the focal length decreases.

Solved Numerical Problems

1. A lens of focal length 40 cm forms Fraunhofer's diffraction pattern of slit 0.5 cm wide. Calculate distance of the first dark band and the next bright band from the axis when wavelength of light used is 4890 $\mathring{A}$.

Sol. The position of first dark band is

$$(y_1)_{dark} = \pm \frac{\lambda}{e} f$$

The position of the first bright band next to dark band is

$$(y_1)_{bright} = \frac{3}{2}\frac{\lambda}{e} f$$

$$(y_1)_{bight} - (y_1)_{dark} = \frac{3}{2}\lambda/_{e}\, f - \frac{\lambda}{e} f$$

$$(y_1)_{bright} - (y_1)_{dark} = \frac{1}{2}\frac{\lambda}{e} f$$

Given

$$\lambda = 4890\ \mathring{A}$$

$$e = 0.5 \times 10^{-2}$$

$$f = 40 \times 10^{-2}$$

$$\Delta y = \frac{1}{2}\frac{\lambda}{e} f$$

$$\Delta y = \frac{4890 \times 10^{-10} \times 40 \times 10^{-2}}{2 \times 0.5 \times 10^{-2}}$$

$$\Delta y = 1.956 \times 10^{-5} m$$

2. The light of wave length 5000 $\mathring{A}$ is incident normally on a list. The first minimum of diffraction pattern is observed to lie at a distance 5 mm from the central maximum an a screen placed at a distance of 2 m from the slit. Calculate the width of the slit. [UPTU]

Sol. we know that

$$(y_1)_{dark} = \frac{\lambda}{e} f$$

$$\therefore \quad e = \frac{\lambda f}{(y_1)_{dark}} = \frac{5000 \times 10^{-10} \times 2}{5 \times 10^{-3}}$$

$$e = 2 \times 10^{-4} \text{ m}$$

$$e = 0.2 \text{ mm}$$

3. Microwaves as 6000 MH_z frequency are incident normally on a slit of variable width and large length. A receiver is placed at a large distance on the other side at angle 30° with normal. For what width of slit, would receiver show zero intensity.

Sol. the minimum positions are given by

$$e \sin \theta = n \lambda$$

$$e = \frac{n\lambda}{\sin \theta}$$

Given $\theta = 30^\circ$ $\quad \therefore \ \sin \theta = \frac{1}{2}$

and $\lambda = \frac{3 \times 10^8}{6000 \times 10^6} = \frac{1}{20}$ m

$$e = n \frac{1}{20} \times 2$$

$$e = \frac{n}{10} \text{ i.e.,}$$

0.1 m, 0.2 m, 0.3 m……

4. A screen is placed 200 cm away from a narrow slit which is illuminated with light of wavelength 6×10^{-5}cm of the first minima lies 5 mm on either side of central maximum, calculate the slit width.

Sol. The situation is illustrated in figure.

The distance of first minima from the central maxima.

$$y = 5\text{mm} = 0.5\text{cm}$$

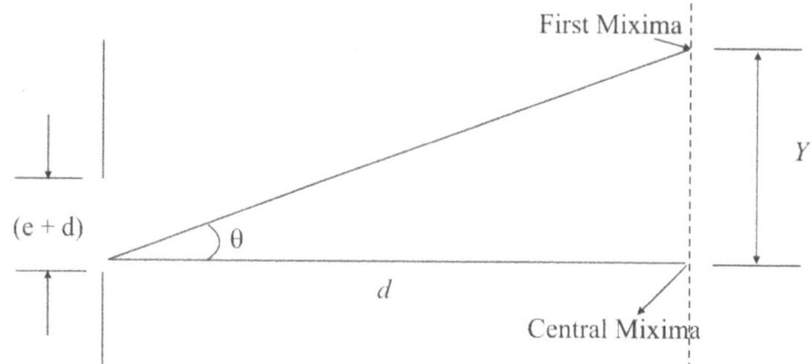

The angular position of the first minima is given by

$$\sin\theta = \frac{\lambda}{e} \qquad (1)$$

As θ is very small, therefore

$$\theta = \sin\theta = \frac{\lambda}{e} \qquad (2)$$

For small θ, we have

$$\theta = \tan\theta = \frac{y}{d}$$

$$\therefore \quad \frac{y}{d} = \frac{\lambda}{e}$$

$$e = \frac{\lambda d}{y}$$

Given d = 200 cm; $\lambda = 6 \times 10^{-5}$ cm; y = 0.5 cm

$$e = \frac{200 \times 6 \times 10^{-5}}{0.5}$$

$$e = 0.024 \text{ cm}$$

5. Plane wave ($\lambda = 6 \times 10^{-5}$cm) falls normally on a straight slit of width 0.2 mm. Calculate the total angular width of the central diffraction maximum and also the linear width as observed on a screen placed 2 meters away

Sol. The angular spread of the central maximum on either side is given by

$$\theta = \frac{\lambda}{e}$$

∴ total angular width

$$2\theta = 2\frac{\lambda}{e} = \frac{2\times(6\times10^{-5}\ \text{cm})}{0.02\ \text{cm}}$$

$$2\theta = 6 \times 10^{-3}\ \text{radian}$$

The linear width at a distance of 2 metres

$$= \text{angle} \times \text{distance}$$

$$= (6 \times 10^{-3}) \times (200\ \text{cm})$$

$$= 1.2\ \text{cm}$$

6. In a double slit Fraunhofer diffraction pattern, the screen is 160 cm away from the slits. The slit width are 0.08 mm and they are 0.4 mm apart. Calculate the wave length light of the fringe spacing is 0.25 cm. Also deduce the missing orders. [UPTU]

Sol. The fringe spacing is given by

$$\beta = \frac{\lambda D}{d}$$

Where d is separation between the slits

$$\lambda = \frac{\beta d}{D}$$

$$\lambda = \frac{(0.25\ \text{cm})\times(0.04\ \text{cm})}{160\ \text{cm}}$$

$$\lambda = 6.25 \times 10^{-5}\ \text{cm}$$

$$\lambda = 6250\ \text{A}^{\circ}$$

If e be the slit width and d be the separation between the slits, then the condition of missing orders is given by

$$\frac{e+d}{e} = \frac{n}{m}$$

Here

$$d = 0.4\ \text{mm}$$

$$e = 0.08\ \text{mm}$$

$$d = 5e \qquad \text{so that}$$

$$n = 6\ m$$

$$n = 6, 12, 18, \ldots\ldots$$

Thus 6th, 12th, 18th,...... orders will be missing

7. Light of wavelength 5500 A° falls normally on a slit of width 22.0×10^{-5} cm. Calculate the angular position of the first two minima on either side of the central maximum.

Sol. The angular position of minima in a single slit diffraction pattern are given by

$$e \sin \theta = \pm n \lambda$$

or $$\sin \theta = \frac{n \lambda}{e}$$

whence $n = 1, 2, 3$ - - -

Given $e = 22.0 \times 10^{-5}$ cm

$$\lambda = 5500 \times 10^{-8} \text{ cm}$$

For first order minimum, n = 1

$$\sin \theta_1 = \frac{\lambda}{e} = \frac{5500 \times 10^{-8}}{22.0 \times 10^{-5}}$$

$$\sin \theta_1 = 0.25$$

$$\theta_1 = \sin^{-1}(0.25) = 14^\circ 29'$$

For second order minimum, n = 2

$\therefore$ $$\sin \theta_2 = \frac{2\lambda}{e} = 2 \times 0.25 = 0.5$$

$$\theta_2 = \sin^{-1}(0.5)$$

$$\theta_2 = 30^\circ$$

Therefore, the first two minima will occur at an angle $14^\circ 29'$ and 30° an either side of central maximum.

8. Calculate the angle at which the first dark band and the next bright band are formed in the Fraunhofer diffraction pattern of slit 0.3 mm wide ($\lambda = 5890$ Å)

[UPTU]

Sol. In Fraunhofer diffraction pattern due to single slit of width e, the directions of minima are given by

$$e \sin \theta = n\lambda$$

where n = 1, 2, 3, - - -

For first dark band

$$n = 1$$

$\therefore$ $$e \sin \theta = \lambda$$

$$\sin \theta = \frac{\lambda}{e} = \frac{5890 \times 10^{-8}}{0.03}$$

$$\sin \theta = 0.00196$$

or $$\theta = \sin^{-1}(0.00196)$$

$$\theta = 0.112^{\circ}$$

The angle of diffraction θ' corresponding to the first bright band on either side of the central maximum is approximately given by

$$e \sin \theta' \simeq \frac{3\lambda}{2} = 1.5\lambda$$

$$\sin \theta' = \frac{1.5\lambda}{e}$$

since $$\frac{\lambda}{e} = 0.00196$$

$\therefore$ $$\sin \theta' = 1.5 \times 0.00196$$

$$\sin \theta' = 0.00294$$

$$\theta' = \sin^{-1} = (0.00294)$$

$$\theta' = 0.168^{\circ}$$

9. A single slit of width 0.14 mm is illuminated normally by monochromatic light and diffraction bands are observed on a screen 2 m away. If the centre of second dark band is 1.6 cm from the middle of the central bright band, deduce the wavelength of light. [UPTU]

Sol. For single slit diffraction pattern, the direction of nth minima is given by

$$e \sin \theta = n\lambda$$

or $$\sin \theta = \frac{n\lambda}{e}$$

For small θ, $\sin \theta \simeq \theta$

$\therefore$ $$\theta = \frac{n\lambda}{e}$$

If D is the distance of the screen from the slit and x is the distance of centre of second dark band from the middle of the central bright band, then

$$\theta = \frac{x}{D} \qquad \therefore \qquad \frac{x}{D} = \frac{n\lambda}{e}$$

or $$\lambda = \frac{x\,e}{n\,D}$$

Given $x = 1.6 \text{ cm} = 1.6 \times 10^{-2} \text{ m}$

$e = 0.14 \text{ mm} = 0.14 \times 10^{-3} \text{ m}$

$n = 2$ and

$D = 2 \text{ m}$

$$\lambda = \frac{1.6 \times 10^{-2} \times 0.14 \times 10^{-3}}{2 \times 2}$$

$$\lambda = 5.6 \times 10^{-7} \text{ m}$$

$$\lambda = 5600 \text{ A}^{o}$$

10. A single slit is illuminated by light composed of two wavelengths λ_1, and λ_2. One observes that due to Fraunhofer diffraction, the first minima obtained for λ_1 coincides with the second diffraction minima of λ_2. What is the relation between λ_1 and λ_2.

Sol. In the Fraunhofer diffraction pattern due to single slit of width e, the direction of minima are

$$e \sin\theta = \pm n\,\lambda$$

For wavelength λ_1, the direction of first minima (n = 1) is

$$e \sin\theta_1 = \lambda_1 \qquad \text{.....(1)}$$

similarly, for wavelength λ_2, the direction of second minima (n = 2) is

$$e \sin\theta_2 = 2\lambda_2 \qquad \text{.....(2)}$$

The direction of first minima (θ_1) due to λ_1 and the second minima due to λ_2 coincides, that is, $\theta_1 = \theta_2 = \theta$. Therefore from (1) and (2), we have

$$e \sin\theta = \lambda_1 = 2\lambda_2$$

$$\lambda_2 = \frac{\lambda_1}{2}$$

11. The intensity distribution due to single slit diffraction is given by

$$I = I_o\left(\frac{\sin\alpha}{\alpha}\right)^2 \qquad \text{where } \alpha = \frac{\pi}{\lambda} e \sin\theta$$

Show that successive maxima bear the intensity ratio $\left(\frac{1}{3\pi}\right)^2 : \left(\frac{1}{5\pi}\right)^2 : \left(\frac{1}{2\pi}\right)^2$

Sol. The maxima shall occur when

$$\alpha = \left(n + \frac{1}{2}\right)\pi$$

$$I_{maxima} = I_o \left[\frac{\sin\left(n + \frac{1}{2}\right)\pi}{\left(n + \frac{1}{2}\right)\pi}\right]^2$$

$$I_{maxima} = \frac{I_o}{\left(n + \frac{1}{2}\right)^2 \pi^2}$$

Therefore,

$$I_1 : I_2 : I_3 \text{ - - -} = \left(\frac{1}{3\pi}\right)^2 : \left(\frac{1}{5\pi}\right)^2 : \left(\frac{1}{7\pi}\right)^2 \text{- - - - - - -}$$

12. Show that intensities of first and second maximum in Fraunhofer's diffraction at single slit are $\frac{1}{22}$ and $\frac{1}{61}$ of principal maxima

Sol. We know that

$$I_{maxima} = I_o \left[\frac{\sin\left(n + \frac{1}{2}\right)\pi}{\left(n + \frac{1}{2}\right)\pi}\right]^2$$

$$I_{maxima} = \frac{I_o}{\left(n + \frac{1}{2}\right)^2 \pi^2}$$

Therefore,

$$I_1 : I_2 : I_3 \text{ - - -}$$

$$= \left(\frac{1}{3\pi}\right)^2 : \left(\frac{1}{5\pi}\right)^2 : \left(\frac{1}{7\pi}\right)^2 : \text{- - - - - - -}$$

The intensities of successive maxima are in the ratio

$$I_2 = \frac{I_o}{\left(\frac{3\pi}{2}\right)^2} = \frac{I_o \times 4}{9\pi^2}$$

$$I_2 = \frac{I_o}{22}$$

and
$$I_3 = \frac{I_o}{\left(\frac{5\pi}{2}\right)^2}$$

$$I_3 = \frac{4\,I_o}{25\,\pi^2}$$

$$I_3 = \frac{I_o}{61}$$

13. In relation to plane transmission grating having 5000 lines per cm, answer the following

(i) The largest wavelength of light for which spectrum can be observed

(ii) The highest order of spectrum that can be observed with light of wavelength 6000 A°

(iii) If opaque spaces are exactly two times the transparent spaces, which order of the spectra will be missing

(iv) If 90% of grating space is covered, what shall happen to spectrum

Sol. From the grating equation, we have

$$(e + d)\sin\theta = n\,\lambda$$

$$\lambda = \frac{(e+d)\sin\theta}{n}$$

(i) $\lambda = \lambda_{max}$ when $n = 1, \theta = 90^\circ$

$$\therefore \quad \lambda = (e + d)$$

$$e + d = \frac{10^{-2}}{5000}$$

$$\lambda = 2 \times 10^{-6}\ m$$

(ii) $n = \dfrac{(e+d)\sin\theta}{\lambda}$, $n = n_{max}$ when $\theta = 90^\circ$

or

$\sin\theta = 1$

therefore

$$n = \frac{(e+d)}{\lambda} = \frac{2\times10^{-6}}{6\times10^{-7}} = \frac{10}{13} = 3.33 = 3$$

(iii) n^{th} order spectrum will be missing when

$$\frac{(e+d)}{e}=\frac{n}{m}; \quad m=1,2,3,\ldots$$

$$n=m\frac{(e+d)}{e}=\frac{e+2e}{e}=3$$

Thus, 3, 6, 9,... orders will be missing

(iv) when 90° of grating space is covered, the number of lines exposed to light are

$$500\times\frac{9}{10}=4500=0.99N$$

Since the intensity of lines $\propto N^2$

The intensity of lines becomes 81%

Width of spectral line $\propto \frac{1}{N}$

Therefore, the width of the lines be comes

$$\frac{10}{9}\text{ times}$$

14. A parallel beam of sodium light inlident on plane transmission grating having 4250 line per cm and a second order spectral line is observed at an angle of 30°. Calculate the wavelength of light

Sol. from the grating equation

$$(e+d)\sin\theta=n\lambda$$

$$(e+d)=\frac{10^{-2}}{4250}$$

Since $n=2$ and $\theta=30^\circ$

$$\therefore \quad \lambda=\frac{(e+d)\sin\theta}{n}=\frac{10^{-2}}{4250}\times\frac{1}{2}$$

$$\lambda=5.882\times10^{-7}\text{ m}$$

$$\lambda=5882\ \overset{o}{A}$$

15. How many lines per cm are there in a grating which gives an angle of diffraction of 30° in first order of light of wavelength 6×10^{-5} cm.

Sol. From the grating equation

$$(e + d) \sin\theta = n\lambda$$

$$(e + d) = \frac{n\lambda}{\sin \theta}$$

$$\frac{1}{e+d} = \frac{\sin \theta}{n\lambda}$$

Given $\theta = 30°$, $\sin 30° = \frac{1}{2}, \lambda = 6 \times 10^{-7}$ m; $n = 1$

Therefore

$$\frac{1}{e+d} = \frac{1}{2} \times \frac{1}{1 \times 6 \times 10^{-7}} = \frac{10^7}{12} \text{ per meter}$$

$$\frac{1}{e+d} = \frac{10^5}{12} = 8333 \text{ liner per cm}$$

16. A diffraction grating used at normal incidence gives green line $\lambda_1 = 5400\ \mathring{A}$ in a certain order coincided on a violet line $\lambda_2 = 4050\ \mathring{A}$ of the next higher order. If angle of diffraction is 30°, how many lines are there in one cm of grating.

Sol. From the grating equation

$$(e + d) \sin \theta = n_1 \lambda_1 = n_2 \lambda_2$$

Given $n_1 = n, \ \lambda_1 = 5400\ \mathring{A}$

$n_1 = (n + 1)$

$\lambda_2 = 4050\ \mathring{A}$

$\theta = 30°$

$\sin 30° = \frac{1}{2}$

Hence, $n.\ 5400 = (n + 1)\ 4050$

$n = 3$

Therefore

$$(e + d) = \frac{n_1 \lambda_1}{\sin \theta}$$

$$= \frac{3 \times 5400 \times 10^{-10}}{0.5}$$

$$= 3.24 \times 10^{-6} \text{ m}$$

$$= 3.24 \times 10^{-4} \text{ cm}$$

Thus, the no. of lines per cm is

$$\frac{1}{3.24 \times 10^{-4}} = 3086$$

17. Two spectral lines have wavelength λ and λ + dλ respectively. Show that if dλ << λ, their angular separation dθ in a grating spectrometer is given by

$$d\theta = \frac{d\lambda}{\left[\{(e + d/n)\}^2 - \lambda^2\right]^{1/2}}$$

where (e +d) is the grating element and n is the order of the spectrum

Sol. The plane grating equations is

$$(e + d) \sin \theta = n \lambda \qquad .....(1)$$

Differentiating eq. (1) w.r.t to λ we get

$$(e + d) \cos \theta \frac{d\theta}{d\lambda} = n$$

or

$$d\theta = \frac{n\, d\lambda}{(e + d) \cos \theta}$$

$$d\theta = \frac{n\, d\lambda}{(e + d)\sqrt{1 - \sin^2 \theta}}$$

substituting the value of sin θ from (1) we get

$$d\theta = \frac{n\, d\lambda}{(e + d)\sqrt{1 - n^2\lambda^2/(e + d)^2}}$$

$$d\theta = \frac{n\,d\lambda}{\sqrt{(e+d)^2 - n^2\lambda^2}}$$

$$d\theta = \frac{n\,d\lambda}{\left[\left\{\frac{e+d}{n}\right\}^2 - \lambda^2\right]^{1/2}}$$

18. Monochromatic light from a Helium-neon laser (λ = 6328 A°) is incident normally on a diffraction grating 6000 lines/cm. Find the angles at which one would observe the first order maximum, the second order maximum etc. [UPTU]

Sol. In case of diffraction grating, the angles at which the interference maxima occur are given by

$$(e + d) \sin\theta = n\,\lambda$$

or $$q = \sin^{-1}\left(\frac{n\ \lambda}{e+d}\right)$$

Hence $$\frac{1}{e+d} = 6000$$

and $$\lambda = 6328 \text{ A}^{\circ}$$

If θ_1 and θ_2 are the angles at which the first and second maxima are observed, then

$$\theta_1 = \sin^{-1}(1 \times 6328 \times 10^{-8} \times 6000)$$
$$= \sin^{-1}(0.38) = 22^{\circ}\,20'$$
$$\theta_2 = \sin^{-1}(2 \times 6328 \times 10^{-8} \times 6000)$$
$$= \sin^{-1}(0.76) = 49^{\circ}\,28'$$

19. A diffraction grating at normal incidence gives a yellow line (λ = 6000 A°) in a certain spectral order super imposed on a blue line (λ = 4800 A°) of next higher order. If the angle of diffraction is $\sin^{-1}(3/4)$, calculate the grating element. [UPTU]

Sol. The directions of principal maxima for normal incidence are expressed as

$$(e + d) \sin\theta = n\,\lambda \qquad .....(1)$$

Let n^{th} order maximum of n_1 coincide with $(n + 1)^{th}$ order maximum of λ_2, then we have

$$(e + d) \sin\theta = n\,\lambda_1 = (n + 1)\,\lambda_2$$

or $$n = \frac{\lambda_2}{\lambda_1 - \lambda_2}$$

Thus equation (1) becomes

$$(e + d) \sin \theta = \frac{\lambda_1 \lambda_2}{\lambda_1 - \lambda_2}$$

or $$(e + d) = \frac{\lambda_1 \lambda_2}{(\lambda_1 - \lambda_2) \sin \theta}$$

given $$\lambda_1 = 6000 \text{ A}^\circ = 6000 \times 10^{-8} \text{ cm}$$

$$\lambda_1 = 4800 \text{ A}^\circ = 4800 \times 10^{-8} \text{ cm}$$

and $$\theta = \sin^{-1}(3/4) \text{ or } \sin \theta = 3/4$$

$$\therefore \text{ grating element } (e + d) = \frac{6000 \times 10^{-8} \times 4800 \times 10^{-8}}{(6000 - 4800) \times 10^{-8} \times 3/4}$$

$$\therefore \quad (e + d) = \frac{6000 \times 10^{-8} \times 4800}{1200 \times (3/4)} = 3.2 \times 10^{-4} \text{ cm}$$

20. How many orders will be visible if the wavelength of incident radiation is 5000 $\overset{o}{A}$ and the number of lines on the grating is 2620 to an inch. [UPTU]

Sol. We know that (e + d) sin θ = nλ, the maximum value of θ is 90°, that is, sin θ = 1. Therefore the number of orders visible with the grating is given as

$$(e + d) \sin 90^\circ = n \lambda$$

$$n = (e + d)/\lambda$$

here $$(e + d) = \frac{2.54}{2620} \text{ cm}$$

and $$\lambda = 5000 \text{ A}^\circ = 5000 \times 10^{-8} \text{ cm}$$

$$n = \frac{2.54}{2620 \times 5000 \times 10^{-8}}$$

$$n = 19.4$$

or $$n > 19$$

hence, the maximum number of orders visible in the spectrum are 19

21. How many orders will be observed by a grating having 4000 lines per cm, if it is illuminated by light of wavelength in the range 5000 A° to 7500 A°

Sol. Number of orders visible with grating is

$$n = \frac{(e+d)\sin\theta}{\lambda}$$

For maximum order visible with grating, $\sin\theta = 1$ or $\theta = 90°$, therefore

$$n_{max} = \frac{(e+d)}{\lambda}$$

Here, $$e + d = \frac{1}{4000}$$

Therefore $$n_{max} = \frac{1}{4000\,\lambda}$$

For $\lambda = 5000$ A° $= 5000 \times 10^{-8}$ cm

$$n_{max} = \frac{1}{4000 \times 5000 \times 10^{-8}} = \frac{10}{2} = 5$$

For $\lambda = 7500$ A° $= 7500 \times 10^{-8}$ cm

$$n_{max} = \frac{1}{4000 \times 5000 \times 10^{-8}} = \frac{10}{3} = 3.3$$

therefore, in the wavelength range 5000 Å to 7500 Å, the observed number of orders range between three to five.

22. Show that only first order spectra is possible if the width of grating element is less than twice the wavelength of light. [UPTU]

Sol. The grating equation is,

$$(e + d)\sin\theta = n\lambda$$

The maximum possible order will be available with the grating if $\sin\theta$ is maximum or $\theta = 90°$

$$(e + d) = n_{max}\lambda$$

or $$n_{max} = \frac{e+d}{\lambda}$$

here, grating element $(e + d) < 2\lambda$

$$n_{max} < \frac{2\lambda}{\lambda}$$

or

$$n_{max} < 2$$

Hence, only first order spectra is possible if the width of the grating element is less than twice the wavelength of light.

23. What other spectral lines in the range 4000 $\overset{o}{A}$ to 7000 $\overset{o}{A}$ will coincide with the fifth order line of 6000 $\overset{o}{A}$ in a grating spectrum? [UPTU]

Sol. A line of wavelength λ_1 (unknown) in the order n_1 coincides in position with a line of known wavelength λ in the nth order provided.

$$(e + d) \sin \theta = n\lambda = n_1 \lambda_1$$

or $$\lambda_1 = \frac{n\lambda}{n_1}$$

Here $n = 5$, $\lambda = 6000\ \overset{o}{A}$ and λ_1 lie between 4000 $\overset{o}{A}$ and 7000 $\overset{o}{A}$, that is 4000 $\overset{o}{A}$ $< \lambda_1 < 7000\ \overset{o}{A}$

$\therefore$ $$\lambda_1 = \frac{6000 \times 5}{n_1} = \frac{30000}{n_1}$$

when $n_1 = 4$, $$\lambda_1 = \frac{30000}{4} = 7500\ \overset{o}{A}$$

when $n_1 = 5$, $$\lambda_1 = \frac{30000}{5} = 6000\ \overset{o}{A}$$

when $n_1 = 6$, $$\lambda_1 = \frac{30000}{6} = 5000\ \overset{o}{A}$$

when $n_1 = 7$, $$\lambda_1 = \frac{30000}{7} = 4285.7\ \overset{o}{A}$$

when $n_1 = 8$, $$\lambda_1 = \frac{30000}{8} = 3750\ \overset{o}{A}$$

Hence, in the grating spectrum the spectral lines with wavelength 5000 A° and 4285.7 A°, in the range 4000 A° to 7000 A° will coincide with fifth order line of 6000 A°.

On Resolving Power

24. Find the minimum number of lines in a plane diffraction grating required to just resolve the sodium doublet (5890 Å and 5896 Å) in the [UPTU]

(i) First order

(ii) Second order

Sol. The resolving power of a grating is given by

$$\frac{\lambda}{d\lambda} = nN$$

Where n is the order and N is the total number of lines on the grating

$$\therefore \qquad N = \frac{1}{n}\left(\frac{\lambda}{d\lambda}\right)$$

(i) For the first order, n = 1

$$\text{Mean wavelength } (\lambda) = \frac{\lambda_1 + \lambda_2}{2} = \frac{5890 + 5896}{2}$$

$$= 5893 \text{ Å}$$

and $$d\lambda = 5896 - 5890 = 6 \text{ Å}$$

$$\therefore \qquad N = \frac{5893}{1 \times 6} = 982$$

(ii) For the second order, n = 2, the minimum number of lines

$$N = \frac{5893}{2 \times 6} = 491$$

25. Calculate the least width that a grating must have to resolve the components of sodium D-lines in the second order, the grating having 1000 lines per cm. The wavelength of two components of sodium lines are 5890 Å and 5896 Å respectively [UPTU

Sol. The resolving power of a grating is given by

$$\frac{\lambda}{d\lambda} = nN$$

For second order, n = 2,

$$\lambda = \frac{\lambda_1 + \lambda_2}{2} = \frac{5890 + 5896}{2} = 5893 \text{ Å}$$

and $d\lambda = 5896\ A^o - 5890\ A^o = 6\ A^o$

∴ the total number of lines on the grating

$$N = \frac{1}{n}\frac{\lambda}{d\lambda} = \frac{1}{2}\left(\frac{5893}{6}\right) = 491$$

∴ width of grating,

$$N(e+d) = \frac{\text{Total lines on grating}}{\text{lines per cm on grating}}$$

$$N(e+d) = \frac{491}{1000} = 0.491 \text{ cm.}$$

26. Can D_1 and D_2 lines of Na (Sodium) light be resolved for ($\lambda_1 = 5890\ A^o$, $\lambda_2 = 5896\ A^o$) in second order. Number of lines in grating of 2.0 cm wide = 4500. [UPTU]

Sol. The resolving power of the grating required to resolve the lines in the second order in given by

$$\frac{\lambda}{d\lambda} = N.n$$

Here $$\lambda = \frac{5890+5896}{2} = 5893\ A^o$$

$$d\lambda = 5896\ A^o - 5890\ A^o = 6\ \overset{o}{A} \text{ and } n = 2$$

∴ $$\frac{\lambda}{d\lambda} = \frac{5893}{6} = 2N$$

or $$N = \frac{5893}{12} = 491$$

Given N= 4500 which is very very greater than the required value, hence lines can be resolved in second order.

27. Find the minimum number of lines that a diffraction grating would need to have in order to resolve in first order the red doublet given by a mixture of hydrogen and deuterium. The wave length difference is 1.8 A^o at $\lambda = 6553\ A^o$. [UPTU]

Sol. The resolving power of a grating in given by

$$\frac{\lambda}{d\lambda} = n\,N$$

where n is the order and N the total number of lines on the grating.

Here $\lambda = 6553\ A^o; \quad d\lambda = 1.8\ A^o$ and $n = 1$

$$N = \frac{1}{n}\left(\frac{\lambda}{d\lambda}\right) = \frac{6553}{1\times1.8} = 3641$$

28. A diffraction grating is just able to resolve two lines of wave length 5140.34 A° and 5140.85 A° in the first order. Will it resolve the lines 8037.20 A° and 8037.50 A° in the second order. [UPTU

Sol. the resolving power of a grating is given by

$$\frac{\lambda}{d\lambda} = n\,N$$

Therefore

$$N = \frac{1}{n}\left(\frac{\lambda}{d\lambda}\right)$$

where $d\lambda$ is the smallest wavelength difference at the mean wave length λ, N the number of the lines on grating and n is the order of the speculum.

In the given Question,

$$\lambda = \frac{5140.34 + 5140.85}{2}\ 5140.595\ \overset{o}{A}$$

$$d\lambda = 5140.85 - 5140.34 = 0.51\ \overset{o}{A} \text{ and } n = 1$$

$$\therefore \qquad N = \frac{1}{1}\left(\frac{5140.595}{0.51}\right) = 10080$$

Hence the resolving power $\left(\frac{\lambda}{d\lambda}\right)$ of a gating in second order should be

$$n\,N = 2 \times 10080 = 20160.$$

The resolving power require to resolve the lines 8037.20 A° and 8037.50 A° in the second order is $\left(\frac{\lambda}{d\lambda}\right)$

In this case

$$\lambda = \frac{8037.20 + 8037.50}{2}$$

$$\lambda = 8037.50 \text{ A}^{o}$$

And $$d\lambda = 037.50 - 8037.50 = 0.30$$

$\therefore$ $$\text{Resolving power} = \frac{8037.35}{0.30} = 26791.17$$

Thus, the grating will not be able to resolve the lines 8037.20 $\overset{o}{\text{A}}$ in the second order because the required resolving power (26791.17) is greater than the actual resolving power 20160.

29. A plane transmission grating has 15000 lines per inch. Find the resolving power of the grating and the smallest wavelength difference that can be resolved with a light of wavelength 6000 A^{o} in the second order. [UPTU]

Sol. the resolving power of grating is

$$\frac{\lambda}{d\lambda} = nN$$

Here $$N = 15000 \text{ and } n = 2$$

$\therefore$ $$\frac{\lambda}{d\lambda} = 2 \times 15000 = 30000$$

Here $$\lambda = 6000 \text{ A}^{o}$$

$\therefore$ the smallest wave length difference

$$d\lambda = \frac{\lambda}{d\lambda} = \frac{6000 \overset{o}{\text{A}}}{30000} = 0.2 \text{ A}^{o}$$

30. A parallel beam of light of wavelength 5460 A^{o} is incident at an angle of 30° on a plane transmission grating which has 6000 lines/cm. Find the highest order spectrum that can be observed.

Sol. From the grating under oblique incidence

$\therefore$ $$(e + d)\,[\sin\theta + \sin i] = n\,\lambda$$

Here $$\theta = i$$

$\therefore$ $$(e + d)\,[2 \sin i] = n\lambda$$

$\therefore$ $$(e + d) = \left[\frac{1}{6000}\right] \text{cm}$$

$$\sin 30^{o} = 0.5$$

$$\lambda = 5460 \text{ A}^{o} = 5460 \times 10^{-8} \text{ cm}$$

$$n = \frac{(e+d)\ (2\sin 30^\circ)}{\lambda}$$

$$n = \frac{2 \times 0.5}{6000 \times 5460 \times 10^{-8}}$$

$$n = 3.05$$

$$n \approx 3.0$$

31. A plane transmission grating having the grating element 18000 $\mathring{A}$, obtain the dispersions in the first order spectra around $\lambda = 5000\,\mathring{A}$, assuming normal incidence.

Sol. the angular separation is given by

$$\frac{d\theta}{d\lambda} = \frac{1}{\sqrt{\left[\left(\frac{e+d}{n}\right)^2 - \lambda^2\right]}}$$

Given $(e + d) = 18000\ \mathring{A}$

$$n = 1$$

$$\lambda = 5000\ \mathring{A}$$

$$\therefore \quad \frac{d\theta}{d\lambda} = \frac{1}{\sqrt{\left[(18000)^2 - (5000)^2\right]}}$$

$$\therefore \quad \frac{d\theta}{d\lambda} = \frac{1}{10^3\sqrt{(18)^2 - (5)^2}}$$

$$\frac{d\theta}{d\lambda} = \frac{1}{10^3\sqrt{324 - 25}}$$

$$\frac{d\theta}{d\lambda} = \frac{1}{10^3 \times 17.3}$$

$$\therefore \quad \frac{d\theta}{d\lambda} = 5.78 \times 10^{-5}\ \text{radian}\ /\ \mathring{A}$$

32. What will be the diameter of a telescope objective required to resolve two stars separated by an angle of 10^{-3} degree? Assume $\lambda = 5000\ \mathring{A}$.

Sol. As $$\theta = \frac{1.22\,\lambda}{d} \text{ radian}$$

$$d = \frac{1.22\,\lambda}{\theta}$$

Given $$\lambda = 5000\ \overset{o}{A} = 5 \times 10^{-5} \text{ cm}$$

and $$\theta = 10^{-3} \text{ degree} = \frac{\pi}{180^o} \times 10^{-3} \text{ radian}$$

$$d = \frac{1.22 \times 5 \times 10^{-5} \text{ cm}}{\left(\frac{3.14}{180} \times 10^{-3}\right) \text{radian}} = 3.5 \text{ cm}$$

33. A source containing a mixture of hydrogen and deuterium atoms emits a red doublet at λ = 656 A° where separation in 1.8 A°. Find the minimum number of lines required in a plane transmission grating which can resolve the doublet in the first order.

Sol. As the resolving power is given by

$$\frac{\lambda}{d\lambda} = n\,N$$

$\therefore$ $$N = \frac{1}{n}\frac{\lambda}{d\lambda}$$

Here $$n = 1,\ \lambda = 6563\ \overset{o}{A}, \quad \text{and } d\lambda = 1.8\ \overset{o}{A}$$

$\therefore$ $$N = \frac{1}{1} \times \frac{6563 \text{ A}^o}{1.8 \text{ A}^o} = 3646$$

34. Two gratings A and B have the same width of ruled surface but A has greater number of lines. Giving reasons compare in two cases (i) intensity (ii) width of principal maxima (iii) dispersive power (iv) resolving power

Sol. (i) Intensity $\propto N_2$ and so $I_A > I_B$

(ii) The width of principal maxima $\approx \frac{1}{N}$

Therefore $W_B > W_A$

(iii) The dispersive power

$$\frac{d\theta}{d\lambda} = \frac{n}{(e+d)\cos\theta}$$

$$\frac{d\theta}{d\lambda} \propto \frac{1}{(e+d)}$$

since $(e + d)_B > (e + d)_A$

$$\Rightarrow \quad \left(\frac{d\theta}{d\lambda}\right)_A > \left(\frac{d\theta}{d\lambda}\right)_B$$

(iv) Resolving power = Nn

$$N_A > N_B$$

Therefore $(RP)_A > (RP)_B$

35. The wavelength of sodium D lines are 589.593 mμ and 588.996 mμ. What is the minimum number of lines a grating must have in order to resolve these lines in the first order.

Sol. Given
$$\lambda = \frac{588.996 + 589.593}{2} \times 10^{-9}$$

$$d\lambda = (589.593 - 588.996) \times 10^{-9} \text{ and } n = 1$$

Therefore
$$N = \frac{\lambda}{nd\,\lambda} = \frac{589.295}{0.597} = 987.09$$

The desired number of lines is 988.

Conceptual Questions and Answers

1. **What is diffraction of light?**

Ans. The phenomenon of bending of light round the sharp corners and its spreading into the regions of the geometrical shadow is called diffraction.

2. **State the condition for diffraction of light to occur?**

Ans. The size of the obstacle should be of the order of the wavelength of light used.

3. **Yellow light is used in a single slit diffraction experiment with slit width of 0.6 mm. If yellow is replaced by x-rays, how will the diffraction pattern be affected?**

Ans. Since wavelength of x-rays is very small as compared to the size of the slit, no diffraction pattern will be observed.

4. **What is the basic difference between interference and diffraction of light?**

Ans. Interference is the result of superposition of light waves from two sources, whereas diffraction takes place due to interference of light waves from the different parts of the same source of light.

5. **Diffraction is common in sound best not common in light waves, why?**

Ans. For diffraction to take place, the size of the obstacle should be of the order of the wavelength of the waves. Since the wavelength of light $\simeq 10^{-6}$ m is very small as compared to the size of the obstacle around us, light cannot be diffracted readily. On the other hand, as wavelength of sound is of the order of such objects, it gets diffracted easily.

6. **Radio waves diffract pronouncedly around buildings while light waves, which are electro magnetic waves do not why?**

Ans. For a wave to undergo diffraction, its wavelength should be of the size of the obstacle. The radio waves particularly short radio waves have wavelengths of the order of the size of the building and other obstacles coming in their way and hence they easily diffracted.

Since wavelength of the light waves is very small, they are not diffracted by buildings.

7. **Coloured spectrum is observed, when we see through muslin cloth. But no such spectrum is observed when we see through coarse cloth. Why?**

Ans. Muslin cloth is made of very fine threads such and as such fine slits are formed. White light passing through these slits gets diffracted giving rise to coloured spectrum. The central maximum will be white, while the secondary maxima will be coloured. It is because, position of all the secondary maxima (except central) depends upon wavelength of the light.

In a coarse cloth, the slits formed due to threads are very wide and hence no such spectrum is seen.

8. **Two students are separated by 6 m partition wall in a room 10 m high. If both light and sound waves can bend around obstacles, how is it that the students are unable to see each other, even through they can converse easily.**

Ans. For pronounced diffraction effect, the size of the obstacle should be comparable to wavelength of the waves. The size of partition is very large as compared to the wavelength of light and hence it is not diffracted and the two students cannot see each other.

The wavelength of sound waves is of the order of the height of partition. It causes sound to diffract and hence they can converse with each other.

9. Stars are often photographed through a blue filter. What is the advantage of this?

Ans. The resolving power of a telescope

$$= \frac{1.22\,D}{\lambda}$$

where D is diameter of the telescope's objective of the light. If shorter wavelength is used, the resolving power of the telescope will be more and like wise the starts which subtend a smaller angle at the earth, will also be resolved and seen distinctly as separate ones. The use of blue filter allows only blue light (shorter wavelength) from the star to enter the relescope.

10. A slit is illuminated with monochromatic light. The light emerging from it is collimated by a lens L_1 and focussed on a screen by a lens L_2. A large diffraction grating is interposed midway between the lenses with grating lines parallel to the slit. What will be the effect on diffraction pattern, when

1. The grating is moved forward or backward between the lenses

2. The grating is shifted in its own plane and perpendicular to the slit

3. The grating plane is rotated round the axis parallel to the slit

Ans.

1. When the grating is moved forward or backward between the lenses, there will be no change on the diffraction pattern because light is still falling normally on the grating and thus the direction of maxima will remain unchanged.

2. When the grating is shifted in its own plane and perpendicular to the slit, again there will be no change on the diffraction pattern till a distance equal to the diameter of lens L_1 of the ruled surface of the grating. If the parallel beam emerging from L_1 and falling on the grating covers a small ruled surface of the grating, then the total number of lines used will decrease and consequently the spectral lines will become wider and fainter (intensity $\propto N^2$)

3. When the grating plane is rotated round the axis parallel to the slit, the incidence on the grating remains no more normal. In such a case position of maxima will change.

Objective Questions

1. In Fraunhofer diffraction pattern due to single slit, the width of the central maximum is

(a) $2x = \frac{2 f \lambda}{a}$ (b) $2x = \frac{2 f}{a}$

(c) $x = \frac{2 f \lambda}{a}$ (d) None of these

2. With white light the central maximum obtained in Fraunhofer diffraction pattern of a single slit is

(a) dark (b) bright (c) coloured (d) white

3. The intensity distribution in the Fraunhofer diffraction pattern due to a single slit is given as

(a) $I = A^2$ (b) $I = \frac{\sin^2 \alpha}{\alpha}$

(c) $I = I_o \sin^2 \alpha$ (d) $I = \frac{I_o \sin^2 \alpha}{\alpha^2}$

4. The Fraunhofer diffraction pattern due to a double slit is

(a) Interference pattern due to the secondary waves emanating from the corresponding points of the two slits

(b) diffraction pattern due to the secondary waves from the two slits individually

(c) only (b)

(d) both (a) and (b)

5. The angular separation between any two consecutive minima (or maxima) of interference in Fraunhofer diffraction due to double slit is

(a) $\frac{\lambda}{ab}$ (b) $\frac{\lambda}{a-b}$

(c) $\frac{\lambda}{a+b}$ (d) None of these

where a is the width of each slit and b is width of opaque portion.

6. The direction of diffraction minima in Fraunhofer diffraction due to double slit is given as

(a) $\sin \phi_n = n\lambda$ (b) $\phi_n = n\lambda$

(c) $a \sin \phi_n = n\lambda$ (d) $\sin \phi_n = na\lambda$

7. The ratio of intensities of the central maximum due to double slit and single slit is

(a) 1:2 (b) 2:1

(c) 4:1 (d) 3:1

8. Condition for the orders 2, 4, 6 etc., of the interference maxima to be missing in the diffraction pattern of double slits is

(a) $a = b$ (b) $a = \frac{b}{2}$

(c) $a = 2b$ (d) $a = 4b$

where a is the width of each slit and b is the separation between the slits.

9. Condition for all the orders of the interference maxima to be missing in the diffraction pattern of double slit in

(a) $a + b = 0$ (b) $a + b = b$

(c) $a + b = 1$ (d) $a + b = a$

10. The missing orders of the interference maxima for a double slit Fraunhofer diffraction if the slit width are 0.16 mm and they are 0.8 mm apart is

(a) 1, 2, 3 etc., (b) 3, 6, 9 etc.,

(c) 5, 10, 15 etc., (d) 6, 12, 18 etc.,

11. The maximum intensely by Fraunhofer diffraction and to N slits is given as

(a) $I = I_o \frac{\sin^2\alpha}{\alpha^2}$ (b) $I = I_o \frac{\sin^2 N\beta}{\sin^2\beta}$

(c) $I = I_o \sin^2\alpha \sin^2 N\beta$ (d) $I = I_o \left(\frac{\sin^2\alpha}{\alpha^2}\right)\left(\frac{\sin^2 N\beta}{\sin^2\beta}\right)$

Whore α, β are related to diffraction and interference respectively

12. How many narrow slits are there is a diffraction grating?

(a) about 5000 in one cm (b) about 6000 in one cm

(c) about 7000 in one cm (d) about 10,000 in one cm

13. The intensity for principal maxima in diffraction grating is given by

(a) $I = I_o^2 \left(\frac{\sin^2\alpha}{\alpha^2} \right)$ (b) $I = NI_o^2 \left(\frac{\sin^2\alpha}{\alpha^2} \right)$

(c) $I = N^2 I_o \left(\frac{\sin^2\alpha}{\alpha^2} \right)$ (d) none of these

14. The equation which tells that which order of the spectrum is absent in diffraction grating is

(a) $\frac{a+b}{b} = n$ (b) $\frac{a+b}{a} = n$

(c) $a + b = n$ (d) $a + b = \frac{n}{2}$

15. The higher order spectrum which may be seen with monochromatic light of wavelength 6000 A° by means of diffraction grating with 5000 lines/cm is

(a) $n = 1$ (b) $n = 2$

(c) $n = 3$ (d) $n = 4$

16. Two nearly images are said to be resolved if the position of the central maximum of one coincides with the first secondary maximum of the other and vice versa. The criterion was proposed by

(a) Fraunhofer (b) Einstein

(c) Newton (d) Rayleigh

17. The Linear distance between two just resolvable point objects is given by (for a convex lens)

(a) $h = \frac{0.71\,\lambda_o}{\mu \sin\theta}$ (b) $h = \frac{0.71\mu}{\lambda_o \sin\theta}$

(c) $h = \frac{0.61\lambda_o}{\mu \sin\theta}$ (d) $h = \frac{0.61\mu}{\lambda_o \sin\theta}$

18. The human eye in most sensitive to a wavelength

(a) 7500 A° (b) 6500 A°

(c) 5500 A° (d) 4500 A°

19. The resolving power of microscope is given as

(a) $\frac{\lambda_o}{2 \sin \alpha}$ (b) $\frac{\lambda_o}{2 \mu \sin \alpha}$

(c) $\frac{\lambda_o}{2 \mu \cos \alpha}$ (d) $\frac{\lambda_o}{\mu \cos \alpha}$

20. The resolving power of a grating is
   (a) directly proportional to the order of spectrum
   (b) directly proportional to the total number of lines on the grating surface
   (c) both (a) and (b)
   (d) None of these.

21. Ray optics is valid, when characterise dimensions are
   (a) much smaller than the wavelength of light
   (b) much larger than the wavelength of light
   (c) of the same order as the wavelength of light
   (d) of the order of one millimetre

22. The frequency of e.m wave which in best suited to observe a particle of radius 3×10^{-4} cm, in the order of

(a) 10^{15} (b) 10^{13} (c) 10^{14} (d) 10^{12}

23. The interference phenomenon differs from diffraction that unlike diffraction
   (a) interference fringes are of varying intensity
   (b) interference cannot be observed with white light
   (c) interference minima may be perfectly dark
   (d) interference fringes are of unequal width

24. If I_o is the intensity of the principal maximum in the single slit diffraction pattern, then what will be its intensity when the slit width is doubled.

(a) $2 I_o$ (b) $4 I_o$ (c) I_o (d) $I_o/2$

25. The limit of resolution of one meter radio telescope for radio waves of wavelength 10 cm is about

(a) 7° (b) few seconds of arc

(c) 18° (d) one minute of arc

26. The area of each half period zone is

(a) $b\lambda$ (b) b^2 (c) λ^2 (d) $\pi b\lambda$

27. Each half period differs from its neighbour by a phase difference of

    (a) $\pi/2$ (b) π (c) 2π (d) 3π

28. Zone phase is similar to

    (a) concave lens (b) convex lens

    (c) glass (d) none of these

Theoretical Questions

1. Explain the difference between interference and diffraction?
2. What is meant by diffraction of light? Distinguish between Fresnel and Fraunhofer classes of diffraction?
3. Discuss Fraunhofer diffraction pattern due to a single slit. Find the expression for the width of the central maximum?
4. Derive expression for the intensity distribution due to Fraunhofer diffraction at a single slit and show that the intensity of the first subsidiary maximum is about 4.5% of that of the principal maximum?
5. Discuss the phenomenon of Fraunhofer diffraction at a single slit and show that the relative intensities of the successive maximum are nearly

$$1 : \frac{4}{9\pi^2} : \frac{4}{25\pi^2} : \frac{4}{49\pi^2} : - - -$$

6. Describe Fraunhofer diffraction due to a single slit and derive the positions of the maxima and minima. Show that the relative intensities of successive maxima are nearly

$$1 : \frac{1}{22} : \frac{1}{61} : - - -$$

7. Describe the Fraunhofer diffraction pattern due to a double slit. What are the missing orders?
8. Describe the feature of a double slit Fraunhofer's diffraction pattern. What is the effect of increasing the

   (a) slit width

   (b) slit separation and

   (c) wavelength

9. What do you understand by missing order spectrum? What particular spectra would be absent if the width of transparencies and opacities of grating are equal?

10. Show that only first order of spectra is possible if the width of a grating element is less than twice the wavelength of light.

11. Explain the formation of spectra by a plane transmission grating. What are its chief characteristics?

12. Show that the rays forming spectrum in a grating suffer minimum deviation when angle of incidence equals the angle of diffraction.

13. Define dispersive power of a grating and obtain an expression for it.

14. Explain Rayleigh criterion of resolution. Define limit of resolution and resolving power.

15. What is meant by the resolving power of an optical instrument

16. What is meant by resolving power of a plane transmission grating? Derive an expression for it. Show that the maximum resolving power is proportional to the width of the grating.

Numerical Problems

1. In Fraunhofer diffraction due to narrow slit, a screen is placed 2 m away from the lens to obtain the pattern. If the slit width is 0.2 mm and first minima lie 5 mm on either side of the central maximum, find the wavelength of light

[5000 Å]

2. Light of wavelength 5000 Å is incident normally on a single slit. The central maximum falls out at 30° on both sides of the direction of the incident light. Calculate the slit width. For what width of the slit the central maximum would spread out to 90° from the direction of the incident light.

[5000 Å]

3. Plane wave of $\lambda = 6.0 \times 10^{-5}$ cm fall normally on a slit of width 0.20 mm. Calculate (i) the total angular width of the central maximum (ii) the linear width of the central maximum on a screen placed 2 m away.

[6×10^{-3} rad, 1.2 cm]

4. In a single slit diffraction pattern the distance between the first minimum as the right and the first minimum on the left is 5.2 mm. The screen on which the pattern is displaced is 80 cm from the slit and the wavelength is 5460 Å. Calculate the slit width.

   [168×10^{-4} cm]

5. Two parallel slits having widths 0.15 mm each and separation between them is 0.30 mm. They are illuminated normally by light $\lambda = 6000$ Å and the emergent light is focussed by a convergent lens of 100 cm focal length. Deduce the positions of the first four interference maxima an one side in the focal plane of the lens.

   [x = 1.33 mm, 2.66 mm, 5.32 mm, 6.65 mm]

6. A parallel beam of monochromatic light is allowed to be incident normally on a plane grating having 1250 lines per cm and a second order spectral lines is observed to be deviated through 30°. Calculate the wavelength of the spectral line.

   [20000 Å]

7. White light (400 nm < λ < 700 nm) is incident on a grating. Show that no matter what is the value of grating spacing d, the second and third order spectra overlap.

8. Light of wavelength 5×10^{-5} cm is incident normally on a plane transmission grating of width 3 cm and 15000 lines. Find the angle of diffraction in first order

   [14°29']

9. The limits of visible spectrum are approximately 400 mμ to 700 mμ. Find the angular width of the first order visible spectrum produced by a plane grating having 15000 lines per inch when light is incident normally as the grating (Given sin 13°14' = 0.237 and sin 24°30' = 0.415)

   [10°50']

10. A plane transmission grating produces an angular separation of 0.01 radian between two wavelengths observed at an angle of 30°. If the mean value of wavelength is 5000 Å and the spectrum is observed in the second order (86.6 Å), calculate the difference in two wavelengths.

11. A set of 10 parallel equidistant slits of width 0.50 cm and opaque space 1.4 cm are used to study Fraunhofer diffraction of $\lambda = 0.6$ cm, falling normally on the planes of the slits. Calculate

    (i) angular position and half width of first maximum.

(ii) The effect of covering up alternative slits on angular position and width of first maximum

[18.41°; 3.66°]

12. Light which is a mixture of the two wavelengths 5000 Å and 5200 Å is incident normally on a plane transmission grating having 10,000 lines per cm. A lens of focal length 150 cm is used to observe the spectrum on a screen. Calculate the separation in cm of the two lines in the first order spectrum

[4.59 cm]

13. Show that for a transmission grating with one inch ruled space, the resolving power cannot exceed 5×10^4 at normal incidence for $\lambda = 5080$ Å

14. A plane transmission grating has 16,000 lines to an inch over a length of 5 inches. Find in the wavelength region of 6000 Å, in the second order.

(a) the resolving power of the grating and

(b) the smallest wavelength difference that can be resolved

[80,000; 16×10^4]

15. A monochromatic light from He-Ne laser $\lambda = 632.8$ nm is incident normally on a diffraction grating containing 6000 lines per cm. Find angles at which first and second order maxima are observed

[22.31°, 49.39°]

16. A diffraction grating has total ruled width 5 cm. For normal incidence it is found that a line of wavelength 6000 Å in a certain order super impose on another line of wavelength 4500 Å of the next order. If angle of diffraction is 30°, how many lines are there on grating

[13,889]

17. A microscope objective has focal length of 2.2 cm and diameter of its aperture is 10 mm. Estimate resolving limit assuming that object is placed close to it ($\lambda = 5500$ Å)

[1.7×10^{-6} cm]

18. A source emits 531.62 nm and 531.81 nm light,

    (a) What minimum number of lines is required for a grating that resolves the two wavelengths in the first order spectrum?

    (b) Determine the slit spacing for a grating 1.32 cm wide that has required minimum number of lines

[2799; 4.72 μm]

19. A diffraction grating of width 4.0 cm has 300 lines/cm rules over it,

    (a) What is the resolving power of this grating in first three order?

    (b) If two monochromatic waves incident on this grating have a mean wavelength of 400 nm, what is their wavelength separation if they are just resolved in third order?

[12000, 24000, 36000; 0.133 Å]

20. A light of wavelength 5000 Å falls on a transmission grating normally. Two adjacent principal maximum occur at $\sin \theta_1 = 0.2$ and $\sin \theta_2 = 0.3$. If grating is 2.5 cm wide, find its resolving power.

[10,000]

Answers to Objective Questions

1. (a)	2. (d)	3. (d)	4. (d)	5. (c)	6. (c)
7. (c)	8. (a)	9. (d)	10. (d)	11. (d)	12. (b)
13. (c)	14. (b)	15. (c)	16. (d)	17. (c)	18. (c)
19. (b)	20. (c)	21. (b)	22. (c)	23. (c)	24. (c)
25. (a)	26. (d)	27. (b)	28. (b)		

CHAPTER 3A

Polarisation

"As to the other emanation which should produce the irregular refraction, I wished to try what elliptical waves, or rather, spherical waves would do, and these I suppose spread differently both in the ethereal matter diffused throughout the crystal and in the particles of which it is composed......"

"Christian Huygens"

3A. Introduction

According to Maxwell, light is electromagnetic in nature, an electromagnetic wave consists of varying electric and magnetic fields, such that the two fields are mutually perpendicular to each other and to the direction of propagation of waves. The optical phenomenon i.e., the phenomenon concerning light are described by the electric field vector in electromagnetic wave. Therefore, the transverse nature of light may be attributed to the vibration of electric field vector in a direction perpendicular to the direction of propagation of light.

The wave in which the electric vector is confined to the x-y plane is said to be linearly polarised or also called plane polarised as shown in Fig. 3A.1(a). Generally in light obtained from a source such as the sun or an incandescent lamp, the orientation of electric vector is not confined to one plane only. However, when viewed from along the direction of propagation, the electric vector appears to be randomly oriented as shown in Fig. 3A.1(b), although confined to a plane normal to the direction of propagation. Such a light wave is transverse but unpolarised; that is there is no preferred plane of polarisation.

It may be noted that the phenomenon of interference and diffraction illustrate the wave nature of light. However the transverse nature of light waves is demonstrated only by the phenomenon of polarisation.

Fig. 3A.1 (a) A linearly polarised wave, viewed from along the direction of propagation. The wave is moving out of the plane of the page, only the direction of E vector is shown. (b) Unpolarised light

An unpolarised wave, which can be considered to be a random superposition of many polarised waves.

3A.1 Transverse Nature of Light

In general, the waves are of two types.

1. ***Longitudinal waves:*** The waves in which the particles, oscillates along the direction of propagation of the waves are called longitudinal waves.
2. ***Transverse waves:*** The waves in which the direction of oscillation of particles is perpendicular to the direction of propagation of the waves, are called transverse waves.

Both types of these waves exhibit the phenomenon of reflection, refraction, diffraction and interference but polarisation of the waves can only be exhibited by the transverse wave. This is the only phenomenon, where the two types of waves essentially differ from one another.

To understand the polarisation of the waves, let us first consider the following mechanical experiment.

Two vertical slits S_1 and S_2 are placed parallel to each other and a string is passed through them (Fig. 3A.2(a)). The end B of the string is fixed and the end A is given jerks in any direction perpendicular to its length. Then, transverse wave, where vibrations are parallel to the length of the slit, will only pass through the slit S_1. If the slit S_2 is parallel to S_1, then the amplitude of vibration will remain unaffected.

On the other hand, when the slit S_2 is at right angle to the slit S_1, no vibrations will pass through it i.e., amplitude of the vibration will become zero (Fig. 3A.2(b).

If instead of producing transverse vibration, if we produce longitudinal vibration in the string at the end A, the vibration will pass through the slit S_2 irrespective of the fact that the slit S_1 and S_2 are parallel to or at right angles to each other. From the above mechanical experiment it follows that if on rotating the slit S_2, the amplitude of vibration in the string decreases and becomes zero, when the two slits are at right angles, then the wave motion is transverse in nature, on the other hand if the amplitude of vibration remains unaffected, the wave motion is longitudinal one.

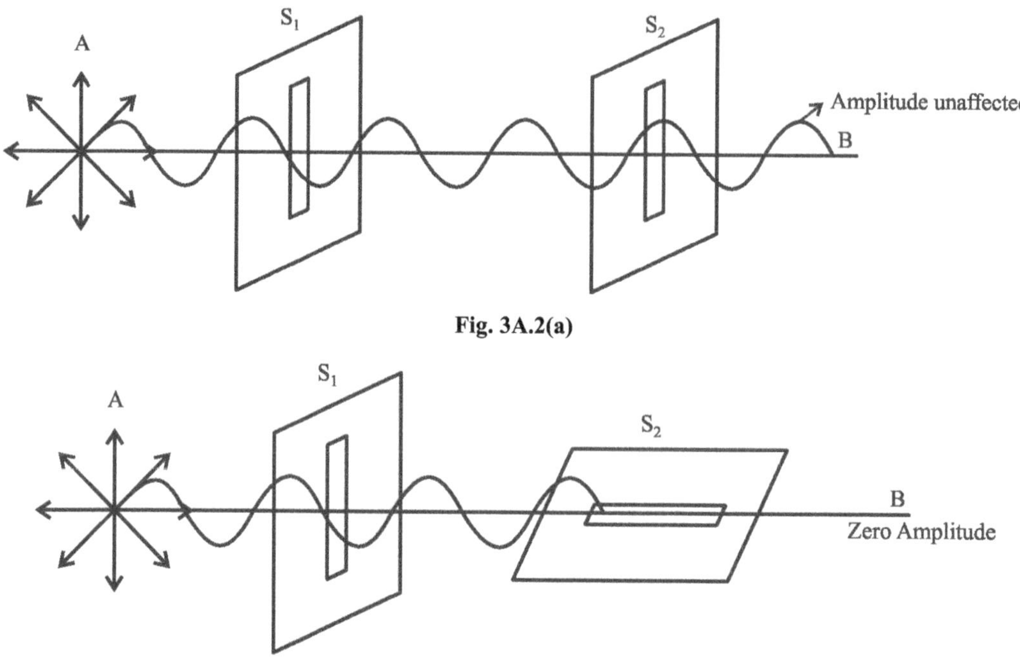

Fig. 3A.2(a)

Fig. 3A.2(b)

A Nicol prism or tourmaline crystal cut parallel to its crystallographic axis behaves in a similar manner for light waves as the slit does in case of the wave motion produced in the string. Suppose that the ordinary light in incident normally on a pair of such crystal C_1 and C_2. It is observed that the intensity of the emergent light is maximum, when the axis of the two crystals are parallel as shown in Fig. 3A.3(a) below, and minimum when the axis of the two crystals become at right angles to each as shown in Fig. 3A.3(b). In view of the mechanical experiment discussed earlier, it follows that light is transverse in nature.

3A.2 Polarisation of Light

The phenomenon due to which the vibration of light are restricted in a particular plane is called the polarisation of light.

When ordinary light i.e., unpolarised light passes through a tourmaline crystal, out of all the vibrations which are symmetrical about the direction of propagation, only those pass through it, which are parallel to its crystallographic axis AB, therefore on emerging through the crystal, the vibrations no longer remain symmetrical about the direction of propagation but are confined to a single plane as shown in Fig. 3A.4.

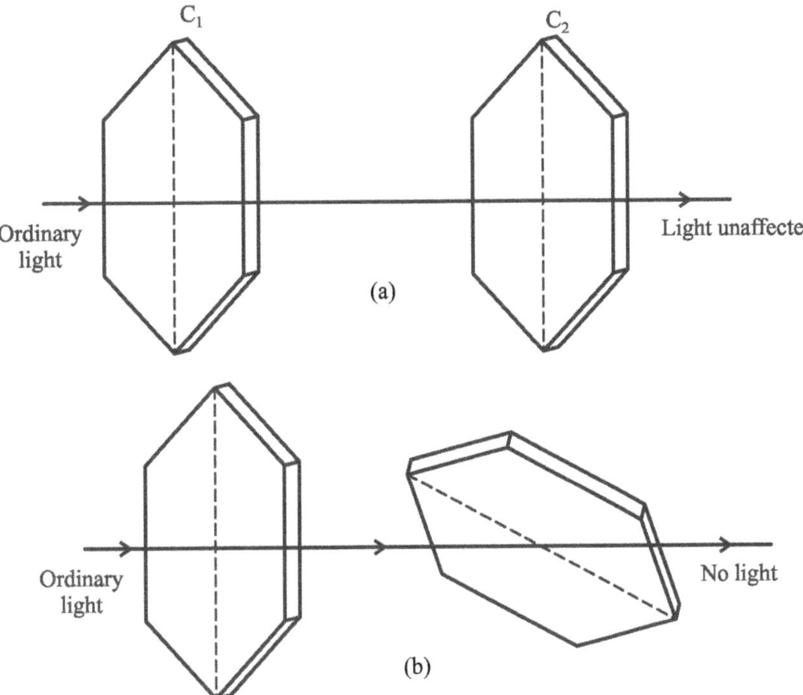

Fig. 3A.3

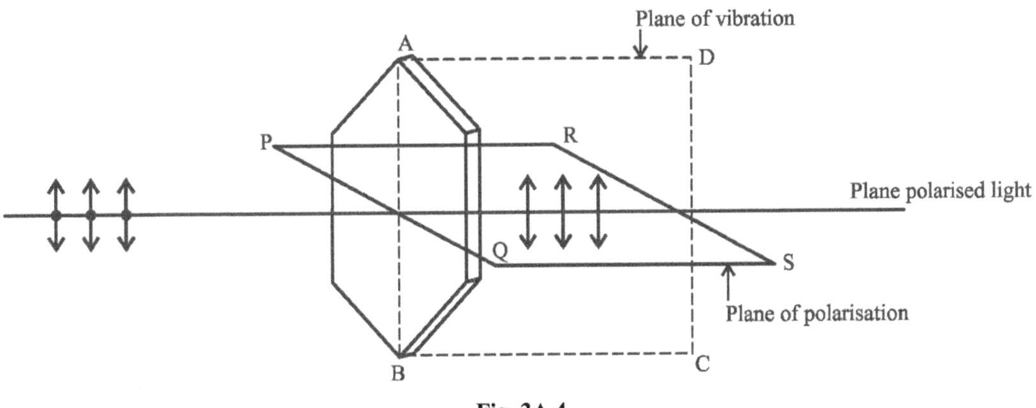

Fig. 3A.4

Plane of vibration: The plane ABCD which contains the vibration of plane polarised light is called plane of vibration.

Plane of polarisation: The plane PQRS, perpendicular to the plane of vibration is called the plane of polarisation.

3A.3 Representation of Polarized and Unpolarised Light

3A.3.1 Unpolarized Light

Light waves are transverse and so have vibrations at right angles to the direction of propagation. As seen, since light emitting atoms are oscillating independently, the waves emitted have their own planes of vibration. Hence the waves vibrate in all possible planes with equal probabilities. If we consider a beam propagating in a direction perpendicular to the plane of the paper it will have the vibrations along the plane of the paper in all possible directions with the same amplitude (Fig. 3A.5(a).

But for the purpose of easy representation it is considered to consist of vibrations in two mutually perpendicular direction xx' and yy' (since vibrations in all other directions can have their resolved components along these two directions. If the direction of propagation of light is considered to be along the plane of the paper then, for unpolarised light the vibrations are indicated by dots (to represent vibration perpendicular to the plane of the paper) and arrows to represent vibration along the plane of the paper (Fig. 3A.5(b).

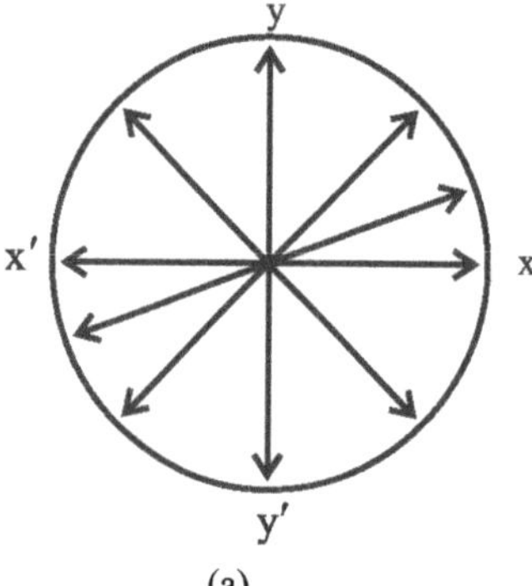

(a)

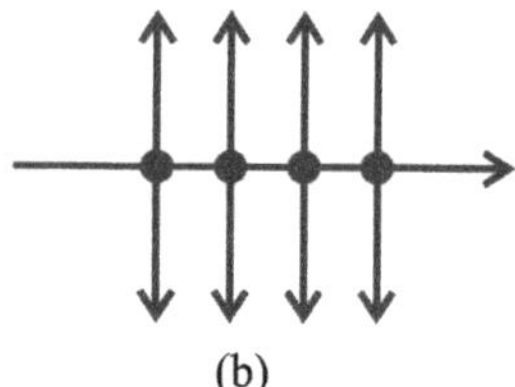

(b)

Fig. 3A.5

3A.3.2 Plane polarised light

If the vibrations are confined to a single plane either in the direction along the plane of the paper or in the direction perpendicular to the plane of the paper, then it is called plane polarised or linearly polarised light. The vibrations along the plane of the paper are represented by arrows Fig. 3A.6(a) while the vibration perpendicular to the plane of the paper are represented by dots (Fig. 3A.6(b)).

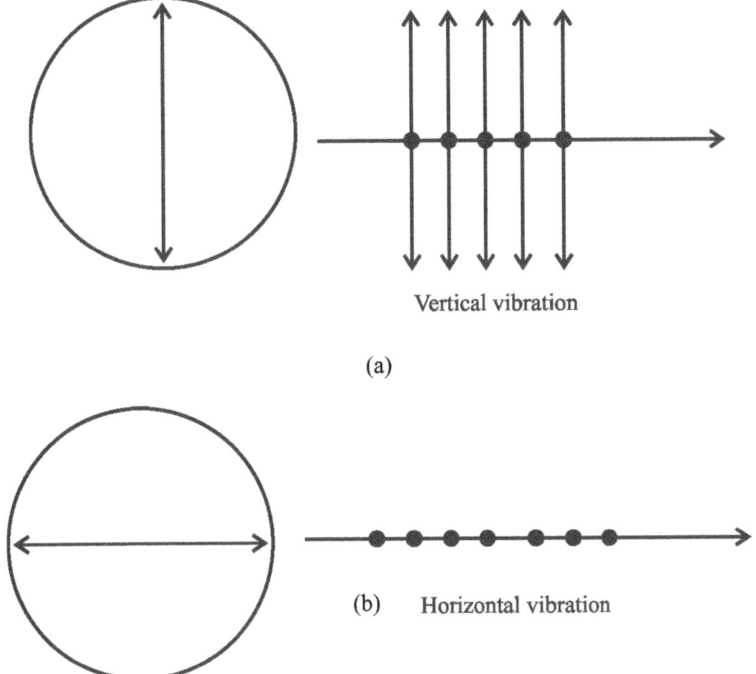

Fig. 3A.6

3A.3.3 Partially plane polarized light

If the linearly polarised light contains small additional component of unpolarised light, it becomes partially polarised light. Then it is represented by either more arrows and less dots as shown in Fig. 3A.7(a) below or less arrow and more dots (as shown in Fig. 3A.7(b)).

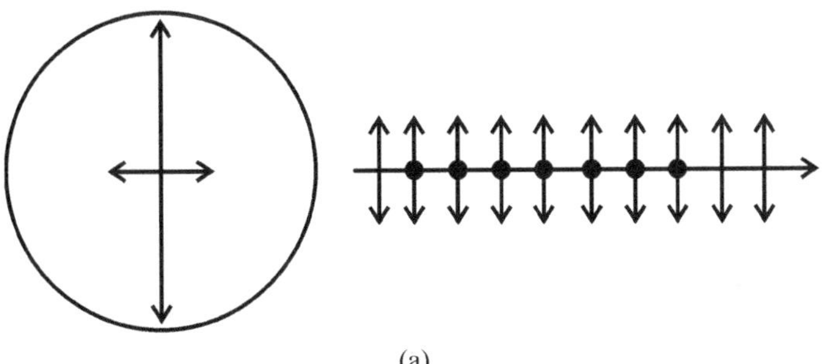

(a)

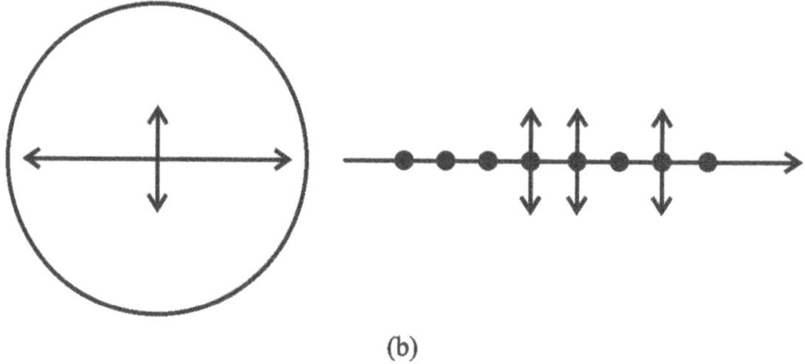

(b)

Fig. 3A.7

3A.3.4 Circularly polarised light

In circular polarization, the electric vector of constant amplitude no longer oscillates but rotates while proceeding in the form of a helix. The projection of wave on a plane intercepting the axis of propagation gives a circle with the amplitude of the vector remaining constant as shown in Fig. 3A.8. If the vector rotates in the clockwise direction with respect to the direction of propagation, it results in right circularly polarised light while the rotation in the anticlockwise direction results in left circulating polarised light.

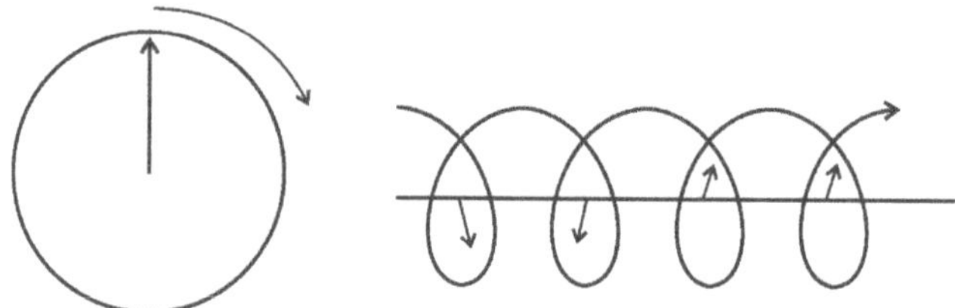

Fig. 3A.8 Representation of circularly polarised light

3A.3.5 Elliptically polarised light

We have seen that in circular polarization, the electric vector of constant amplitude rotates while proceeding. If the amplitude of the electric vector is not a constant but varies periodically then it results in elliptically polarised light. For example, if the electric vector has minimum amplitude while oscillating vertically and rotates while propagating to have maximum amplitude when oscillating horizontally as shown in Fig. 3A.9.

3A.4 Production of Plane Polarised Light

The light can be polarised by the following processes:

1. Polarisation by scattering
2. Polarisation by reflection
3. Polarisation by refraction
4. Polarisation by double refraction
5. Polarisation by selective absorption

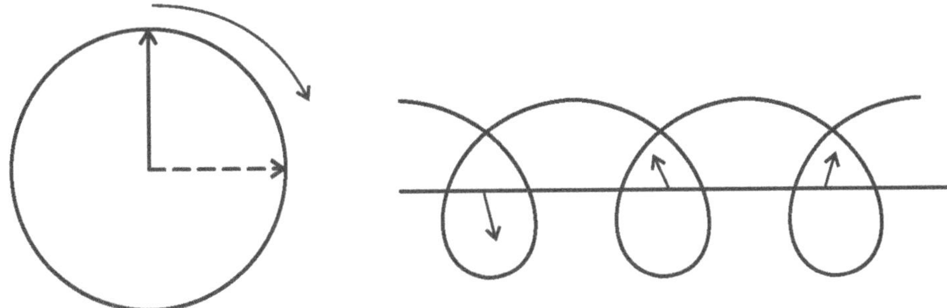

Fig. 3A.9 Representation of elliptically polarised light

3A.4.1 Polarisation by scattering

The light from a clear blue portion of the sky shows a rise and fall of intensity when viewed through a polaroid which is rotated. This is nothing but sunlight, which has changed its direction (having been scattered) on encountering the molecules of the earth's atmosphere. As shown in Fig. 3A.10 the incident sunlight is unpolarised. The dots stand for polarisation perpendicular to the plane of the Figure. There is no phase relation between these two in unpolarised light.

If μ is the refractive index of the medium, then

Brewster's discovered the relation

$$\mu = \tan i_p$$

i.e., the tangent of angle of polarisation for a given medium is numerically equal to the refractive index of the medium. This is called "Brewsters law.

He also found an important result that "At the polarising angle the reflected and refracted rays are perpendicular to each other".

Let a beam PQ of unpolarised light be incident on the surface XY of a transparent substance at the polarising angle i_p. Let QR be the reflected beam and QS the refracted beam as shown in Fig. 3A.11. Let r be the angle of refraction. Then from Brewster's law

$$\mu = \tan i_p = \frac{\sin i_p}{\cos i_p} \qquad \text{.....(3A.1)}$$

By Snell's law we have

$$\mu = \frac{\sin i_p}{\sin r} \qquad \text{.....(3A.2)}$$

Comparing eq. (3A.1) and (3A.2) we get

$$\frac{\sin i_p}{\sin r} = \frac{\sin i_p}{\cos i_p}$$

Under the influence of the electric field of the incident wave the electrons in the molecules acquire components of motion in both these directions. We have drawn an observer looking at 90° to the direction of the sun. Clearly charges accelerating parallel to the double arrows do not radiate energy towards this observer since their acceleration has no transverse component. The radiation scattered by the molecule is therefore represented by dots. It is polarised perpendicular to the plane of the figure. This explains the polarisation of scattered light from the sky.

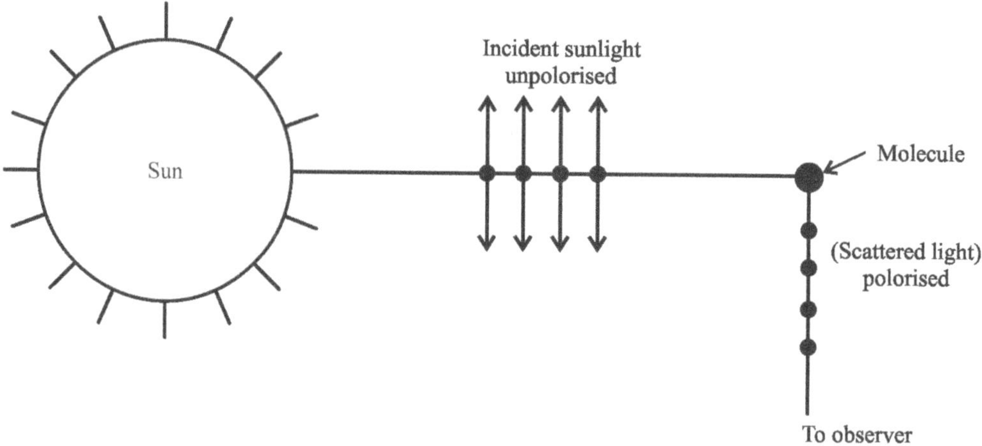

Fig. 3A.10

3A.4.2 Polarisation by Reflection

In 1808 Malus discovered the polarisation of light by reflection from the surface of glass. He found that when unpolarised light is incident on the surface of any transparent material, the reflected beam is partially plane polarised. The vibrations of the reflected beam are parallel to the reflecting surface. When this reflected light is passed through a tourmaline crystal and the later is rotated, the intensity of the reflected beam is minimum (not zero) for one particular position of the crystal. If now the angle of the incidence is changed, we see that intensity of the reflected beam is zero for one particular position of the crystal. This means that the reflected beam is completely plane polarised for a particular angle of incidence. This angle of incidence is called the polarising angle.

The polarising angle depends upon the reflecting surface and slightly up on the wavelength of light. The value of polarising angle is 57° for air-glass reflection, 33° for glass-air reflection and 53° for air-water reflection.

3A.4.3 Brewster's Law

(a) In 1881, sir David Brewster observed that when a beam of ordinary light is incident on the surface of a given transparent medium, the reflected light is completely plane-polarized. This particular angle of incidence is called polarising angle or Brewster's angle. It is denoted by i_p and is defined as the particular angle of incidence for which the reflected beam is completely plane polarised (Fig. 3A.11).

$$\sin r = \cos i_p = \sin (90 - i_p)$$

$$r = 90^\circ - i_p$$

$$\boxed{r + i_p = 90^\circ} \qquad \text{.....(3A.3)}$$

If θ is the angle between reflected and refracted beam, then

$$i_p + \theta + r = 180^\circ$$

$$\theta = 180 - (i_p + r)$$

$$\theta = 180 - 90$$

$$\theta = 90^\circ$$

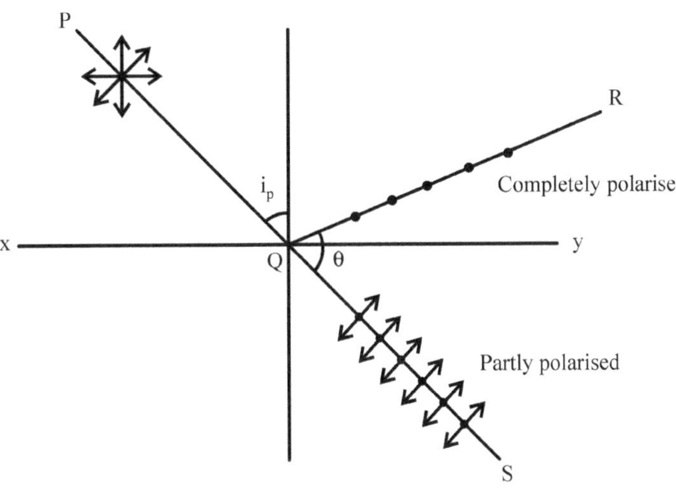

Fig. 3A.11

This shows that the reflected and refracted beams are at right angle to each other. Hence, when light is incident at the polarising angle, the reflected beam is at right angle to the refracted beam.

3A.4.4 Polarisation by Refraction through Piles of Plates

When ordinary unpolarised light is incident on the upper surface of a single glass plate at polarising angle i_p, only a small fraction of the light is reflected e.g. for glass ($\mu = 1.5$) out of incident light, 100% of the light having vibration perpendicular to the plane of incidence is transmitted; where as only 15% of light having vibration perpendicular to the plane of incidence is reflected. Thus the reflected light is completely plane polarised with its vibrations perpendicular to the plane of incidence. While the transmitted or reflected light is partially polarised having its vibration both in the plane of incidence as well as perpendicular to the plane of incidence. The refracted light is incident at the lower surface of the glass plate at an angle r where r is the angle of incidence at the upper face of the plate. We have

$$\tan r = \frac{\sin r}{\cos r}$$

$$\tan r = \frac{\sin r}{\sin(90 - r)}$$

$$\tan r = \frac{\sin r}{\sin i_p} \qquad \text{since } i_p + r = 90^o$$

$$\tan r = g^{\mu_a}$$

Thus from Brewster's law r is the angle of polarisation for a beam incident from glass to air. Therefore the light reflected at the lower face of the plate is completely plane polarised; while the light refracted in to air is partially polarised (Fig. 3A.12). Thus the light refracted through a single glass plate is partially polarised. In order to obtain completely polarised light, it is refracted through pile of plates.

Note : When a beam of light, polarised by reflection at one plane surface is allowed to fall on another similar reflecting surface at polarising angle, of plate, its vibration are parallel to the plane of incidence. Obviously the polarisation depends upon the number of the plates. If I_p and I_n are the intensities of the parallel and normal (perpendicular) components in refracted light. The degree of polarisation is given by

$$P = \frac{I_p - I_n}{I_p + I_n} = \frac{m}{m + \left(\frac{2\mu}{1 - \mu^2}\right)}$$

where m is the number of plates and μ is refractive index of their material.

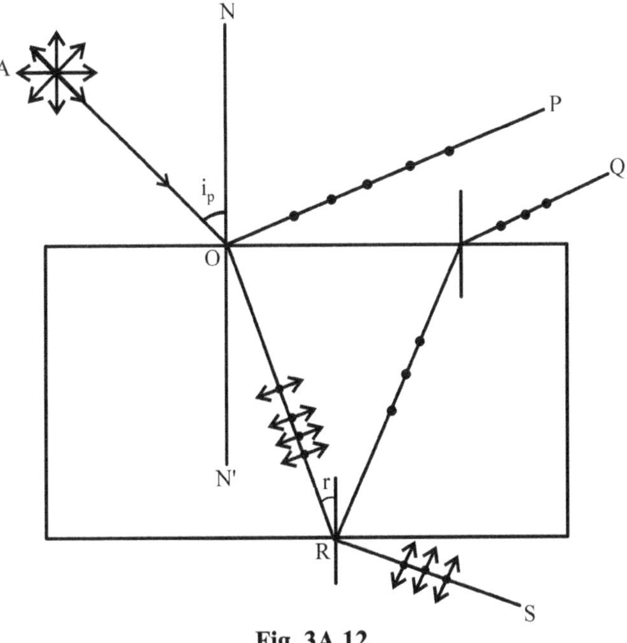

Fig. 3A.12

3A.4.5 Malus Law

(a) "According to Malus,

When a completely plane polarised light beam is incident on a polarising sheet the intensity of the emergent light varies as the square of the cosine of the angle between the direction of electric field vector of the incident light and the polarising direction of the sheet i.e.,

$$I = I_o \cos^2 \theta$$

where I_o being the intensity of plane polarised light.

Proof: Let A represent the amplitude transmitted by the polariser for which the plane of transmission intersects the plane of figure as shown in Fig. 3A13, the vertical dotted line. When the light strikes the analyser, set at the angle θ, one can resolve the incident amplitude into components A cos θ and A sin θ.

The vibration of the A cos θ are parallel to the plane of transmission of the analyser and the vibration of A sin θ are perpendicular to the plane of transmission of the analyser, therefore the component A cos θ is transmitted while the component A sin θ is reflected.

Therefore the intensity of light emerging from the analyser is given by

$$I = (A \cos \theta)^2 = A^2 \cos^2 \theta$$

$$I = I_o \cos^2 \theta$$

where $I_o = A^2$ is the intensity of the incident plane polarised light.

1. If $\theta = 0°$ or $180°$ then $\cos^2\theta = 1$

   therefore,

   $$I = I_o$$

2. If $\theta = 90°$ then $\cos^2\theta = 0$

   therefore, $I = 0$

This law does not hold when the incident light is unpolarised, because in unpolarised light the angle θ, made by the electric field vector with the polarising direction of the plate is not constant. Therefore we have to take average value of $\overline{\cos^2\theta} = \frac{1}{2}$

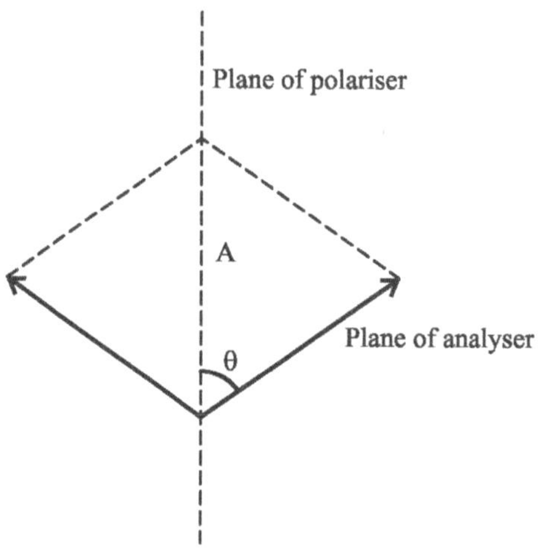

Fig. 3A.13

Thus the intensity of the beam transmitted from the polarising plate is given by

$$I = I_o\,\overline{\cos^2\theta}$$

$$I = \frac{I_o}{2}$$

Thus the plate reduces the intensity of the unpolarised light by 50%. This means that even an ideal polariser can have maximum transmission of 50%.

3A.5 Uniaxial and bi-axial Crystals

When a ray of unpolarised light is incident on a certain special crystal known as doubly refracting crystal, two refracted rays are produced. Such crystals are of two types.

1. Uniaxial double refracting crystal. 2. Biaxial double refracting crystal

In uniaxial double refracting crystal there is one direction known as optic axis along which the two refracted rays travel with the same velocity. The examples of uniaxial crystals are calcite, tourmaline and quartz.

In biaxial crystals there are two optic axis and none of the refracted rays follows the ordinary laws of refraction. The examples of Biaxial crystals are topaz and aragonite.

3A.6 Geometry of Calcite Crystal

Chemically it is crystalline calcium carbonate ($CaCo_3$). Iceland spar or calcite crystal is colourless transparent crystal belonging to hexagonal system. In nature it is found in several forms. A piece of Iceland spar of any shape readily breaks into simple rhombohedron or the rhomb is bounded by six faces each of which is a parallelogram of angles 102° and 78° (more correctly 101° 55' and 78° 5').

At two diametrically opposite corners, all the angles of the faces meeting there are obtuse. These two corners A and B are termed as blunt corners of the crystal. At the rest of six corners, two angles are acute and the remaining is obtuse (Fig. 3A.14).

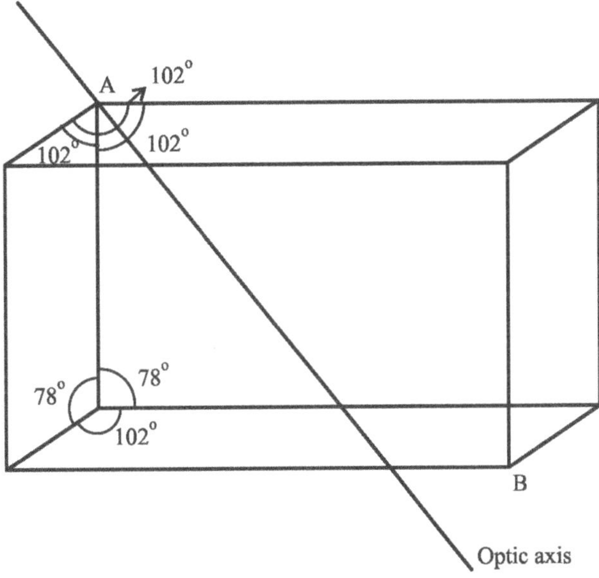

Fig. 3A.14

3A.7 Optic Axis

A lines passing through any one of the blunt corners and making equal angles with each of three edges which meet there gives the direction of the optic axis. In fact any line parallel to this line represents the optic axis.

1. Optic axis is not a line, but it is a direction.
2. Optic axis is not obtained by joining the two blunt corners.
3. In special case when the three edges of the crystal are equal, the line joining the two blunt corners coincides with the axis of the crystal and gives the direction of the optic axis.
4. The crystal is symmetrical about it.
5. The velocities of the two refracted beams along the direction of optic axis are same.

Note:

(a) When a ray of light is incident along the optic axis or in a direction parallel to the optic axis, the ray is not split up into ordinary and extra ordinary components. In this case both the rays travel along the same direction with the same velocity.

(b) When a ray of light is incident perpendicular to the optic axis, the ray of light is not split up into ordinary and extra ordinary rays. In this case both the rays travel along the same direction but with different velocities.

3A.8 Principal Plane or Principal Section

A plane containing the optic axis and perpendicular to the face of the crystal is called the principal plane or principal section for that face. In Fig. 3A.15(a), PQRS is the principle plane for the top and the bottom faces of the calcite crystal. For every point there are three principal planes, one for each pair of opposite faces of the crystal.

The principle plane for calcite crystal is a parallelogram with angles 71° and 109° as shown in Fig. 3A.15(b). Since the opposite faces of a calcite crystal are always parallel, the two refracted beam emerge out parallel to each other.

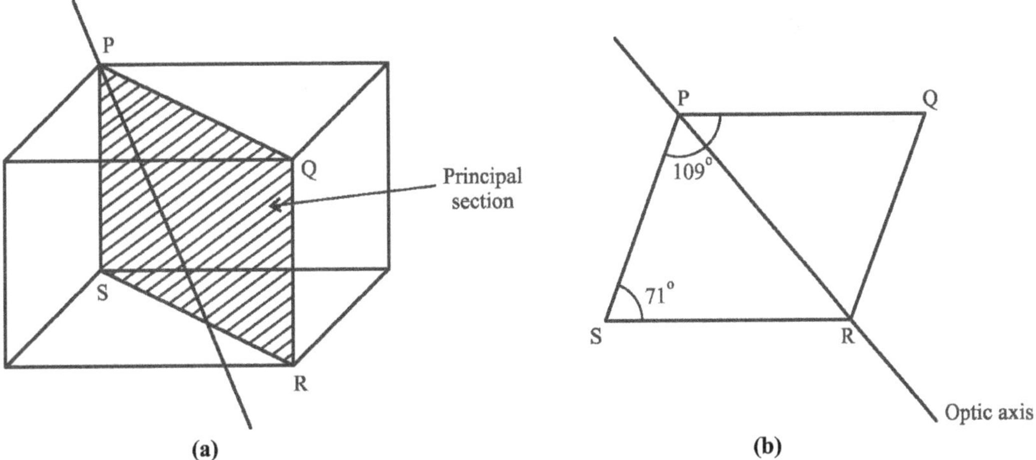

Fig. 3A.15

3A.9 Ordinary and Extraordinary Rays

When a ray of unpolarised light is passed through a doubly refracting crystal like calcite or quartz, it splits up into two refracted rays, of the two refracted rays, one of the rays obeys snell's law and is called the ordinary ray or O-ray. The other ray does not obey snell's law and is not even in the plane of incidence and is called the extraordinary ray orE-ray.

1. O-ray is plane polarised in the principal plane i.e., it has vibration perpendicular to the principal plane. These have been shown by dots.
2. E-ray is also plane polarised in a plane perpendicular to the principal plane i.e., the vibrations takes place in the principal plane. These have been shown by lines with arrow heads.
3. O-ray travels, in the crystal with the same speed in all the directions. That is, refractive index for the O-ray is having single value.
4. E-ray travels in the crystal with a speed that varies with direction. That is refractive index for E-ray varies with the direction.
5. The difference between the refractive index for O and E-rays is called birefringe

   $$(\mu_o - \mu_E = \text{birefringes})$$

6. Along the optic axis, both O-ray and E-ray travel with the same speed.
7. In uniaxial crystal the wavefront of O-ray is spherical in nature and the wavefront of E-ray is ellipsoidal in nature.

3A.10 Double Refraction

When an un polarised light is passed through a doubly-refracting crystal such as calcite or quartz it splits up into two refracted rays. One of the rays obeys the ordinary laws of refraction and is called 'ordinary rays' or o-ray. The other however behaves in an extraordinary way and is therefore called extraordinary ray (E-ray). Hence if an object is viewed through such a crystal, two images of the object are seen, one due to O-ray and the other due to E-ray. The phenomenon is known as double refraction.

Fig. 3A.16 represents the double refraction phenomenon. A ray of light SA from a point sources is incident on a calcite crystal making an angle of incidence i, it is refracted along two paths AB and AC making angle of refraction r_1 and r_2 respectively. These rays emerges out along BO and CE parallel to each other. Thus, two images O and E can be obtained on the screen. If the crystal is rotated about the incident beam as axis one image remains stationary while the other image E rotates around the image O.

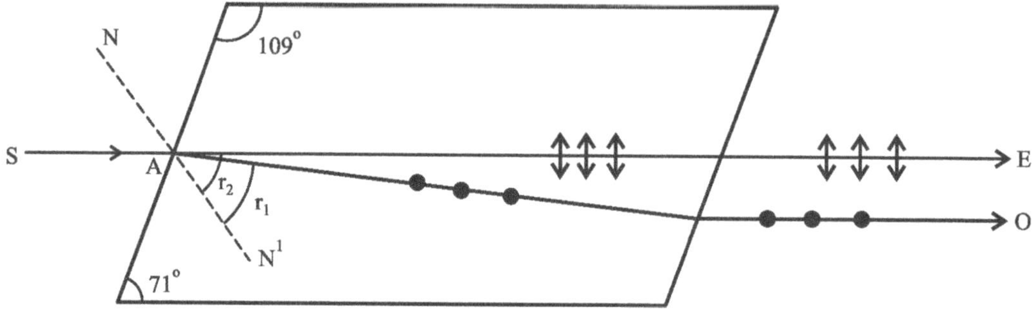

Fig. 3A.16

In calcite crystal $r_2 > r_1$, therefore $\mu_o > \mu_E$, hence $v_o < v_E$ that is, inside the calcite crystal, the E-ray travels faster as compared to O-rays. Such crystal are called uniaxial negative crystal.

In quartz crystal $r_2 < r_1$, therefore $\mu_E > \mu_o$ hence $v_o > v_E$ that is inside the quartz the O-ray travel faster as compared to E-ray. Such crystals are called uniaxial positive crystals.

3A.11 Huygen's Theory of Double Refraction

Huygen 's explained the phenomenon of double refraction by extending his theory of secondary wavelets in a suitable manner. According to his theory:

1. Every point on the crystal surface is disturbed by the incident wavefront and become the origin of two wavefronts, one the ordinary and the other extraordinary to account for ordinary and extraordinary rays.

2. The wavefront corresponding to the ordinary wavelets is spherical because O-ray travels with the same velocity in all directions.
3. The wavefront corresponding to the extraordinary wavelets is ellipsoid because E-ray travels with different velocity in different directions.
4. The sphere and the ellipsoid touch each other at points which lie on the optic axis of the crystal.
5. In a negative uniaxial crystal like calcite the sphere lies inside the ellipsoid as shown in Fig. 3A.17(a).

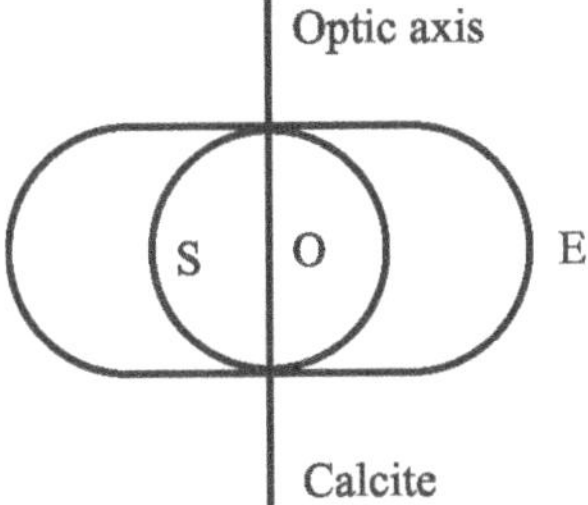

Fig. 3A.17(a) Uniaxial negative crystal

6. While in a positive uniaxial crystal like quartz the ellipsoid lies inside the sphere as shown in Fig. 3A17(b).

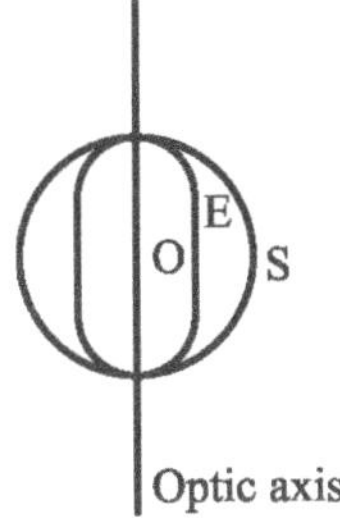

Fig. 3A.17(b) Uniaxial positive crystal (quartz)

7. The geometric construction of wavefronts immediately explains double refraction. From Fig. 3A.18(a), OA is incident wave plane wavefront. By the time primary disturbance from A reaches point B, ordinary disturbance expands on sphere with O as centre. The tangent from point B on to the sphere touches it at B'. The OB' is ordinary refracted ray similarly OB" is extraordinary refracted ray.

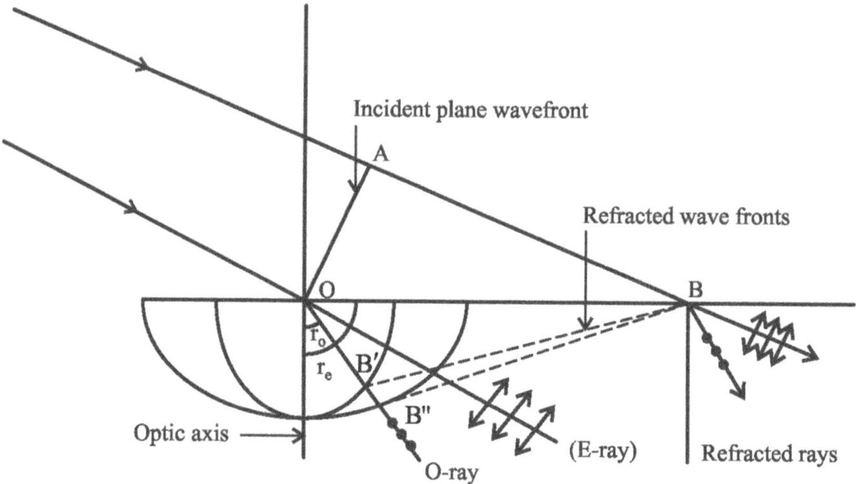

Fig. 3A.18(a) Huygen's construction of wavefronts for double refracting negative crystal (calcite). Since e-disturbance expands on ellipsoid where as o-disturbance on sphere. The tangent from B on to these disturbances touch at different points giving two refracted rays.

8. Fig. 3A.18(b) and (c) below shows the propagation of light wave along the optic axis and perpendicular to optic axis respectively. It turns out that light when travels perpendicular to optic axis in principal plane, ordinary and extraordinary rays travel with different velocities but in the same direction. This result is of great importance and allows us to create desired phase difference between two transverse components of light, namely E_y and E_z which are travelling in the same direction. In other words, it allows us to produce circularly and elliptically polarised light.

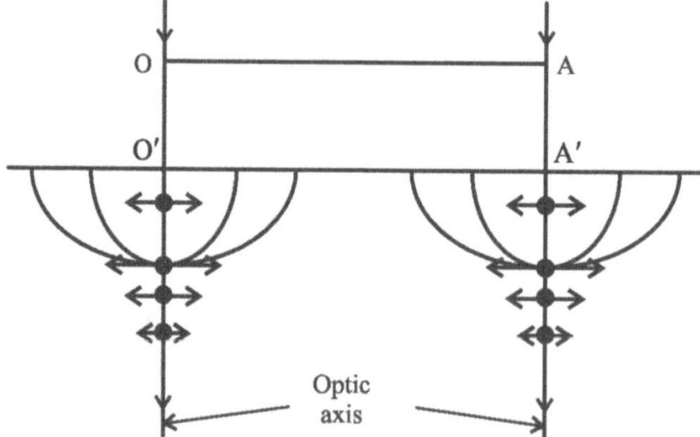

Fig. 3A.18(b) When light travels in the direction of optic axis, there is no double refraction.

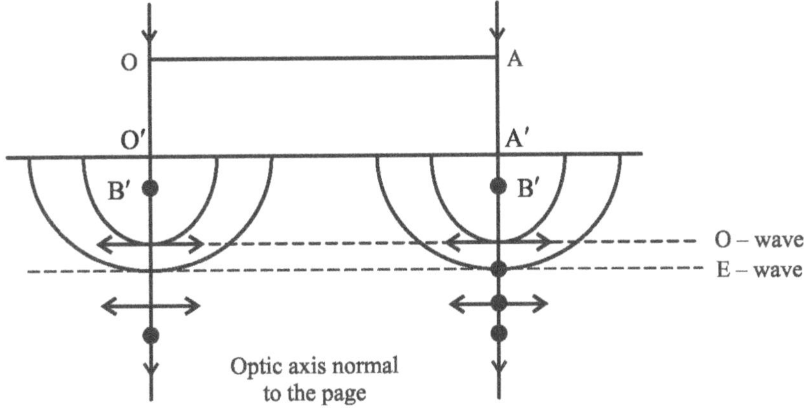

Fig. 3A.18(c) When light travels perpendicular to optic axis, the O and e-rays travel in the same direction but with different speeds.

3A.12 Principal Refractive Indices

The velocity of O-ray is the same in all directions while the velocity of E-ray is different in different directions, but both the rays travel with the same velocity along the direction of optic axis. Thus in unaxial crystals there are two principal refractive indices, one corresponding to the velocity of O-ray and the other corresponding to the velocity of E-ray.

$$\mu_o = \frac{\text{Velocity of light in vacuum}}{\text{Velocity of O - ray}}$$

The principal refractive index of E-ray is defined as the ratio of velocity of light in vacuum to the velocity of E-ray in a direction perpendicular to the optic axis.

$$\mu_E = \frac{\text{Vecloty of light in vacuum}}{\text{Velocity of E - ray}}$$

3A.13 Nicol Prism

The Nicol prism is an optical device made from a calcite crystal and used in many optical instruments for producing and analysing plane polarised light.

3A.13.1 Construction

Nicol prism is constructed from a calcite crystal whose length is nearly three times its width. The end faces of the crystal are cut down so as to reduce the angle in the principal section to a more acute angle of 68° instead of 71° Fig. 3A.19. The crystal is then cut along a diagonal and the two cut surfaces after polishing, cemented back together with a special cement called Canada Balsam, which is a transparent substance. It is optically more denser than calcite for E-ray and less dense for the O-ray (for sodium light).

$$\mu_o = 1.65836$$

$$\mu_{cb} = 1.55$$

$$\mu_E = 1.4864$$

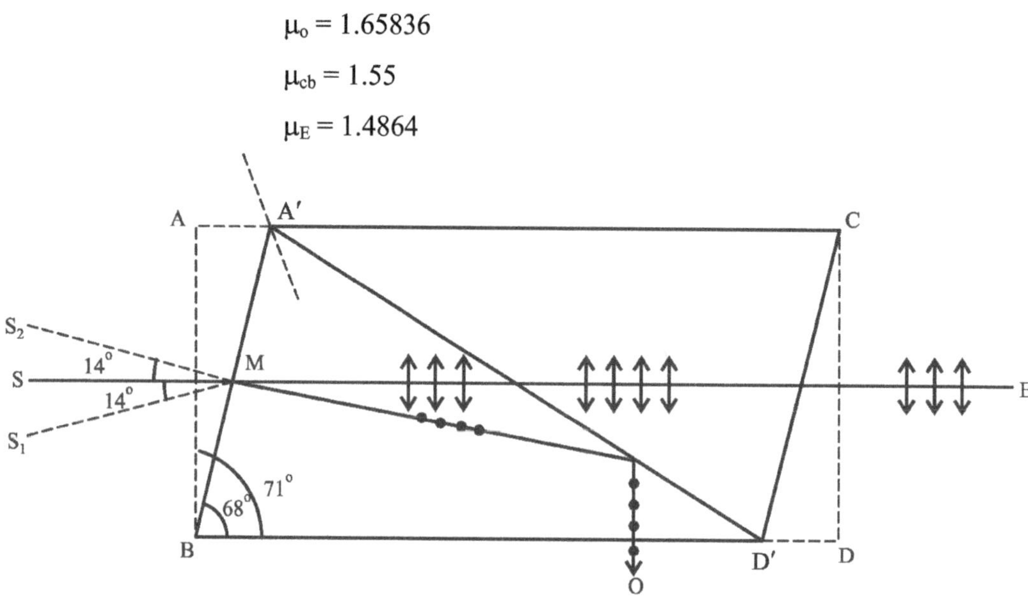

Fig. 3A.19

3A.13.2 Action

A ray of light SM is incident nearly parallel to BD' on the face A'B of the Nicol Prism. It splits in to E-ray and O-ray, whose vibrations are respectively perpendicular and parallel to the principal section of the nicol prism as shown in Fig. 3A.19.

The O-ray suffers total internal reflection at the Canada Balsam surface for nearly normal incidence because Canada balsam is optically more dense than calcite for the E-ray and lens dense for the O-ray. The E-ray refracted through Canada balsam and is transmitted but the O-ray, moving from a denser calcite medium to the rarer canada Balsam medium, is totally reflected for angle of incidence greater than the critical angle.

The value of critical angle i_c for the O-ray for calcite to Canada Balsam is

$$i_c = \sin^{-1}\left[\frac{1.550}{1.658}\right] = 69.2^\circ$$

Thus starting with ordinary unpolarised light, a Nicol prism transmits only the plane polarised light.

Limitations

The Nicol Prism works only when the incident ray is slightly convergent or slightly divergent. If the incident ray makes an angle much smaller than BMS with the surface A'B, the O-ray will strike the Canada Balsam layer at an angle less than the critical angle and hence will be transmitted and the light emerging from the Nicol will not be plane polarised.

If the incident ray makes an angle greater than BMS, the E-ray will become more and more parallel to the optic axis and hence its refractive index will become nearly equal to that of calcite for the O-ray. This will then also offer total internal reflection like O-ray. Hence no light will emerge out of the Nicol prism.

Thus to get the plane polarised light the angle between the extreme rays of the incoming beam < S_1 M S_2 is limited to about 28°.

3A.13.3 Nicol Prim as a Polariser and Analyser

Nicol prism can be used as 'polariser' and 'Analyser' when an unpolarised light is incident on a Nicol prism P Fig. 3A.20(a), the ray emerging from P is plane polarised with vibrations in the principal section of P. If this ray falls on a second Nicol prism A, whose principal section is parallel to that of P, its vibration will be in the principal section of A. Hence the ray will be completely transmitted. The intensity of the emergent light will be maximum.

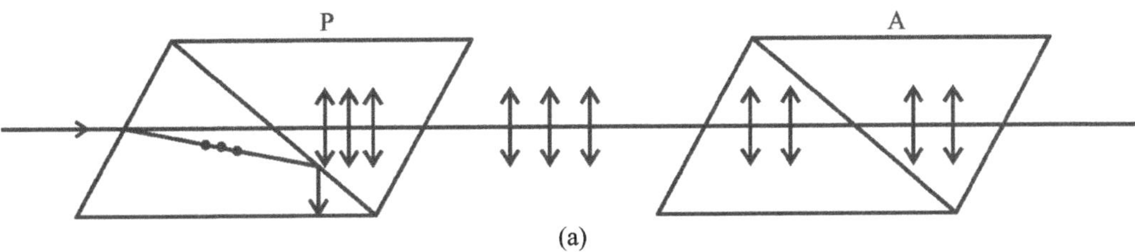

(a)

Now, if the Nicol A be rotated such that its principal section becomes perpendicular to that of P, the vibration in the plane-polarised ray incident on A will be perpendicular to the principal section of A. Hence the ray will behave as O-ray inside A and will be lost by total reflection at the Canada Balsam surface. Therefore no light will emerge from A. In this position the two Nicols are said to be crossed as shown in Fig. 3A.20(b).

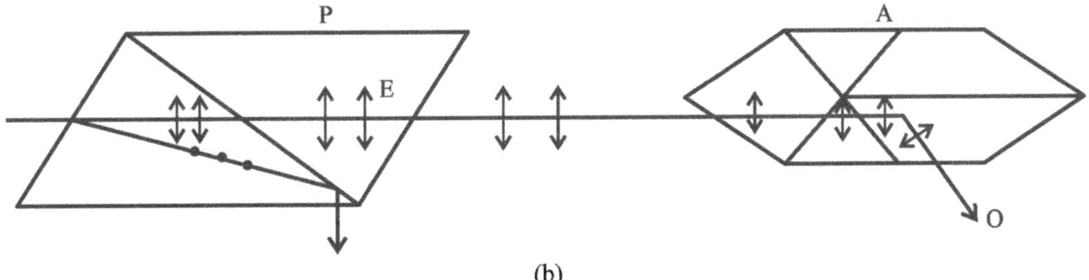

(b)

If the Nicol A is further rotated to have its principal section again parallel to that of P Fig. 3A.20(c), the intensity of emergent light will again be maximum. Therefore P is called the polariser and the prism A is called Analyser.

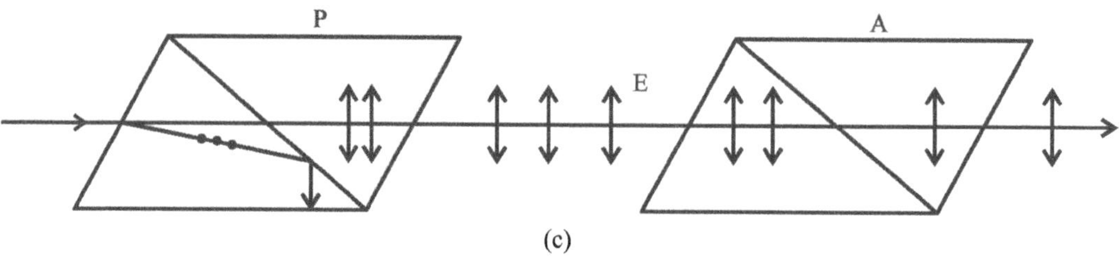

(c)

Fig. 3A.20

3A.14 Plane, Circularly and Elliptically Polarised Light

Plane Polarised Light

When light travels along a certain direction the vibrations takes place in the direction at right angles to the direction of propagation. If the vibrations of the ether particles are linear and take place parallel to a plane through the axis of the beam or the direction of propagation, light is said to be plane polarised.

Elliptically polarised light

When the vibrations of the ether particles are elliptical, having a constant period and take place in a plane perpendicular to the direction of propagation, light is said to be elliptically polarised. In elliptically polarised light, the amplitude of the vibrations changes in magnitude as well as in direction.

Circularly polarised light

When the vibrations of the ether particles are circular having constant period and take place in the transverse plane, light is said to be circularly polarised. In circularly polarised light, the amplitude of the vibrations remains constant but the direction only changes.

3A.15 Superposition of Two Linearly Polarised Sinusoidal Light Waves

Lat a beam of linearly polarised light be incident normally on a calcite plate cut with its optic axis parallel to its faces. Let the linear vibration in the incident light be along PA, making an angle θ with the optic axis. Let A be the amplitude of vibration in the incident light (Fig. 3A.21).

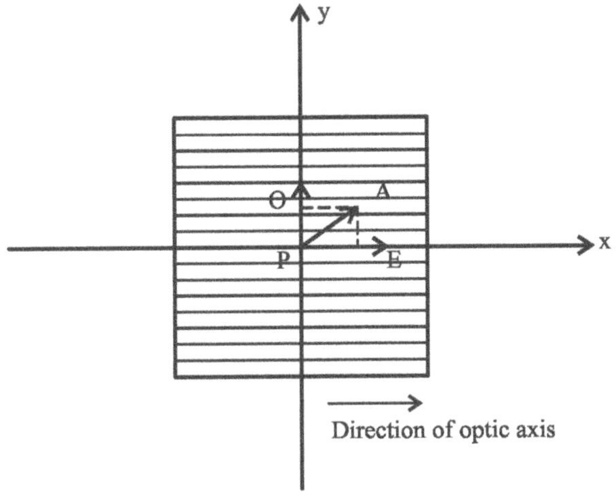

Fig. 3A.21

If A sin ωt is the incident wave, then the two emergent linearly polarised E and O-waves would be respectively

E-ray ⇒

$$x = A\cos\theta\sin(\omega t + \delta) \quad \text{.....(3A.4)}$$

O-ray ⇒

$$y = A\sin\theta\sin\omega t \quad \text{.....(3A.5)}$$

Let $A\cos\theta = a$

$A\sin\theta = b$

Thus

$$x = a\sin(\omega t + \delta) \quad \text{.....(3A.6)}$$

$$y = b\sin\omega t \quad \text{.....(3A.7)}$$

∴

$$\frac{x}{a} = \sin\omega t\cos\delta + \cos\omega t\sin\delta$$

$$\frac{x}{a} = \sin\omega t\cos\delta + \sin\delta\sqrt{1-\sin^2\omega t} \quad \text{.....(3A.8)}$$

Since

$$\sin\omega t = \frac{y}{b}$$

∴

$$\frac{x}{a} = \frac{y}{b}\cos\delta + \sqrt{1-{y^2}/{b^2}}\ \sin\delta$$

On solving we get

$$\frac{x^2}{a^2} + \frac{y^2}{b^2} - \frac{2xy}{ab}\cos\delta = \sin^2\delta \quad \text{.....(3A.9)}$$

Eq. (3A.9) in general represents an ellipse. Hence the light merging from the crystal plate is, in general elliptically polarised.

(a) If the thickness of the plate is such that

$$\delta = 0, 2\pi, 4\pi, \ldots..$$

Then $\cos \delta = 1$ and $\sin \delta = 0$

Then eq. (3A.9) reduces to

$$\frac{x^2}{a^2} + \frac{y^2}{b^2} - \frac{2xy}{ab} = 0$$

$$\left(\frac{x}{a} - \frac{y}{b}\right)^2 = 0$$

$$\therefore \qquad y = \pm \frac{b}{a} x$$

This represents a pair of coincident straight line i.e., the emergent light is plane polarised with the same direction of vibration as the incident light.

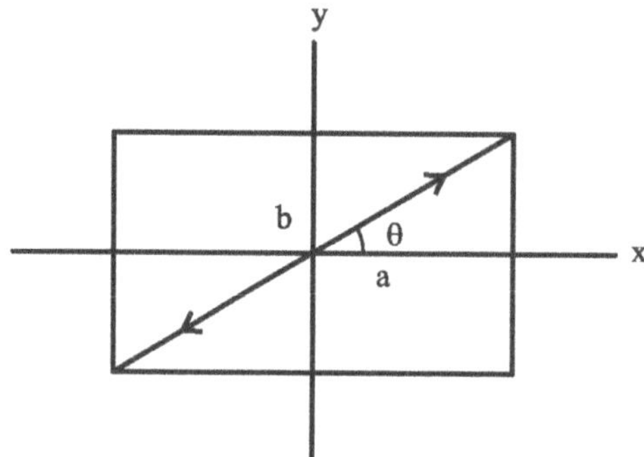

(b) If $\delta = \pi, 3\pi, 5\pi, \ldots.$

Then $\cos \delta = -1$ and $\sin \delta = 0$

Then eq. (3A.9) becomes,

$$\frac{x^2}{a^2} + \frac{y^2}{b^2} + \frac{2xy}{ab} = 0$$

or

$$\left(\frac{x}{a} + \frac{y}{b}\right)^2 = 0$$

$$\pm y = \mp \frac{b}{a} x$$

Again this represents a plane-polarised light

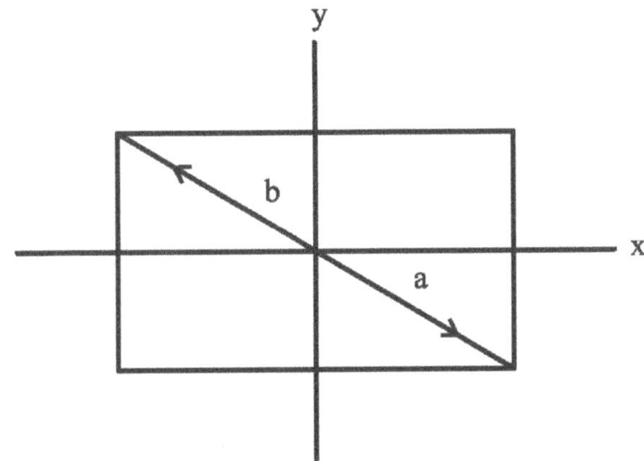

(c) If $\delta = \frac{\pi}{2}, \frac{3\pi}{2}, \frac{5\pi}{2}, \frac{7\pi}{2}$...... then $\cos\delta = 0$ and

$$\sin^2\delta = 1$$

Then eq. 3A.9 becomes

$$\frac{x^2}{a^2} + \frac{y^2}{b^2} = 1$$

Hence the emergent light elliptically polarised

(i) If $\delta = \frac{\pi}{2}, \frac{5\pi}{2}, \frac{9\pi}{2}$,..... then emergent light is left handed elliptically polarised.

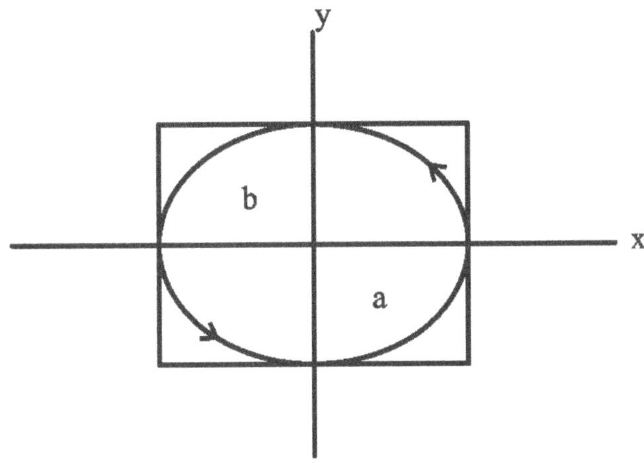

(ii) If $\delta = \frac{\pi}{2}, \frac{5\pi}{2}, \frac{9\pi}{2}, \ldots$ Then emergent light is right handed elliptically polarised.

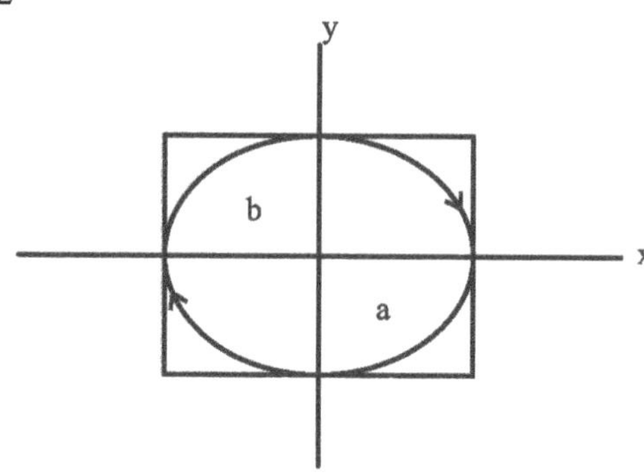

(iii) If $\theta = 45°$, then $a = A \cos 45°; b = A \sin 45°$

then $a = b$

$\therefore$ $\frac{x^2}{a^2} + \frac{y^2}{a^2} = 1$

$\Rightarrow$ $x^2 + y^2 = a^2$

Hence in this case the emergent light is circularly polarised.

(a) If $\delta = \frac{\pi}{2}, \frac{5\pi}{2}, \frac{9\pi}{2}, \ldots$ then the light obtained is left handed circularly polarised.

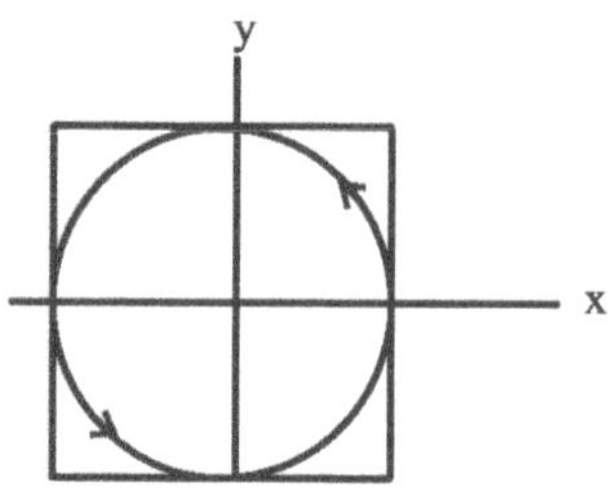

(b) If $\delta = \frac{3\pi}{2}, \frac{7\pi}{2}, \frac{11\pi}{2}, \ldots$ then the light obtained is right handed circularly polarised.

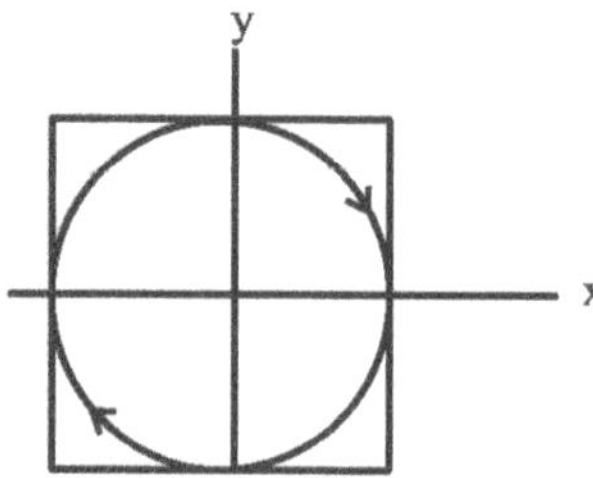

Hence, we can conclude that

1. When the phase difference is 0, 2π or any even multiple of π, the results is a linear vibration at 45° to both original vibrations.
2. When the phase difference π, 3π or any odd multiple of π, the result is also a linear vibration, but at right angles to those corresponding to even multiple of π.
3. When the phase difference is $\frac{\pi}{2}, \frac{3\pi}{2}$ or any odd multiple of $\frac{\pi}{2}$, the resulting vibration is a circle.

   Thus we conclude that the plane polarised and circularly polarised lights are the special cases of elliptically polarised light.

3A.16 Retardation Plates

The crystal plate of doubly refracting crystal, that retards the motion of one of the refracted beams (O-rays or E-rays) are known as retardation plates. These are of two types.

1. Quarter wave plate
2. Half wave plate

3A.16.1 Quarter Wave Plate

It is a crystal plate of a doubly refracting material of a uniaxial crystal cut with its optic axis parallel to the refracting faces. The thickness of the plate is such as to produce a path difference of $\lambda/4$ or phase difference of $\pi/2$ between the O-ray and the E-ray, when monochromatic plane polarised light of wavelength λ is incident normal to the surface.

In calcite (negative crystal)

As $\mu_o > \mu_E$, hence the velocity of E-ray is greater than the velocity of O-ray if t is the thickness of the crystal plate, then the path difference between the two components on coming out = $(\mu_o - \mu_E)\,t$

Hence, $$(\mu_o - \mu_E)\,t = \frac{\lambda}{4}$$

$$t = \frac{\lambda}{4(\mu_o - \mu_E)} \quad \text{.....(3A.10)}$$

In quartz (a positive crystal)

As $\mu_E > \mu_o$

The path difference is given by $(\mu_E - \mu_o)t$

$$\therefore \quad (\mu_E - \mu_o)t = \frac{\lambda}{4}$$

$$t = \frac{\lambda}{4(\mu_E - \mu_o)} \qquad .....(3A.11)$$

A quarter wave plate is used for the production and detection of circularly polarised light.

Action of Quarter Wave Plate (QWP):
Production of Circular and Elliptic Polarizations

Let a linearly polarised wave travelling in the x-direction be incident normally on a QWP (Quarter wave plate) located at origin in the y-z plane. If the incident wave amplitude E_o makes an angle θ with the optic axis of the plate and that the optic axis is in the y-direction, then the amplitude of O- and E-waves are respectively,

$$a = E_o \cos\theta$$

$$b = E_o \sin\theta$$

The electric field components of O- and E-waves at incidence are

$$E_y = a \sin \omega t$$

$$E_z = b \sin \omega t \qquad .....(3A.12)$$

The electric field components of O- and E-waves passing out of the QWP are

$$\text{O-wave} \qquad E_y' = a \sin\left(\omega t \pm \frac{\pi}{2}\right) = \pm a \cos \omega t$$

$$\text{E-wave} \qquad E_z' = b \sin \omega t \qquad .....(3A.13)$$

The locus of electric field in transverse plane is obtained by eliminating t in eq. 3A.12, then

$$\frac{E_y'^2}{a^2} + \frac{E_z'^2}{b^2} = 1 \qquad .....(3A.14)$$

This is the equation of an ellipse. Thus, when plane polarised light is incident on a QWP the out going light is in general elliptically polarised. However if $\theta = 0°, 90°, 180°$ and $270°$, the phenomenon of double refraction does not occur and light passes as linearly polarised light.

The sign of + or –ve in E_y' components decides whether elliptic polarization is left-handed or right handed. This can be done by considering electric field components at t = 0 and a little later say after Δt. At t = 0, $E_y' = +a$ and $E_z' = 0$ but at t = Δt, $E_y' = a \cos \omega\Delta t$ and $E_z' = b \sin \omega\Delta t$. As time increases y-component decreases ($\cos \omega\Delta t < 1$) and that z-component increases. This means electric vector rotates from y to z i.e., rotation is left-handed. The plus (+) sign in E_y' indicates left-handed rotation while minus (–) sign indicate right-handed rotation.

Circular polarization

Consider the case when $\phi = 45^\circ$ or 135°

When incident electric field makes an angle of 45° or 135° with the optic axis of quarter wave plate, the amplitudes of O- and E-wave are equal.

$$E_o \cos 45^\circ = E_o \sin 45^\circ = \frac{E_o}{\sqrt{2}} = a \text{ (say)}$$

The outgoing components of electric field are

$$E_y' = \pm a \cos \omega t$$

$$E_z' = \pm a \sin \omega t \qquad \text{.....(3A.15)}$$

The locus of electric field in the transverse plane s

$$E_y'^2 + E_z'^2 = a^2 \qquad \text{.....(3A.16)}$$

which is an equation of a circle.

Thus when a plane polarised light falls normally on a QWP with its vibrations (E_o) making an angle of 45 ° or 135° to the optic axis, then the outgoing light is circularly polarised. The plus sign in E_y' indicates left-handed and minus sign indicates right-handed circular polarization (Fig. 3A.22).

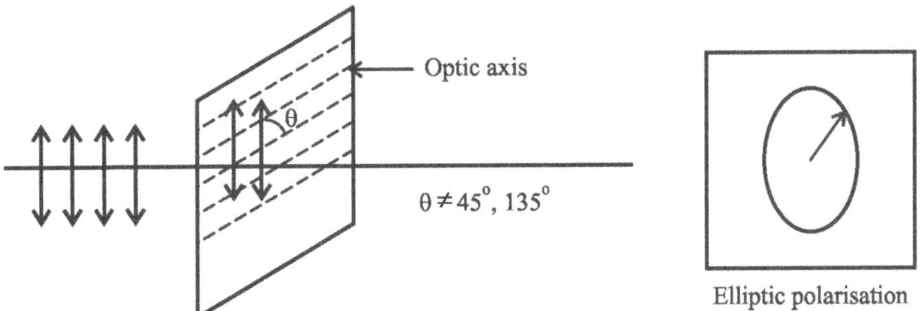

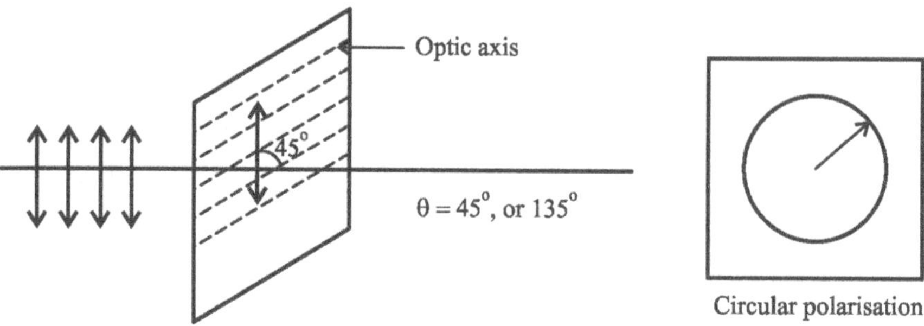

Fig. 3A.22

3A.16.2 Elliptically or Circularly Polarised Light Incident on a QWP

When circularly or elliptically polarised light is incident normally on a QWP an additional phase difference of $\pm \pi/2$ is introduced between O and E-components of the wave. As a result, outgoing light field E_y' and E_z' shall either be in phase of π or cut of phase. The light thus becomes plane polarised. When circularly or elliptically polarised light falls normally on a QWP outgoing light is plane polarised.

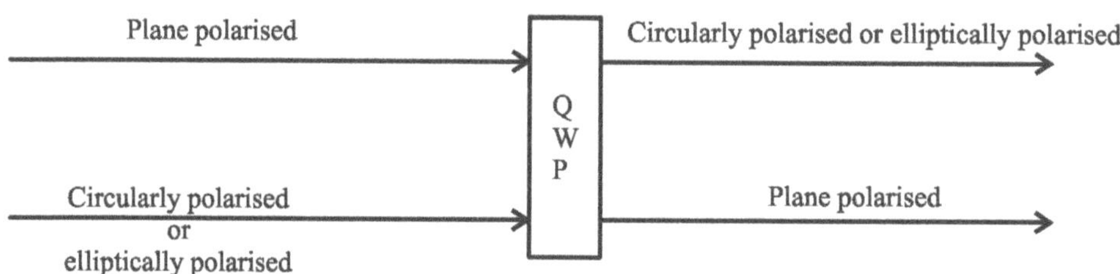

3A.16.3 Half Wave Plate (HWP)

It is crystal plate of a doubly refracting material of a uniaxial crystal cut with its optic axis parallel to the refracting faces. The thickness of the plate is such as to produce a path difference of $\lambda/2$ or phase difference of π, between the O-ray and the E-ray when monochromatic plane polarised light if wavelength λ is incident normal to the surface.

In Calcite (negative crystal)

As $\mu_o > \mu_E$, hence the velocity of E-ray is greater than the velocity of O-ray. If t is the thickness of the crystal plate, then the path difference between the two components on coming out = $(\mu_o - \mu_E)$ t

Hence $$(\mu_o - \mu_E)\, t = \frac{\lambda}{2}$$

$$t = \frac{\lambda}{2(\mu_o - \mu_E)}$$

In Quartz (a positive crystal)

As $\mu_E > \mu_o$

Therefore, the path difference = $(\mu_E - \mu_o)\, t$

Hence $(\mu_E - \mu_o)t = \frac{\lambda}{2}$

3A.16.4 Action of HWP on Plane Polarised Light

Consider a plane polarised light is incident on a HWP whose y-axis coincides with optic axis.

1. If incident polarisation is parallel to y-axis so it travels as O-wave and suffers a phase difference of π. Thus outgoing light is plane polarised.
2. If incident polarisation is parallel to z-axis, it travels through HWP as E-wave and so passes as plane polarised light.
3. If incident polarisation makes an angle θ with the optic axis which is along y-axis, then E_y suffers a phase change of π but E_z passes without change in phase. Consequently as shown in Fig. 3A.23(a), the outgoing light is plane polarised but its polarisation is rotated by an angle 2θ towards the optic axis.

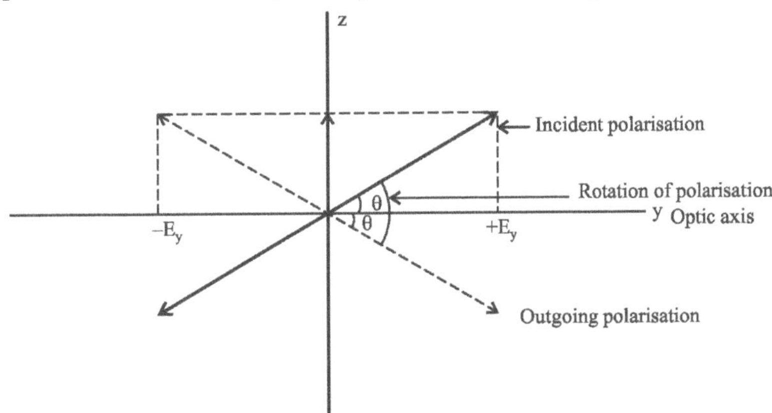

(a) Action of HWP, E_z is E-component and it remains unchanged but E_y, the O-component changes in phase by π. Consequently the outgoing polarisation is rotated towards optic axis by an angle 2θ.

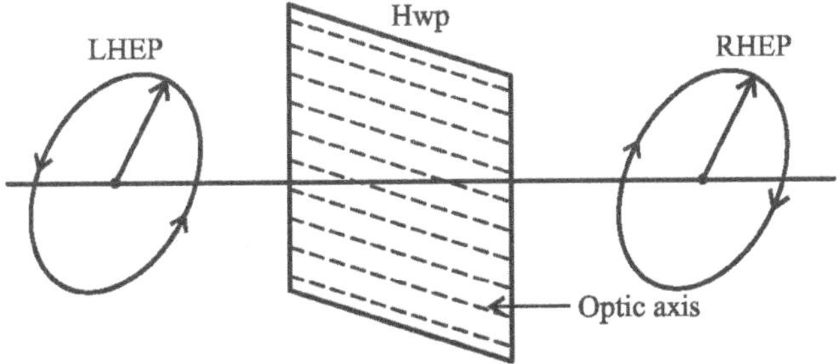

(b)

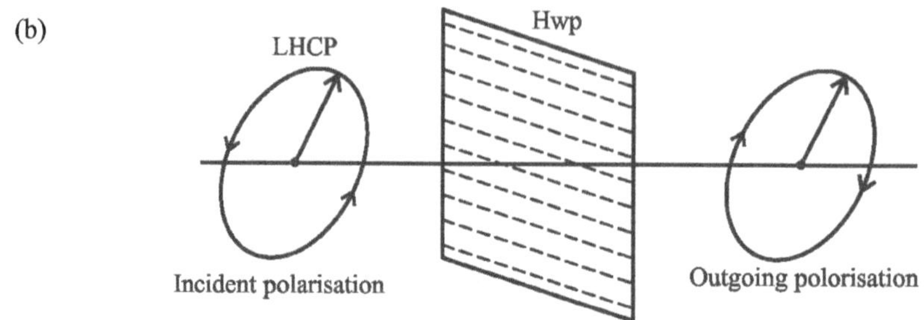

Fig. 3A.23

3A.16.5 Action of HWP on Circularly or Elliptically Polarised Light

When circularly or elliptically polarised light is incident on HWP, the O-component wave suffers a phase change of π but E-component wave remains unaffected. Consequently the left handed wave becomes right-handed and vice-versa but it passes as circularly or elliptically polarised light as shown in Fig. 3A.23(b) above.

3A.17 Production of Plane, Circularly and Elliptically polarised Light

(a) Plane polarised light

The light in which the vibration lie only along a plane passing through the axis of beam is known as plane polarised light.

Production: A beam of unpolarised light when passed through a Nicol prism it splits in to ordinary ray i.e., O-ray and extraordinary ray i.e., E-ray. The O-ray and E-ray are both plane polarised, their plane of polarisation are perpendicular to each other. The O-ray is totally internally reflected or absorbed, while E-ray passes through the prism. This ray is plane polarised.

(b) Circularly polarised light

When the vibrations of the light in transverse plane are circular having constant period, the light is said to be circularly polarised. In circularly polarised light the amplitude of vibrations remains constant but direction changes regularly.

Production: When a beam of unpolarised light is allowed to pass through a Nicol prism it becomes plane polarised. Now when this plane polarised light is passed through a Quarter wave plate such that the angle between the optic axis and the vibration of the light is 45°, the light emerging from the plate will be circularly polarised.

(c) Elliptically polarised light

When the vibrations in the transverse plane are elliptical having constant period, the light is said to be elliptically polarised. In this case the amplitude of vibrations changes in magnitude as well as in direction.

Production: Unpolarised light after passing through a Nicol prism becomes plane polarised. When this plane polarised light is allowed to pass through a quarter wave plate in such a manner that the angle between the optic axis and plane of vibration is other than 0°, 45° and 90°, elliptically polarised light is obtained.

3A.18 Process of Detection of Plane, Circularly and Elliptically Polarised Light

The process of detection of plane, circular and elliptically polarised light is as follows:

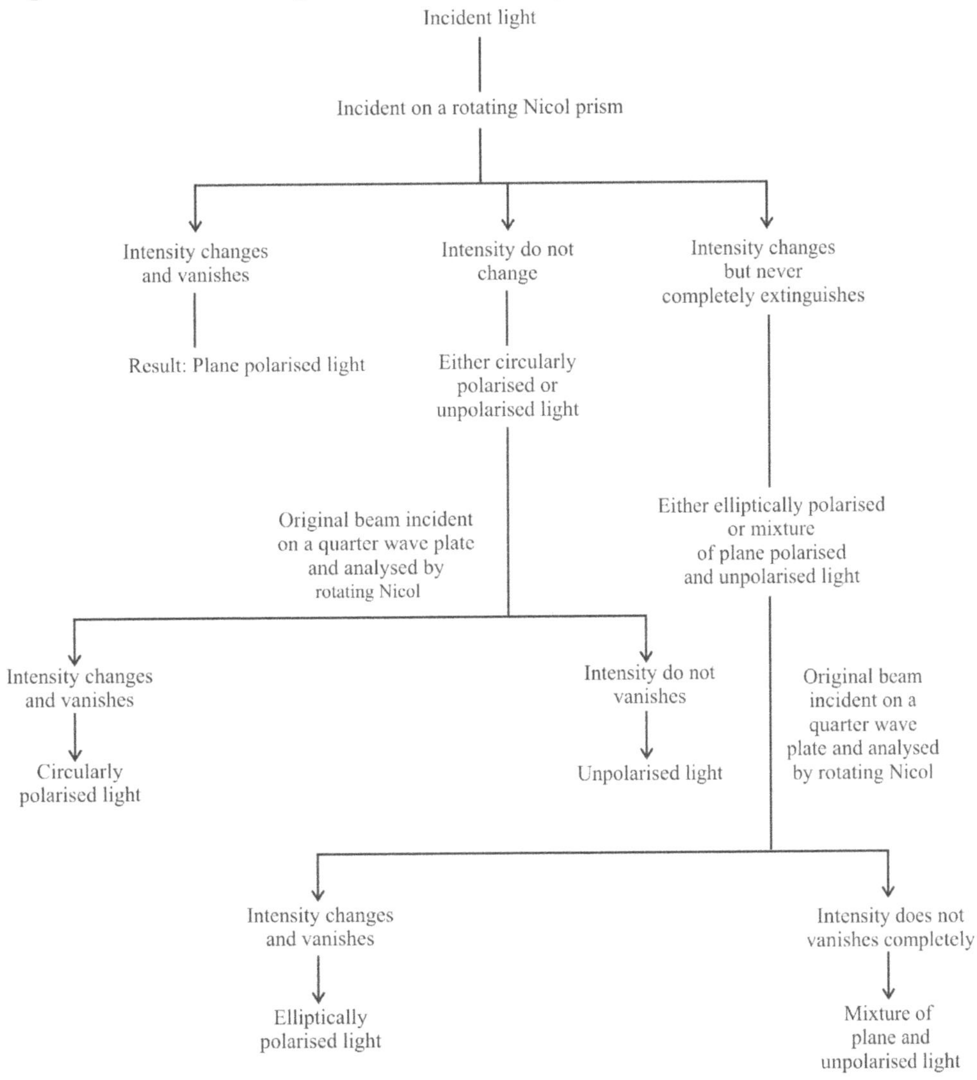

3A.19 Distinction Between QWP, HWP and Plane Glass Plate (GP)

The procedure is as follows:

1. Two Nicols N_1 and N_2 are arranged coaxially and seeing through these N_2 is rotated till no light comes out; Nicols N_1 and N_2 are crossed.
2. Plates are inserted one by one and observed through N_2 when plate is rotated. The following may happen:
   (a) Light never comes out; If this happens then plate is glass plate which does not effect polarization.
   (b) Light does come at certain positions. Set the plate so that no light comes out.

This shall happen when optic axis of QWP or HWP coincides with incident polarization. Now rotate the plate slightly from no light position. The QWP shall convert PP into EPL and so rotating Nicol N_2 shall record $I_{max} \rightarrow I_{min} \neq 0$. But HWP will just rotate the direction of polarisation and so rotating Nicol N_2 shall register $I_{max} \rightarrow I_{min} = 0$ in one complete rotation.

Thus all the three plate are identified. The procedure is made clear in the following flow chart

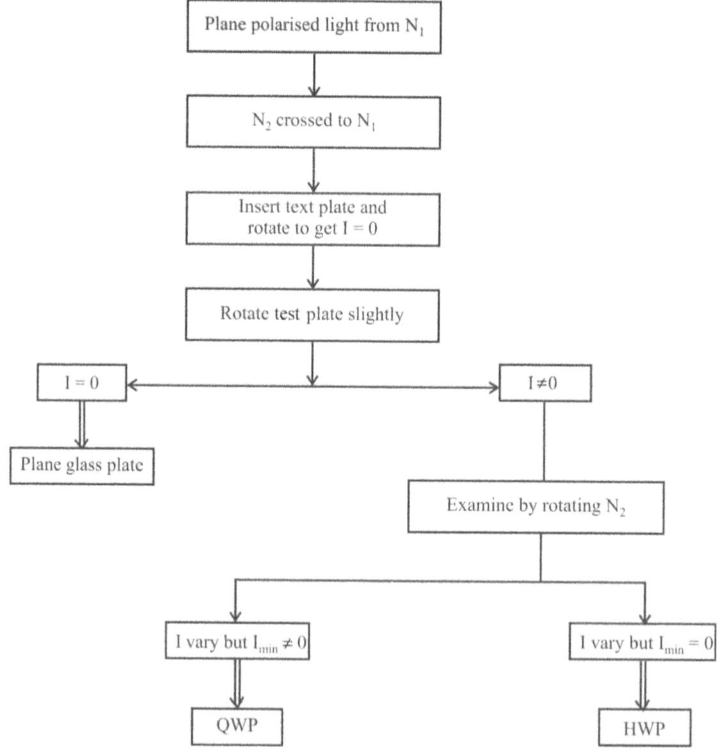

3A.20 Optical Activity

When plane-polarised light passes through certain substances, the plane of polarisation of the light is rotated about the direction of propagation of light through a certain angle. This property is known as "Optical Activity" and the substances are said to be "Optically Active". This phenomenon is called "Optical rotation".

When a light falls on Nicol N_1, then after emerging it becomes plane polarised and it remains plane polarised even after passing through the quartz plate, but the direction of vibration and, therefore, the plane of polarisation is rotated through a certain angle. This angle is equal to the angle through which the second Nicol N_2 has to be rotated to completely stop the light. This shows that quartz plate is acting as an optically-active substance (Fig. 3A.24).

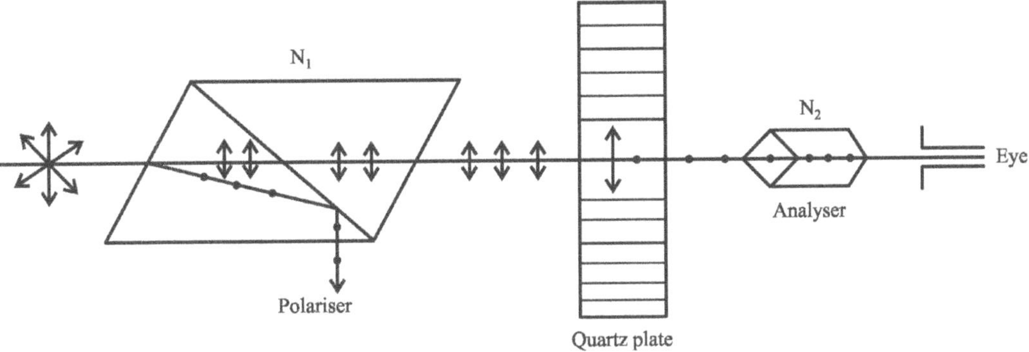

Fig. 3A.24

It has been found that calcite does not produce any change in the plane of vibration of the plane polarised light. Therefore calcite is not optically active substance. The examples of optically active substances are sugar crystals, sugar solution sodium chlorate etc.

The substance which rotate the plane of vibration to the right they are called dextro-rotary (or right handed). The substance that rotates the plane of vibration to the left are called Laevo-rotary (or left handed).

3A.21 Biot's Law for Rotary Polarisation or Optical Rotation

Experimentally, it is found that the optically active substance follow the following laws.

1. The amount of optical rotation produced by an optically active substance is proportional to path length travelled by light in the medium

$$\theta \propto l$$

2. In the case of solutions and vapours, in addition to path length, rotation depends on concentration, C or density d,

$$\theta \propto c \text{ or } d$$

3. The total optical rotation produced by mixture of a number of optically active and chemically inactive substances put together is algebraic sum of rotations produced by individual sample

$$\theta = \theta_1 + \theta_2 + \theta_3 + \ldots\ldots..$$

Dextro rotary, d-type is taken as positive and laevo rotary or l-type is taken as negative.

4. The rotation is found to decrease with temperature for a given substance.

5. The optical rotation is inversely proportional to the square of wavelength of light used

$$\theta \propto \frac{1}{\lambda^2}$$

Thus the rotation is least for red and maximum fir violet; and so, when white light passes through an optically active substance like Quartz cut parallel to optic axis, different colours undergo different rotations. This leads to dispersion called rotary dispersion.

3A.21.1 Fresnel's Theory of Optical Rotation

Fresnel's theory is based on the fact that a rectilinear vibration is equivalent to two opposite circular vibrations of same frequency.

Assumptions of Fresnel's Theory

1. When a plane polarised beam enters a substance, it is broken up into two oppositely directed circularly polarised beams, one clockwise and the other anticlockwise.

2. In an optical inactive substance these two circularly polarised beams travel with the same velocity; while in an optically active substance they travel with different velocities.

   (a) In a dextro-rotary substance the velocity of right-handed circularly polarised light is greater than that of left-handed circularly polarised light.

   (b) While in laevo rotary substance the velocity of left-handed circularly polarised light is greater than that of right-handed.

   Hence in an optically active substance a phase difference is developed between the two beams while traversing through the substance.

3. On emergence from the optically active substance, the two circularly polarised beam recombine to form a plane polarised beam whose plane of polarisation has rotated with respect to that of the incident light through a certain angle which depends on the phase difference between the two beams.

Derivation of the relation

$$\theta = \frac{\pi l}{\lambda}(\mu_L - \mu_R)$$

Let a plane polarised light beam be incident normally on a doubly refracting crystal like quarter plate with its faces perpendicular to the optic axis. Let the equation of the incident beam is given by

$$y = 2a \sin \omega t$$

$$x = 0$$

where 2a is the amplitude of the incident vibrations.

We can rewrite the above equation as

$$y = a \sin \omega t + a \sin \omega t \qquad .....(3A.17)$$

$$x = a \cos \omega t - a \cos \omega t \qquad .....(3A.18)$$

But the principle of superposition states that equations 3A.17 and 3A.18 may be taken as resultant of two vibrations x_1, x_2 and y_1, y_2; where x_1 and y_1 are the components of clockwise circular motion and x_2 and y_2 are the components of anticlockwise circular motion.

$$\left.\begin{aligned} x_1 &= a \cos \omega t \\ y_1 &= a \sin \omega t \end{aligned}\right\} \text{Clockwise}$$

$$\left.\begin{aligned} x_2 &= -a \cos \omega t \\ y_2 &= a \sin \omega t \end{aligned}\right\} \text{Anti-clockwise}$$

Since these two circular motions travelling along the optic axis have different velocities, a phase difference δ is developed between them.

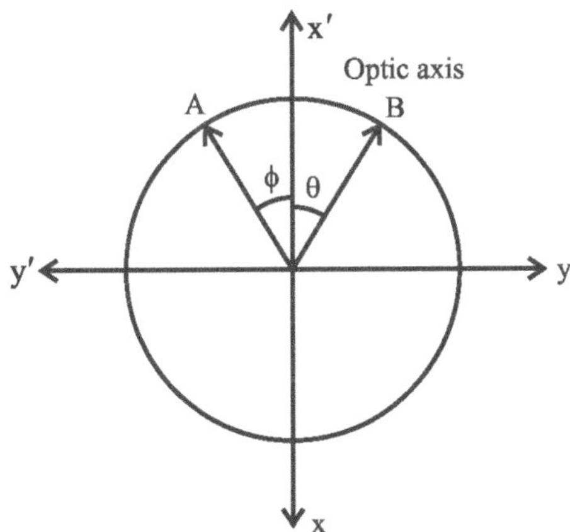

Thus we have,

$$\left.\begin{aligned} x_1 &= a\cos\omega t \\ y_1 &= a\sin\omega t \end{aligned}\right\} \text{Clockwise}$$

$$\left.\begin{aligned} x_2 &= -a\cos(\omega t+\delta) \\ y_2 &= a\sin(\omega t+\delta) \end{aligned}\right\} \text{Anti-clockwise}$$

The resultant vibrations along x and y axis may be given by

$$X = x_1 + x_2 = a\cos\omega t - a\cos(\omega t + \delta)$$

$$= 2a\sin\frac{\delta}{2}\sin\left(\omega t + \frac{\delta}{2}\right) \qquad .....(3A.19)$$

$$Y = y_1 + y_2 = a\sin\omega t - a\sin(\omega t + \delta)$$

$$= 2a\cos\frac{\delta}{2}\sin\left(\omega t + \frac{\delta}{2}\right) \qquad .....(3A.20)$$

Dividing eq. 3A.19 by equation 3A.20, we get

$$\frac{X}{Y} = \tan\frac{\delta}{2}$$

This equation is similar to a straight line equation y = mx. Therefore the resultant is a plane polarised light with vibrations inclined at an angle $\delta/2$ with the y-axis.

Let μ_R and μ_L be refractive indices for the right and left handed circularly polarised light and l is the thickness of the crystal, then the path difference will be

$$\Delta = (\mu_L \sim \mu_R)\, l$$

Corresponding phase difference, $\delta = \frac{2\pi}{\lambda}(\Delta)$

$$\delta = \frac{2\pi}{\lambda}(\mu_L \sim \mu_R)\, l$$

∴ the angle of rotation θ is given by

$$\theta = \frac{\delta}{2} = \frac{\pi l}{\lambda}(\mu_L \sim \mu_R)$$

$$\theta = \frac{\pi l}{\lambda}(\mu_L \sim \mu_R)$$

Results

1. If V_L and V_R are the velocities of left-handed and right-handed circular vibration respectively, then

$$\mu_L = C/V_L \text{ and } \mu_R = \frac{C}{V_R}$$

where C is the velocity of light

In terms of velocities the angle of rotation is

$$\theta = \frac{\delta}{2} = \frac{\pi C l}{\lambda}\left(\frac{1}{V_L} \sim \frac{1}{V_R}\right)$$

2. If T is the time period, then

$$\frac{1}{T} = \frac{C}{\lambda}$$

$$\therefore \qquad \theta = \frac{\pi l}{T}\left(\frac{1}{V_L} \sim \frac{1}{V_R}\right)$$

3. In case of left handed substances $V_L > V_R$

$$\theta_L = \frac{\pi l}{\lambda}(\mu_R - \mu_L) = \frac{\pi l}{T}\left(\frac{1}{V_R} - \frac{1}{V_L}\right)$$

4. In case of right handed substances $V_R > V_L$

$$\theta_R = \frac{\pi l}{\lambda}(\mu_L - \mu_R)$$

$$= \frac{\pi l}{T}\left(\frac{1}{V_L} - \frac{1}{V_R}\right)$$

5. In case of optically inactive substances, like calcite

$$V_R = V_L$$

$\therefore$ $\theta = 0$, that is, the direction of vibrations remain unaffected. Thus, calcite does not exhibit the phenomenon of optical rotation.

3A.22 Specific Rotation

The angle through which the plane of polarisation is rotated depends upon:

1. the thickness of the substance
2. concentration of the solution
3. wavelength of the light,
4. temperature

"Specific rotation for a given wavelength of light at a given temperature is defined as the rotation produced by one decimetre length of the solution containing one gram of optically active substance per c.c of solution".

If θ is the rotation produced by l decimetre length of a solution and C is the concentration in grams per c.c then the specific rotation S at a given temperature for given wavelength of light is given by

$$S = \frac{\theta}{l \times c}$$

$$S = \frac{\text{Rotation (degree)}}{\text{Length (decimeter)} \times \text{Concentration (gm/cc)}}$$

Generally, specific rotation is expressed in $(dm)^{-1}$ $(gm/cc)^{-1}$

$$S = \frac{10\theta}{l\,c} \quad \text{if } l \text{ is in cm}$$

3A.23 Polarimeter

Polarimeter is an optical instrument which is used to measure the angle through which the plane of polarisation is rotated by the optically active substance. When the instrument is used to determine the quantity of sugar in a solution. It is known as saccharimeter.

There are basically three types of polarimeters:

1. Laurent's half shade polarimeter
2. Biquartz polarimeter
3. Lippich polarimeter

3A.23.1 Laurent's Half shade Polarimeter

It is an instrument used for finding the optical rotation of certain solutions. When used for finding the optical rotation of sugar it is called a saccharimeter. If the specific rotation of sugar is known, the concentration of the sugar solution can be determined.

Construction: The optical parts of a polarimeter are shown in Fig. 3A.25. Monochromatic light from a source, usually a sodium lamp, after passing through a narrow slit s is rendered in to a parallel beam by the lens L. This light is rendered plane polarised by the Nicol P and after passing through the half-shade device H (called laurent's plate), a glass tube T containing the solution is made to fall on the analysing Nicol A. The light is viewed through a telescope E. The analysing Nicol A can be rotated about the axis of the tube and its rotation can be measured with the help of a vernier scale on the graduated circular scale C divided in degrees.

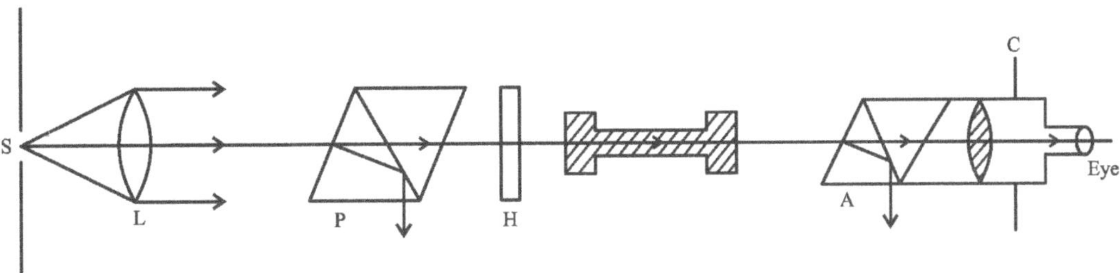

Fig. 3A.25

Action of half-shade device

It consists of a semi-circular plate ADB of glass cemented to a semi-circular plate ACB of quartz. The quartz plate is cut with its optic axis parallel to the line of separation AOB. The thickness of the quartz plate is such that it introduces a phase difference of π between the O and the E-vibrations. In other words, it is a half-wave plate. The thickness of the glass plate is such that it absorbs the same amount of light as is done by the quartz half wave plate.

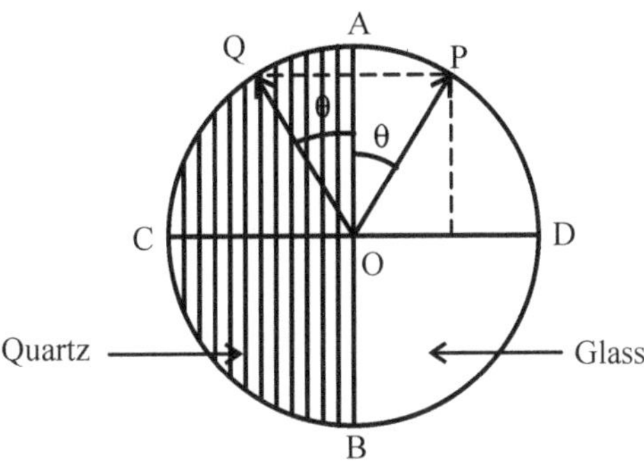

Fig. 3A.26

Suppose light after passing through the polariser P is incident normally on the half-shade plate and has vibrations along OP. On passing through the glass half the vibrations will remain along OP, but on passing through the quartz half of these will be split up into E and O components. The vibrations of the O-component are along OD and those of E-component along OA (Fig. 3A.26).

On passing through the quartz plate a phase difference of π is introduced between these two vibrations. The O-vibrations will advance in phase by π and will occur along OC instead of OD on emergence. The resultant vibration on emerging from the quartz plate will, therefore, be along OQ, such that

$$\angle POA = \angle QOA$$

If the analysing Nicol is fixed with its principal plane parallel to OP, the plane polarised light through glass half will pass and hence it will appear brighter than the quartz half from which light is partially obstructed.

If the principal plane of the Nicol is parallel to OQ the Quartz half will appear brighter than the glass half due to the same reason.

When the principal plane of the analysing Nicol is parallel to AOB, the two halves will appear equally bright. It is because the vibrations emerging out of the two halves are equally inclined to its principal plane and hence two components will have equal intensity.

When the principal plane of the analyser is at right angle to AOB again the components of OP and OQ are equal. The two halves are again equally illuminated, but as the intensity of the components passing through is small as compared to that in the previous case, the two halves are said to be equally dark.

The eye can easily detect a slight change when the two halves are equally dark. The reading are, therefore, taken for this purpose.

3A.23.2 Merits and Demerits

1. The instrument is suitable only for one wavelength for which the path difference between O-ray and E-ray is $\lambda/2$. It is usually constructed for sodium source.
2. When position is adjusted for equally dark halves, it gives fairly accurate observation as slight rotation of the analyser changes the intensity of two halves.

3A.23.3 Biquartz Polarimeter

It is an instrument used for finding the optical rotation of certain solutions. The arrangement of bi-quartz polarimeter is same as that of a Laurent's half shade polarimeter with the difference that it consists of a biquartz plate in place of Laurent's half-shade plate and, this is used for white light instead of monochromatic light.

Construction: The optical parts of bi-quartz polarimeter are shown in Fig. 3A.27.

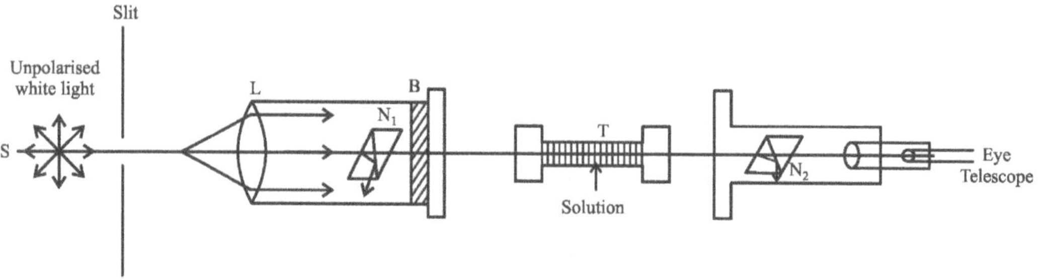

Fig. 3A.27

According to the figure, the light from a white source s which is just an ordinary bulb after passing through a slit is incident on a convex lens L which converts the emergent light in to a parallel beam. This parallel beam of light falls on a Nicol N_1 (polariser) which makes this light plane polarised when it comes out of it.

This plane polarised light coming out of polariser N_1 to be incident on a solution of an optically active solution filled in tube T.

The light coming out of T is made to fall on another nicol N_2 used as analyser and finally the emergent light from N_2 is received by a Telescope which is focused on the bi-quartz plate B, which is arranged between N_1 and T. The analyser N_2 can be rotated about a horizontal axis and its rotation can be recorded by a circular scale.

Action of Bi-quartz Plate B

A bi-quartz plate essentially is a combination of two semicircular plates out of which one is left handed quartz and other of right handed quartz both of which are cut perpendicular to their optic axes and cemented along a diameter (xy) as shown in Fig. 3A.28.

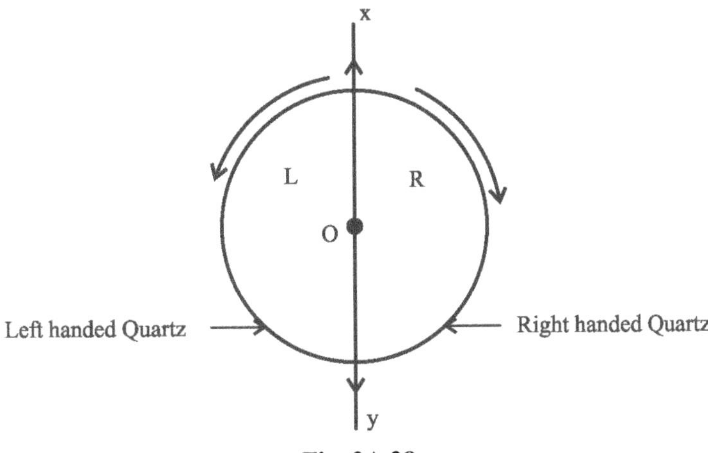

Fig. 3A.28

Since the optic axes of both semi circular plates are perpendicular to their planes, hence double refraction does not take place in the quartz plate. The thickness of each plate is nearly 3.75 mm which is so much that each plate can rotate the plane of polarisation of yellow light through 90°. One rotation is clockwise and other is anticlockwise.

When a beam of white light is incident on Nicol N_1, it emerges in a parallel direction and falls on the bi-quartz plate B normally and it travels in the direction of optic axis. In the plate the rotatary dispersion of every colour takes place in opposite direction

$$\because \qquad \theta \propto \frac{1}{\lambda^2}$$

If now the vibrations in incident light are in xy direction Fig. 3A.29(a) a, then the emergent light from the bi-quartz plate for yellow colour the vibrations will be perpendicular to xy because the vibrations are rotated through 90° by bi-quartz whereas the red light is rotated through the smallest amount and that of violet light is rotated through the largest angle.

If the principal section of N_2 is placed parallel to XOY, the two yellow vibrations ceases and both the halves appear to the similar and has greyish violet tint, called the sensitive tint or tint of passage. Fig. 3A.29(b). This is due to mixture of red and blue.

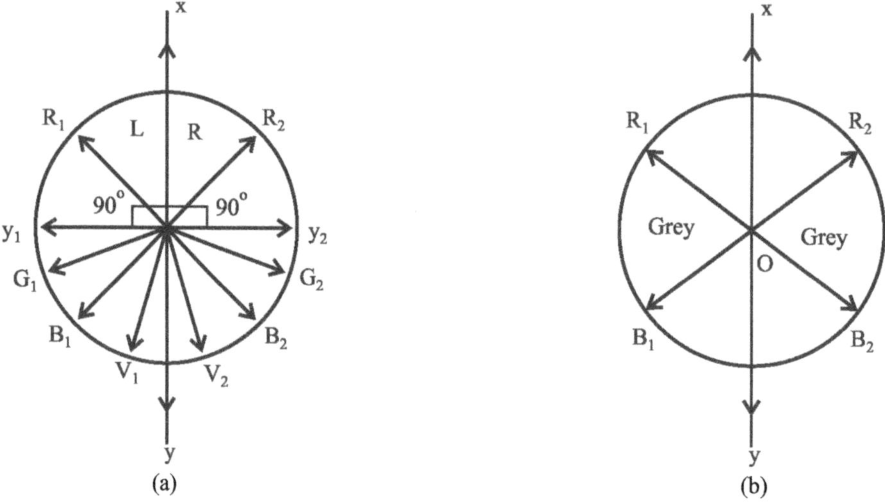

Fig. 3A.29

Now, if the principal section of N_2 is rotated through a small angle from this position as shown R_1B_2 in Fig. 3A.30(a), then right half appears more bluish and left more red. If the principal section of Nicol N_2 is rotated in opposite direction R_2B_1 Fig. 3A.30(b) then we see an observation just opposite to the above. The position of the sensitive tint is very sensitive and is used for accurate determination of optical rotation.

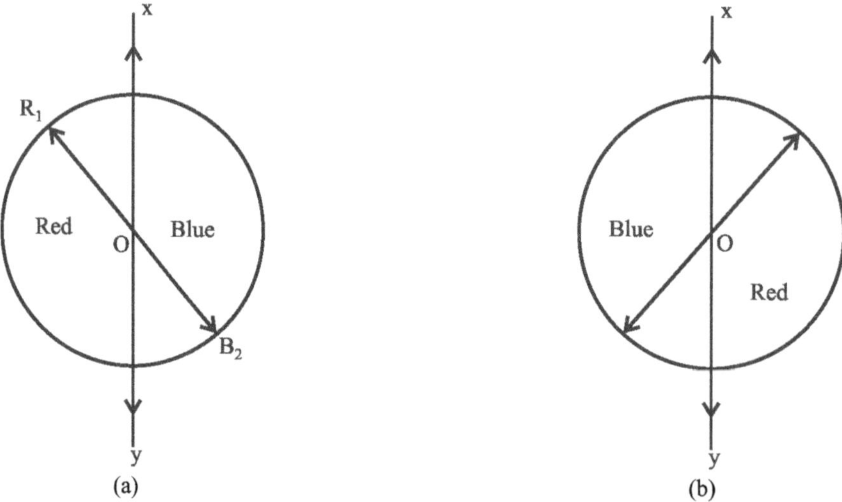

Fig. 3A.30

3A.23.4 Merits and Demerits

(a) *Merits*

1. As the transition from red to blue is very rapid, the zero position can be obtained very accurately.
2. For this instrument, white light can also be used.
3. It is highly sensitive device for measuring optical rotation.

(b) *Demerits*

1. This instrument does not give accurate result for colourless optically active substance.
2. It is not possible for a colour blind person to use this instrument.

Solved Numerical Problems

On Brewster's law

1. The refractive indices of glass and water are 1.54 and 1.33 which will be greater, the polarising angle for a beam incident from water to glass or that for a beam incident from glass to water.

Sol. The Brewster's law is given by

$$\mu = \tan i_p$$

For water to glass

$$\mu = \frac{1.54}{1.33}$$

$$\mu = 1.158$$

$$i_p = \tan^{-1}(1.158)$$

$$i_p = 49.2^\circ$$

For glass to water

$$\mu = \frac{1.33}{1.54} = 0.864$$

$$i_p = \tan^{-1}(0.864) = 40.8^\circ$$

Thus i_p is larger from water to glass.

2. The polarising angle of a piece of glass for green light is 60°. What is the angle of minimum deviation for a 60° prism made of the same glass?

Sol. Using Brewster's law

$$\mu = \tan i_p$$

$$\mu = \tan 60^\circ$$

$$\mu = 1.73.2$$

We know that

$$\mu = \frac{\sin\left(\frac{A+\delta_m}{2}\right)}{\sin\left(\frac{A}{2}\right)}$$

Here $\mu = 1.732,\ A = 60^\circ$ and $\delta_m = ?$

$$1.732 = \frac{\sin\left(\frac{60^\circ + \delta m}{2}\right)}{\sin\frac{60^\circ}{2}}$$

$$1.732 = \frac{\sin\left(\frac{60^\circ + \delta m}{2}\right)}{0.5}$$

$\therefore$

$$\sin\left(\frac{60^\circ + \delta m}{2}\right) = 0.5 \times 1.732 = 0.86605$$

$$\frac{60^\circ + \delta m}{2} = \sin^{-1}(0.86605) = 60^\circ$$

or

$$60^\circ + \delta_m = 120^\circ$$

$$\delta_m = 60^\circ$$

On Malus Law

3. Two polarising sheets have their polarising directions parallel so that the intensity of the transmitted light is a maximum. Through what angle must either sheet be turned so that the intensity becomes one half of the initial value?

Sol. From Malus law, we have

$$I = I_o \cos^2\theta$$

Given that

$$I = I_o/2$$

$$\frac{I_o}{2} = I_o \cos^2\theta$$

$$\cos^2\theta = \frac{1}{2}$$

$$\cos\theta = \pm\frac{1}{\sqrt{2}}$$

$$\theta = \cos^{-1}\left(\pm\frac{1}{\sqrt{2}}\right)$$

$$\theta = \pm 45^\circ,\ \pm 135^\circ$$

4. Discuss the state of polarisations of the following light waves

1. $E = \hat{j}A\cos(kx - wt) + \hat{k}\,2A\cos(kx - wt + \pi/4)$

2. $E = \hat{j}A\cos(kx - wt) + \hat{k}\,A\sin(kx - wt)$

Sol. 1. The two transverse components are

$$E_y = A\cos(kx - wt)$$

$$E_z = 2A\cos(kx - wt + \pi/4)$$

The components have different amplitudes and a phase difference of $\pi/4$ hence light is elliptically polarised.

At $t = 0$, and $x = 0$, $E_y = A$ and $E_z = \sqrt{2}\,A$, as t increases, cos wt decreases and so Ey decreases and E_z increases i.e., electric vector rotates from y to z i.e., left handed polarisation.

2. The two transverse components are

$$E_y = A\cos(kx - wt)$$

$$= A\sin(kx - wt + \pi/2)$$

$$E_z = A\sin(kx - wt)$$

The two components have equal amplitudes and a phase difference of $\pi/2$ hence light is circularly polarized.

Next, at $t = 0$ and $x = 0$, $E_y = A$ and $E_z = 0$, as time increases, E_y decreases and E_z increases in negative direction hence it is right handed circular polarization.

5. A partially polarized beam is passed through a nicol. In 90° rotation of Nicol, the intensity changes by 75%. Calculate the degree of polarisation

Sol. Let I_o be the maximum intensity, then the minimum intensity will be

$$I_{min} = I_o - \frac{75}{100}\,I_o$$

$$= 0.25\,I_o$$

$$\text{Degree of polarisation} = \frac{I_{max} - I_{min}}{I_{max} + I_{min}}$$

$$= \frac{I_o - 0.25\ I_o}{I_o + 0.25\ I_o}$$

$$= \frac{0.75}{1.25} = 0.6 \text{ or } 60\%$$

6. Two polarizers are orthogonally placed giving extinction. A third polarizer is placed in between them with its transmission axis making an angle θ with the first polarizer. Calculate the intensity of transmitted light if incident intensity of light is I_o.

Sol. The intensity of light transmitted by the first polarizer is

$$\left(\frac{I_o}{2}\right)\cos^2\theta$$

The intensity transmitted by the second polarizer is

$$I_t = \left(\frac{I_o}{2}\right)\cos^2\theta\cos^2(90 - \theta)$$

$$I_t = \left(\frac{I_o}{2}\right)\cos^2\theta\sin^2\theta$$

$$I_t = \frac{I_o}{2}\sin^2 2\theta$$

7. Light from qausi monochromatic sodium lamp of intensity I_o falls on two polaroids oriented at an angle of 30°. Find the intensity of light passing out of the polarizers.

Sol. When light beam of intensity I_o falls on the first polaroid; it is plane polarized and its intensity reduces to $\frac{I_o}{2}$. Therefore from Malus law

$$I = \left(\frac{I_o}{2}\right)\cos^2 30^\circ$$

$$I = \frac{I_o}{2} \cdot \frac{3}{4}$$

$$I = \frac{3}{8} I_o$$

8. Two polarising sheets are so arranged that the amount of light transmitted through them is maximum. What will be the percentage reduction in the intensity of the incident light when the analyser is rotated through 30°?

Sol. From the Malus law,

$$I = I_o \cos^2 \theta$$

$$\frac{I}{I_o} = \cos^2 \theta$$

∴ Percentage reduction of intensity

$$= \left(\frac{I_o - I}{I_o}\right) \times 100$$

$$= \left(1 - \frac{I}{I_o}\right) \times 100$$

$$= (1 - \cos^2 \theta) \times 100$$

Given that $\theta = 30°$

∴ $$\cos \theta = \frac{\sqrt{3}}{2} = \frac{1.732}{2} = 0.866$$

∴ % reduction of intensity

$$= [1 - (0.866)^2] \times 100$$

$$= 25\%$$

On Retardation Plates

9. Calculate the thickness of a quarter wave plate of quartz for sodium light of wavelength 5893A°. The ordinary and extraordinary refractive indices for sodium are 1.54425 and 1.55336 respectively. [UPTU]

Sol. The thickness of a quarter wave plate for positive crystal like quartz is

$$t = \frac{\lambda}{4(\mu_E - \mu_o)}$$

Given that

$$\mu_o = 1.54425$$

$$\mu_E = 1.55336$$

and for sodium light

$$\lambda = 5893\ A^{o} = 5.893 \times 10^{-5}\ cm$$

$$t = \frac{5.893 \times 10^{-5}\ cm}{4\,(1.55336 - 1.54425)}$$

$$t = \frac{5.893 \times 10^{-5}}{4 \times 911 \times 10^{-5}} = 1.62 \times 10^{-3}\ cm$$

10. Calculate the thickness of a doubly refracting plate capable of producing a path difference of $\lambda/4$ between ordinary and extraordinary rays with light of wavelength 5890 $\overset{o}{A}$. The refractive indices for ordinary and extraordinary rays are 1.53 and 1.54 respectively. [UPTU]

Sol. The thickness of a doubly refracting plate capable of producing a path difference of $\lambda/4$ or of positive quarter wave plate is given by

$$t = \frac{\lambda}{4(\mu_e - \mu_o)}$$

Give that

$\lambda = 5890\ \overset{o}{A} = 5890 \times 10^{-10}$ m,

$\mu_e = 1.54$ and

$\mu_o = 1.53$

$$\therefore \quad t = \frac{5890 \times 10^{-10}}{4(1.54 - 1.53)} = \frac{5890 \times 10^{-10}}{4 \times 0.1} = 1.47 \times 10^{-5}\ m$$

11. Find the thickness of a quarter wave plate for the wavelength of 589 nm and $\mu_o = 1.55$, $\mu_e = 1.54$. [UPTU]

Sol. The thickness of a quarter wave plate for negative crystal is

$$t = \frac{\lambda}{4(\mu_o - \mu_e)}$$

Given

$\lambda = 589$ nm; $\mu_o = 1.55$; and $\mu_e = 1.54$

$$t = \frac{589\ nm}{4 \times (1.55 - 1.54)}$$

$$t = \frac{589\ \text{nm}}{4 \times 0.1}$$

$$t = 147.25 \times 10^2\ \text{nm}$$

12. The value of μ_E and μ_O for quartz are 1.5508 and 1.5418 respectively. Calculate the phase retardation for $\lambda = 5000\ \overset{o}{A}$ when the plate thickness is 0.032 mm. [UPTU]

Sol. Foe a quartz crystal of the thickness t, the path difference between the O and E rays is given by

$$\text{Path difference } (\Delta) = (\mu_E - \mu_O)t$$

$$\text{The phase retardation} = \frac{2\pi}{\lambda}(\Delta)$$

$$= \frac{2\pi}{\lambda}(\mu_E - \mu_O)t$$

Given that

$\mu_E = 1.5508$

$\mu_o = 1.5418$

$t = 0.032\ \text{mm} = 0.0032\ \text{cm}$

and $\lambda = 5000\ \overset{o}{A} = 5000 \times 10^{-8}$ cm

$$\therefore \quad \text{Phase retardation} = \frac{2 \times 3.14}{5000 \times 10^{-8}}(1.5508 - 1.5418) \times 0.0032$$

$$\therefore \quad \text{Phase retardation} = 3.617\ \text{rad}$$

13. Calculate the thickness of doubly refracting crystal required to introduce a path difference of $\lambda/2$ between the O and E rays when $\lambda = 6000\ \overset{o}{A}$, $\mu_o = 1.65$ and $\mu_E = 1.48$.

Sol. The required thickness of a doubly refracting crystal to introduce a path difference of $\lambda/2$ is given by

$$t = \frac{\lambda}{2(\mu_o - \mu_E)}$$

Given that

$$\mu_o = 1.65;\ \mu_E = 1.48;\ \text{and } \lambda = 6000\ \mathring{A} = 6000 \times 10^{-8}\ \text{cm}$$

$$\therefore \qquad t = \frac{6000 \times 10^{-8}}{2(1.65 - 148)} = 0.000176\ \text{cm}$$

$$= 1.76 \times 10^{-4}\ \text{cm}$$

14. The values of μ_E and μ_o for quartz are 1.5508 and 1.5418 respectively. Calculate the phase retardation for $\lambda = 5 \times 10^{-5}$ cm where the plate thickness is 0.035 mm.

Sol. The path difference between E-ray and O-ray is given by

$$\text{Path difference } (\Delta) = (\mu_E - \mu_O)t$$

$$\therefore \qquad \text{The path difference} = \frac{2\pi}{\lambda} \times (\text{path difference})$$

$$= \frac{2\pi}{\lambda} \times (\mu_E - \mu_O)t$$

$$\therefore \qquad \text{Phase difference} = \frac{2 \times \pi \times (1.5508 - 1.5418) \times 0.0035}{5 \times 10^{-5}}$$

$$\therefore \qquad \text{Phase difference} = 1.26\ \pi \text{ radians}$$

15. A quartz plate 0.850 mm thick is illuminated at normal incidence by green light of wavelength 5461 $\mathring{A}$. The optic axis is parallel to the surface. Find

(a) The optical paths of the two rays in travelling through the plate and

(b) The phase difference between these two

(Given that $\mu_E = 1.555$ and $\mu_o = 1.546$)

Sol. (a) The optical paths for

(i) E-ray $\mu_E\, t = 1.555 \times 0.850$ mm

$= 1.32175$ mm

(ii) For O-ray $\mu_o\, t = 1.546 \times 0.850$ nm

$= 1.3141$ mm

(b) $$\text{Phase difference} = \frac{2\pi}{\lambda} \times (\text{path difference})$$

$$= \frac{2\pi}{\lambda} \times (1.32175 - 1.3141)$$

$$= \frac{2\pi}{\lambda} \times (0.00765 \text{ mm})$$

$$= \frac{2\pi \times 0.00765 \times 10^{-3}}{5461 \times 10^{-10}}$$

Phase difference = 87.91 radian

16. A phase retardation plate of quartz has thickness 0.1436 mm. For what wavelength in visible region ($\lambda = 4500\,\text{Å}$ to $8000\,\text{Å}$) will it act as

(i) a quarter wave plate (ii) a half wave plate

Sol. (i) For quarter wave plate

$$(\mu_e - \mu_o)\, t = \frac{\lambda}{4} \text{ or any odd multiple of } \lambda/4$$

or $$\lambda = \frac{4t\,(\mu_e - \mu_o)}{(2n+1)}$$

where $n = 0, 1, 2, 3, \ldots\ldots$

Given $t = 0.1436 \times 10^{-3}$ n

$$(\mu_e - \mu_o) = (1.5533 - 1.5443)$$

$$= 0.009$$

$\therefore$ $$\lambda = \frac{4 \times 0.1436 \times 10^{-3} \times 0.009}{(2n+1)}$$

$$\lambda = \frac{51696\,\text{Å}}{(2n+1)}$$

$\therefore$ $$\lambda_{visible} = 7385\,\text{Å},\ 5774\,\text{Å},\ 4700\,\text{Å}$$

(ii) For Half Wave Plate

$$(\mu_e - \mu_o)\ t = \left(n + \frac{1}{2}\right)\lambda$$

$\therefore$ $$\lambda = \frac{2t\,(\mu_e - \mu_o)}{2n+1}$$

$$\lambda = \frac{2 \times 0.1436 \times 10^{-3} \times 0.009}{(2n+1)}$$

$$\lambda = 25848\ \mathring{A},\ 8616\ \mathring{A},\ 5170\ \mathring{A},\ 3693\ \mathring{A}$$

$$\lambda_{visible} = 5170\ \mathring{A}$$

17. The thickness of a calcite plate to produce polarised light is 8.56×10^{-7} m. The principal refractive indices are $\mu_o = 1.658$ and $\mu_E = 1.486$ at the wavelength of light used as $5890\ \mathring{A}$. Calculate the types of retardation plate.

Sol. The path difference is given by

$$(\mu_o - \mu_E)\ t = \frac{\lambda}{x}$$

$$(1.658 - 1.486)\ t = \frac{5890 \times 10^{-10}\ m}{x}$$

$$(0.172 \times 8.56 \times 10^{-7}\ m)\ t = \frac{5890 \times 10^{-10}\ m}{x}$$

$$x = \frac{5890 \times 10^{-10} m}{(0.172)\ (8.56 \times 10^{-7} m)}$$

$$x = 4$$

Thus it is a quarter wave plate.

18. A left circularly polarised beam ($\lambda = 5893\ \mathring{A}$) is incident normally on a calcite crystal (with its optic axis cut parallel to the surface) of thickness 0.005441 mm. What will be the state of polarisation of the emergent beam? Given that $\mu_o = 1.65836$ and $\mu_E = 1.48641$.

Sol. The phase difference

$$\delta = \frac{2\pi}{\lambda} \times (\mu_o - \mu_E)t$$

$$\delta = \frac{2\pi\ (1.65836 - 1.48641) \times 0.005141}{5893 \times 10^{-7}}$$

$$\delta = 3\pi$$

The equation of polarised beams are

$$x = a \cos \omega t$$

$$y = b \sin (\omega t - \delta)$$

Since light is circularly polarised

$$\therefore \qquad a = b$$

$$\therefore \quad y = a \sin(\omega t - 3\pi)$$

$$y = -a \sin \omega t$$

which represents a right circularly polarised beam.

On Optical Rotation

19. A tube of sugar solution 20 cm long is placed between crossed Nicols and illuminated with light of wavelength 6×10^{-5} cm. If the optical rotation produced is 13° and the specific rotation is 65°/dm/gm/cm^3, determine the strength of the solution.

Sol. For specific rotation we have

$$S = \frac{\theta}{l \times C}$$

$$C = \frac{\theta}{l \times S}$$

Given

$$\theta = 13^\circ$$

$$l = 2.0 \text{ dm}$$

$$S = 65^\circ \, (\text{dm})^{-1} \, (\text{gm/CC})^{-1}$$

$$C = 0.1 \text{ gm/CC} = 10\%$$

20. A certain length of 5% solution causes the optical rotation of 20°. How much length of 10% solution of the same substance will cause 35° rotation.

Sol. We have

$$S = \frac{\theta_1}{l_1 \times C_1} = \frac{\theta_2}{l_2 \times C_2}$$

$$\therefore \quad \frac{20^\circ}{l_1 \times 5\%} = \frac{\theta_2}{l_2 \times 10\%}$$

Given $\theta_1 = 20^\circ$; $\theta_2 = 35^\circ$

$$l_2 \times 10 \times 20 = 35 \times l_1 \times 5$$

$$l_2 = \frac{5 \times 35 \times l_1}{10 \times 20}$$

$$l_2 = \frac{7}{8} l_1$$

21. The specific rotation of quartz at 5086 $\overset{\circ}{A}$ is 29.73 deg/mm. Calculate the difference in the refractive indices.

Sol. The optical rotation θ is given by

$$\theta = \frac{\pi l}{\lambda}(\mu_L - \mu_R)$$

Thus,
$$\mu_L - \mu_R = \frac{\lambda}{\pi}\frac{\theta}{l}$$

The specific rotation,

$$\frac{\theta}{l} = 29.73 \text{ deg/mm}$$

and

$$\lambda = 5086\,\text{Å} = 5.086 \times 10^{-4} \text{ mm}$$

$$\therefore \quad \mu_L - \mu_R = \frac{5.086 \times 10^{-4}\text{mm}}{180^\circ} \times \frac{29.73^\circ}{\text{mm}}$$

$$\therefore \quad \mu_L - \mu_R = 8.4 \times 10^{-5}$$

22. The refractive indices of quartz for the right handed and left handed circularly polarised light of wavelength 7620 Å are 1.53914 and 1.53920 respectively. Calculate the angle of rotation produced by its plate of thickness of 0.5 mm.

Sol. Given

$$l = 0.5 \text{ mm} = 5 \times 10^{-4} \text{ m}$$

$$\mu_L = 1.53920$$

$$\mu_R = 1.53914$$

$$\lambda = 7620\,\text{Å} = 7.62 \times 10^{-7} \text{ m}$$

The angle of rotation is given by

$$\theta = \frac{\pi l}{\lambda}(\mu_L - \mu_R)$$

$$\theta = \frac{3.14 \times 5 \times 10^{-4}}{7.62 \times 10^{-7}}(1.53920\text{–}1.53914)$$

$$\theta = 0.1236 \text{ radian}$$

or
$$\theta = \left(0.1236 \times \frac{180}{\pi}\right)^\circ$$

$$\theta = 7.1^\circ$$

23. 80 gm of impure sugar when dissolved in a litre of water gives an optical rotation of 9.9°, when placed in a tube of length 20 cm. If the specific rotation of sugar is 66°, find the percentage purity of the sugar sample. [UPTU]

Sol. The rotation is given by

$$S = \frac{\theta}{l \times C}$$

∴ The strength of the solution is given by

$$C = \frac{\theta}{l \times S}$$

$$C = \frac{9.9^{o}}{2.0 \times 66^{o}} = 0.075 \text{ gm/cc} = 75 \text{ gm/litre}$$

The sugar sample dissolved in a litre of water is 80 gm in which 75 gm is pure sugar. Therefore purity is $\frac{75}{80} \times 100 = 93.75\%$.

24. A 20 cm long tube containing 48 cm^3 of sugar solution rotates the planes of polarisation by 11°. If the specific rotation of sugar is 66°, calculate the mass of sugar in the solution. [UPTU]

Sol. The specific rotation, S of a sugar is given by

$$S = \frac{\theta}{l \times C},$$

where θ is the rotation, C is the concentration in gm/cm^3 and l the length in decimetres.

Hence $S = 66^{o}$, $l = 20$ cm $= 2.0$ decimeter

$$\theta = 11^{o}$$

$$C = \frac{\theta}{l \times S} = \frac{11}{2 \times 66} = \frac{1}{12} \text{ gm/cm}^3$$

1 cm^3 of sugar solution contains $= \frac{1}{12}$ gm of sugar

∴ 48 cm^3 of sugar solution contains $= \frac{1}{12} \times 48 = 4$ gm

Thus the mass of sugar in solution is 4 gm.

25. Calculate the specific rotation, which rotates the plane of polarisation 15.2° in 20% sugar solution of 25 cm length. [UPTU]

Sol. The specific rotation S of a sugar solution of concentration C is given by

$$S = \frac{\theta}{l \times C}$$

Here

$$\theta = 15.2^\circ$$

$$l = 25 \text{ cm} = 2.5 \text{ decimeter}$$

$$C = 20\% = \frac{20}{100} = 0.2 \text{ gm/cc}$$

$$S = \frac{15.2}{2.5 \times 0.2} = 30.4^\circ \text{ (dm)}^{-1} \text{ (gm/CC)}^{-1}$$

26. Determine the specific rotation of the given sample of sugar solution if the plane of polarisation is turned through 13.2°. The length of the tube containing 10% sugar solution is 20 cm. [UPTU]

Sol. The specific rotation S of a sugar solution of concentration C is given by

$$S = \frac{\theta}{l \times C}$$

Here $\theta = 13.2^\circ$

$$l = 20 \text{ cm} = 2.0 \text{ decimeter}$$

$$C = 10\% = \frac{10}{100} = 0.1 \text{ gm/cc}$$

$$S = \frac{13.2}{2.0 \times 0.1}$$

$$S = 66^\circ \text{ (dm)}^{-1} \text{ (gm/cc)}^{-1}$$

27. The refractive indices of quartz for right handed and left handed circularly polarised light of wavelength $6300 \overset{\circ}{A}$ are 1.53915 and 1.53921 respectively. Calculate the angle of rotation produced by quartz plate of thickness 0.5 mm. [UPTU]

Sol. In case of right handed substance, the rotation of plane of polarisation is given by

$$\theta = \frac{\pi l}{\lambda} (\mu_L - \mu_R)$$

Given

$$l = 0.5 \text{ mm } = 0.5 \times 10^{-3} \text{ m}$$

$$\lambda = 6300 \text{ Å} = 6300 \times 10^{-10} \text{ m}$$

$$\mu_L = 1.53921$$

$$\mu_R = 1.53915$$

$$\therefore \quad \theta = \frac{3.14 \times 0.5 \times 10^{-3}}{6300 \times 10^{-10}} (1.53921 - 1.53915)$$

$$\theta = \frac{3.14 \times 0.5 \times 10^{-3} \times 0.00006}{6300 \times 10^{-10}}$$

$$\theta = 0.1495 \text{ rad}$$

$$\theta = \frac{0.1495 \times 180}{3.14} = 8.57^\circ$$

28. The specific rotation of quartz at 5086 Å is 29.73 deg/mm. Calculate the difference in the refractive indices. [UPTU]

Sol. The optical rotation is given by

$$\theta = \frac{\pi l}{\lambda} (\mu_L - \mu_R)$$

$$(\mu_L - \mu_R) = \frac{\theta \lambda}{\pi l}$$

Given $\quad \frac{\theta}{l} = 29.73^\circ/\text{mm}$

$$\lambda = 5086 \text{ Å} = 5.086 \times 10^{-4} \text{ mm}; \quad \text{since } 180^\circ = \pi \text{ radian}$$

$$(\mu_L - \mu_R) = \frac{5.086 \times 10^{-4} \times 29.73^\circ}{180^\circ}$$

$$(\mu_L - \mu_R) = 8.4 \times 10^{-5}$$

Conceptual Questions and Answers

1. Sound waves cannot be polarised? Why?

Ans. It is because, sound waves are longitudinal in nature.

2. Will ultrasonic waves show any polarisation? Give reason for your answer.

Ans. No, it is because, ultrasonic waves are longitudinal in nature.

3. What is plane polarised light.

Ans. The light in which the vibrations of the light (vibrations of electric field vector) are restricted in a particular plane is called plane polarised light.

4. Which plane is defined as the plane of polarisation in a plane polarised electromagnetic wave.

Ans. The plane perpendicular to the plane of vibrations in a plane polarised light is called plane of polarisation.

5. When light is polarised by reflection, what is the plane of vibration of the electric field vector in polarised light.

Ans. The electric field vector is perpendicular to the plane of incidence.

6. What is a Polaroid.

Ans. A polaroid is a thin and large sheet of a crystalline polarising material capable of producing a plane polarised beam of light of large cross-section.

7. How do sunglass reduce the glare of intense light ? Explain the physical principle involved.

Ans. The sunglasses are made of polaroids, when ordinary light passes through sunglasses, the ordinary light is rendered in to plane polarised light. Due to this, the glare of the intense light, gets reduced.

8. What do you mean by the term crossed polaroids.

Ans. Two polaroids are said to be crossed, when the planes of polarisation of the two polaroids are perpendicular to each other.

9. In light waves, the vibrations are perpendicular to the direction of propagation of light. But it is said that light require no material medium for its propagation, then what vibrates transversely in light waves.

Ans. Light is electromagnetic in nature and it travels in the form of varying electric and magnetic fields. The optical phenomenon are described by the electric field vector of the electromagnetic waves. Hence, the electric field vector vibrates transversly in light waves.

10. How will you distinguish between circularly polarised and unpolarised light.

Ans. Circularly polarised light and unpolarised light has electric field same in all directions and so rotating Nicol shows uniform intensity in all positions. To distinguish these a quarter wave plate is put in front of the source and examined by rotated Nicol. Since quarter wave plate converts circularly polarised light in to plane polarised light, the rotating Nicol shall record $I_{max} \rightarrow I_{min} = 0 \rightarrow I_{max} \rightarrow I_{min} = 0$. On the other hand unpolarised light remains unaffected by the quarter wave plate and so rotating Nicol shall register uniform illumination.

11. What will be observed if white light is incident on a birefringent plate switched between crossed Nicols? Examine all possible case. Take thickness of plate to be d.

Ans. (i) Wavelength satisfying the condition

$$(\mu_o - \mu_e)\, d = n\lambda$$

where n = 0, 1, 2, 3, ….. will be stopped as Nicols are crossed

(ii) All wavelengths satisfying the condition

$$(\mu_o - \mu_e)\, d = (n + \tfrac{1}{2})\lambda$$

where n = 0, 1, 2, … shall come out parallel to Nicol N_2 hence transmitted to the maximum.

(iii) Wavelength not satisfying the condition (i) and (ii) shall emerge as elliptically polarised light

The outgoing light is coloured and it is a mixture of plane polarised and elliptically polarised light.

12. How will you distinguish elliptically polarised light from partially polarised light.

Ans. To distinguish elliptically polarised light from partially polarised light we examine light coming out of Quarter wave plate by rotating Nicol as shown in the figure below.

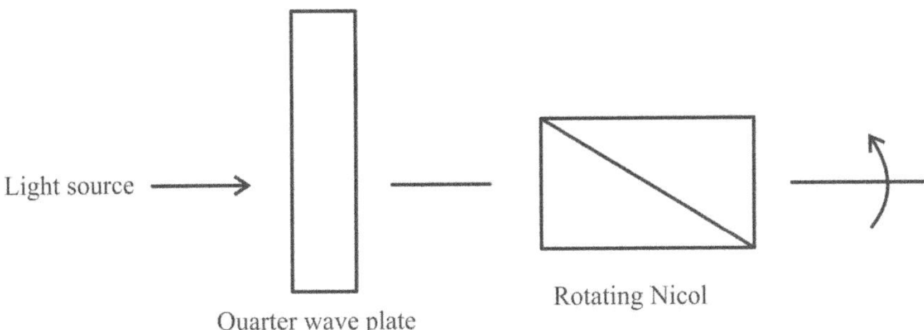

Since quarter wave plate converts elliptically polarised light in to linearly polarised light so rotating Nicol shall show $I_{max} \rightarrow I_{min} = 0 \rightarrow I_{max} \rightarrow I_{min} = 0$. On

the other hand, unpolarised light remains unaffected but plane polarised light is converted to either circularly polarised or elliptically polarised. Consequently, the rotating Nicol shall give either uniform illumination if $I_{max} \rightarrow I_{min} \neq 0 \rightarrow I_{max} \rightarrow I_{min} \neq 0$.

13. If a Quarter wave plate is inserted between crossed Nicols and then rotated slowly about light beam as axis, the light transmitted by the second Nicol will go successively through maxima and minima. How many maxima and minima will be observed in one complete rotation.

Ans. (i) Nicols are crossed and so when the quarter wave plate passes incident electric field without modification i.e., polarization is retained then we shall get minimum. This happens when incident electric field or polarization is parallel or perpendicular to optic axis i.e., incident polarisation makes an angle 0, $\pi/2$, π, $3\pi/2$. Thus in one rotation four such positions shall occur.

(ii) As the quarter wave plate is rotated from the minimum position, the light becomes elliptically polarised. As a result outgoing component increases and becomes maximum when light becomes circularly polarised. This happens when incident polarization makes the following angles with optic axis of the quarter waveplate viz, $\pi/4$, $3\pi/4$, $5\pi/4$ and $7\pi/4$.

Objective Questions

1. The light waves are transverse in nature establish beyond doubt by
   (a) Fresnel's diffraction (b) Fraunhofer diffraction
   (c) Phenomenon of polarization (d) None of these
2. The plane of polarization is that plane in which
   (a) Few vibration occur (b) Number of vibrations occurs
   (c) No vibrations occur (d) None of these
3. The plane of vibration is that plane in which
   (a) Vibrations occurs at right angles to the plane of polarization
   (b) No vibrations occur
   (c) Few vibrations occur
   (d) None of these

4. The angle of polarization is related to

(a) Brewester's law (b) Snell's law
(c) Both above are same (d) None of these

5. When a monochromatic light is passed through nicol prism, it splits in two rays, one ray through the nicol prism is

(a) Ordinary ray (b) Extra ordinary ray
(c) Both (a) and (b) (d) None of these

6. The plane polarized light can be produced by using

(a) Rochon prism (b) Wallaston prism
(c) Nicol prism (d) Some other prism

7. When a light incident on a rotating nicol prism a plane polarized light is obtained if and only if

(a) Intensity does not change
(b) Intensity changes but never completely extinguished
(c) Intensity changes and vanishes
(d) None of the above

8. When an extraordinary ray come out from the nicol prism, it can be represented as simple harmonie motion as

(a) A sin θ sin wt (b) A cos θ sin wt
(c) A sin wt (d) A cos θ sin (wt + δ)

9. The general equation which represents all the cases of the polarized light can be written as

(a) $\frac{x^2}{a^2}+\frac{y^2}{b^2}-\frac{2xy}{ab}=\sin^2\delta$ (b) $\frac{x^2}{a^2}+\frac{y^2}{b^2}-\frac{2xy}{ab}\cos\delta=\sin^2\delta$
(c) $\frac{x^2}{a^2}+\frac{y^2}{b^2}-\frac{2xy}{ab}\cos\delta=1$ (b) None of these

10. Elliptically polarized light can be produced by using

(a) Single nicol prism (b) Two crossed nicol prisms
(c) Three crossed nicol prisms (d) None of these

11. In case of circularly polarized light, the two waves vibrating at right angles to each other have

(a) Un equal amplitude (b) Equal amplitude
(c) Both (a) and (b) (d) None of these

12. The circularly polarized can be detected by using
    - (a) Nicol prism
    - (b) Rotating nicol prism
    - (c) Quarter wave plate
    - (d) Both (b) and (c)
13. When elliptically polarized light is allowed to fall on a nicol prism, then its intensity
    - (a) Is maximum
    - (b) is minimum
    - (c) Varies from a maximum to a minimum value
    - (d) None of these
14. Which is true about the detection of the plane polarized light
    - (a) The light is completely extinguished in each rotation of the nicol prism
    - (b) The light is completely extinguished twice in each rotation of the nicol prism
    - (c) The light is not completely extinquished in each rotation of the nicol prism
    - (d) The intensity of the light remains constant in each rotation of the nicol prism
15. Double refraction is a phenomenon is which
    - (a) When a ray of light is reflected by a crystal of calcite it gives two reflected rays
    - (b) When a ray of light is refracted by a crystal of calcite it gives two refracted rays
    - (c) Both (a) and (b)
    - (d) None of the above
16. Calcite is
    - (a) Other name of Iceland
    - (b) Crystallized calcium carbonate
    - (c) Crystallized sodium carbonate
    - (d) Both (a) and (b)
17. Optic axis is a
    - (a) Line
    - (b) Direction
    - (c) Both (a) and (b)
    - (d) None of these

18. The phenomenon of double refraction is absent when light is allowed to enter the crystal
    (a) Perpendicular to the optic axis
    (b) Along the optic axis
    (c) In any direction
    (d) None of these

19. The refractive index of Canada Balsam
    (a) Greater than the ordinary ray
    (b) Less than the extraordinary ray
    (c) Equal to the ordinary ray
    (d) Lies in between the ordinary and extraordinary ray

20. The negative crystals are those in which
    (a) The extraordinary wave surface lies within the ordinary wave surface
    (b) The ordinary wave surface lies within the extraordinary wave surface
    (c) There is no optic axis
    (d) None of these

21. The Quarter wave plate made of Quartz undergoing double refraction process has a thickness given by
    (a) $t = \frac{\lambda}{4(\mu_E - \mu_o)}$
    (b) $t = \frac{\lambda}{4(\mu_o - \mu_E)}$
    (c) $t = \frac{\lambda}{2(\mu_E - \mu_o)}$
    (d) None of these

22. The thickness of half wave plate made of calcite in
    (a) $t = \frac{\lambda}{4(\mu_E - \mu_o)}$
    (b) $t = \frac{\lambda}{4(\mu_o - \mu_E)}$
    (c) $t = \frac{\lambda}{2(\mu_E - \mu_o)}$
    (d) None of these

23. When an unpolarised light intensity I_o is incident on a polarizing sheet, the intensity of the light which does not get transmitted is
    (a) $\frac{1}{2} I_o$
    (b) $\frac{1}{4} I_o$
    (c) Zero
    (d) I_o

24. The specific rotation of a liquid of length 10 cm, concentration 2 g/cm^3 is 40° dm^{-1} g^{-1} cm^3. The angle of rotation is
    (a) 10°
    (b) 70°
    (c) 80°
    (d) 90°

25. The Quartz is a

(a) Negative crystal (b) Positive crystal

(c) Can act both as positive and negative (d) None of these

Theoretical Questions

1. What do you understand by plane-polarised light?
2. Light wave is transverse, but ordinary light is symmetric towards propagation direction, explain?
3. Distinguish between polarised and unpolarised light?
4. Explain 'plane of polarisation' and 'plane of vibration' for plane polarised light?
5. How would you obtain plane-polarised light by reflection?
6. State and explain Brewater's law.
7. Using Brewster's law show that when light is incident on a transparent substance at the polarising angle, the reflected and refracted rays are at right angles.
8. State and explain Malus law.
9. Give three methods for producing plane polarised light.
10. What are doubly-refracting crystals? Explain the geometry of calcite crystal.
11. Define optic axis and principle section of a crystal.
12. What do you understand by double refraction? What are ordinary and extra-ordinary rays in a uniaxial crystal?
13. What is the difference between negative and positive crystals?
14. Describe the construction and action of Nicol prism.
15. Describe the construction of a Nicol prism and show how it can be used as a polarised or as an analyser?
16. Give Huygen's theory of double refraction.
17. What is meant by plane, elliptically and circularly polarised light?
18. Discuss theorectically the super position of two linearly polarised sinusoidal light waves of the same frequency travelling the same direction when the optical vectors are

    (i) Parallel and (ii) Mutually perpendicular

19. Show that the plane-polarised and circularly polarised lights are the special cases of elliptically polarised light?

20. What is retardation plate? Draw a ray diagram for extra-ordinary and ordinary rays before and after passing through a quarter wave plate.

21. Explain how the state of polarisation of a given beam of light can be determined with the help of a Nicol prism and quarter wave plate?

22. How would you distinguish between

    (a) Unpolarised and partially polarised light

    (b) Circularly and elliptically polarised light

23. What is meant by optical rotation? Explain optical activity.

24. Discuss the phenomena of the rotation of plane of polarisation of light by optically active materials. Give the necessary theory, show that the rotation of plane of polarisation is given by

$$\theta = \frac{\pi l}{\lambda}(\mu_A - \mu_C)$$

where μ_A and μ_C are the refractive indices of the crystal in the direction of optic axis for anticlockwise and clockwise circularly polarised light respectively and l is the thickness of the crystal.

25. Explain Fresnel's theory of optical activity.

26. Define specific rotation. How it is determined experimentally?

27. Describe a polarimeter and explain how it is used to measure the strength of sugar solution?

28. Describe the construction and working of a "Laurent's half-shade polarimeter".

29. Explain the working of a biquartz polarimeter. How would you use it to find the specific rotation of an optically active substance?

Numerical Problems

1. Two Nicols A and B are placed in the path of light beam. In between these a third Nicol C is placed whose plane of transmission makes an angle of 30° with the plane of transmission of Nicol A. If the intensity of unpolarised light incident on Nicol A is 32 watt/m^2 then what is the intensity of light emerging out of Nicol B.

$[\frac{I_o}{8}(\sin 2\theta)^2]$

2. Light from Quasi monochromatic sodium lamp of intensity I_o falls on two polaroids oriented at an angle of 30°. Find the intensity of light parsing out of the polarizers.

   [$\frac{3}{8} I_o$]

3. Unpolarised light falls on two polarising sheets placed one on top of the other. What must be the angle between their characteristic directions if intensity of transmitted light is one-third of intensity of the incident beam

   [$\theta = 35.26°$]

4. A plane polarised light is incident on a plate of double refracting uniaxial crystal cut with its faces parallel to optic axis. If the light is incident with vibrations at an angle of 30° with optical axis on the crystal, calculate ratio of intensities of extra-ordinary and ordinary waves

   [3 : 1]

5. Find the thickness of a quarter wave plate when the wavelength of light is equal to 5890 $\overset{o}{A}$ and $\mu_o = 1.55$ and $\mu_E = 1.54$.

   [1.47×10^{-3} cm]

6. Calculate the thickness of a quarter wave plate of quartz for sodium light of wavelength 5893 $\overset{o}{A}$. The ordinary and extraordinary refractive indices for sodium are 1.54425 and 1.55336 respectively

   [1.62×10^{-3} cm]

7. Calculate the thickness of a calcite plate which would convert plane polarised light into circularly polarised light. The principal refractive indices are $\mu_o = 1.658$ and $\mu_E = 1.486$ at the wavelength of light used as 5890 $\overset{o}{A}$.

   [8.56×10^{-5} cm]

8. A quarter wave plate is meant for $\lambda_o = 5.893 \times 10^{-5}$ cm. What phase retardation ϕ will show for $\lambda = 4.358 \times 10^{-5}$ cm (neglect changes of μ_o and μ_E with λ)

   [$0.67\ \pi$ rad]

9. A beam of linearly polarised light is changed into circularly polarised light by passing it through a slice of crystal 0.003 cm thick. Calculate the difference in the refractive index of the two rays in the crystal assuming this to be minimum thickness that will produce the effect and that the wavelength is 6×10^{-5} cm

[5×10^{-3}]

10. Plane polarised light, is incident on a plate of quartz cut with faces parallel to optic axis. Calculate the thickness for which the phase difference between the two rays is 60°, where $\mu_o = 1.5442$, $\mu_e = 1.5583$ and $\lambda = 5000\ \text{Å}$.

[0.00091 cm]

11. The faces of a quartz plate are parallel to the optic axis of the crystal.
    (a) What is the thinnest possible plate that would serve to put the ordinary and extraordinary rays of $\lambda = 5890\ \text{Å}$ a half wave apart on their exist.
    (b) What multiples of this thickness would give the same result? The indices of refraction of quartz are $\mu_E = 1.553$ and $\mu_o = 1.554$.

[$t = 3.27 \times 10^{-3}$ cm; 3.27×10^{-3} cm; $3 \times 3.27 \times 10^{-3}$ cm, $5 \times 3.27 \times 10^{-3}$ cm]

12. The plane of polarisation of plane polarised light is rotated through 6.5° in passing through a length of 2.0 decimeter of sugar solution of 5% concentration. Calculate the specific rotation of the sugar solution.

[$65^\circ\ (\text{dm})^{-1}\ (\text{gm/cc})^{-1}$]

13. A tube 20 cm long filled with a solution of 15 gm of cane sugar in 100 cc of water is placed in the path of polarised light. Find the angle of rotation of the plane of polarisation, if the specific rotation of cane sugar is 66°.

[19.8°]

14. If 1 mm of the quartz rotates the plane of polarisation of the light of wavelength λ by 18°, find the thickness of quartz, which if placed between two parallel nicols then no light of this wavelength λ transmits through the system

[5 mm]

15. Calculate the required thickness of a quartz crystal whose faces are perpendicular to the optic axis in order to produce a rotation of 43.5° for sodium light of wavelength 5893 Å. The specific rotation in quartz for this light is 21.72°/mm.

[2 mm]

16. 20 cm path of solution A shows 70° clockwise optical rotation, and 10 cm path of solution B shows 60° anticlockwise rotation. Compute the rotation by 10 cm path of a mixture of A and B in the ratio 1 : 2.

[28.3° right handed]

17. The optical rotation produced by a particular material is found to be 30° per mm at $\lambda = 5000\,\text{Å}$ and 50° per mm at $\lambda = 4000\,\text{Å}$. In the visible region of the spectrum the rotation of the plane of the polarisation is given by

$$\theta = a + b\,\lambda^{-2}$$

Find the values of the constants a and b for the given material.
Calculate the possible thickness of the given material which when interposed between two crossed Nicols will produce maximum transmission for $\lambda = 5500\,\text{Å}$.

[$b = 8.88 \times 10^{-7}$ deg-cm; $a = -55.55$ deg-m; $\theta = 238.30$ deg/cm]

Answers to Objective Questions

1. (c)	2. (c)	3. (a)	4. (a)	5. (b)	6. (c)
7. (c)	8. (d)	9. (b)	10. (b)	11. (b)	12. (d)
13. (c)	14. (b)	15. (b)	16. (d)	17. (b)	18. (b)
19. (d)	20. (b)	21. (a)	22. (c)	23. (a)	24. (c)
25. (b)					

CHAPTER 3B

Lasers

"In the war of worlds, written before the turn of the century H.G. Wells told a fanciful story of how Martians invaded and almost conquered the earth. Their weapon was a mysterious 'sword of beat' from which flickered' a ghost of a beam of light; it felled men in their tracks, made lead run like water and flashed anything combustible into masses of flame. Today wells's sword of beat comes close to reality in the Laser...."

Thomas Meloy

3B.1 Introduction

The word "Laser" is an acronym for "Light Amplification by stimulated emission of Radiation". Albert Einstein in 1917 theoretically proved that the process of stimulated emission must exist. But only in 1954, a group of scientists at Columbia university headed by Charles H. Townes operated a microwave device for "Microwave Amplification by Stimulated Emission of Radiation (MASER)". In 1960, T.H. Maiman of the Huges Research laboratories first achieved laser action at optical frequency in ruby. Since 1960 the development of lasers has been extremely rapid. Though it was called as "an invention in search of an application" at the time of its invention, the variety of lasers and the wealth of laser applications developed is enormous.

3B.2 Absorption of Radiation

If an atom is initially in a lower state 1, it can rise to a higher state 2 by absorbing a quantum of radiation (photon) of frequency ν given by

$$\nu = \frac{E_2 - E_1}{h}$$

where E_1 and E_2 are the energies of the atom in the state 1 and 2 respectively. (Fig. 3B.1). This process is known as absorption of radiation.

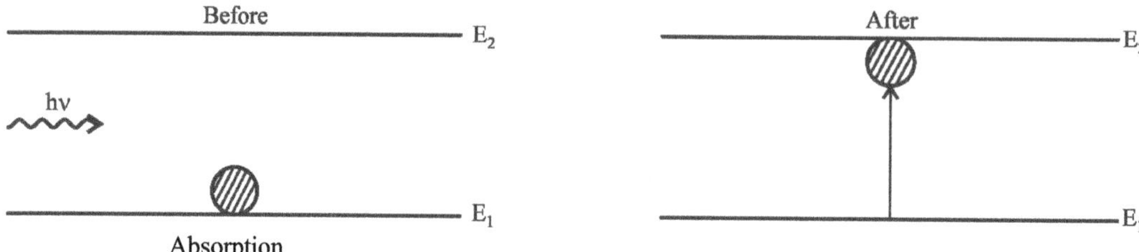

Fig. 3B.1

The probable rate of occurrence of the absorption transition $1 \rightarrow 2$ depends on the properties of states 1 and 2 and is proportional to the energy density u(ν) of the radiation of frequency ν incident on the atom.

Thus

$$P_{12} = B_{12}\, u(\nu) \qquad \text{.....(3B.1)}$$

where B_{12} is proportionality constant and is known as Einstein's coefficient of radiation.

3B.3 Spontaneous Emission

Let us now consider an atom initially in the higher excited state 2. Excited state with higher energy is inherently unstable, hence atom in excited state does not stay for longer time and it jumps to the lower energy state 1 emitting a photon of frequency ν. This is spontaneous emission of radiation (Fig. 3B.2). If there is an assembly of atoms, the radiation emitted spontaneously by each atom has a random direction and a random phase and is therefore incoherent from one atom to another.

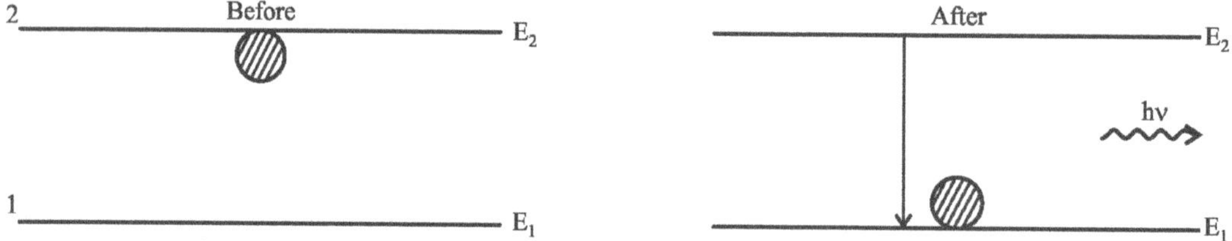

Fig. 3B.2 Spontaneous emission

The probability of spontaneous emission $2 \rightarrow 1$ is determined only by the properties of states 2 and 1. This is denoted by

$$A_{21}$$

which is known as 'Einstein's coefficient of spontaneous emission of radiation'. In this case the probability of spontaneous emission is independent of it.

3B.4 Stimulated or Induced Emission

According to Einstein, an atom in an excited energy state may, under the influence of the electromagnetic field of a photon of frequency ν incident upon it, jump to a lower energy state, emitting an additional photon of same frequency (ν). (Fig. 3B.3).

Hence two photons, one original and the other emitted move together. This is stimulated (or induced) emission of radiation. The direction of propagation, phase, energy and state of polarisation of the emitted photon is exactly same as that of the incident stimulating photon, so the result is an enhanced beam of coherent light.

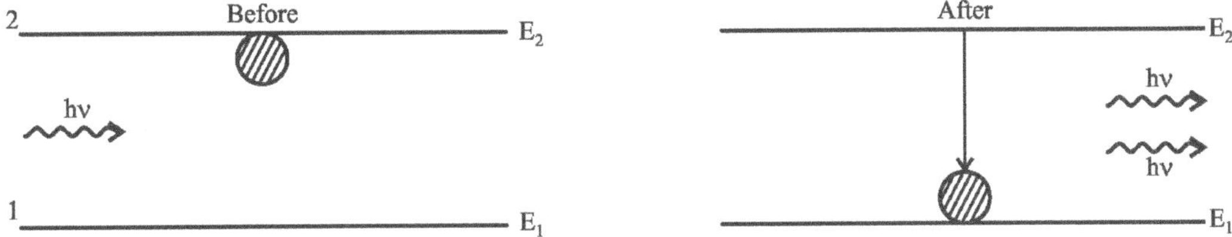

Fig. 3B.3

The probability of stimulated emission transition 2 → 1 is proportional to the energy density u(ν) of the stimulating radiation and is given by

$$B_{21}\, u(\nu)$$

where B_{21} is the 'Einstein coefficient of stimulated emission of radiation'.

The total probability for an atom in state 2 to drop to lower state 1 is therefore

$$P_{21} = A_{21} + B_{21}\, u(\nu)$$

3B.5 Distinction Between the Spontaneous and Stimulated Emission Processes

In the case of spontaneous emission, the atom emits an electromagnetic wave which has no definite phase or directional relation with that emitted by another atom.

In the case of stimulated emission, since the process is forced by the incident electromagnetic wave, the emitted light by the atom is in phase with that of the incident electromagnetic wave. The emitted wave is also in the same direction as that of the incident wave.

3B.6 Relation between Spontaneous and Stimulated Emission Probabilities: Einstein Coefficient Relationship

Let us consider an assembly of atoms in thermal equilibrium at temperature T with radiation of frequency ν and energy density u(ν). Let N_1 and N_2 be the number of atoms in states 1 and 2 respectively at any instant. The number of atoms in state 1 that absorb a photon and rise to state 2 per unit time is

$$N_1 P_{12} = N_1 B_{12} u(\nu) \quad .....(3B.2)$$

The number of atoms in state 2 that drop of state 1, either spontaneously or under stimulation, emitting a photons per unit time is

$$N_2 P_{21} = N_2 [A_{21} + B_{21} u(\nu)] \quad .....(3B.3)$$

For equilibrium, the absorption and emission must occur equally. Thus

$$N_1 P_{12} = N_2 P_{21}$$

or

$$N_1 B_{12} u(\nu) = N_2 [A_{21} + B_{21} u(\nu)]$$

$$[N_1 B_{12} - N_2 B_{21}] u(\nu) = N_2 A_{21}$$

$$u(\nu) = \frac{N_2 A_{21}}{N_1 B_{12} - N_2 B_{21}}$$

$$u(\nu) = \frac{A_{21}}{B_{21}} \frac{1}{\frac{N_1}{N_2}\left(\frac{B_{12}}{B_{21}}\right) - 1} \quad .....(3B.4)$$

Einstein proved thermo dynamically that the probability of stimulated absorption is equal to the probability of stimulated emission

i.e., $B_{12} = B_{21}$

Then, we have

$$u(\nu) = \frac{A_{21}}{B_{21}} \frac{1}{\left(\frac{N_1}{N_2} - 1\right)}$$

The equilibrium distribution of atoms among different energy states is given by using Boltzmann's Distribution law, according to which

$$\frac{N_2}{N_1} = \frac{e^{\frac{-E_2}{KT}}}{e^{\frac{-E_1}{KT}}}$$

or
$$\frac{N_2}{N_1} = e^{\frac{-(E_2-E_1)}{KT}}$$

$$\frac{N_2}{N_1} = e^{\frac{-h\nu}{KT}}$$

Consequently
$$u(\nu) = \frac{A_{21}}{B_{21}} \frac{1}{e^{h\nu/KT} - 1} \qquad \text{.....(3B.5)}$$

This is the energy density of photon of frequency ν in equilibrium with atoms in energy states 1 and 2, at temperature T. Comparing it with the Planck's radiation formula, according to which the energy density of the black body radiation of frequency ν at temperative T is given as

$$u(\nu) = \frac{8\pi h\nu^3}{C^3} \frac{1}{e^{h\nu/KT} - 1} \qquad \text{.....(3B.6)}$$

We get
$$\frac{A_{21}}{B_{21}} = \frac{8\pi h\nu^3}{C^3} \qquad \text{.....(3B.7)}$$

This shows that the ratio of Einstein's coefficient of spontaneous emission to the Einstein's coefficient of stimulated emission of radiation is proportional to cube of the frequency ν^3. This means that the probability of spontaneous emission increases rapidly with the energy difference between two states.

Units of Einstein Coefficients

From eq. (3B.7) we have
$$\frac{A_{21}}{B_{21}} = \frac{8\pi h\nu^3}{C^3}$$

$$= \frac{\text{Joule} - \text{sec} \times \text{sec}^{-3}}{(\text{meters sec}^{-1})^3}$$

$$= \text{Joule} - \text{sec} - \text{meters}^{-3}$$

$$= \text{J} - \text{s} - \text{m}^{-3}$$

These are the units of Einstein coefficients.

3B.7 Properties of a Laser

1. ***Coherence:*** From the very process of emission, lasers are coherent sources. Those have high degree of spatial and temporal coherence. Coherent length is in kilometre.
2. ***Directivity:*** In stimulated emission, photons of the same state are added. The state of a photon is defined by momentum and polarization. Hence laser is highly directive, non-diverging beam as all photons added in phase have the same momentum.
3. ***Polarised:*** As photons add in the same polarization so it is highly plane polarised light.
4. ***Monochomaticity:*** Lasers are truly, practically available monochromatic light. As photons of same frequency are added, These have high Q and low band width $\Delta\nu$.
5. ***Intensity:*** Since wave trains are added in phase and hence amplitudes are added. Suppose n wave trains each of amplitude a and intensity I are added in stimulated emission. Then,

$$I_{laser} = n^2 a^2 = n^2 I$$

For ordinary light, being incoherent intensities are added

$$I_{inc} = n I$$

Therefore, $$I_{coherent} = n I_{inc}$$

Since n is very large so lasers have extremely high intensity. The intensity can further be increased for short duration by using laser in 'Q'-switching mode.

3B.8 Population Inversion

Under ordinary conditions of thermal equilibrium the number of atoms in higher energy level 2, (N_2) is considerably smaller than the number in lower energy level 1, (N_1), so that there is very little stimulated emission compared to absorption. Hence under ordinary condition an incident photon is more likely to be absorbed rather than emitted. Hence laser action will not take place. If however, the larger number of atoms are made available in the higher energy level ($N_2 > N_1$) then stimulated emission will take place easily. This process of achieving the larger number of atoms in the higher energy level than the lower energy level is known as population inversion.

The term population inversion describes an assembly of atoms in which the majority are in energy levels above ground state (lower energy level).

The process of achieving population inversion is known as "pumping" of atoms. Most commonly used methods are as follows:

1. Optical pumping (used in Ruby laser)
2. Electric discharge (used in He-Ne)
3. Inelastic-atom-atom collision (used in He-Ne)
4. Direct conversion (used in semiconductor laser)
5. Chemical reaction (used in CO_2 laser)

3B.9 Need of Three Levels

Suppose there are only two energy levels, a metastable state hν above the ground state. The more photons of frequency ν we pump in to the assembly of atoms, the more upward transitions there will be from the ground state to the metastable state. However, at the same time the pumping will induce or stimulate downward transition from the metastable state to the ground state. When half the atoms are in each state, the rate of stimulated emission will equal the rate of induced absorptions, so the assembly cannot even have more than half its atoms in the metastable state. Hence a condition of population inversion will not achieve because $N_1 = N_2$ and therefore in this situation stimulated emission will not take place and laser amplification will not occur.

3B.10 Theory of laser: components of a Laser

For laser action, the stimulated emission must dominate the other two processes. Comparing rates of induced emission and absorption, it is seen that since $B_{12} = B_{21}$, the stimulated emission will dominate absorption only when

$$N_2 > N_1 \quad \text{.....(3B.8)}$$

There should be atoms/molecules in the excited state than in the ground state. This situation is known as population inversion.

Therefore, the first requirement for laser action is that energy is supplied from outside to breakdown the thermal equilibrium. This energy is used to pump atoms/molecules to higher energy state and so it is also called pumping energy. Different types of lasers use different pump mechanisms. Under normal situation, the excited atoms/molecules shall spontaneously transit to ground state as life time of the excited state is quite low (10^{-8} sec). This shows that two level laser is not possible because what is at the most achievable ($N_2 = N_1$).

Consider a three-level system, ground state energy E_1, excited state at energy E_3 and there is metastable state in between at energy E_2. The energy states that have substantially larger life time such as 10^{-3} second, 10^{-5} second against 10^{-8} seconds are called metastable states. These are schematically shown in Fig. 3B.4. The pump energy pumps atoms/molecules from E_1 to E_3. Since E_3 has life 10^{-8} seconds, these quickly transits to either E_1 or E_2. Since E_2 is metastable as pumping continues, the population of E_2, i.e., N_2 increases and a stage is reached when N_2 becomes greater than N_1. The population inversion is thus achieved. The second requirement for laser action is therefore, that system must have three or more energy levels and at least one excited must be metastable state.

Now, comparing rate of stimulated emission with the spontaneous emission, it turns out that with population inversion, if the photon density ρ_v is made quite high, stimulated emission may dominate spontaneous emission. This may be achieved by having standing waves of photons. Two parallel mirrors enclose the lasing medium with separation equal to $n\lambda$. One of the mirrors is perfectly reflecting and another 99% reflecting. The to and fro reflections of light waves stimulate more and more photons to add coherently and ultimately photon density reach to threshold value. Out of which 1% of photon density come out as laser. This system of parallel mirrors is called optical resonator.

The requirements of laser are

1. Input pump energy which breaks thermal equilibrium and lead to population inversion $N_2 > N_1$
2. Active medium of desired energy levels, minimum three levels with one metastable state.
3. Optical resonator to provide very high photon density. Parallel mirrors with separation $n\lambda$, λ being laser wavelength.

3B.11 Threshold condition for Laser Action

When atoms from metastable state transit spontaneously to ground or lower state, these emit photons in random directions. The photons which are emitted normal to the optical resonator mirror bounce back and forth. These photons induce stimulated emission and laser action grows.

The laser field or laser intensity I, grow exponentially in one bounce (back and forth)

$$I = I_o\, e^{2\beta l} \qquad \text{.....(3B.9)}$$

where β is the gain coefficient and can be defined as the growth in laser intensity per unit length, l being length of medium and I_o initial intensity.

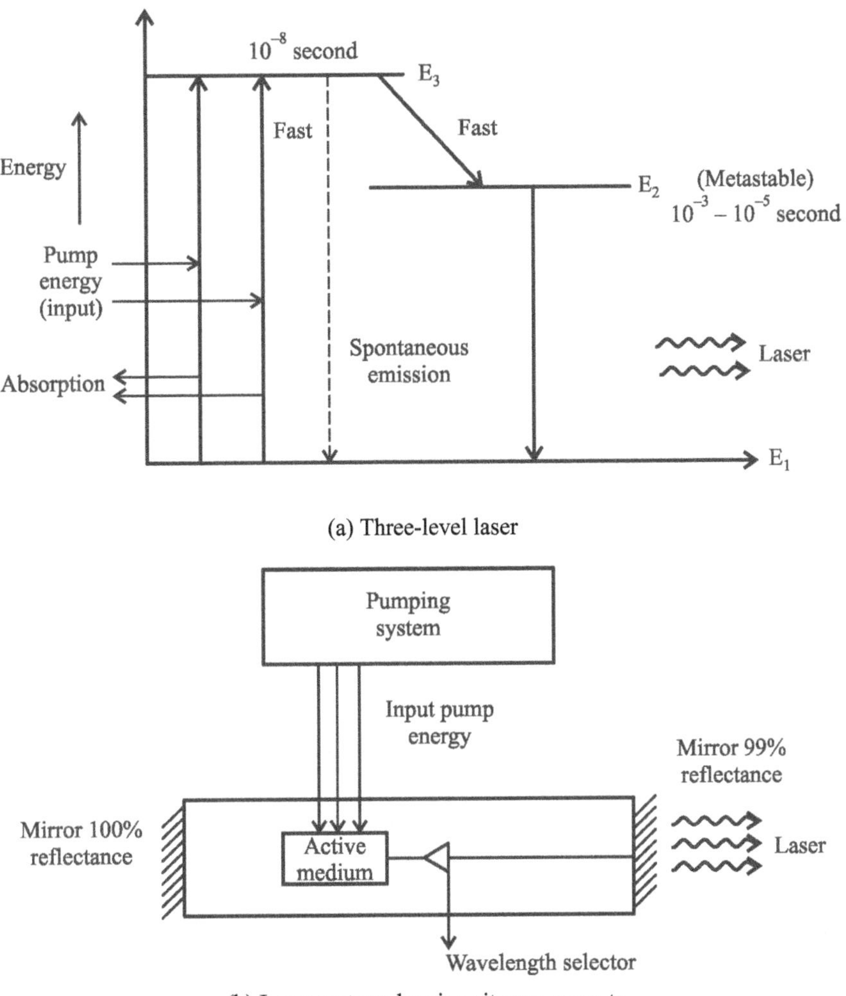

(a) Three-level laser

(b) Laser system showings its components

Fig. 3B.4

Besides gain, there are losses in the resonator when light bounces back and forth. These include loss due to diffraction because mirrors have finite size, absorption by mirrors, absorption of light by the medium causing undesired transitions, scattering due to inhomoginities and due to finite reflectivities of mirrors. Front mirror is used intentionally to have reflectance. 0.99 to get laser output. All the losses except reflectance losses may be grouped into a single loss constant α, the intensity loss per unit length.

Let the reflectivities of two front and back mirrors be R_1 and R_2, then

$$I = I_o\, e^{2\beta l} . e^{-2\alpha l}\, R_1\, R_2$$

The gain $I/I_o = G$ is given by

$$G = e^{2\beta l} . e^{-2\alpha l} R_1 R_2 \quad .....(3B.10)$$

For sustained laser action

$$G \geq 1 \quad \text{or}$$

$$e^{2\beta l} e^{-2\alpha l} R_1 R_2 \geq 1 \quad .....(3B.11)$$

$$2\beta l - 2\alpha l + \log_e R_1R_2 \geq 0 \quad .....(3B.12)$$

$$\beta \geq \left\{ \alpha - \frac{\log_e R_1R_2}{2l} \right\} \quad .....(3B.13)$$

Eq. (3B.13) is the threshold condition for sustained laser oscillations. The second term of the equation is the laser output.

The gain constant β depends on various factors, such as:

1. $\beta \propto (N_2 - N_1)$ population inversion
2. $\beta \propto B_{21}$, the probability of stimulated emission
3. $\beta \propto v^2 = \left(\frac{C}{n_o}\right)^2$, square of the velocity of light in the medium, Higher is velocity, more is the number of bounces per second. n_o is refractive index.
4. $\beta \propto \frac{1}{\Delta v}$, $h\Delta v$ is the energy width of metastable state.
5. $\beta \propto \frac{1}{v^2}$, v is the frequency of operation

Thus

$$\beta = \frac{(N_2 - N_1)(C/n_o)^2 B_{12}}{4v^2 (2\pi\Delta v)}$$

or

$$\beta = \frac{(N_2 - N_1)(C/n_o)^2}{4v^2 (2\pi\Delta v)\tau_m} \quad .....(3B. 14)$$

where $\tau_m = \frac{1}{B_{21}}$ is transition time of metastable state involved in laser action.

Grouping all the losses including reflection losses into a single constant τ_r, the characteristic time of resonator. In the absence of laser action, the energy of the resonator decays exponentially so

$$W = We^{-\frac{t}{\tau_r}} \qquad .....(3B.15)$$

Thus, ζ_r is the time in which energy of the optical resonator reduces to $1/e$ of its initial value. Then,

$$e^{-2\alpha l}R_1R_2 = e^{-\frac{t}{\tau_r}} = \exp\left(\frac{-2ln_o}{C\tau_r}\right) \qquad .....(3B.16)$$

where $t = \frac{2l}{v} = \frac{2ln_o}{C}$ is the time taken in one complete bounce or

$$-2\alpha l + \log_e R_1R_2 = \frac{2ln_o}{C\tau_r}$$

$$\tau_r = \frac{2ln_o}{C(2\alpha l - \log_e R_1R_2)} \qquad .....(3B.17)$$

From eq. (3B.12) and (3B.16), we have

$$e^{2\beta l}.e^{-\frac{2ln_o}{C\tau_r}} \geq 1$$

or

$$\beta \geq \frac{n_o}{C\tau_r} \qquad .....(3B.18)$$

Using value of β from eq. (3B.14), we have

$$\frac{(N_2 - N_1)\left(\frac{C}{n_o}\right)^2}{4v^2(2\pi\Delta v)\tau_m} \geq \frac{ln_o}{C\tau_r}$$

or

$$(N_2 - N_1) = \frac{8\pi v^2\Delta v\left(\frac{\tau_m}{\tau_r}\right)}{\left(\frac{C}{n_o}\right)^3} \qquad .....(3B.19)$$

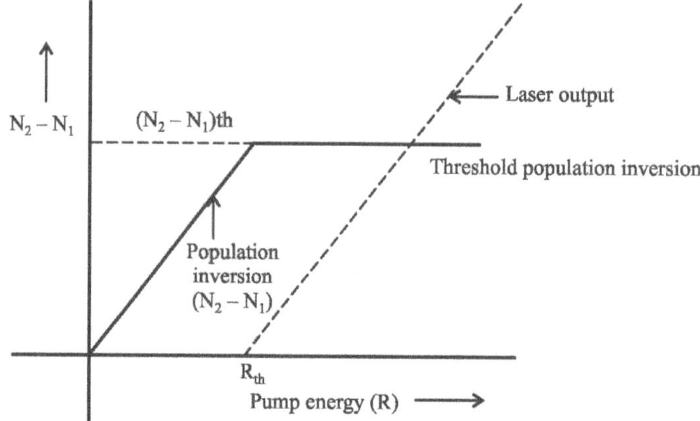

(Population inversion $(N_2 - N_1)$ versus pump energy rate (R) and laser output versus pump rate)

Eq. 3B.19 gives the required threshold population for sustained laser oscillations. Higher is $(N_2 - N_1)$ more difficult it is to have sustained laser action. The requirement for this are:

1. Losses in the resonator should be low, τ_r high
2. Life time of metastable state is low
3. Small width of metasable state. Also centre of the mode shall excite first
4. Low frequency lasers (IR) are easy to design as compared to UV lasers.

Neglecting losses, α and taking $R_1 = R_2 = 1$, it is found that gain saturation occurs after m bounces given by

$$m = \frac{12}{(\beta l)}$$

I
l
D
I_{th}
Threshold intensity
Linear growth
Exponential growth
$e^{\beta l} = 16(L/D)^2$
Length of medium

(Variation in laser intensity with length of active medium for given diameter. The length corresponding to saturation intensity be chosen to have optimum cost)

3B.12 Q-Switching

Q-switching is a technique for obtaining short, intense bursts of pulses from lasers. We know that to get laser action mere population inversion is not sufficient. Since there are losses, the inversion density must be high enough to produce gain compensating the losses. The inversion density for which gain is equal to the loss is called threshold inversion density. If we deliberately introduce some losses inside the cavity, the threshold inversion density increases. Hence we can build up the inversion density to a high value and correspondingly gain also increases but since gain is less than the deliberately introduced loss, no oscillation takes place. Now suddenly removing the introduced loss, the threshold inversion density switches back to its original value. Under this condition the gain is very much higher than the loss and intense amplification takes place and all the available energy is emitted as a single large pulse. Since in this process the Quality Q of the cavity is switched to a high value (on removing the deliberately introduced loss) this technique is called Q-switching. There are different methods of Q-switching.

In the ordinary pulsed mode, the output of an insulating crystal laser such as NdYAG consists of many random "spikes" of about 1 μs duration with a separation of about 1 μs and the length of train of spikes may be about 1 ms peak powers within the spikes are typically of the order of kilowatts. When the laser is Q-switched the result is a sigle pulse of high power typically in the mega watt range, with a duration of 10 – 100 ns.

3B.13 Mode Locking

The output from laser cavity may support oscillations in many modes simultaneously. The output of such a laser varies with time and depends on the relative phase, frequencies and amplitudes of the modes. The output irradiance may exhibits small fluctuations. Suppose we force the modes to maintain the same relative phase with respect to one another, then the output of the laser consists of a sequence of short pulses, separated in time by $2L/C$, where L is the length of the cavity, each of peak power equal to N times the average power (where N is the number of modes in the cavity). Thus mode locking is a technique of producing periodic, high power, short duration laser pulses. Using this technique pulses of pico second, duration with nano second pulse separation and peak powers of the order of mega watts can be produced.

3B.14 Coherence

The term coherence basically refers to the degree of correlation between the phases at different points in a beam of light. It is of two types:

(a) Temporal or time coherence

(b) Spatial coherence

3B.14.1 Temporal or time coherence

If the phase difference of waves crossing the two points lying along the direction of propagation of the beam is time dependent, then a beam of light is said to possess temporal or time coherence. This coherence is also known as longitudinal coherence. It refers to spectral purity of the degree of two waves of laser light remains in phase over many wavelengths.

The beam is said to possess temporal coherence if the phase difference of the waves crossing P and Q at any instant is always constant.

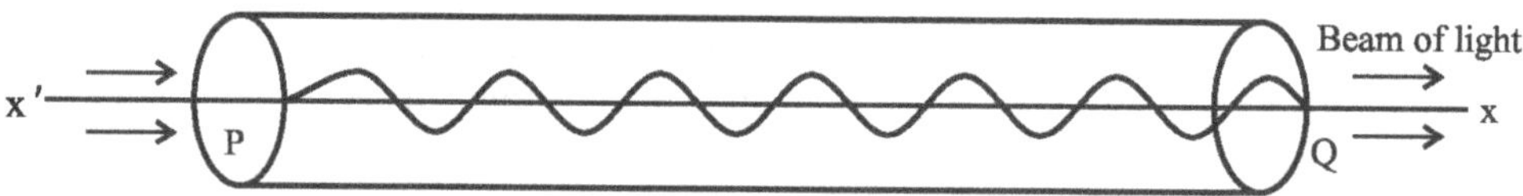

Accordingly, the light wave train emitted from each source can be characterized by an average time T_o. It is called coherent Time. Also the band width of the spectral distribution is inversely proportional to T_o.

$$\therefore \quad \Delta\omega = \frac{2\pi}{T_o} \qquad .....(3B.20)$$

The average length of the wave trains is called coherence length. If the velocity of light is C then coherence length L_C is given by

$$L_C = C\,T_o \qquad .....(3B.21)$$

Since $\Delta\omega = 2\pi\Delta\nu$

where $\Delta\nu$ is the frequency in Hz

$$\therefore \quad 2\pi\Delta\nu = \frac{2\pi}{T_o}$$

or

$$\Delta\nu = \frac{1}{T_o} \qquad .....(3B.22)$$

Again $\nu = \frac{C}{\lambda}$

$$\therefore \quad |\Delta\nu| = +\frac{C}{\lambda^2}\Delta\lambda \qquad .....(3B.23)$$

From equation (3B.21) and (3B.22), we get

$$\Delta\nu = \frac{C}{L_C} \qquad .....(3B.24)$$

From equation (3B.23) and (3B.24), we get

$$\frac{C}{L_C} = \frac{C}{\lambda^2}\Delta\lambda$$

$$\therefore \qquad \Delta\lambda = \frac{\lambda^2}{L_C} \qquad .....(3B.25)$$

Hence $\Delta\lambda$ is called the natural line width

Hence temporal coherence depends upon the value of coherent length and coherent time.

3B.14.2 Spatial Coherence

A laser beam is said to possess spatial coherence, if the phase difference of the waves crossing the two points lying on a plane perpendicular to the direction of propagation of the beam is time-independent. This coherence is also termed as transverse or lateral coherence. It is a measure of the minimum separation across the wave form where two waves remain coherent.

Fig. 3B.5 shows a beam of light travelling along the line X'X. Here abcd is a plane perpendicular to X'X. The beam is said to possess spatial coherence if the phase difference of the waves crossing P and P' at any instant is always constant.

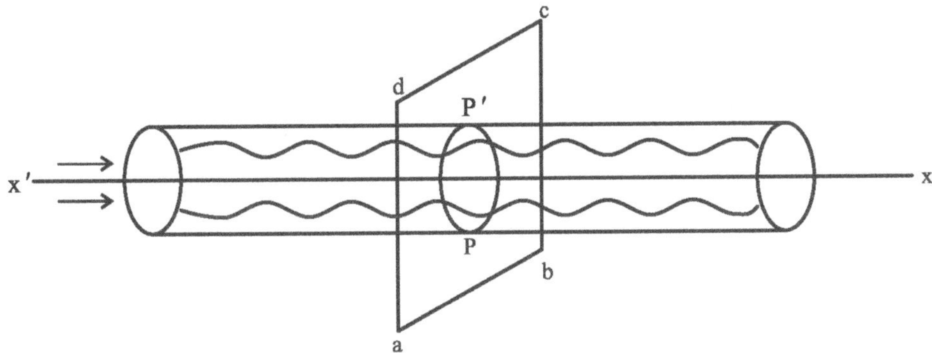

Fig. 3B.5

In Fig. 3B.6, S is a source and S_1, S_2, are two slits of separation l. If interference occurs for $l_{max} = l_\omega$ then this dimension is called the spatial coherence length of the source.

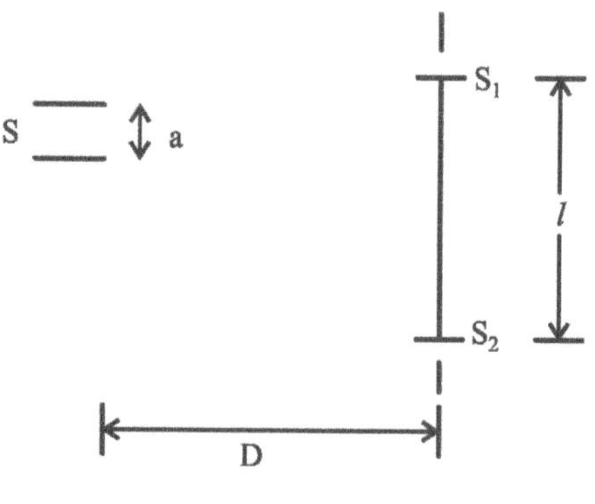

Fig. 3B.6

In Fig. 3B.6, the source S is at a distance D from slits S_1 and S_2. The condition for coherence of S_1 and S_2 is

$$\frac{\lambda}{a} > \frac{l}{D}$$

$$l < \frac{\lambda}{\frac{a}{D}}$$

Since $\frac{a}{D} = \theta$, the angle subtended by the width of the slits. Thus we get

$$l_w = \frac{\lambda}{\theta}$$

3B.15 Types of Lasers

Lasers are mainly divided in to four categories

1. Solid state lasers
2. Gas lasers
3. Liquid dye lasers
4. Semiconductor lasers

1. ***Solid state lasers:*** The first solid state lasers, the ruby laser was developed by T.H. Maiman in 1960. In solid state lasers, the bulk of the materials acts as a host to the active centres which are present in concentrations of less then 1%. The common solid state lasers are

   (a) Ruby laser (b) Nd: YAG laser

3B.15.1 Ruby Laser

Ruby laser is based on three energy levels and this laser consists of a pink ruby cylindrical rod whose end are optically flat and parallel as shown in Fig. 3B.7. One end is fully silvered and the other end is only partially silvered. Upon the rod is wound a coiled flash lamp filled with Xenon gas.

The ruby rod is basically a Al_2 O_3 (aluminium oxide) crystal doped with 0.05% by weight of chromium oxide (Cr_2O_3). The Al^{+3} ions are replaced by Cr^{+3}. These “impurity” Cr^{+3} ions are responsible for the pink colour (or red colour) of ruby laser.

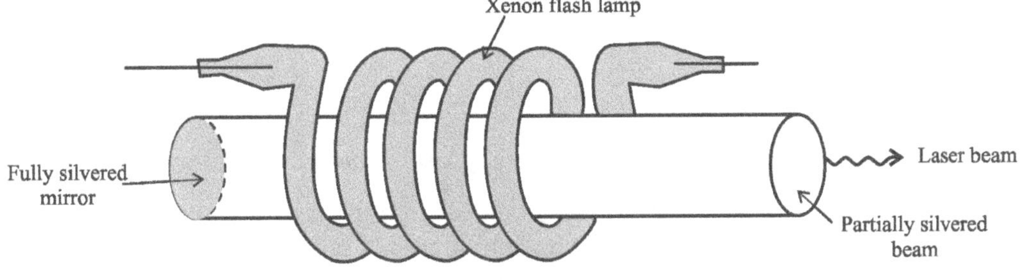

Fig. 3B.7

Working: The energy levels of Cr^{+3} ions on the crystal lattice are shown in Fig. 3B.8. below. It consists of three-level system. Upper energy level is short lived state E_3 above the ground state energy level E_1. There is an intermediate excited state level E_2 which is metastable having a life-time of 3×10^{-3} sec.

Normally, most of the chromium ions (Cr^{+3}) are in the ground state E_1. When a flash of light falls upon the ruby rod, the 5500 A° radiation photons are absorbed by the Cr^{+3} ions which are pumped to the excited state E_3. The transition from E_1 to E_3 is the optical pumping transition. Now the Cr^{+3} ions on the excited ions give a part of their energy to the crystal lattice and decay to the metastable state E_2. Hence the transition from E_3 to E_2 is radiationless transition. Metastable state E_2 is a long-lived state (life time 10^{-3} sec), hence the number of Cr^{+3} ions goes on decreasing. Thus population inversion is established between the metastable state E_2 and the ground state E_1.

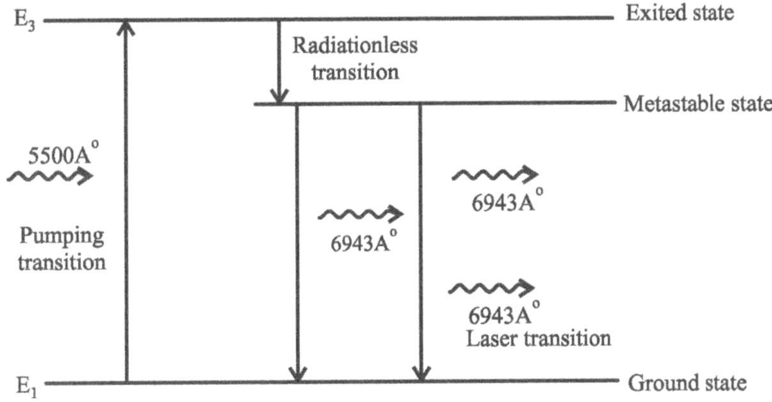

Fig. 3B.8

A spontaneous photon emitted by a Cr^{+3} ion at E_2 level initiates the stimulated emission by the other Cr^{+3} ions in the metastable state. The wavelength of the photon (beam) is 6943 A°. This photon travels through the ruby rod and if it is moving along the axial direction and repeated by reflection. This results in amplified strong laser beam of wavelength 6943 A°. This stimulated transition is the laser transition.

The laser beam is sufficiently intense, part of it emerges through the partially silvered end of the crystal.

3B.15.2 Drawbacks

There are some drawbacks in the three-level ruby laser, which are as follows:

1. The laser requires high pumping power because the laser transition terminates at the ground state and more than one-half of ground state atoms must be pumped to the higher state to achieve population inversion.
2. The efficiency of ruby laser is very low because only the green component of the pumping light is utilized while the rest of the components of incident light are left unused.
3. The laser output is not continuous but occurs in the form of pulses of micro second duration.
4. The defects due to crystalline imperfection are also present in this laser.

3B.15.3 Gas Lasers

In gas lasers, gases work as the active medium. Commonly used gases are helium, argon neon, xenon, nitrogen and carbon dioxide. In gas lasers population inversion is achieved by electric discharge. The gas lasers radiate in the visible and infrared region of spectrum. Among the gas lasers, helium neon (He-Ne) laser is most widely used laser and another important gas laser is CO_2 laser.

The Helium-Neon laser: In the helium-neon laser, a mixture of Helium (He) and Neon (Ne) gases are used as active medium. Lasing levels are provided by the excited states of the Ne atoms, whereas the He-atoms play an important role in pumping Ne atoms to the excited states. The He-Ne laser is shown schematically in Fig. 3B.9.

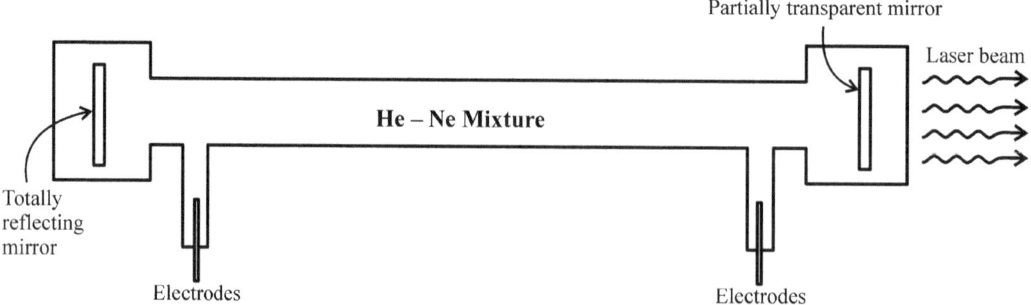

Fig. 3B.9 The He-Ne laser

The pumping is done by a stationary glow discharge fired by a direct current. When the potential difference between the anode and cathode is about 1000 V, a glow discharge is initiated in the working capillary (containing He-Ne mixture) of a few millimetre diameter.

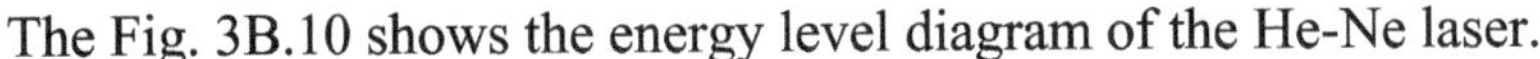
The Fig. 3B.10 shows the energy level diagram of the He-Ne laser.

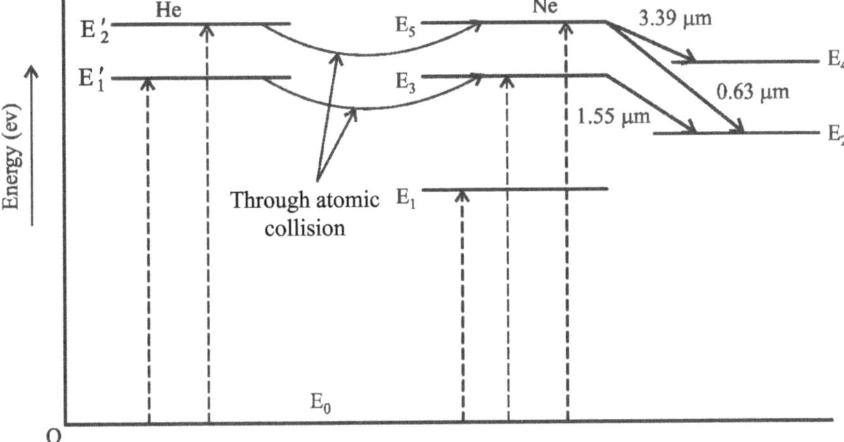

Fig. 3B.10

1. When free electrons produced during the gas discharge pass through the He-Ne mixture, they collide with the He and Ne atoms and excite them by impact energy transfer. Such absorptive transitions due to electron impacts are shown by dashed arrows. These excited states of the He i.e., E_1' and E_2' are metastable. Thus, He-atoms excited to these states stay there for a long time before losing energy by collision.

2. The interesting feature of the He-Ne energy diagram is that the excited states of Ne, namely E_3 and E_5 have approximately same energy as that of E_1' and E_2' of He atom. Therefore, when He-atoms in E_1' and E_2' collide with Ne-atoms in ground state, the He-atom transfer their energy to Ne-atoms and raise them to the states E_3 and E_5. Such an exchange of energy is known as resonant collision energy transfer.

3. Due to this energy transfer, He-atoms fall back to ground state. As a result, the excited states E_3 and E_5 of Ne-atoms have a sizable population which is much more than that of states E_2 and E_4.

4. Thus a condition of population inversion is achieved between the upper lasing levels E_5 or E_3 and lower lasing levels E_4 or E_2. In such a situation, any spontaneously emitted photon can trigger laser transition between these levels. The Ne-atoms then drop down from the lower lasing levels E_2 and E_4, to the level E_1 through spontaneous emission.
5. The wavelength of transition between levels $E_5 \rightarrow E_4$, $E_5 \rightarrow E_2$, $E_3 \rightarrow E_2$ are 3.39 μm, 0.63 μm and 1.55 μm respectively. We know that radiations corresponding to 3.39 μm and 1.55 μm fall in the infrared region of the electromagnetic spectrum.
6. The radiation corresponding to 0.63 μm however gives the red light – characteristic light of He-Ne laser.

3B.16 Semiconductor Laser Design and Working

Material: The material used for design of semiconductor diode laser are direct band gap semiconductors. There are semiconductors whose bottom of conduction band and top of valence bond are in the same K-axis. Such semiconductors are GaAs, InP, BeTl etc. Highly doped p-n junction is made from these materials and ends are polished flat to form an optical resonator.

Pumping and Resonator: In semiconductor diode laser the current causes pumping and forward bias provides population inversion. The third requirement optical resonator is formed by cleaving (polishing flat) the ends of the semiconductor crystal perpendicular to optical axis of the crystal. Since the refractive index is quite high $n \simeq 3.5$, the reflectivities of the cleaved planes is small ~ 30%. To increase the reflectivity, the dielectric coating is used at these cleaved surfaces. The optical resonator so formed is commonly called as Fabry-Perot resonator.

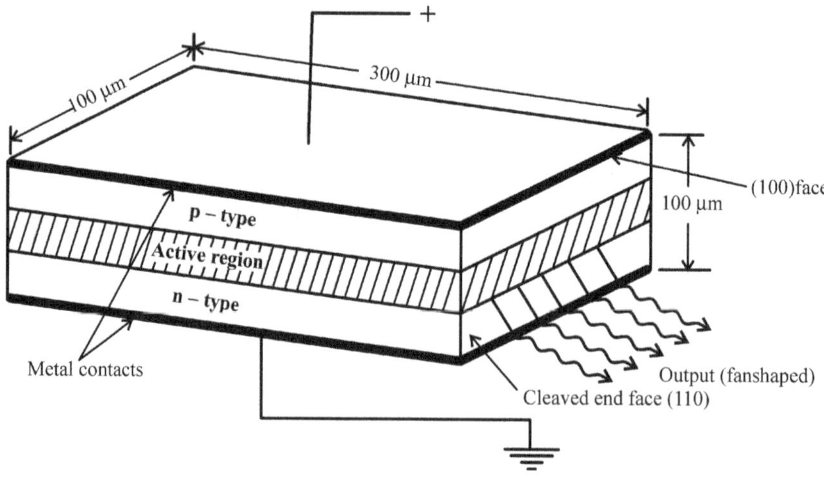

Fig. 3B.11

Homo-junction laser: The radiation generated within active region spreads out in surrounding lossy GaAs region. No doubt, the high refractive index of the active junction region as compared to outside acts as a waveguide to the radiation. Since the difference in refractive indices between active and passive regions is small ($\Delta n \sim 0.02$) so the guiding is not sufficient. This, in turn, requires high threshold density, $J_{th} = 400\ A/mm^2$ and consequently high heat dissipation. Such lasers have to be operated at low temperatures. Typical design of such a laser called homo-junction laser is shown in Fig. 3B.11.

In Fig. 3B.12 from a to c shows the working of the laser. Unbiased junction has almost no electron and holes to combine. When forward biased, there are large number of electrons and holes to recombine. This is population inversion. The current causes pumping or recombination of these electrons and holes. The carrier concentration, refractive index profile and optical field leading to gain are also shown in figure d and e respectively.

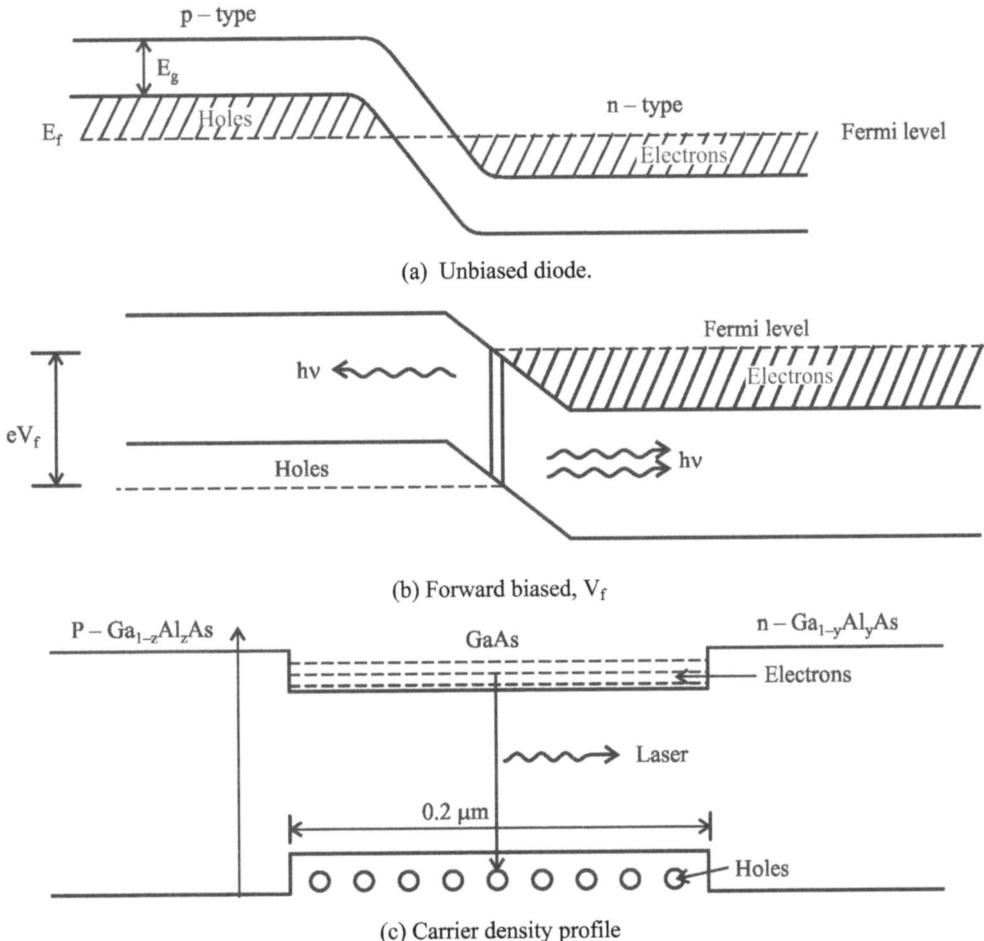

(a) Unbiased diode.

(b) Forward biased, V_f

(c) Carrier density profile

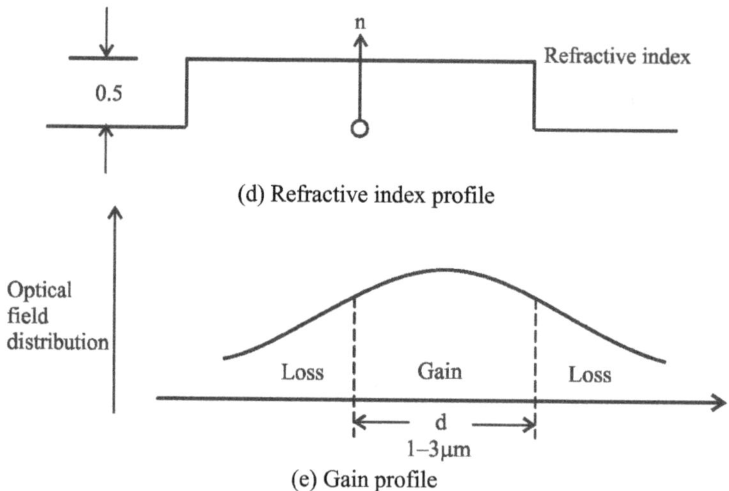

(d) Refractive index profile

(e) Gain profile

Fig. 3B.12

Heterojunction lasers

To operate semiconductor lasers at room temperature, the losses have to be drastically reduced so that the threshold current is low. This is done by adding cladding layers semiconductors having low refractive index and high energy band gap. Such structures are called heterogeneous semiconductor lasers. Fig. 3B.13 shows structure, energy band diagram, refractive index profile and the gain of heterogenous laser. It consists of an upper p-type GaAs followed by layers of P AlGaAs, P GaAs, n AlGa As, nGaAs. Basic heterojunction is P AlGaAs – PGaAs – nGaAs, the other two layers provide desired refractive index profile so that undesired power dissipation is avoided.

Because the charge carrier in heterojunction lasers are confined to a much smaller region, the heat deposition is much lower while threshold current density or recombination radiation are still at sufficient high level for laser output.

The active region may be further narrowed via lithographic technique reducing the threshold current further. Normally the thickness of active region is from 0.1 to 0.2 μ. If the thickness is reduced further say 5 to 10 nm, then energy levels exhibit quantum behaviour and such lasers are called quantum well lasers. These lasers allow desired laser wavelength.

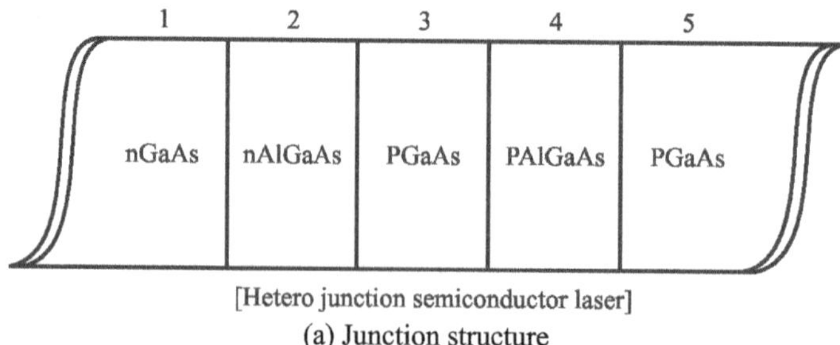

[Hetero junction semiconductor laser]

(a) Junction structure

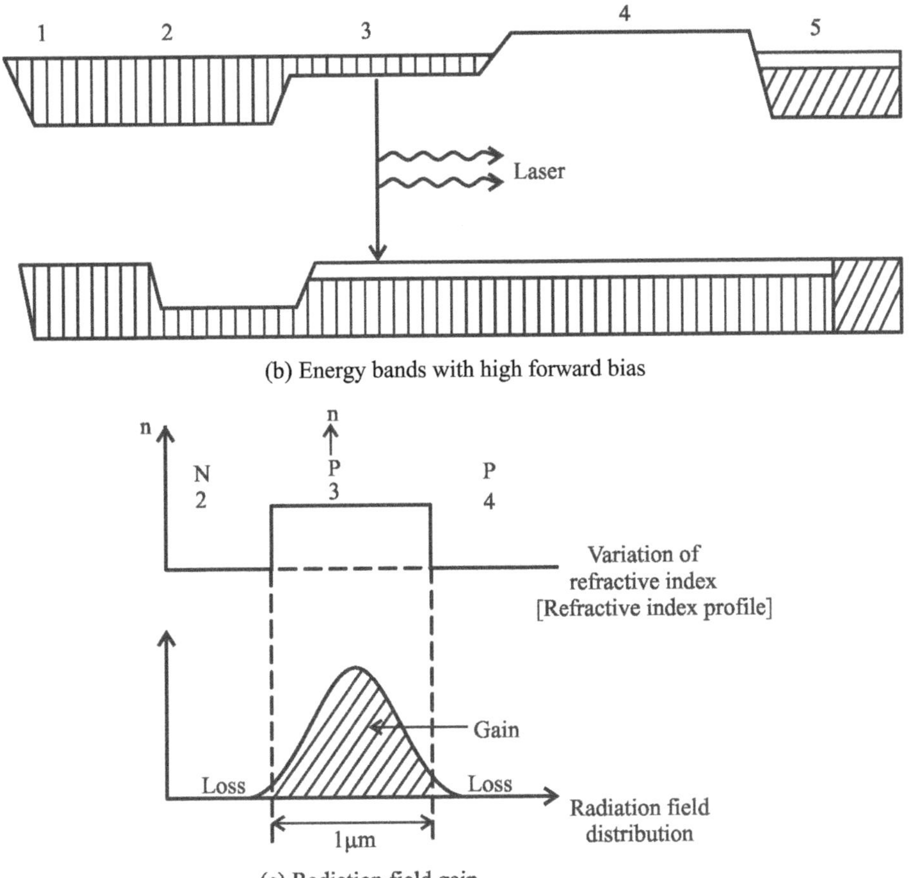

(b) Energy bands with high forward bias

(c) Radiation field gain

Fig. 3B.13

3B.17 Applications

The common applications of lasers are as follows.

1. Metallic rods can be melted and joined by means of a laser beam (laser welding).
2. The laser beam is used to vapourise unwanted material during the manufacture of electronic circuits on semiconductor chips.
3. CO_2 gas lasers of about 100 W output are helpful in surgery because they seal small blood vessels while cutting through tissue by vapourising water in the path of their IR beams.
4. Lasers are used to detect and destroy enemy missiles during warfare.

5. Low power semiconductor lasers are used in CD (compact disc) players, laser printers, laser copiers, facsimile machine etc.
6. Semiconductor lasers are ideal for fibre-optic transmission lines in which the electric signals that would normally be sent along copper wires are first converted into a series of pulses according to a standard code. Lasers then turn the pulses in to flashes of IR light that travel along thin glass fibers and at the other end are changed back into electric signals.
7. High power lasers are used to bring about thermo nuclear reactions which would become the ultimate inexhaustible power source for human civilization.
8. Lasers are also being employed for separating the various isotopes of an element.
9. Laser beams have also been used in the "inertial confinement" of plasma.
10. Lasers are used in the production of three-dimensional images of an object in holography.

Solved Numerical Problems

1. In a ruby laser, the total number of Cr^{+3} ions is 2.8×10^{19}. If the laser emits radiation of wavelength $7000\,\mathring{A}$. Calculate the energy of the laser pulse. [UPTU]

Sol. Energy of the laser pulse = total number of ions (n) ×energy of one photon

$$E = nh\nu$$

$$E = n\frac{hc}{\lambda}$$

Given

$$h = 6.62 \times 10^{-34} \text{ J.sec}$$

$$c = 3 \times 10^{8} \text{ m/sec}$$

$$\lambda = 7000\,\mathring{A} = 7 \times 10^{-7} \text{ m}$$

$$n = 2.8 \times 10^{19}$$

$$E = 2.8 \times 10^{19} \times \frac{6.62\times10^{-34}\times3\times10^{8}}{7\times10^{-7}}$$

$$E = 7.94 \text{ Joule}$$

2. Calculate the energy and momentum of a photon of a laser beam of wavelength $6328\,\mathring{A}$. [UPTU]

Sol. The energy of the photon is

$$E = h\nu = \frac{hc}{\lambda}$$

Given $h = 6.62 \times 10^{-34}$ J.sec

$C = 3 \times 10^{8}$ m/sec

$\lambda = 6328\ \text{Å} = 6.328 \times 10^{-7}$ m

$$E = \frac{6.62 \times 10^{-34} \times 3 \times 10^{8}}{6.328 \times 10^{-7}}$$

$$E = 3.14 \times 10^{-19} \text{ Joule}$$

$$E = \frac{3.14 \times 10^{-19}}{1.6 \times 10^{-19}} = 1.96 \text{ eV}$$

The momentum of photon

$$P = \frac{h}{\lambda} = \frac{6.62 \times 10^{-34}}{6.328 \times 10^{-7}}$$

$$= 1.05 \times 10^{-27} \text{ Kg m/sec}$$

3. A laser beam has a wavelength of 8×10^{-7} m and aperture 5×10^{-3} m. The laser beam is sent to moon. The distance of moon is 4×10^{5} km from the earth. Calculate. [UPTU]

(a) The angular spread of the beam

(b) The axial spread when it reaches the moon.

Sol. (a) The angular width or angular spread of the beam is given by

$$\Delta\theta \approx \frac{\lambda}{d} \text{ rad}$$

where λ is the wavelength and d is the diameter of the aperture.

Given $\lambda = 8 \times 10^{-7}$ m

$d = 5 \times 10^{-3}$ m

$$\therefore \quad \Delta\theta \approx \frac{8 \times 10^{-7}}{5 \times 10^{-3}} = 1.6 \times 10^{-4} \text{ rad}$$

(b) The axial spread is given by

$$W = (L\Delta\theta)^2$$

Here $L = 4 \times 10^{5}$ Km

$$W = (4 \times 10^{5} \times 1.6 \times 10^{-4})^2$$

$$W = 4.096 \times 10^{3} \text{ Km}^2$$

4. A laser beam has a power of 50 mw. It has an aperture of 5×10^{-3} m and it emits light of wavelength $7200\,\mathring{A}$. The beam is focussed with a lens of focal length 0.1 m. Calculate the area and the intensity of the image. [UPTU]

Sol. (a) The angular spread is given by

$$\Delta\theta \approx \frac{\lambda}{d}$$

where λ is the wavelength

and d is the aperture diameter

Given $\lambda = 7200\,\mathring{A} = 7.2 \times 10^{-7}$ m

$d = 5 \times 10^{-3}$ m

$$\therefore \quad \Delta\theta \approx \frac{7.2 \times 10^{-7}}{5 \times 10^{-3}}$$

$$\approx 1.44 \times 10^{-4} \text{ rad}$$

The area of the image or areal spread

$$= (\Delta\,\theta\,f)^2$$

Given $f = 0.1$ m

$$= (1.44 \times 10^{-4} \times 0.1)^2$$

or $A = 2.074 \times 10^{-10}\ m^2$

(b) The intensity of the image is given by

$$I = \frac{P}{A}$$

where P is the power of the laser beam

$$I = \frac{50 \times 10^{-3}\ w}{2.074 \times 10^{-10}\ m^2}$$

$$I = 2.41 \times 10^8\ w/m^2$$

5. The coherence length of sodium light is 2.945×10^{-2} m and its wavelength is $5890\,\mathring{A}$. Calculate (a) the frequency (b) the number of oscillations corresponding to the coherence length and (c) the coherence time. [UPTU]

Sol. We know that

$$\nu = \frac{C}{\lambda}$$

Given $C = 3 \times 10^8$ m/sec

$$\lambda = 5890\,\overset{\circ}{A} = 5890 \times 10^{-10}\ \text{m}$$

$$\nu = \frac{3 \times 10^8\ \text{m/sec}}{5890 \times 10^{-10}\ \text{m}} = 5.09 \times 10^{12}\ \text{Hz}$$

(b) The number of oscillations in any length l may be given as

$$n = \frac{l_c}{\lambda}$$

Given $l_c = 2.945 \times 10^{-2}$ m

$$\lambda = 5.89 \times 10^{-7}\ \text{m}$$

$$n = \frac{2.945 \times 10^{-2}}{5.89 \times 10^{-7}}$$

$$n = 5 \times 10^4$$

(c) The coherence time τ_c and coherence length l_c of a laser light are related as

$$\tau_c = C = l_C$$

or $$\tau_c = \frac{l_C}{C}$$

Here $l_c = 2.945 \times 10^{-2}$ m

and $C = 3 \times 10^8$ m

$$\tau_c = \frac{2.945 \times 10^{-2}}{3 \times 10^8}$$

$$\tau_c = 9.816 \times 10^{-11}\ \text{sec}$$

6. A laser beam ($\lambda = 6000\,\overset{\circ}{A}$) on earth is focussed by lens of diameter 2m on to a crater on the moon. The distance of the moon is 4×10^8 m. How big is the spot on the moon. Neglect the effect of the atmosphere. [UPTU]

Sol. If d is the diameter of the focussing lens, then the angular spread of the laser beam is

$$\Delta\theta \approx \frac{\lambda}{d}$$

Here $\lambda = 6000\,\overset{\circ}{A} = 6.0 \times 10^{-7}$ m

$$d = 2.0\ \text{m}$$

$$\Delta\theta \approx \frac{6.0\times10^{-7}}{2.0} = 3.0 \times 10^{-7} \text{ rad}$$

If L is the distance of moon, then the area of the spot on the moon or the areal spread is given as

$$A = (L\,\Delta\theta)^2$$

Given $L = 4 \times 10^8$ m

$$A \approx (4 \times 10^8 \times 3.0 \times 10^{-7})^2$$

$$A = 1.44 \times 10^4 \text{ m}^2$$

7. The coherence length for sodium D_2 line is 2.5 cm. Deduce

(a) The coherence time

(b) The spectral width of the line

(c) The purity factor Q.

Sol. (a) The coherence length l_c and coherence time τ_c are related as

$$\tau_c \times C = l_c$$

or

$$\tau_c = \frac{l_C}{C}$$

Given $l_C = 2.5 \text{ cm} = 2.5 \times 10^{-2}$ m

$$C = 3 \times 10^8 \text{ m/sec}$$

$$\tau_c = \frac{2.5\times10^{-2}}{3\times10^8}$$

$$\tau_c = 0.83 \times 10^{-10} \text{ sec}$$

(b) The spectral width $\Delta\lambda$ is given by

$$\Delta\lambda = \frac{\lambda^2}{l_c}$$

Here $\lambda = 5896\ \text{Å} = 5.896 \times 10^{-7}$ m

$$\Delta\lambda = \frac{(5.896\times10^{-7})^2}{2.5\times10^{-12}}$$

$$\Delta\lambda = 13.90 \times 10^{-12} \text{ m}$$

(c) The purity factor Q is

$$Q = \frac{\lambda}{\Delta\lambda} = \frac{5.896 \times 10^{-7}}{13.90 \times 10^{-12}}$$

$$Q = 4.24 \times 10^{4}$$

8. Find the relative population of the two states in a ruby laser that produces a light beam of wavelength 6943 Å at 300 k.

Sol. The population ratio is given by

$$\frac{N_2}{N_1} = \exp\left[-\frac{(E_2 - E_1)}{KT}\right]$$

Since $$E_2 - E_1 = h\nu = \frac{hc}{\lambda}$$

$$= \frac{6.6 \times 10^{-34} \times 3 \times 10^{8}}{1.6 \times 10^{-19} \times 6943 \times 10^{-10}}$$

$$E_2 - E_1 = 1.79 \text{ eV}$$

$$\frac{N_2}{N_1} = \exp\left[-\frac{1.79 \text{ eV}}{8.61 \times 10^{-5} \times 300}\right]$$

$$\frac{N_2}{N_1} = \exp\ (-69.3)$$

$$\frac{N_2}{N_1} = 8 \times 10^{-31}$$

9. For a He-Ne laser at 1 m and 2 m distances from the laser, the output beam spot diameters are 4 mm and 6 mm respectively. Calculate the divergence.

Sol. Divergence $$\theta = \frac{a_2 - a_1}{2(d_2 - d_1)}$$

$$= \frac{(6-4) \times 10^{-3}}{2(2-1)}$$

$$= 10^{-3} \text{ radian}$$

$$= 1 \text{ milli radian}$$

10. In a DVD player laser of 650 mm and 1 mm beam size is focussed with a lens of focal length 1 mm, calculate the spot size.

Sol. Spot size is given by

$$\text{Spot size} = \frac{0.6\,\lambda}{\text{NA}}$$

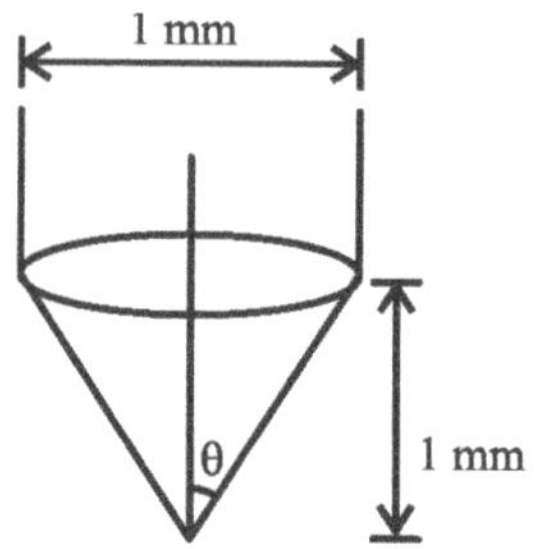

where $\text{NA} = n \sin\theta$

$$\tan\theta = \frac{0.5\times10^{-3}}{1\times10^{-3}}$$

$$\theta = 26.57$$

$\therefore$ $\sin\theta = 0.45$

Hence numerical aperture NA = 0.45

Since for air n = 1

$$\therefore \quad \text{spot size} = \frac{0.6\times650\times10^{-9}}{0.45}$$

$$= 0.867 \times 10^{-6}\ \text{m}$$

$$\text{spot size} = 0.867\ \mu\text{m}$$

11. Interference is to be observed in a typical double slit experiment with the help of sunlight with $\lambda = 5000\ \text{Å}$ filter. What should be separation of two slits so that fringes of good contrast are seen. The sun subtends an angle of about 32' on the surface of the earth.

Sol. The coherence length $= \dfrac{\lambda}{\theta}$

Given $\theta = 32' = \dfrac{32\pi}{180\times60} = 0.01$ rad

$$\therefore \quad l = \frac{5000\times10^{-10}}{10^{-2}} = 5\times10^{-5}\ \text{m}$$

12. Two wavetrains overlap to 29% of their length. If maximum in the resulting interference pattern receive 20 units of light, how much do minima receive?

Sol. The degree of coherence is equal to the overlapping of wave trains. Hence visibility is 0.29.

$$0.29 = \frac{I_{max} - I_{min}}{I_{max} + I_{min}}$$

Given $\qquad I_{max} = 20$

$\therefore \qquad I_{min} = 11$ units

13. An individual atom emits coherence light for 10^{-8} sec(a) what is coherent length of wave train (b) If this wavetrain is separated in two parts with a partially reflecting mirror and later reunited after one beam travels 5 m and other 10 m. Do the waves produce interference fringes observable by human eye?

Sol. (a)
$$l_c = c\,\tau_c = 3 \times 10^8 \times 10^{-8}$$
$$= 3 \text{ m}$$

(b) The path difference between two beam is

$$\Delta x = 10 - 5 = 5\text{m}$$

which is greater than coherent length, hence interference is not observable by human eye.

14. In a ruby laser, the flash lamp operates due to discharge of 1000 μF capacitor bank charged to 4KV. If it produces a flash of energy 10 J, what is efficiency of laser.

Sol.
$$\eta = \frac{\text{Optical power output}}{\text{Input power}}$$

$$= \frac{10}{\frac{1}{2}\left(1000 \times 10^{-6}\right)\left(4 \times 10^{3}\right)^{2}}$$

$$= 1.25 \times 10^{3}$$

$$\eta = 0.125\%$$

15. A ruby laser (n_o = 1.76) is 10 cm long and 0.5 cm in diameter. The ends of the crystal are highly polished and form the mirrors of the resonant cavity. Does diffraction contribute seriously to energy loss at lasing wavelength 694.3 nm?

Sol. So long as the first minimum lies well within the periphery of the mirror, diffraction is not considered to be important. The diffraction angle

$$\theta = \frac{\text{Half mirror width}}{\text{Cavity length}}$$

$$\theta = \frac{a}{L} >> \frac{\lambda}{2a}$$

$$\frac{a^2}{\lambda} >> L$$

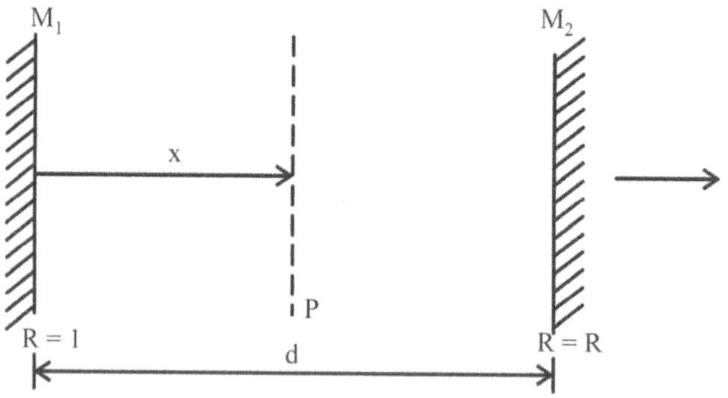

Substituting the values, we get

$$\frac{a^2}{\lambda} = \frac{(0.1)^2}{694.3 \times 10^{-9}}$$

$$\frac{a^2}{\lambda} = 6250 \text{ m} >> 0.1 \text{ m}$$

16. For He-Ne laser assuming mirrors to be 100% reflecting and gain medium length is 0.2 m, how many pass must occur through the amplifier before saturation is reached? Given that threshold gain $\beta_{th} = 0.15/\text{m}$

Sol. We know that

$$m = \frac{12}{\beta_{th} l} = \frac{12}{0.15 \times 0.2}$$

$$= 400$$

Thus, radiation has to make 400 passes before the threshold.

Conceptual Questions and Answers

1. What does LASER stand for?

Ans. The word LASER stands for light amplification by stimulated emission of radiation.

2. What are the main characteristics of a laser beam?

Ans. A laser beam is highly coherent, highly monochromatic and perfectly parallel.

3. The laser beams are highly parallel. Explain?

Ans. The extend of parallelism of the laser beam can be judged from the fact that if a laser beam of 1 cm diameter is sent to moon, the at the moon its diameter may not increase to more 500 m. Thus, the angle between the two sides of the beam is of the order of

$$\frac{500\text{ m}}{3\times10^{8}\text{ m}} = 1.3\times10^{-6}\text{ rad.}$$

4. What is stimulated emission?

Ans. The emission of photons in an atomic system by shining photons of energy equal to the difference in energies associated with the higher and lower energy states along with the incident photons is called stimulated emission.

5. Can we obtain light amplification in the absence of stimulated emission?

Ans. No. only in case of the stimulated emission, we obtain photons of same phase, energy, frequency, state of polarization and travel parallel to each other, they constitute the coherent radiation which super impose to produce radiation of very high intensity.

6. What do you mean by pumping?

Ans. The process of raising atoms from the lower energy state to the higher energy state by induced absorption is known as pumping.

7. What is a metastable state?

Ans. The energy state in an atomic system, which has a mean life time much larger than the life time of a normal higher energy state, is called metastable state.

8. What is the need of three levels in laser?

Ans. In case of two level laser (with a metastable state above the ground state), when half the atoms are in each state, the rate of stimulated emission will equal the rate of induced absorption and as a result the metastable state will not be having more atoms than ground state. Hence, a condition of population inversion will not be achieved and stimulated emission will not take place and laser amplification will cease. That is the reason, why we prefer three level laser or higher over two level laser.

9. What is population inversion in laser?

Ans. Population inversion is the phenomenon of having more atoms occupy an excited energy level E_2, than a lower energy level E_1. So that the normal equilibrium distribution is reversed i.e., $N_2 > N_1$. Population inversion occurs temporarily as a result of the excitation of a medium, called pumping.

10. Why population inversion is an essential condition for stimulated emission?

Ans. We know that the rate of stimulated absorption of photons is given by

$$P_a = N_1 B_{12} u(\nu)$$

and the rate of stimulated emission of photons is given by

$$P_e = N_2 B_{21} u(\nu)$$

where N_1 and N_2 are the numbers of atoms in the ground energy state and higher energy state respectively, B_{12} and B_{21} are Einstein's coefficient of stimulated absorption and emission respectively.

Thermodynamically

$$B_{12} = B_{21}$$

$$\therefore \quad \frac{P_e}{P_a} = \frac{N_2}{N_1}$$

In order to achieve dominance of stimulated emission over stimulated absorption, we must have $N_2 > N_1$ i.e., the population of higher energy state is more than ground energy state which is called population inversion.

11. What is mode locking and Q-switching mechanism is lasers?

Ans. Mode locking is technique for producing periodic, higher power, short duration laser pulses.

Whereas Q-switching is a techniques for producing short, intense bursts of laser beam.

12. Why existence of metastable state is an essential requirement for achieving population inversion?

Ans. Orderly, an atom may remain in an excited state for about 10^{-8} sec, after which it drops to some lower energy state along with emission of a photon. This emission of photon is governed by some selection rules. These rules do not permit certain transitions from a higher energy state to the ground state. Such energy levels are known as metastable states. Once the valence electron finds itself in the metastable state, it continues to be there for long time and the atom is unable to return to the ground state instantaneously. As time goes on, more and more atoms will land up in the metastable and ultimately the number of atoms in the metastable state are more than that in the ground state, which results into population inversion. Thus the existence of metastable state is an essential condition for achieving population inversion.

Objective Questions

1. The basic principle involved in the lasing action is
   (a) Population inversion (b) Spontaneous emission
   (c) Stimulated emission (d) None of these
2. The rate of spontaneous emission is proportional to
   (a) Number of atoms in excited state
   (b) Number of atoms in the lower energy state
   (c) Number of atoms making transitions from excited to lower state
   (d) All the above
3. The rate of stimulated emission depends on
   (a) Intensity of the external field
   (b) Number of atoms in the upper state
   (c) Number of atoms in the lower state
   (d) Both (a) and (b)
4. The main component of any laser device are
   (a) Active medium and pumping source
   (b) The optical resonator
   (c) Only (a)
   (d) Both (a) and (b)
5. A permanent transition between two states can be described by relation
   (a) $E_1 - E_2 = -\frac{1}{2}h\nu$ (b) $E_1 - E_2 = +\frac{1}{2}h\nu$
   (c) $E_1 - E_2 = \pm h\nu$ (d) None of these
   where E_1, E_2 corresponds to excited state and ground state respectively.
6. The intensity of stimulated emission in the modern ruby laser is
   (a) Less than 10^6 watts (b) More than 10^6 watts
   (c) More than 10^8 watts (d) More than 10^9 watts
7. Which is not an application of laser
   (a) It is used to carry signals by amplitude modulation
   (b) It is useful in photographic and spectrographic research
   (c) It is used in certain surgical and metallurgical purposes
   (d) All are correct.

8. The Helium-Neon laser is a

(a) First gas laser (b) Semiconductor laser

(c) Liquid laser (d) Second gas laser

9. The He-Ne laser consist of a mixture of the He-Ne in a ratio of about

(a) 10:1 (b) 1:10 (c) 3:10 (d) 10:3

10. The relation between the Einstein's coefficients A_{21} and B_{21} is

(a) $\frac{A_{21}}{B_{21}} = \frac{n_o^3}{\pi^2 C^3}$ (b) $\frac{A_{21}}{B_{21}} = \frac{\hbar^3}{\pi^2 C^3}$

(c) $\frac{A_{21}}{B_{21}} = \frac{\hbar \omega^3}{\pi^2 C^3}$ (d) $\frac{A_{21}}{B_{21}} = \frac{\hbar \omega^3 n_o^3}{\pi^2 C^3}$

11. The distance for which the beam may be assumed to be spatially coherent is

(a) $l_\omega = \lambda\theta$ (b) $l_\omega = \theta/\lambda$

(c) $l_\omega = \lambda/\theta$ (d) None of these

12. The lateral coherence width is

(a) $l_\omega = \lambda\theta$ (b) $l_\omega = \theta/\lambda$

(c) $l_\omega = \lambda/\theta$ (d) $l_\omega = \lambda^2/\theta$

13. Which of the following properties of radiation emitted under stimulated emission is not correct

(a) The emitted radiation is of a very precisely defined frequency

(b) The emitted radiation is in phase

(c) The stimulating and emitted radiation are coherent

(d) All are correct

14. A resonance phenomenon is associated with

(a) Stimulated emission (b) Spontaneous emission

(c) Both (a) and (b) (d) None of these

15. The directionality of a laser beam is measured by

(a) The visibility of interference fringes

(b) The size and aperature of the laser source

(c) The divergence angle of the beam with the distance from the source

(d) Nature of the lasing medium.

16. The spontaneous emission is a
    (a) Negative process
    (b) Non-equilibrium process that takes place when atoms are excited
    (c) Process that requires population inversion
    (d) Natural and equilibrium process that takes place when atoms are excited
17. Population inversion can be achieved in atoms if there is
    (a) Transfer of energy from laser substance to the reflecting mirrors
    (b) Continuous pumping of energy in the system by reflectors and amplification
    (c) An extra discharge of the gases in the gas discharge tube
    (d) Infinite amplification of light in the source by high discharge process
18. In a ruby laser
    (a) The chromium ions collide with Al ions to give out energy
    (b) A flash lamp is used to excite chromium ions
    (c) Very low peak power pulses of radiation are given out
    (d) A flash lamp is used to excite Al atoms
19. The useful material for semiconductor laser source is
    (a) Silicon (b) Organic semiconductor
    (c) Germanium (d) Gallium arsenide
20. In semi conductor laser diode the population inversion takes place
    (a) At the junction on n side only
    (b) Near the junction as the p side
    (c) Far away from the junction on the neutral n side
    (d) Far away from the junction on the neutral p-side
21. For stimulated emission to takes place in an active medium, the condition is
    (a) The production of population inversion in energy levels
    (b) Large volume of active medium
    (c) Sufficient number of atoms in ground stable
    (d) Coherent incident radiation on the medium

Theoretical Questions

1. What is laser? Explain its principle.
2. What are the characteristic properties of a laser beam? Describe its important applications.

3. Explain the term 'spontaneous emission' and 'stimulated emission' of radiation. Obtain a relation between transition probabilities of spontaneous and stimulated emissions.
4. What are the differences between spontaneous and stimulated emission? Why is spontaneous radiations are incoherent?
5. What are the differences between spontaneous and stimulated emissions? Which are, out of these two processes, is maximized in case of laser action, give suitable justification to your answer?
6. What are Einstein's coefficients A and B? Establish a relation between them.
7. Laser action is some times called "inverted absorption" explain which situation may A/B be small enough for laser action.
8. Explain coherence length, spatial coherence and temporal coherence.
9. Explain the spontaneous and stimulated emission of radiation. Describe the working principle of ruby laser.
10. Explain spontaneous and stimulated emission of radiation. How stimulated emission taken place with the exchange of energy between Helium and Neon atoms?
11. Explain the action of Helium-Neon laser. How is it superior to a Ruby laser.
12. What is a laser? What are its application? What is the difference between a continuous and a pulse laser? Give one example of each type and discuss the basic principle and working of any one of them.
13. What is laser? Explain its principle.
14. Draw a neat diagram of He-Ne laser and describe its method or working. What are the characteristic of laser beam? Discuss its important applications.
15. In He-Ne laser, what is the functions of the atoms? Explain the answer with the help of energy level diagram for He-Ne. Describe with a neat sketch the working of He-Ne laser.
16. Obtain the threshold condition for sustained laser action. On what factors does threshold population inversion depend?
17. What are semiconductor lasers? How is population inversion achieved in these lasers?
18. What are the disadvantages in a homojunction semiconductor laser and how are they overcomed in heterojunction.
19. With a suitable diagram explain the construction and working of a semiconductor laser.
20. What is 'Q-switching'? How is it achieved? Out line the applications of Q-switched lasers.

Numerical Problems

1. In a CO_2 laser, the energy difference between two levels is 0.121 eV. Calculate the frequency of radiation

   [0.029×10^{-15} sec^{-1}]

2. Calculate the population ratio of two states in He-Ne laser that produces light of wavelength 6000 Å at 300 K

   [e^{-80}]

3. The Ruby laser has two states at 27 °C and 227 °C. If is emits radiation of wavelength 7000 Å, then calculate the relative population.

   [$e^{-41.1}$]

4. Find the intensity of a laser beam of 1 mw power and having a diameter of 1.4 mm. Assume the intensity to be uniform across the beam.

   [6.5×10^2 w/m^2]

5. A laser beam has a wavelength of 8×10^{-7} m and aperture 5×10^{-3} m. The laser beam is sent to moon. The distance of the moon is 4×10^5 km from the earth. Calculate (a) the angular spraed of the beam (b) The axial spread when it reaches the moon

   [(a) 1.6×10^{-4} rad; (b) 4.096×10^3 Km^2]

6. A laser beam of wavelength 6328 Å and aperture 5×10^{-3} m from He-Ne laser can be focussed on an area equal to the square of its wavelength. If the laser radiates energy at the rate of 10 mw find (a) the angular spread of the beam and (b) the intensity of focussed beam

   [(a) 1.265×10^{-4} rad; (b) 2.5×10^{16} w/m^2]

7. A light of wavelength 4800 Å has wave train length of 25 waves. What are coherent length and coherent time.

   [12 μm; 4×10^{-14} sec]

8. A laser operates at wavelength 6000 Å and its spectral width $\Delta\nu$ is 10^{-2} Hz. For this laser calculate (a) coherence length (b) quality factor

   [5×10^{14} Hz; 3000 km]

9. The red cadmium line has wavelength 643.8 nm. If coherence time be $\tau_c \sim 10^{-9}$ second. Find spectral purity on the line.

$$[\Delta\lambda = 0.1 \text{ nm}; \frac{\lambda}{\Delta\lambda} = 6438]$$

10. A continuous laser of wavelength λ = 650.0 nm is Q-switched into 0.5 n sec pulses. Compute its coherent length, band width and line width?

[0.15 m; 2×10^9 Hz; 0.28 Å]

11. A ruby laser emits 0.1 J pulses of light of wavelength 720 nm. What is minimum number of Cr^{+3} ions are there in ruby?

[3.625×10^{17}]

12. A semiconductor heterojunction laser is grown with GaAsP active region layer with lattice constant 0.56 nm. Assume that laser transiton occurs from the bottom of conduction band involving band energy gap 1.85 eV. Determine wavelength of laser emitted out?

[6707 Å]

Answers to Objective Questions

1. (c)	2. (a)	3. (d)	4. (d)	5. (c)	6. (b)
7. (d)	8. (a)	9. (a)	10. (d)	11. (c)	12. (c)
13. (d)	14. (a)	15. (c)	16. (d)	17. (b)	18. (b)
19. (d)	20. (b)	21. (a)			

CHAPTER 4

Radiation

The spectral density of black body radiation... represents something absolute, and since the search for the absolutes has appeared to me to be the highest form of research, I applied my self vigorously to its solution.

–Max Planck

4.1 Introduction

Radiation is the phenomenon of propagation of heat through vacuum or a material medium without the medium playing an active part. However, it does not mean that the intervening medium will not be heated when the heat radiation fall on it or pass through it. Some materials like glass, wood, metal etc, get heated to less or more extent while materials like quartz, rock slat etc., allow the radiation to pass through them without being appreciably heated. It must also be understood that the molecules of intervening material owing to the heat vibrations act as source of radiations which are in the form of waves that carry the thermal energy.

The statement that 'no intervening medium is required for the radiation phenomenon' simply means that unlike conduction and convection, in radiation phenomenon it is not necessary for the intervening medium to get itself heated. Also since heat is a form of energy, the mode of propagation is in the form of waves just as light waves travel even in empty space.

4.2 Emissive and Absorptive Power

Emissive power

When the temperature of a body is higher than that of its surroundings, then due to this temperature difference, radiant energy is continuously 'emitted' from the surface of the body. At a given temperature, the rate of emission depends upon the nature of the surface, the area of the surface and the wavelength of the emitted radiation. The total amount of radiant energy emitted per unit area of a surface per second is called the 'emissive power' e of that surface. Its unit is Joule per metre^2 per second (J m^{-2} sec^{-1}).

Radiant energy has different wavelengths and its spectrum can be obtained. The amount of radiant energy emitted per unit area of a surface per second per unit wavelength range at wavelength λ is called the 'spectral emissive power' e_λ of the surface at wavelength λ. Its unit is Joule per metre^2 per second per angstrom. (J m^{-2} s^{-1} A^{o-1}).

If the Spectral emissive power of a surface at wavelength λ be e_λ, then the amount of radiant energy emitted per unit area of the surface per second between the wavelengths λ and $\lambda + \Delta\lambda$ will be $e_\lambda \Delta\lambda$.

Absorptive power

When radiant energy falls on a body, it is partly reflected (or scattered) from the surface of the body, partly absorbed by the surface and the rest is transmitted to the other side through the body. At a given temperature, the absorbed part of the radiant energy depends upon the nature of the surface and the wavelength of the radiation. The ratio of the radiant energy absorbed by a surface in a given time to the total radiant energy incident on the surface in the same time, is called the 'absorptive power' or 'absorption coefficient' of the surface, it is denoted by a. Since a is a ratio therefore it has no unit.

If the radiant energy incident on the surface has different wavelengths, then the absorptive power of the surface for these wavelengths will be different. The ratio of the radiant energy absorbed by a surface per unit wavelength range at wavelength λ in a given time to the incident radiant energy in that time is called the 'spectral absorptive power' a_λ of the surface of wavelength λ.

If the spectral absorptive power of a surface at wavelength λ be a_λ, then the incident radiant energy Δ Q between the wavelength λ and $\lambda + \Delta\lambda$, then the energy absorbed by the surface will be $a_\lambda \Delta Q$.

The emissive and absorptive power of some of the source are given as under

Emissive power

1. Lamp black 98%
2. Platinum 11%
3. Copper 5%
4. Silver 2%
5. Indian ink 88%

Absorptive power

1. Lamp black 98%
2. Shellac 72%
3. Copper 13%

4B.3 Black Body

A perfectly black body is that which can absorb radiation of all wavelengths which fall on it. It neither reflects nor refracts and so its appears black, whatsover form of radiations may fall on it. The property of a black body is that when hot, it gives out radiations of all wavelengths or colours because it has the characteristic of emitting full or total radiation at high temperature. A perfect black body being good absorber, and is also perfect radiator, i.e., it gives out maximum emission of radiation for every wavelength that falls on it at any temperature.

4.3.1 Perfectly Black Body

A perfectly Black body is one which absorbs completely all the radiation falling on its surface what ever be the wavelength.

Thus for a perfectly black body, we have

$$a_\lambda = 1$$

Since a_λ is same for all values of λ, so for a perfectly black body.

$$a = 1$$

Since perfectly black body is a perfect absorber hence according to Kirchhoff's law it will also be a perfect radiator (good absorbers are good radiators also). So when a perfectly black body is heated to a high temperature, it emits radiation of all possible wavelengths.

No known body is a perfect black body. Lamp black and platinum black, when absorbs all the visible and near infrared parts of the incident radiation, reflect the far infrared parts.

4.3.2 Pratical Black Body

The radiation in a uniform temperature enclosure has all possible wavelengths. It is called the full radiation. Thus if a perfectly black body be placed inside the enclosure, it will completely absorbs this radiation. But the quantity and quality of radiation inside a uniform-temperature enclosure is not affected by the presence of any body inside the enclosure. Therefore the radiation emitted by the perfectly black body will be identical, in quantity and quality, to the radiation present in the enclosure. Fery, on this basis designed a perfectly black body which is shown in Fig. 4.1.

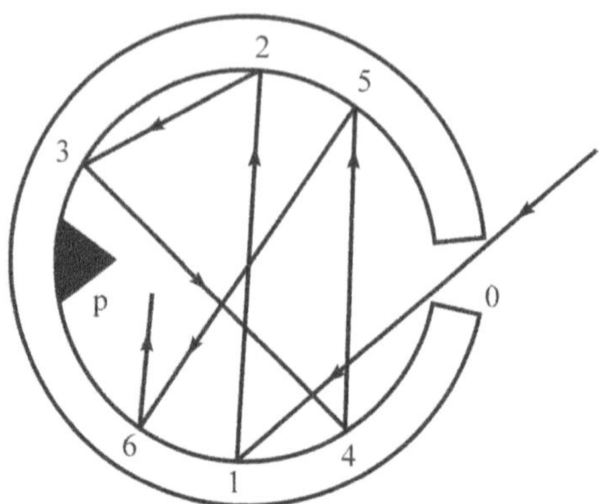

Fig. 4.1

It is a double-walled metallic hollow sphere with a narrow opening O. The inner wall of the sphere is blackened. Just opposite the opening O, a conical projection P is made on the inner wall. Any radiation incident at the hole O suffers multiple reflection at the inner wall of the sphere and is thus absorbed. If the cone P were not there, the radiation reflected from this part of the wall could have come out of the opening O. This possibility is eliminated by the cone P. Thus any radiation incident at O is almost completely absorbed. Hence the opening O behave like a perfectly black body. If the sphere be uniformly heated to a high temperature, then the radiation, coming out of the opening O will be very nearly the black body radiation.

It s not essential that a perfectly black body should appear dark.

4.4 Kirchoff's Law

According to this law, the ratio of the emissive power to the absorptive power for radiation of a given wavelength is the same for all bodies at the same temperature and is equal to the emissive power of a perfectly black body at that temperature.

4.4.1 Importance of Kirchoff's Law

A conclusion from the Kirchoff's law is that if a surface is a good absorber of a particu;ar wavelength of radiation, it is also a good emitter of that wavelength of radiation. This fact is observed in daily life.

We know that red glass appears red in white light, because if reflects red light and absorbs the rest. Since the average effect of the rest of the light i.e., white minus red is greenish, we can say that the red glass absorbs green light strongly.

Hence according to Kirchoff's law, the emissive power of red glass should be maximum for green light. That is why when red glass is heated to a high temperature it gives greenish glow, because now it strongly emits green light. Similarly green glass, which absorbs red light, gives reddish glow on being heated. Fraunhofer lines found in sun's spectrum are the best example of the illustration of Kirchoff's law.

4.4.2 Proof of Kirchoff's Law

Let a body be placed in a uniformly heated enclosure at a constant temperature T. In the equilibrium state, the temperature of the body will be equal to that of the enclosure. Suppose at this temperature, the absorptive power of the surface of the body for the wavelength λ and $\lambda + \Delta\lambda$ be incident per second on the unit surface area of the body. Then an amount of energy $a_\lambda\ \Delta Q$ will be absorbed by unit surface area of the body per second.

Now if at temperature T and wavelength λ, the emissive power of the surface of the body be e_λ, then an amount of energy $e_\lambda\ \Delta\lambda$ between the wavelength λ and $\lambda + \Delta\lambda$ will be emitted by unit surface area of the body per second by virtue of its temperature.

In a uniform-temperature enclosure, the quantity and quality of radiation is not affected by the presence of any body inside it. Hence what ever radiant energy of any wavelength the body absorbs, it emits exactly equal amount of radiant energy of the same wavelength . Thus

$$a_\lambda\ \Delta Q = e_\lambda \Delta\lambda \qquad .....(1)$$

Now, a perfectly black body absorbs whole of the radiant energy incident on it i.e., its absorptive power $a\lambda = 1$. So if a perfectly black-body, whose emissive power is E_λ is placed inside the enclosure, than for it the equation corresponding to equation (1) will be

$$\Delta Q = E\lambda\ \Delta\lambda \qquad .....(2)$$

Dividing (1) by (2) we get

$$a_\lambda = \frac{e_\lambda}{E_\lambda} \qquad .....(3)$$

or

$$\frac{e_\lambda}{a_\lambda} = E_\lambda \qquad .....(4)$$

Since for a given temperature, E_λ is constant, so for all materials at that temperature, we have

$$\frac{e_\lambda}{a_\lambda} = E_\lambda \qquad \text{(constant)}$$

$$\left(\frac{e_\lambda}{d_\lambda}\right)_1 = \left(\frac{e_\lambda}{a_\lambda}\right)_2 = \left(\frac{e_\lambda}{a_\lambda}\right)_3 \ldots = E_\lambda \qquad \text{(for perfectly black body)}$$

where 1, 2, 3, ... represents different surfaces.

Thus at a definite temperature and for a given wavelength, the ratio of the emissive power to the absorptive power for different surface is same. If a surface A radiates 10 times more energy than surface B, then at the same wavelength the surface A will absorb 10 times more energy than B.

4.5 Uses and Effects of Good and Bad absorber of Thermal Radiations

1. *In deserts, days are very hot and night are very cold:* Sand is good absorber of heat, so in day time it absorbs heat and becomes hot. In the night, it loses heat by radiation and becomes cool.
2. *Cloudy nights are hotter than clear sky nights:* Earth becomes hot in day time due to solar heat and becomes cool in night by radiation. When the sky is clear, heat from the earth goes to the sky by radiation. But when sky is clouded, heat radiation strikes the clouds. But since the clouds are bad absorber of heat, the thermal radiation returns back to the earth, which therefore remains hot.
3. *Hills are cooler in summer:* Because of uneven surfaces of Hills, most of its part remain under shadow. So most of the hill's land is not heated up by the sun. Besides this, sun-rays fall slanting on the hills and are spread over a larger area. So the heat received by the hills per unit area is less and they are less heated compared to planes.

After sun-set, earth starts cooling by radiation. For a given mass, the surface area of hills is larger compared to planes, so hills become cool more rapidly. That is why nights at hills are even more colder.

4.6 Stefan's Law

As the temperature of a surface rises, more and more radiant energy is emitted by it. In this connection, Stefan in 1879, enunciated a law called 'Stefan law'.

According to this law, the total radiant energy emitted per second per unit surface area of a black body is proportional to the fourth power of the absolute temperature of the body.

Thus if the absolute temperature of a black body is T, then the energy E emitted per second per unit surface area of the body is given by

$$E \propto T^4$$

$$R = \sigma T^4$$

Where σ is a constant, called the Stefan's constant. The unit of σ is (Joule/metre^2–sec–k^4) or watt/meters^2 – k^4

According to this law, on raising the absolute temperature of a body to double its initial value, the rate of energy emitted by it becomes 2^4 or 16 times, and when the absolute temperature is raised to three times, the energy emitted becomes 3^4 or 81 times.

In 1884, Boltzmann gave its theoretical proof and proved that stefan's law is applicable for a perfectly black body. Hence this law is also called 'Stefan-Boltzmann law'.

Stefan's law does not state the rate of cooling of a body, because the rate of cooling also depends upon the temperature of the surroundings.

Suppose a black body at absolute temperature T is surrounded by a black enclosure at temperature T'. It will emit energy $E = \sigma T^4$ per second per unit surface area.

Beside this, it will also absorb energy $E' = s\ E' = \sigma T'^4$ emitted by the enclosure area. So, the net energy emitted per second per unit area of the black body is

$$E - E' = \sigma\left(T^4 - T'^4\right)$$

It is important to note that the rate of radiant energy emitted does not depend upon the shape, size material etc., of the body. It depends only on the temperature. Though it is practically impossible to obtain a 100% perfectly black body, even then its concept is important. Even more important is the fact that the wavelength distribution in the radiation of black body depends only on the temperature of the body.

If the emissive power of a body at absolute temperature T be e, then the energy emitted by its unit area per second will be $\sigma T^4 e$. If A be the surface area of the body, then the energy emitted per second by the whole body will be

$$\sigma T^4 eA$$

4.7 Newton's Law of Cooling

Newton studied experimentally the cooling of hot bodies in air and established a law, called Newton's law of cooling. According to this law, when a hot body is cooled in air, the rate of loss of heat by the body is proportional to the temperature difference between the body and its surroundings that is

$$\text{Rate of loss of heat} \propto \text{temperature difference}$$

Clearly as the body cools down, its rate of cooling slows down. For the validity of this law, the temperature difference between the body and its surroundings should not be large.

4.8 Derivation of Newton's Law of Cooling from Stefan's Law

Suppose the absolute temperature of a hot body is T and that of its surrounding medium is T'. Suppose the emissive power of the body is e (< 1). According to Stefans' law, the net rate of energy emission by the body, or the net rate of loss of heat by the body is

$$e\,\sigma\,(T^4 - T'^4)$$

Suppose the temperature T of the body is only slightly higher than the temperature $T'$of its surrounding medium and that the difference between these temperature is $(T - T') = \Delta T$

Then $T = T' + \Delta T$ and so the rate of loss of heat is

$$e\,\sigma\left[(T'+\Delta T)^4 - T'^4\right] = e\,\sigma\left[T'^4\left(1+\frac{\Delta T}{T'}\right)^4 - T'^4\right]$$

Since ΔT is very small compared to T'. So neglecting higher power of ΔT in the binomial expansion, we have

$$\left(1+\frac{\Delta T}{T'}\right)^4 = 1 + 4\frac{\Delta T}{T'}$$

$$\therefore \text{ rate of loss of heat} = e\,\sigma\left[T'^4\left(1+4\frac{\Delta T}{T'}\right) - T'^4\right]$$

$$= e\,\sigma\left[4T'^3\right]\Delta T$$

$$\propto \Delta T$$

Thus the radiant energy (heat) emitted per second by a body is proportional to the difference in temperature of the body and that of the surrounding medium. The rate of cooling of the body depends upon the energy radiated by it. Hence, the rate of cooling of a body is proportional to the temperature difference between the body and its surrounding medium. Thus, Newton's law is only a special case of Stefan's law and is applicable only at small temperature difference.

4.9 Spectral Distribution of Black Body Radiation

A perfect black body is a full radiation, i.e., it emits radiation of all possible wavelengths. Lummer and Pringsheim in 1899, studied the spectral distribution of energy (i.e., energy distribution among different wavelengths) in the radiation of a black body at different temperatures. They used an electrically-heated chamber with a narrow hole as the black body. The temperature of the black body was measured by a thermocouple. The experimental arrangement is shown in Fig. 4.2. Radiation coming from the black body O is allowed to fall on a slits S by means of a concave mirror M_1. The slit S is in the focal plane of another concave mirror M_2, and so the radiation falling on M_2 is reflected in the form of a parallel beam and falls on a fluorspar prism P. After refraction through the prism P, the radiation falls on the mirror M_3 which focuses it on a thermopile B placed in its focal plane. By rotating M_3 about a vertical axis, radiant energy of different wavelengths is made to fall on the thermopile one after the other. Each time the deflection in the galvanometer G connected in the thermopile circuit is read. These deflections are the measure of radiant energies of different wavelengths falling on the thermopile. Now a graph is plotted between energy and wavelength. The graph so obtained is called 'spectral distribution curve'.

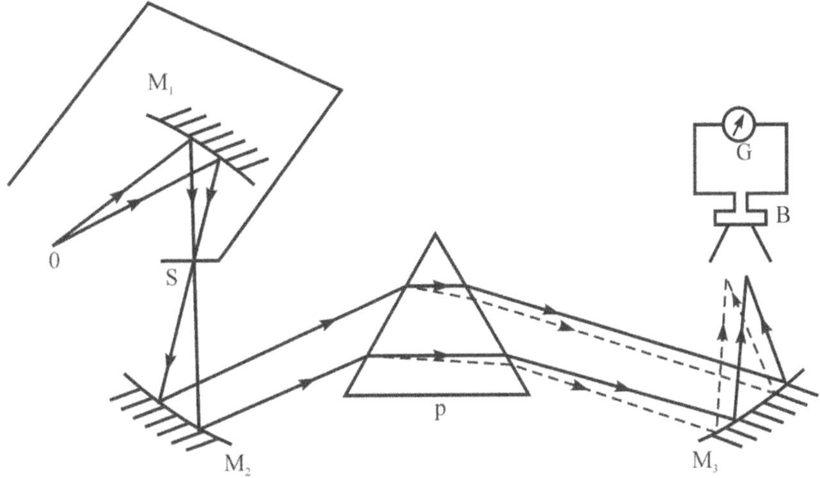

Fig. 4.2

Lummer and Pringsheim heated the black body to different temperatures and drew distribution curves for all temperatures on the same graph paper. The curves so obtained are shown in Fig. 4.3. These curves whose general shape for all temperatures is the same give following information regarding the characteristic of black-body radiation.

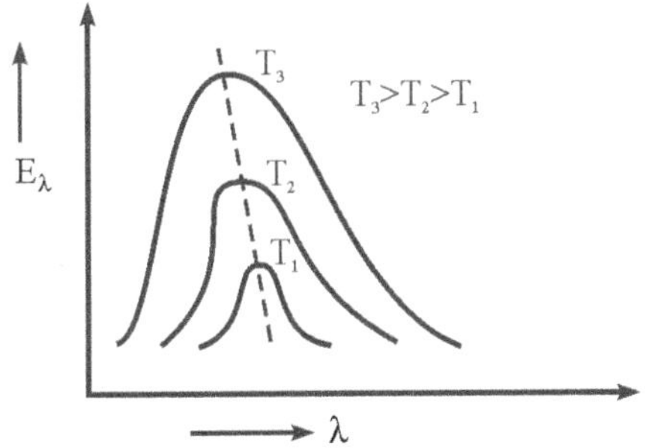

Fig. 4.3

1. At a given temperature T, with increase in wavelenth λ, the energy E_λ first increases, reaches a maximum and then decreases. This means that for a given temperature, the radiant energy emitted by a black body is maximum for a particular wavelength.
2. As the temperature increases, the peak of the curve shifts toward shorter wavelength (λ) i.e., the maximum value of E_λ is obtained at smaller value of λ.

   In 1896, Wein established the following relation between temperature T and λ_m (for which E_λ is maximum)

   $$\lambda_m T = \text{constant} = b$$

   This is called 'Weins displacement law'. It shows that λ_m is inversely proportional to the absolute temperature T, that is, as the temperature of the back body rises, the maximum energy of the emitted radiation shifts towards shorter wavelength. This means that on heating a body to ordinary temperature, only long-wavelength radiation is emitted. But as the temperature of the body is raised, radiation of shorter and shorter wavelength is also emitted in increasing quantity. That is why when iron is heated, it first becomes light-red, then dark-red, then yellow and ultimately it becomes white.

   From Wein's displacement law, we have

   $$\lambda_m = \frac{b}{T}$$

   A graph between λ_m and T will be a rectangular hyperbola as shown in Fig. 4.4.

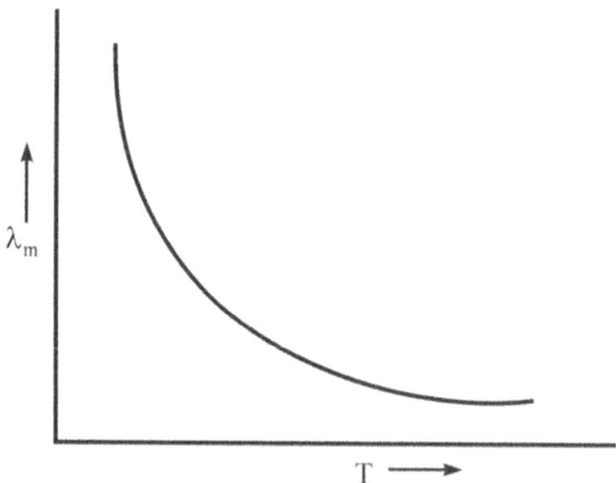

Fig.4.4

3. Further, as the temperature rises, the area enclosed by the curve goes on increasing. The area enclosed by a curve represents the total radiant energy E (of all wavelength) emitted by the black body at that temperature. When the areas

enclosed by different curves are measured, they are found to be proportional to the fourth power of the corresponding absolute temperature that is

$$E \propto T^4$$

Thus these curves verify stefan's law

4. Wein also proved that at a temperature T, the maximum emitted energy $(E_\lambda)_m$ corresponding to a wavelength λ, is proportional to the fifth power of that temperature (T^5) i.e.,

$$(E_\lambda)_m \propto T^5$$

It means that on raising temperature, the maximum energy $(E_\lambda)_m$ emitted corresponding to wavelength λ decreases very rapidly.

4.10 Planck's Hypothesis: Quantum Model of Radiation

The black-body radiation has, in general, all wavelengths from zero to infinity. At a certain temperature, the total radiant energy is distributed among different wavelengths. This distribution was experimentally studied by Lummer and Pringheim. But no formula based on mechanics and thermodynamics could be obtained which could express the distribution of radiant energy among different wavelength at a given temperature.

In 1900, Planck put forward a revolutionary idea that the emission of radiation is not continuous, but takes place in the form of small bundles or packets of definite amount of energy which are called Quanta. Later on, these bundles were called 'photons'. The energy associated with each photon (quantum) is $h\gamma$, where γ is the frequency of radiation and h is Planck's constant. Thus, emitted energies may be $h\gamma$, $2h\gamma$, $3h\gamma$,... but not in between.

4.11 Momentum of Photon

The rest-mass of a photon is zero, but its kinetic mass is not zero. The photon travels with the speed of light C.

According to Einstein's mass-energy relation ($E = mC^2$), the energy of a photon is

$$E = h\nu = mC^2$$

Hence the kinetic mass of the photon is

$$m = \frac{h\nu}{C^2}$$

But $\nu = C/\lambda$, where λ is the wavelength of the photon

$$m = \frac{h}{C^2}\left(\frac{C}{\lambda}\right) = \frac{h}{C\lambda}$$

Thus, the kinetic mass of the photon is

$$m = \frac{h\nu}{C^2} = \frac{h}{C\lambda}$$

The momentum of the photon is

P = Kinetic mass × velocity of photon

P = mC

$$P = \frac{h\nu}{C^2} \times C = \frac{h\nu}{C}$$

Again

$$\nu = C/\lambda$$

∴

$$P = \frac{h}{C}\left(\frac{C}{\lambda}\right) = \frac{h}{\lambda}$$

Thus, the momentum of the photon is

$$P = \frac{h\nu}{C} = \frac{h}{\lambda}$$

Solved Problems

1. A body having surface area 5 cm^2 is at 727°C and radiates 300 joule energy in one minute. Find out its emissive power. The Stefan-Bolzmann constant σ is 5.67 × 10^{-8} $W/m^2 - k^4$.

Sol. The energy emitted per second from the surface of area A of a body (which is not a black body) is given by

$$E = \sigma T^4 A e$$

∴

$$e = \frac{E}{\sigma T^4 A}$$

Here,

E = 300 J/min = 5 J/s

T = 727 + 273 = 1000 k

$A = 5\ cm^2 = 5 \times 10^{-4}\ m^2$

$$\therefore \qquad e = \frac{5\ J/sec}{5.67 \times 10^{-8} W/m^2 - k^4 \times (1000k)^4 \times 5 \times 10^{-4}\ m^2}$$

$$e = 0.176$$

2. The surface area of a black body is $5 \times 10^{-4}\ m^2$ and its temperature is 727 °C. Calculate the energy radiated by it per minute. Stefan's constant = $5.67 \times 10^{-8}\ J/m^2 - s - k^4$.

Sol. According to Stefan's law

$$E = \sigma T^4$$

If the area of the black body be A, then the energy emitted by it per minute = 60 seconds

$$E = \sigma T^4 \times A \times 60$$

Here

$T = 727\ °C = 727 + 273 = 1000\ k$

$A = 5 \times 10^{-4}\ m^2$

$$\therefore \qquad \text{energy} = (5.67 \times 10^{-8}) \times (1000)^4 \times (5 \times 10^{-4}) \times 60$$

$$= 1.7 \times 10^3\ J$$

3. A black body radiates heat energy at the rate of 1.0×10^6 Joule per second per meter^2 at the temperature of 127 °C. Calculate the temperature of the black body at which the rate of heat radiation is 81×10^6 Joule per second per meter^2.

Sol. According to Stefan's law, the rate of energy emission E by a black body per unit surface area is proportional to the forth power of its absolute temperature.

$$E = \sigma T^4 \qquad .....(i)$$

where σ is the Stefan's constant. If the temperature of the black body is raised from T to T'. Then

$$E' = \sigma T'^4 \qquad .....(ii)$$

From these equations, we have

$$\frac{E'}{E} = \left(\frac{T'}{T}\right)^4$$

$$\frac{T'}{T} = \left(\frac{E'}{E}\right)^{1/4}$$

Here $E = 1.0 \times 10^6$ and

$E' = 81 \times 10^6$ Joule/m^2-sec

$$\therefore \quad \frac{T'}{T} = \left(\frac{81\times10^6}{1.0\times10^6}\right)^{1/4}$$

$$\frac{T'}{T} = (81)^{1/4} = 3$$

But $T = 127 + 273 = 400$ k

$T' = 3T = 3 \times 400 = 1200$ k $= 927$ °C

4. Earth receives an amount of heat radiation 1.4×10^3 w/m^2 from the sun. Assuming that the earth re-emits all the radiation received from the sun, calculate the surface temperature of the earth ($\sigma = 5.67 \times 10^{-8}$ w/m^{2}k^4)

Sol. By Stefan's law

$$E = \sigma T^4$$

$$T^4 = \frac{E}{\sigma}$$

$$T^4 = \frac{1.4\times10^3\,W/m^2}{5.67\times10^{-8}\,W/m^2k^4}$$

$$T^4 = 247 \times 10^8\,k^4$$

$$T = 3.96 \times 10^2\,k \simeq 400\,k$$

5. Sun and moon emit maximum radiation at 5000 A° and 15 μ respectively. If the temperature of the sun is 6000 k, then find the temperature of the moon.

Sol. By Wien's displacement law, we have

$$\lambda_m T = \text{Constant}$$

$$(\lambda_m T)_{sun} = (\lambda_m T)_{moon}$$

$$5000\,A^\circ \times 6000\,k = (15 \times 10^4)A^\circ \times T$$

$$\therefore \quad T = \frac{5000\,\mathring{A} \times 6000k}{(15\times10^4)\,\mathring{A}}$$

$$T = 200\,k$$

Conceptual Questions and Answers

1. **A metallic ball has a black spot. The ball is heated to around 1000 °C and is then taken in to a dark room. It is found that the black sport looks brighter than the rest of the ball. How can you explain this.**

Ans. The black spot is a better absorber of radiation than the ball. Hence at higher temperature it is a better emitter.

2. **Some photographic plates are not affected by red light, but are immediately blackened by white light.**

Ans. The energy of red-light photos is less, so they are unable to affect the photographic plate.

3. **If light propagation takes place through photons then why does the light arriving at the eye not appear discontinuous.**

Ans. The effect of photon on the retina persists for nearly 1/10 second. If during this time another photon arrives, then the combined effect will be continuous. When visible light enter our eye, nearly 10^{18} photons strikes the retina per second. It means that every point on the retina is receiving photons continuously. Hence their effect on the mind is continuous.

4. **How will a white-hot red glass looks in a dark room? If red hot, then?**

Ans. According to Kirchoff's law, the emissive power of red glass will be maximum for green light. Therefore, white hot glass in a dark-room will look green. Red glass, if red-hot, in a dark room will become invisible because it will not emit red colour.

5. **Star A emits green light while star B emits blue light which of these two has a higher temperature**

Ans. Star B

6. **In a dark room, a piece of coal at normal temperature is not visible but a burining coal is visible why?**

Ans. At normal temperature, bodies are visible by reflected light. In a dark room, no light is incident on the coal, no light is reflected and hence the coal is not visible. At high temperatures, the radiation emitted by the bodies have sufficient quantity of visible light which makes the bodies visible in dark. In this state, the bodies are said to be self-luminuous.

7. **On gradually raising the temperature of a body, why does it initially appear red.**

Ans. $\lambda_m T$ = constant on heating up to ordinary temperature. Only long-wavelength (red) radiation is emitted. As the temperature rises, shorter wavelength are also emitted in more and more quantity.

8. If a red glass be heated to a high temperature, with what colour it will glow? Gree glass? Blue glass?

Ans. Green colour, red colour, yellow colour

9. Why does the apparent colour of radiation emitted from a hit body shifts from red to yellow and finally to blue as the temperature of the body increases.

Ans. As the temperature increases, light waves of smaller and smaller wavelengths are also emitted.

10. In a burning-coal oven, the cavities in the coal pieces looks brighter than the surface of the coal , though their temperature is not higher than the coal surface. Why?

Ans. The cavities in the hot coal behave like a black-body. Hence the rate of radiation (σ e T^4) from these cavities, the emissive power e is 1. For the surface of the coal the value of e is less than 1, its rate of radiation is less, so, cavities, being at the same temperature, look brighter.

Objective Questions

1. As compared to a person with white skin, another person dark skin, will experience
   (a) Less heat and less cold (b) More heat and more cold
   (c) More heat and less cold (d) Less heat and more cold
2. The best ideal black body is
   (a) A lump of charcoal heated to a high temperature
   (b) A glass surface coated with coal-fat
   (c) A metal coated with a black dye
   (d) A hollow enclosure blackened inside and having a small hole
3. A hot body will radiate heat most rapidly if its surface is
   (a) White and polished (b) White and rough
   (c) Black and polished (d) black and rough
4. A black body which is at a high temperature TK, radiates energy at the rate of E ω/m^2. When the temperature falls to T/2 k, the radiated energy in ω/m^2 will be
   (a) E/4 (b) E/2 (c) 2E (d) E/16
5. The temperature of a piece of metal is raised from 27 °C to 84 °C. The rate at which the metal radiates energy increases nearly.
   (a) Two times (b) Four times
   (c) Eight times (d) Sixteen times

6. The colour of a star indicates its

   (a) Weight (b) Size (c) Distance (d) Temperature

7. Which of the following is roughly the rate of solar-energy (in kilowatt) falling per meter^2 surface area of the earth

   (a) 1 (b) 100 (c) 0.1 (d) 0.0001

8. A red-glass piece is heated until it glows in dark. The colour of the glow will be

   (a) red (b) Orange (c) Green (d) Violet

9. A spherical black body with a radius of 12 cm radiates 450 ω power at 500 k. If the radius were halved and the temperature doubled, the power radiated in watt would be

   (a) 225 (b) 450 (c) 900 (d) 1800

10. A photon is

    (a) A quantum of radiation

    (b) A quantum of matter

    (c) A positively charged particle

    (d) A negatively charged particles

Theoretical Questions

1. What is meant by 'emissive power' and 'absorptive' power of a surface.
2. Write Kirchoff's law of radiation and prove it.
3. Prove that for a wavelength λ at a definite temperature, the ratio of emissive power e_λ to absorptive power a_λ of all bodies is equal to the emissive power E_λ of a perfectly black body at that temperature.
4. State Stefan's law of thermal radiation. Derive Newton's law of cooling from this law.
5. Define a perfectly black body. Draw spectral distribution curves of black body radiation and write the effect of rise in temperature on it.
6. Explain's Planck's hypothesis. Using this obtain an expression for the mass of a moving photon.

Numerical Problems

1. Two bodies heated to 327°C and 27°C are placed in an evacuated vessel at 27°C. Compare the rates of heat loss from them

[0.52]

2. The rate of radiation from a black-body at 0°C is E Joule/sec. Find the rate of radiation by the same black-body at 273°C

[16 E Joule/sec]

3. Two bodies having temperature of 527°C and 127°C respectively are radiant energy. Compare their rate of radiant energy

[16:1]

4. A black-coated metallic ball of radius 1.0 cm is heated to 327°C of the temperature of the surroundings be 27°C, how much heat will be emitted per second from the ball

[8.65 Joule]

5. Calculate the rate of emission of energy in walt/cm^2 by a surface at 227°C, if its emissive power be 0.60.

[0.21 Watt/cm^2]

6. A black body radiates energy at the rate of 1.6×10^6 J m^{-2} s^{-1} at a temperature of 227°C. Calculate the temperature of the body at which it radiates energy at the rate of 8.1×10^6 J m^{-2} sec^{-1}

[477°C]

7. The temperature of a black-body is 727°C and its surface area is 0.1 m^2. How many calories of heat is this black-body emitting per minute

[8.1×10^4 calories]

8. The temperature of the surface of a silver sphere of radius 5 cm is 527°C. If the emissive power of the surface is 0.04, find out the energy of the emitted radiation per second from the surface of the sphere

[29.2 Watt]

9. If the in the spectral energy distribution of the sun's spectrum, the maximum occurs at 4750A° an the temperature of the sun is 5777°C, then what will be the temperature of that star for which the maximum spectral distribution occurs at 9500 A°

[2752°C]

10. In solar-radiation, the value of λ_m is 4753 A° and the temperature of the sun is 6080.3 K. Calculate the Wein's Constant b.

[2.89×10^{-3} m-k]

Answers to Objective Questions

1. (b)	2. (d)	3. (d)	4. (d)	5. (a)	6. (d)
7. (a)	8. (c)	9. (d)	10. (a)		

CHAPTER 5A

Wave Mechanics

Erwin with his psi can do calculations quite a few.
But one thing has not been seen
Just what does psi really mean?

5A.1 Introduction

At a meeting of the German Physical Society on Dec. 14, 1900, Max Planck read his paper "on the theory of the energy distribution law of the Normal Spectrum". This paper, which first attracted little attention, was the start of a revolution in physics. The date of its presentation is considered to be the birthday of quantum physics, although it was not until a quarter of a century later than modern quantum mechanics, the basis of our present understanding was developed by Schrodinger and others. Many paths converged on this understanding, each showing another aspect of the breakdown of classical physics.

Quantum physics represents a generalization of classical physics that includes the classical laws as special cases, just as relativity extends the range of application of physical laws to the region of high velocities, so quantum physics extends that range to the region of small dimensions.

And just as a universal constant of fundamental significance, the velocity of light c, characterizes relativity, so a universal constant of fundamental significance, now called Planck's constant h, characterizes quantum physics.

5A.2 Origin of Quantum Hypothesis

A black body absorbs all the radiations which fall on it. (since it does not reflect light and appears black, hence the name black body). Usually in the laboratory experiment, a hollow body (cavity) with blackened walls and having a small hole as shown in Fig.5A.1 acts as a black body.

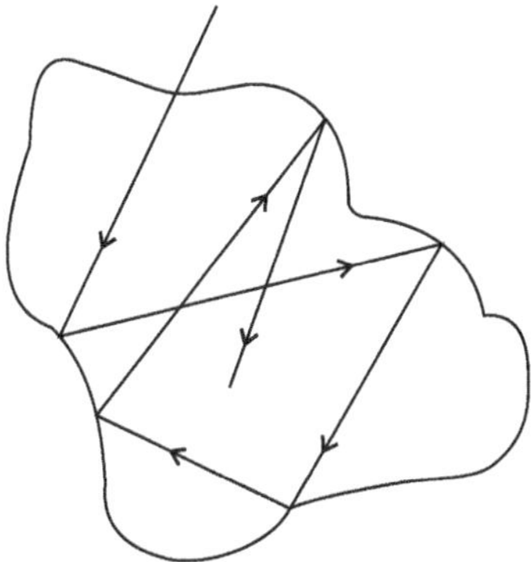

Fig. 5A.1 A black body.

The radiation continued in the body and emitted from the hole produces a black body spectrum. In the last century a number of experiments were carried out to measure the energy per unit volume contained by a black body, denoted by $\rho_T(\lambda)$ at different temperatures. Some of the representative curves showing the variation of $\rho_T(\lambda)$ as a function of λ (black body spectrum) at different temperatures are shown in Fig. 5A.2 below.

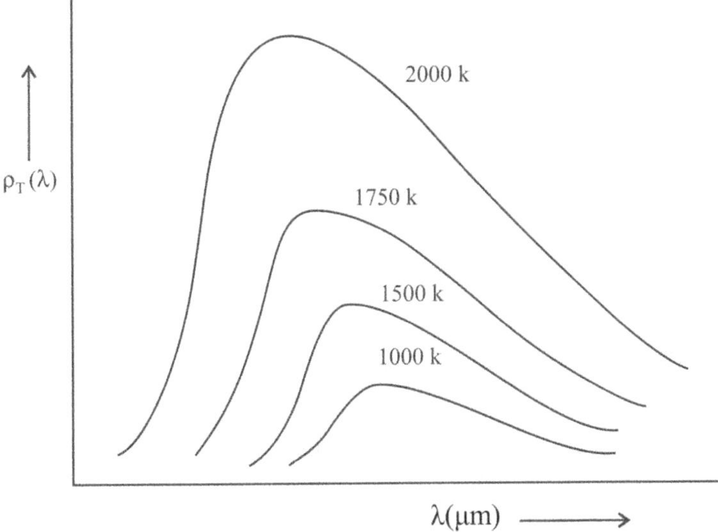

Fig. 5A.2 A black body spectrum.

Various investigation tried to explain the nature of these curves using well established laws of classical physics, including thermodynamics. It was Planck who came up with a theoretical explanation of the black body radiation curve given by

$$\rho_T(\lambda)\, d\lambda = a\, \lambda^{-5} \exp\left[\frac{-b}{KT}\right] d\lambda \qquad \text{.....(5A.1)}$$

Planck found empirically that by the replacement of

$$\exp\left[\frac{-b}{KT}\right] \text{ by } \left[\exp\left(\frac{-b}{KT}\right) - 1\right]$$

a set of parameter a and b could be obtained which would fit equation (5A.1) beautifully with the experimental data over all the values of λ. However, he find it extremely difficult to give a theoretical justification for the above mentioned replacement. Ultimately out of desperation, on 18th Dec 1900 Planck declared that the only way to derive the correct blackbody radiation formula was to postulate that the exchange of energy between matter (walls) and radiation (cavity) could take place only in bundles of certain size.

In classical physics, energy is regarded as an infinitely divisible quantity. Hence the exchange of energy between the walls and the cavity can take place in an infinite number of ways.

However, through his postulate, Planck reduced the number of ways to be finite. The exchange must take place in the units of U_0. Thus he introduced the idea of discreteness in the division of energy (which was thought to be infinitely divisible). If the energy U(λ) is to be exchanged, $U(\lambda)/U_0$ must be an integer. If it is not, Planck suggested that it should be an integer close to $U(\lambda)/U_0$.

Planck further postulated that the units of energy U_0 is directly proportional to its frequency i.e.,

$$U_0 = h\nu \qquad \text{.....(5A.2)}$$

The constant of proportionality h is known as Planck's constant in his honour. He was awarded Nobel prize for physics in 1918 for his work on black body radiation.

This new concept of Planck's gave birth to a new physics, known as Quantum Physics. Hence it is appropriate to take 18th Dec 1900 as the date of birth of quantum physics which later on developed in to quantum mechanics.

5A.3 Bohr's – Correspondence Principle

According to the correspondence principle:

1. The prediction of the quantum theory for the behaviour of any physical system must corresponds to the predictions of classical physics in the limit in which the quantum numbers specifying the state of the system become very large.
2. A selection rule holds true over the entire range of the quantum number concerned. Hence at large quantum numbers of any selection rule is required to obtain correspondence between classical and quantum physics then the same selection rule holds at low quantum number also.

5A.4 Wave Particle Duality

The optical phenomena, such as interference, diffraction and polarisation of light could be explained by wave theory of light. But wave theory of light failed to explain the photoelectric effect, Compton effect, absorption and emission of radiation by substances. These phenomenon could be easily explained by quantum theory of light. According to Einstein, the energy of light is concentrated in to small regions. This represents the smallest quantity of energy known as photon. This photon is an energy particle. Hence light shows itself wave nature at one hand and particle nature on the other hand. This shows that light is having a wave nature as well as particle nature. This nature of light is known as dual nature and the property is known as wave-particle duality.

5A.5 de-Broglie Concept of Matter Waves

In 1924, Louis de-Broglie proposed that the matter also possess dual character like light. His concept about the dual nature of matter was based on the following facts:

1. Matter and light, both are forms of energy and each of them can be transformed from one form to another.
2. Both are governed by the space-time symmetries of the theory of relativity.

According to Louis de-Broglie, a moving particle is surrounded by a wave whose wavelength depends upon the mass of the particle and its velocity. These waves associated with the matter particles are known as matter wave or de-Broglie waves.

de-Broglie equation provided a connection between a wave like property of radiation, i.e., the wavelength and a particle like property i.e., the momentum

$$\lambda = \frac{h}{p} \quad .....(5A.3)$$

or $$\lambda = \frac{h}{mv} \quad .....(5A.4)$$

where, m is the mass of the particle and v is its velocity.

1. For photon

We know that energy of a photon is given by

$$E = h\nu$$

or $$E = \frac{hc}{\lambda}$$

$$.....(5A.5)$$

By Einstein mass-energy equivalence

$$E = mc^2 \quad .....(5A.6)$$

From (5A.5) and (5A.6) we have

$$mc^2 = \frac{hc}{\lambda}$$

$\therefore$ $$\lambda = \frac{h}{mc} \quad .....(5A.7)$$

or $$\lambda = \frac{h}{p} \quad \text{where} \quad P = mc$$

This is the wavelength of de-Broglie wave, P is the momentum associated with photon.

2. For a material particle of mass m is moving with velocity v, then the momentum of the particle is

$$P = mv$$

Therefore, the wavelength of wave associated with the moving particle is given

$$\lambda = \frac{h}{mv} \quad .....(5A.8)$$

This wavelength is called de-Broglie wavelength. The greater the particle momentum, the shorter is its wavelength.

3. For a particle having Kinetic energy E

$\therefore$ $$E = \frac{1}{2} mv^2$$

$$E = \frac{1}{2m} m^2v^2$$

$$\therefore \quad E = \frac{p^2}{2m}$$

$$\Rightarrow \quad p = \sqrt{2m\,E}$$

.....(5A.9)

Thus the de-Broglie wavelength is given by

$$\lambda = \frac{h}{p}$$

$$\lambda = \frac{h}{\sqrt{2mE}} \qquad .....(5A.10)$$

4. According to kinetic theory of gases, the average kinetic energy of the material particle is given by

$$E = \frac{1}{2} mv^2 = \frac{3}{2} KT$$

Where $K = 1.38 \times 10^{-23}$ J/K

$$\lambda = \frac{h}{\sqrt{2mE}} = \frac{h}{\sqrt{2m.\frac{3}{2}KT}}$$

$$\lambda = \frac{h}{\sqrt{3mKT}}$$

5. Suppose an electron accelerated through a potential difference of V volts

Then work done by electric field = gain in kinetic energy

$$eV = \frac{1}{2} mv^2$$

$$m^2v^2 = 2\text{ me }V$$

$$mv = \sqrt{2meV}$$

$$\lambda = \frac{h}{mv} = \frac{h}{\sqrt{2meV}}$$

For an electron,

$m = 9.1 \times 10^{-31}$ kg

$e = 1.6 \times 10^{-19}$ C

$h = 6.62 \times 10^{-34}$ J-sec

$$\therefore \qquad \lambda = \frac{12.27}{\sqrt{V}} \, A^0$$

For a neutron,

$m = 1.67 \times 10^{-27}$ kg

$$\lambda = \frac{6.62 \times 10^{-34}}{\sqrt{2 \times (1.67 \times 10^{-27})\,(1.6 \times 10^{-19})\,E(ev)}}$$

$$\lambda = \frac{0.286}{\sqrt{E\ (ev)}} \, A^0$$

5A.6 Properties of Matter Waves

1. The de-Broglie wavelength of a wave associated with moving light particle is greater than the wavelength associated with heavier particles.
2. The de-Broglie wavelength of a wave associated with a slow moving particle is greater than the wavelength associated with fast moving particle.
3. For a particle at rest i.e., $v = 0$, the de-Broglie wavelength becomes infinite, that is, wave becomes indeterminate and if $v = \infty$ then $\lambda = 0$. This indicates that the matter waves are generated only when the material particles are in motion.
4. The expression for de-Broglie wavelength $\lambda = h / mv$ is independent of charge of the particle therefore the matter waves are generated by moving charged particle as well as by moving neutral particles.
5. The velocity of a matter wave is greater than the velocity of electromagnetic wave

$$v_p = \frac{c^2}{v}$$

5A.7 Experimental Evidence for the Existence of Matter Waves

5A.7.1 Davisson-Germer Experiment

The first experimental demonstration of the matter waves came three years after de-Broglie advanced his hypothesis and was accidentally obtained by two American

Physicists Cliton J. Davisson and his assistant L.H. Germer. They were studying the scattering of electrons by crystals using the apparatus shown in Fig.5A.3.

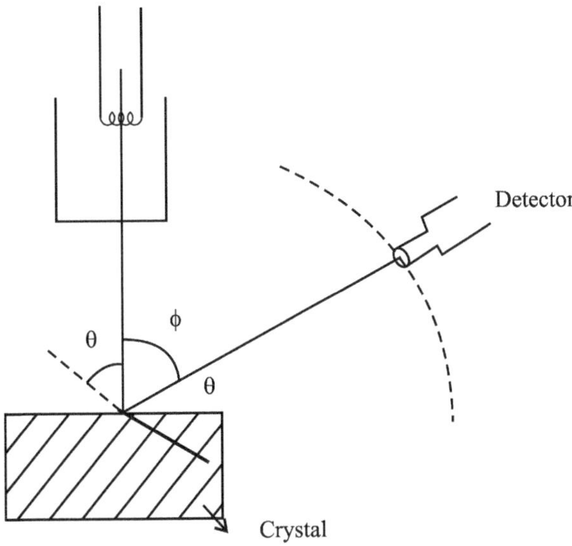

Fig. 5A.3 Davisson - Germer experimental setup.

Electrons from an electron gun were accelerated by a positive electrode maintained at V volts. The accelerated beam of the electrons was incident on a strip of nickel containing many crystals and the number of electrons scattered in various directions was then measured. A smooth variation in the intensity of the scattered electron with the angle was observed. Midway through the experiment an accident occurred which permitted air to enter in to the vacuum tube containing the strip. This resulted in to the formation of an oxide film on the surface. Davisson and Germer were then forced to heat the strip to a very high temperature in order to reduce the oxide.

This heating and slow cooling had the effect of turning the poly crystalline nickel sample into a large single crystal. Again this forced heat treatment of the sample, the experiment was repeated. The experimental results this time were quite different from those obtained before the accident.

The intensity of the scattered electron showed some sharp maxima and minima at certain angle which were found to depend upon the electron energy and hence upon the accelerating voltage. This pattern was similar to a wave diffraction pattern. Some typical results from these experiments are shown in Fig. 5A.4.

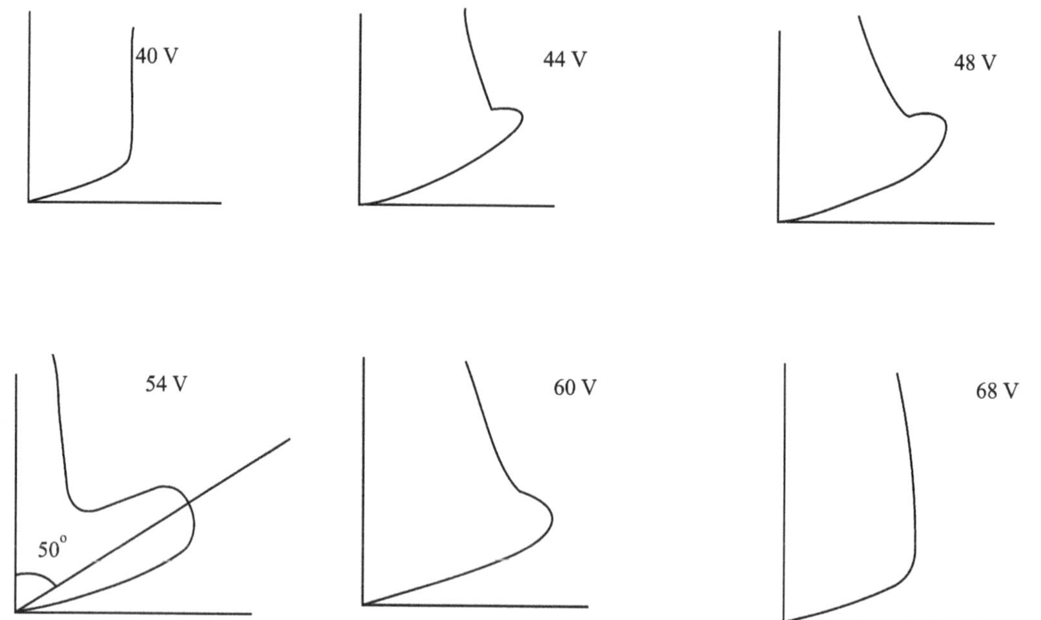

Fig. 5A.4 Polar plots of some typical results of Davisson - Germer experiment.

Davisson and Germer found that the angular peaks of scattered electrons could be explained as diffraction of electrons by atoms if the regular arrays of atoms in the nickel are regarded as a diffraction grating.

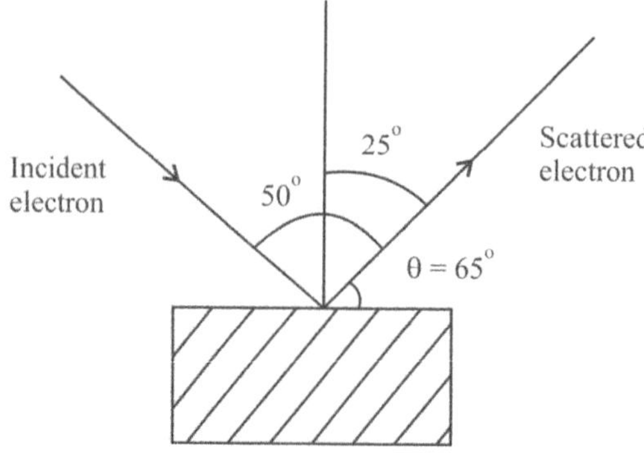

Fig. 5A.5

Using Bragg's equation

$$2\,d \sin\theta = n\,\lambda$$

For nickel $\quad d = 0.911\ A^0$

$$\theta = 65^0$$

$$n = 1$$

$$2 \times 0.911 \times 10^{-10} \times \sin 65^0 = \lambda$$

$$\lambda = 1.65 \text{ A}^0$$

For electron of kinetic energy of 54 eV

$$\therefore \qquad \lambda = \frac{12.27}{\sqrt{V}} \text{ A}^0$$

$$\lambda = \frac{12.27}{\sqrt{54}} = 1.66 \text{ A}^0$$

which agrees well with the observed wavelength of 1.65 A^0. The Davisson-Germer experiment thus directly verifies de-Broglies hypothesis of the wave nature of moving bodies.

5A.7.2 G. P. Thomson Experiment

Within month of the discovery by Davisson and Germer, a British physicist G. P. Thomson also discovered diffraction effect with a beam of highly energetic electrons.

In 1928 Thomson repeated his experiment by using platinum rather than celluloid film. The diffraction rings observed with the poly crystalline metallic foil were found to be exactly similar to those observed for the x-rays of the same wavelength (as that of electron). This experiment of Thomson has provided ever more convincing evidence in support of the de-Broglie hypothesis.

5A.8 Wave Velocity and Group Velocity

Wave velocity is also called phase velocity. Wave motion is a form of disturbance which travels through a medium due to repeated periodic motion of the particles of the medium about their mean positions, the motion being handed over from one particle to the next. The individual oscillations which make up the medium only execute simple harmonic motion about their mean position and do not themselves travels through the medium with the wave.

Energy particle begins its vibration a little later than its predecessor and there is a progressive change of phase as we travel from one particle to the next. It is the phase relationship of these particles that we observe as a wave and the velocity with which the plane of equal phase travels through the medium is known as phase velocity or wave velocity.

Let us assume that a particle like an electron can be described mathematically by a sine wave

$$y = A \sin (wt - kx) \quad .....(5A.11)$$

This wave has no beginning and no end. It is of infinite extent and completely non localised. But particle (electron) is confined to a very small volume. Therefore a mono frequency wave can not represent a particle. It implies that the de-Broglie waves are not harmonic waves but could be a combination of several waves. A superposition of several waves of slightly different frequencies give rise to a wave packet. Such wave packet possesses both wave and particle properties. The regular separation between successive maxima in a wave packet is the characteristic of a wave and at the some time it has a particle like localization in space.

Derivation

The velocity of the wave making up the wave packet is known as the phase velocity, where as the speed with which the wave packet moves is known as group velocity.

Let
$$y_1 = A \cos (wt - kx) \quad .....(5A.12)$$

$$y_2 = A \cos [(w + \Delta w) t - (k + \Delta k) x] \quad .....(5A.13)$$

By the principle of superposition we have

$$y = y_1 + y_2$$

$$y = A \cos (wt - kx) + A \cos [(w + \Delta w)t - (k + \Delta k)x]$$

using the relation

$$\cos C + \cos D = 2 \cos \frac{C+D}{2} \cos \frac{C-D}{2}$$

and
$$\cos (-\theta) = \cos \theta, \quad \text{we get}$$

$$y = 2A \cos \frac{[(2w + \Delta w)t - (2k + \Delta k)x]}{2} \cos \frac{(\Delta wt - \Delta kx)}{2}$$

since Δw and Δk are small

$$2w + \Delta w = 2w \quad \text{and} \quad 2k + \Delta k = 2k$$

$$y = 2A \cos (wt - kx) \cos \left(\frac{\Delta wt}{2} - \frac{\Delta kx}{2}\right)$$

or
$$y = 2A \cos \left(\frac{\Delta wt - \Delta kx}{2}\right) \cos (wt - kx) \quad .....(5A.14)$$

equation (5A.14) represents a wave whose phase is given by the factor cos $(wt - kx)$ and the amplitude of the wave is given by $2A \cos\left[\frac{\Delta wt - \Delta kx}{2}\right]$

Phase Velocity

For the phase of the wave to be constant

$$wt - kx = \text{constant}$$

differentiating, we get

$$wdt - kdx = 0$$

$$\frac{dx}{dt} = \frac{w}{k}$$

$$v_p = \frac{dx}{dt} = \frac{w}{k}$$

$$v_p = \frac{w}{k} \qquad \text{.....(5A.15)}$$

Group Velocity

For amplitude of the wave to be constant

$$\frac{\Delta wt - \Delta kx}{2} = \text{constant}$$

$$\Delta wt - \Delta kx = \text{constant}$$

Differentiating we get

$$\Delta w\, dt - \Delta k\, dx = 0$$

$$\frac{dx}{dt} = \frac{\Delta w}{\Delta k}$$

$$\lim_{w_1 \to w_2} \frac{\Delta x}{\Delta k} = \frac{dw}{dk}$$

$$v_g = \frac{dw}{dk} \qquad \text{.....(5A.16)}$$

Note

1. We know that

$$v_p = \frac{w}{k}$$

$$\therefore \qquad w = kv_p$$

since $$v_g = \frac{dw}{dk}$$

$$v_g = \frac{d}{dk}(kv_p)$$

$$v_g = v_p + k\frac{dv_p}{dk}$$

since $$k = \frac{2\pi}{\lambda}$$

$$\therefore \quad v_g = v_p + \frac{2\pi}{\lambda}\frac{dv_p}{d\left(\frac{2\pi}{\lambda}\right)}$$

$$v_g = v_p - \lambda\frac{dv_p}{d\lambda} \quad \text{.....(5A.17)}$$

(a) If $\frac{dv_p}{d\lambda} = 0$, then $v_g = v_p$ i.e., in a non-dispersive medium phase velocity and group velocity are same.

(b) In dispersive medium $\frac{dv_p}{d\lambda} = \oplus$ve, i.e., in dispersive medium group velocity is less than phase velocity.

5A.8.1 Derivation of the relation $v_p = \frac{v_g}{2}$

The de-Broglie wavelength is given by

$$\lambda = \frac{h}{mv_g} \quad \text{.....(5A.18)}$$

The energy E is given by

$$E = \frac{1}{2}mv_g^2 \quad \text{.....(5A.19)}$$

and the energy of a photon is given by

$$E = h\nu \quad \text{.....(5A.20)}$$

from eq. 5A.19 and 5A.20 we have

$$\frac{1}{2}mv_g^2 = h\nu$$

$$\nu = \frac{1}{2} m \frac{v_g^2}{h} \qquad \text{.....(5A.21)}$$

The phase velocity is given by

$$v_p = \nu \lambda$$

$$\therefore \quad v_p = \frac{1}{2} m \frac{v_g^2}{h} \cdot \nu$$

$$v_p = \frac{1}{2} m \frac{v_g^2}{h} \cdot \frac{h}{m v_g}$$

$$v_p = \frac{v_g}{2}$$

thus

$$\mathbf{v_p = \frac{v_g}{2}} \qquad \text{.....(5A.22)}$$

5A.8.2 Derivation of the relation $v_p . v_g = c^2$

For phase velocity v_p

we know that

$$w = 2\pi\nu$$

by Einstein mass energy relation we have,

$$E = mc^2$$

$$\therefore \quad h\nu = mc^2$$

$$\nu = \frac{mc^2}{h} \qquad \text{.....(5A.23)}$$

$$\therefore \quad w = 2\pi \frac{mc^2}{h}$$

$$w = \frac{2\pi c^2}{h} \left[\frac{m_0}{\sqrt{1 - \frac{v^2}{c^2}}} \right] \qquad \text{.....(5A.24)}$$

we know that

$$k = \frac{2\pi}{\lambda}$$

$$k = \frac{2\pi}{h} . mv$$

$$k = \frac{2\pi v}{h}\left[\frac{m_0}{\sqrt{1-v^2/c^2}}\right] \qquad .....(5A.25)$$

the phase velocity v_p is given by

$$v_p = W/K$$

∴ from 5.24 and 5.25

$$v_p = \frac{\frac{2\pi\, c^2}{h}\left[\frac{m_o}{\sqrt{1-v^2/c^2}}\right]}{\frac{2\pi\, v}{h}\left[\frac{m_o}{\sqrt{1-v^2/c^2}}\right]}$$

$$v_p = \frac{c^2}{v} \qquad .....(5A.26)$$

For group velocity v_g

The group velocity

$$v_g = \frac{dW}{dK} \qquad .....(5A.27)$$

$$v_g = \frac{dW/dv}{dK/dv}$$

we have

$$w = \frac{2\pi\, c^2}{h}\left[\frac{m_o}{\sqrt{1-v^2/c^2}}\right]$$

Differentiation w.r.t v, we get

$$\frac{dW}{dv} = \frac{2\pi\, c^2}{h} m_o \left(1 - \frac{v^2}{c^2}\right)^{-3/2}\left(\frac{v}{c^2}\right) \qquad .....(5A.28)$$

we know that

$$K = \frac{2\pi\, v}{h}\left[\frac{m_o}{\sqrt{1-v^2/c^2}}\right]$$

$$\frac{dK}{dv} = \frac{2\pi\, m_o}{h}\left[v\left(-\frac{1}{2}\right)\left(1-\frac{v^2}{c^2}\right)^{-3/2}\left(\frac{-2v}{c^2}\right)+\left(1-\frac{v^2}{c^2}\right)^{-1/2}\right]$$

$$\frac{dK}{dv} = \frac{2\pi\, m_o}{h}\left[\left(1-\frac{v^2}{c^2}\right)^{-3/2}\left\{\frac{v^2}{c^2}+\left(1-\frac{v^2}{c^2}\right)\right\}\right]$$

$$\frac{dK}{dv} = \frac{2\pi\, m_o}{h}\left[\left(1-\frac{v^2}{c^2}\right)^{-3/2}\right] \quad .....(5A.29)$$

since $$v_g = \frac{dW/dv}{dK/dv}$$

$$\therefore \quad v_g = \frac{\dfrac{2\pi c^2}{h}\, m_o\left(\left(1-\dfrac{v^2}{c^2}\right)^{-3/2}\left(\dfrac{v}{c^2}\right)\right)}{\dfrac{2\pi\, m_o}{h}\left[\left(1-\dfrac{v^2}{c^2}\right)^{-3/2}\right]}$$

$$v_g = v \quad .....(5A.30)$$

from eq. 5.26 and 5.30 we have

$$v_p = \frac{c^2}{v_g}$$

$$\therefore \quad v_p\,.v_g = c^2 \quad .....(5A.31)$$

5A.9 Wave Packet and Uncertainty Velocity

A monochromatic wave is infinite in extent, so instead of associating a single monochromatic de Broglie wave with a moving particle, we associate a wave packet consisting of a group of waves of nearly equal amplitude centred around de-Broglie $\lambda = \frac{h}{p}$ of the particle.

The wave packet has the dimensions of the localised particle and travels with the same velocity as the particle. The association of a group of waves with a moving particle that the position of the particle at any instant of time cannot be specified with any desired degree of accuracy. The particle can be some where within the group of waves i.e with a small region Δ x of space. (Δ x is linear spread of the wave packet)

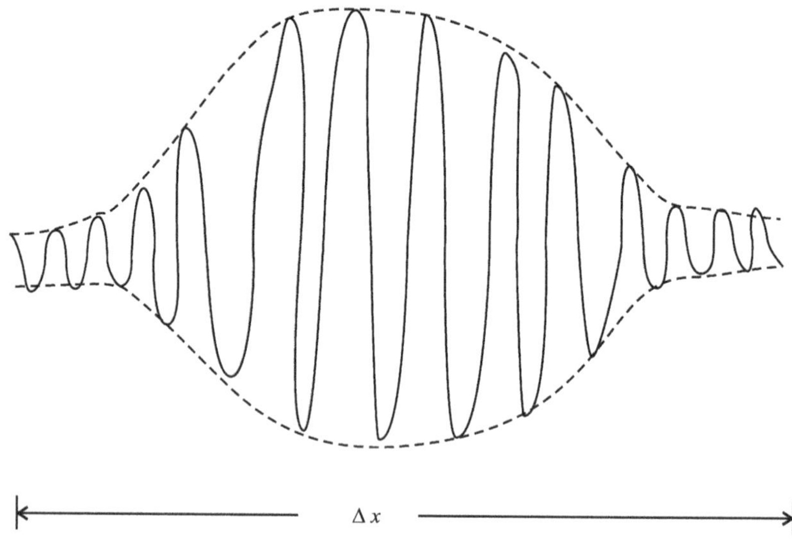

Fig. 5A.6 Wave packet.

The probability of finding the particle is maximum at the centre of the wave packet and falls to zero at its ends. Therefore, there is an uncertainty Δ x in the position of the particle. A wave packet is formed by waves having a range of wavelength. The spread $\Delta \lambda$ related to the spread in dimension Δ x.

As the momentum of the associated particle is related to the wavelength by equation.

$$P = \hbar K \quad \text{.....(5A.32)}$$

$$P = \frac{\hbar}{\lambda} \quad \text{.....(5A.33)}$$

There arises an uncertainty in momentum Δ P. The two uncertainties are interrelated as spread in momentum depends on the spread in wavelength which in turn depends on the length of any wave packet. The mathematical expression governing this situation is called Heisenberg's uncertainty principle which was derived in 1972.

It states that "it is not possible to make simultaneous measurement of the position and momentum of a particle to an unlimited accuracy. The product of uncertainty in the measurement of position and uncertainty in the measurement of momentum is equal to, or greater than Planck's constant $\hbar$.

i.e $$\Delta x.\ \Delta p_x \geq \frac{\hbar}{2\pi} \qquad .....(5A.34)$$

where Δx is the fundamental error or uncertainty in the measurement of the position and Δp_x is the fundamental error or uncertainty in the measurement of momentum (velocity)

The corresponding relation for the other components of position and momentum are

$$\Delta y.\ \Delta p_y \geq \frac{\hbar}{2\pi} \qquad ......(5A.35)$$

$$\Delta z.\ \Delta p_z \geq \frac{\hbar}{2\pi} \qquad .....(5A.36)$$

It means that it is physically impossible to know simultaneously the exact position and exact momentum of particle. It is desired to reduce the error in determination of position i.e., to reduce the error in determination of position i.e to reduce Δx, then this can only be done at the expense of accuracy in determing momentum and the error involved in Δp will increase.

This principle is applicable to all conjugate or complimentary pair of physical variables the product of which has the dimension of Planck's constant e.g time and energy, angular displacement and angular momentum.

$$\Delta E.\ \Delta t \geq \frac{\hbar}{2\pi} \qquad .....(5A.37)$$

and $$\Delta L.\ \Delta \theta \geq \frac{\hbar}{2\pi} \qquad .....(5A.38)$$

It is to be noted that the uncertainty does not apply to a single measurement on a single particle. It is a statement about a statistical average over lots of measurement of position and momentum.

5A.9.1 Experiments for the Verification of Uncertainty Principle

A. Single-Slit Diffraction Method

Let us consider a stream of electrons moving along x direction passes through a narrow slit of width $d = \Delta y$. If Δy is comparable to the wavelength of the electron beam, then the electron will diffract according to single slit diffraction pattern and form a central maximum and two secondary minimum as shown in Fig. 5A.7.

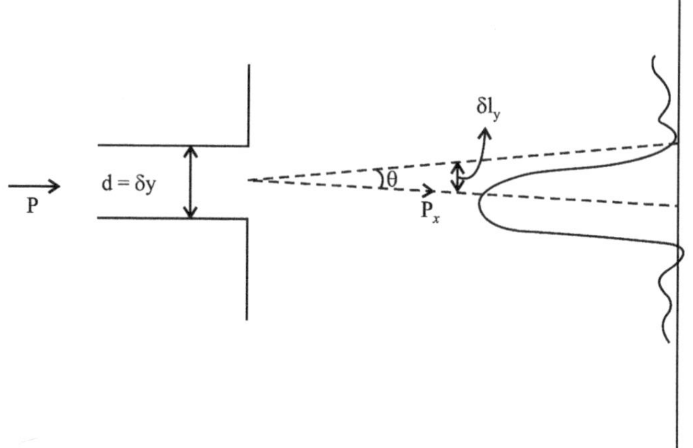

Fig. 5A.7

According to diffraction theory

$$d \sin\theta = \lambda \quad .....(5A.39)$$

$$\Delta y = \frac{\lambda}{\sin\theta} \quad .....(5A.40)$$

d = Δy = uncertainty in position of electron being diffracted. Before diffraction slit, electron has momentum p_x only along x-direction and zero in y direction. Therefore uncertainty of y – component of momentum of $\Delta p_y = 0$). Because the diffraction effect at the slit, the particle acquires a small component of momentum p_y in y direction. The original momentum the of the particle in the y direction was accurately known to be zero. Therefore, Δp_y is the uncertainty introduced in the y component of the momentum.

$\therefore$ Δp_y is the uncertainty introduced in y-component of momentum.

Particles that strike the screen at a point A the first minima, must have left the slit at an angle θ given by

$$\tan\theta = \theta = \frac{\Delta p_y}{p_x} \quad .....(5A.41)$$

equating (5.40) and (5.41) we get

$$\frac{\Delta p_y}{p_x} = \frac{\lambda}{\Delta y}$$

$$\therefore \quad \Delta y \,.\, \Delta p_y = \lambda \,.\, p_x$$

$$\therefore \quad \Delta y \,.\, \Delta p_y = h$$

$$\therefore \qquad \lambda = \frac{h}{p_x} \qquad .....(5A.42)$$

Which is the uncertainty principle i.e. if we try to improve the accuracy in y, we have to reduce Δy using a finer slit which results in turn in a wider pattern. Its leads to larger Δp_y.

B. γ-Ray Microscope Experiment

Consider an experiment for measurement of the position of an electron at rest using microscope. Since the size of the electron is very small light wave cannot be used for measurement as it would 10^4 to 10^5 times smaller than size of electrons. To illuminate an electron γ-rays can be used. To determine the location of an electron one photon must bounce off the electron and pass through the microscope in to our eyes.

A γ-photon carries a very large momentum. When the photon strikes the electron, part of the momentum and energy are transferred to the electron due to the Compton scattering. Thus, when the scattered photon is registered by the microscope, only the earlier position of the electron can be deduced but the momentum of the electron is attend.

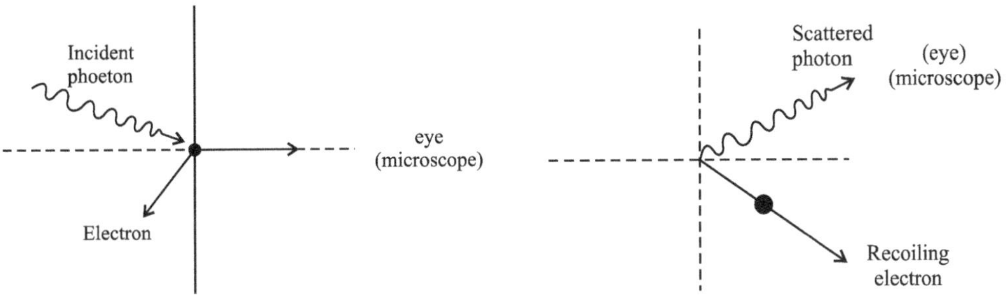

Fig. 5A.8

Let the incoming photon momentum be $\frac{h}{\lambda}$. The uncertainty in the electron momentum after the scattering be Δp. This Δp can have maximum value as the momentum of incident photon i.e $\frac{h}{\lambda}$.

$$\Delta p = \frac{h}{\lambda} \qquad ......(5A.43)$$

The position of the electron can be determined to within one wavelength of photon. The uncertainty in position is $\Delta x = \lambda$.

$$\therefore \quad \Delta x.\ \Delta p = \lambda\ \frac{h}{\lambda} = h \qquad \text{......(5A.44)}$$

Which is uncertainty principle.

5A.9.2 Application of Uncertainty Principle

A. Non-existence of electron and existence of protons and neutrons in nucleus

The radius of nucleus of any atom is of the order of 10^{-14} m. If an electron is confined inside the nucleus then the uncertainty in the position Δx of the electron is equal to the diameter of the nucleus.

$$\text{i.e} \quad \Delta x = 2 \times 10^{-14}\ \text{m} \qquad \text{..... (5A.45)}$$

$$\therefore \quad \Delta x\ \Delta p_x \geq \frac{h}{2\pi}$$

$$\Delta p_x \geq \frac{h}{2\pi.\ \Delta x}$$

since $\frac{h}{2\pi} = 1.055 \times 10^{-34}$ J.sec and $\Delta x = 2 \times 10^{-14}$ m

$$\therefore \quad \Delta p_x \geq \frac{1.055 \times 10^{-34}}{2 \times 10^{-14}}$$

$$\Delta p_x \geq 0.527 \times 10^{-20}\ \text{kg m/sec} \qquad \text{.....(5A.46)}$$

If this is the uncertainty in momentum of the particle then the momentum component p_x and hence the total momentum p of the particle in the nucleus must be at least of the order of magnitude $|p|$ i.e

$$|p| \approx p_x \approx \Delta p_x = 5.27 \times 10^{-21}\ \text{kg m/sec}$$

Since the magnitude of momentum is relativistic as compared to the mass of electron $m_o = 9.1 \times 10^{-31}$ kg, then using the relativistic formula for the energy of the electron.

$$E^2 = p^2 c^2 + m_o^2 c^4 \qquad \text{.....(5A.47)}$$

As the rest energy $m_o c^2$ of the electron is of the order of 0.511 Mev which is much smaller than the value of first term, hence we get

$$E^2 = p^2 c^2$$

$$E = p c$$

$$E = (5.27 \times 10^{-21})\ (3 \times 10^8)\ \text{Joule}$$

$$E = \frac{\left(5.27\times10^{-21}\right)\left(3\times10^{8}\right)}{1.6\times10^{-19}}\ \text{ev}$$

$$E = 10\ \text{Mev} \qquad \text{.....(5A.48)}$$

Which means that if the electron exist inside the nucleus, their energy must be of the order of 10 Mev. However, the electron emitted by radioactive nuclei during beta decay have energies only 3 to 4 Mev. So in general, electron cannot reside inside the nucleus.

For proton and neutrons

$$m_o = 1.67 \times 10^{-27}\ \text{kg}$$

This is non-relativistic problems and v is 3×10^{6} m/sec.

Thus Kinetic energy is given by

$$K.E = \frac{p^2}{2m} = \frac{\left(0.527\times10^{-20}\right)}{2\times1.67\times10^{-27}\times1.6\times10^{-19}}\ \text{ev}$$

$$K.E = 52\ \text{KeV} \qquad \text{.....(5A.49)}$$

Since this energy is less than the energy contained by these particles emitted by nuclei, both these types of particles i.e protons and neutrons exist inside the nucleus.

B. Binding Energy of an electron in atom

In an atom, the electron is under the influence of electrostatic potential of positively charged nucleus. It is confined to the linear dimensions equal to the diameter of electronic orbit. The uncertainty in the position Δx of an electron is of the order of 2R, where R is the radius of the orbit and the corresponding uncertainty in momentum component Δp_x is given by

$$\Delta p_x \geq \frac{h}{2\pi.\Delta x}$$

$$\Delta p_x \geq \frac{h}{2\pi.2R}$$

Which shows that the momentum of electron in an orbit is at least

$$p \sim \Delta p_x \sim \frac{h}{2\pi.2R}$$

$$p \approx \frac{h}{4\pi R} \qquad \text{.....(5A.50)}$$

Kinetic energy of electron $= \frac{p^2}{2m} = \left(\frac{h}{4\pi R}\right)^2 \cdot \frac{1}{2m_o}$

$$K.E = \frac{h^2}{32\pi^2\ m_o\ R^2} \qquad \text{.....(5A.51)}$$

The potential energy of an electron in the field of nucleus with atomic number z is given by

$$V = \frac{-Ze^2}{4\pi\ \epsilon_o\ R} \qquad \text{......(5A.52)}$$

The total energy $E = K.E + P.E$

$$E = \frac{h^2}{32\ \pi^2\ m_o R^2} - \frac{Ze^2}{4\pi\ \epsilon_o\ R}$$

$$E = \frac{\left(1.055\times10^{-34}\right)^2}{32\times(3.14)^2\times9.1\times10^{-31}R^2} - \frac{Z\left(1.6\times10^{-19}\right)^2}{4\times3.14\times8.85\times10^{-12}\times R}$$

where $\frac{h}{2\pi} = 1.055\times10^{-34}$ J.sec

$$E = \frac{10^{-20}}{R^2} - \frac{15\times10^{-10}Z}{R}\ \text{ev}$$

Taking $R = 10^{-10}$ m

we have $E = (1 - 15\ Z)$ ev .....(5A.53)

Now the binding energy of outermost electron is H and H_e are – 13.6 ev and – 24.6 ev respectively. So the value of Binding energy derived on the basis of uncertainty principle is accepted as these are comparable in magnitudes.

C. The Zero-Point Energy

Consider a particle of rest mass m_o is moving under the influence of a potential of linear dimension L. Since the particle can exist any where in this region. Therefore, the maximum uncertainty in the position of the particles is given by

$$(\Delta x)_{mzx} = L \qquad \text{.....(5A.54)}$$

From uncertainty Principle

$$\Delta x.\ \Delta p_x \geq \hbar$$

$$(\Delta p_x)_{min} \geq \frac{\hbar}{L}$$

suppose the value of the momentum of the particle is equal to the uncertainty in the momentum which means that $p = \Delta p_x$. Then the lowest value of Kinetic energy of the particle is given by

$$K.E = \frac{p^2}{2m} = \frac{(\Delta p_x)^2}{2m}$$

$$K.E = \frac{\hbar^2}{2m\ L^2} \qquad .....(5A.55)$$

We know that the energy of the system is minimum when $T \rightarrow 0$ k, thus the Kinetic energy has some finite value even at 0 k. This minimum energy of the system is called zero point energy.

D. Finite width of the spectral lines

According to Heisenberg's uncertainty principle relation between energy and time is

$$\Delta E\ .\ \Delta t \geq \frac{h}{2\pi}$$

or

$$\Delta E\ .\ \Delta t \geq \hbar \qquad .....(5A.56)$$

The average period that takes between the excitation of an atom and time it radiates is 10^{-8} sec. Thus the photon energy is uncertain by the amount

$$\Delta E \approx \frac{\hbar}{\Delta t} = \frac{1.054 \times 10^{-34}}{10^{-8}}$$

$$\Delta E = 1.054 \times 10^{-26}\ J$$

since

$$\Delta E = h\ \Delta \nu$$

$\therefore$

$$\Delta \nu = \frac{1.054 \times 10^{-26}}{h}$$

$$\Delta\ \nu = 1.6 \times 10^{7}\ /sec \qquad .....(5A.57)$$

Thus the energy levels of the atom must have a finite width or in other words the spectral lines can never be sharp but have a natural finite width.

E. The radius of the Bohr's first orbit.

If Δx and Δp_x are the uncertainty in the position and the momentum of the electron in first orbit, then we have

$$\Delta x.\ \Delta\ p_x \geq \hbar$$

$$\Delta\ p_x \geq \frac{\hbar}{\Delta x} \qquad \text{.....(5A.58)}$$

The uncertainty in the kinetic energy of the electron may be written as

$$\Delta\ K = \frac{(\Delta p_x)^2}{2m} = \frac{1}{2m}\left(\frac{\hbar}{\Delta x}\right)^2$$

$$\Delta\ K = \frac{\hbar^2}{2m(\Delta x)^2} \qquad \text{.....(5A.59)}$$

The uncertainty in the potential energy of the electron is

$$\Delta\ V = \frac{-Ze^2}{4\pi\ \epsilon_o\ \Delta x} \qquad \text{.....(5A.60)}$$

So that the uncertainty in the total energy is

$$\Delta\ E = \Delta\ K + \Delta\ V$$

$$= \frac{\hbar}{2m\ (\Delta x)^2} - \frac{1}{4\pi\ \epsilon_o}\frac{Ze^2}{\Delta\ x} \qquad \text{.....(5A.61)}$$

The uncertainty in the energy will be minimum if

$$\frac{d(\Delta E)}{d(\Delta x)} = 0 \text{ and } \frac{d^2(\Delta E)}{d(\Delta x)^2} = \text{positive (+ ve)}$$

$$\therefore \qquad \frac{d(\Delta E)}{d(\Delta x)} = \frac{-\hbar^2}{m(\Delta x)^3} + \frac{1}{4\pi\ \epsilon_o}\frac{Ze^2}{(\Delta x)^2} = 0$$

This gives $$\Delta x = \frac{\hbar^2.4\pi\ \epsilon_o}{mZe^2} \qquad \text{.....(5A.62)}$$

Thus $$\frac{d^2(\Delta E)}{d(\Delta x)^2} = \text{positive (+) ve}$$

Therefore eq. 5.61 represents the conditions of minimum uncertainty in the first orbit. Hence radius of first orbit is

$$r = \Delta x = \frac{\hbar^2.4\pi\ \epsilon_o}{mZe^2}$$

$$r = \Delta x = \frac{h^2 . 4\pi \in_o}{4\pi^2 \ mZe^2}$$

$$r = \Delta x = \frac{h^2 \ \in_o}{\pi \ mZe^2}$$

$$r = \frac{h^2 \ \in_o}{\pi \ mZe^2} \quad \text{.....(5A.63)}$$

5A.10 Wave Function

The amplitude of matter wave is described by wave function represented by a Greek letter ψ consisting of real and imaginary part.

$$\psi = A + i B \quad \text{.....(5A.64)}$$

Conjugate of ψ is

$$\psi^* = A - i B$$

and $$\psi^* \psi = |\psi|^2 = A^2 + B^2$$

$|\psi|^2$ at a particular place at a particular time is proportional to the probability of finding the particle at that time.

$$\therefore \text{ Probability density } = |\psi|^2 = \psi^* \psi \quad \text{......(5A.65)}$$

Let us assume that wave function ψ in specified direction x is given by

$$\psi = A \, e^{-i w (t - x/v)} \quad \text{.....(5A.66)}$$

since

$$w = 2 \, \pi \, \nu \quad \text{.....(5A.67)}$$

and $$v = \nu \, \lambda \quad \text{.....(5A.68)}$$

$$\psi = A \ e^{-2\pi i\left(\nu t - \frac{x\nu}{v}\right)} \quad \text{.....(5A.69)}$$

we know

$$E = h \, \nu$$

$$E = 2\pi \hbar \, \nu$$

$$\therefore \quad \nu = \frac{E}{2\pi \, \hbar} \quad \text{......(5A.70)}$$

and $$v = \nu \ \lambda$$

$$\frac{V}{\nu} = \frac{h}{p}$$

$$\frac{V}{\nu} = \frac{2\pi\,\hbar}{p} \qquad \text{.....(5A.71)}$$

Using eq. 5A.70 and 5A.71 in 5A.69 we get

$$\psi = A\, e^{-i/\hbar\,(Et - px)} \qquad \text{.....(5A.72)}$$

5A. 10.1 Energy Operator

We have

$$\psi = A\, e^{-i/\hbar\,(Et - px)}$$

Differentiating the above equation w.r.t t we get

$$\frac{\partial \psi}{\partial t} = A\, e^{-i/\hbar\,(Et - px)} \left(\frac{-i}{\hbar}E\right)$$

$$\frac{\partial \psi}{\partial t} = \frac{-i}{\hbar}\,E\,\psi$$

$$\frac{-\hbar}{i}\,\frac{\partial \psi}{\partial t} = E\,\psi$$

$$E\,\psi = i\,\hbar\,\frac{\partial \Psi}{\partial t}$$

or

$$\hat{E}\,\psi = i\,\hbar\,\frac{\partial \Psi}{\partial t} \qquad \text{.....(5A.73)}$$

thus energy operator is given by

$$\hat{E} = i\,\hbar\,\frac{\partial}{\partial t} \qquad \text{.....(5A.74)}$$

5A.10.2 Momentum operator

We have

$$\psi = Ae^{-i/\hbar\,(Et - px)}$$

differentiating the above equation w.r.t x we get

$$\frac{\partial \psi}{\partial x} = Ae^{-i/\hbar\,(Et - px)} \left(\frac{i}{\hbar}p\right)$$

$$\frac{\partial \psi}{\partial x} = \frac{i}{\hbar} p\psi$$

$$\frac{\hbar}{i} \frac{\partial \psi}{\partial x} = p\psi$$

$$p\psi = -i\hbar \frac{\partial \psi}{\partial x}$$

$$\hat{p}\psi = -i\hbar \frac{\partial \psi}{\partial x} \qquad \text{.....(5A.75)}$$

Thus momentum operator is given by

$$\hat{p} = -i\hbar \frac{\partial}{\partial x} \qquad \text{.....(5A.76)}$$

***Note*:**

1. *Normalisation*

   The Normalisation condition is given by

$$\int_{-\infty}^{+\infty} |\psi|^2 \, dv = 1$$

as $|\psi|^2 = \psi^* \psi$ = probability density

   The probability between the limit x_1 and x_2 is given by

$$p_{x_1x_2} = \int_{x_1}^{x_2} |\psi|^2 \, dx$$

2. Expectation value

   The expectation value of x is defined as

$$\langle x \rangle = \frac{\int_{-\infty}^{+\infty} \psi^* x \psi \, dx}{\int_{-\infty}^{+\infty} |\psi|^2 \, dx} = \frac{\int_{-\infty}^{+\infty} x|\psi|^2 \, dx}{\int_{-\infty}^{+\infty} |\psi|^2 \, dx}$$

   If ψ is normalised then

$$\langle x \rangle = \int_{-\infty}^{+\infty} x|\psi|^2 \, dx = \int_{-\infty}^{+\infty} \psi^* x \psi \, dx$$

5A. 10.3 Properties of Acceptable Wave Function

The properties of acceptable wave function are:

1. ψ must be continuous and single-valued every where
2. $\frac{\partial \psi}{\partial x} : \frac{\partial \psi}{\partial y}$ and $\frac{\partial \psi}{\partial z}$ must be continuous and single-valued every where
3. ψ must be normlised which means that ψ must go to zero as $x \to \pm \infty$ $y \to \pm \infty$ + and $z \to \pm \infty$
4. $\int |\psi|^2 \, dv$ over all space has a finite constant.

5A.11 Schrodinger's Wave Equations

5A.11.1 Schrodinger's Time dependent Wave Equation

Schrodinger introduced wave function ψ whose periodic variation with space and time constitute matter wave associated with quantum particles such as electron, proton and deuteron etc. $\psi(\bar{r}, t)$ has all the information about the system. In one dimension ψ may be written as

$$\psi = Ae^{-i\,w(t - x/v)} \qquad .....(5A.77)$$

where $w = 2\pi\,\nu$ and $v = \nu\lambda$

$$\therefore \qquad \psi = Ae^{-2\pi i\left(\nu t - \frac{x\nu}{v}\right)} \qquad .....(5A.78)$$

Since

$$E = h\nu$$

$$E = 2\pi\hbar\,\nu$$

$$\nu = \frac{E}{2\pi\hbar} \qquad .....(5A.79)$$

and

$$v = \nu\frac{h}{p}$$

$$\frac{v}{\nu} = \frac{2\pi\,\hbar}{p} \qquad .....(5A.80)$$

Using eq. 5A.79 and eq. 5A.80 in 5A.76 we get

$$\psi = Ae^{-i/\hbar\,(Et - px)} \qquad .....(5A.81)$$

Differentiating eq. 5.81 w.r.t x, we get

$$\frac{\partial \psi}{\partial x} = Ae^{-i/\hbar\,(Et - px)} \left(\frac{i}{\hbar}\right)p$$

Again differentiating we get

$$\frac{\partial^2 \psi}{\partial x^2} = \frac{-p^2}{\hbar^2}\psi$$

or $$p^2\,\psi = -\hbar^2\,\frac{\partial^2 \psi}{\partial x^2} \qquad .....(5A.82)$$

differentiating eq. 5A.81 w.r.t t, we get

$$\frac{\partial \psi}{\partial t} = Ae^{-i/\hbar\,(Et - px)} \left(\frac{-i}{\hbar}\right)E$$

$$\frac{\partial \psi}{\partial t} = -i/\hbar\;E\,\psi$$

or $$E\,\psi = i\,\hbar\,\frac{\partial \psi}{\partial t} \qquad .....(5A.83)$$

at small speeds, the total energy is given by

$$E = K.E + P.E$$

$$E = \frac{p^2}{2m} + V$$

Multiplying both the sides with ψ, we get

$$E\psi = \frac{p^2\psi}{2m} + V\psi \qquad .....(5A.84)$$

Using eq. 5A.82 and 5A.83, we get

$$i\hbar\,\frac{\partial \psi}{\partial t} = \frac{-\hbar^2}{2m}\,\frac{\partial^2 \psi}{\partial x^2} + V\psi \qquad .....(5A.85)$$

Equation 5A.85 is the time dependent form of Schorodiger's equation. In three dimension, the time dependent form of Schordinger's equation can be written as

$$i\hbar\,\frac{\partial \psi}{\partial t} = \frac{-\hbar^2}{2m}\left(\frac{\partial^2 \psi}{\partial x^2} + \frac{\partial^2 \psi}{\partial y^2} + \frac{\partial^2 \psi}{\partial z^2}\right) + V\psi \qquad .....(5A.86)$$

$$i\hbar \frac{\partial \psi}{\partial t} = \frac{-\hbar^2}{2m} \nabla^2 \psi + V\psi$$

When ∇^2 is called Laplacian operator

5A.11.2 Schrodinger's Time Independent Wave Equation or Steady State Equation

The one-dimensional wave function ψ for a free particle is given by

$$\underline{\psi} = Ae^{-i/\hbar\,(Et\,-\,px)} \qquad .....(5A.87)$$

$$\underline{\psi} = Ae^{\left(\frac{-iEt}{\hbar}\right)} \quad e^{(ip/\hbar)x} \qquad .....(5A.88)$$

Thus $\underline{\psi}$ is the product of a time-dependent function $e^{-(iE/\hbar)^t}$ and a position dependent function $\psi = Ae^{-(ip/\hbar)x}$

Thus $$\underline{\psi} = e^{-(iE/\hbar)t}.\ \psi \qquad .....(5A.89)$$

The time-dependent schrodinger equation for wave function $\underline{\psi}$ is given by

$$i\hbar \frac{\partial \underline{\psi}}{\partial t} = \frac{-\hbar^2}{2m} \frac{\partial^2 \underline{\psi}}{\partial x^2} + v\underline{\psi} \qquad .....(5A.90)$$

Differentiating eq. 5.89 with respect to x

$$\frac{\partial \underline{\psi}}{\partial x} = e^{-(iE/\hbar)t} \quad \frac{\partial \psi}{\partial x}$$

Again differentiating we get

$$\frac{\partial^2 \underline{\psi}}{\partial x^2} = e^{-(iE/\hbar)t} \frac{\partial^2 \psi}{\partial x^2} \qquad .....(5A.91)$$

Differentiating eq. 5.89 w.r.t t, we get

$$\frac{\partial \underline{\psi}}{\partial t} = e^{-(iE/\hbar)t} \quad \psi.\left(\frac{-i}{\hbar}\right)E \qquad .....(5A.92)$$

Therefore putting the values from eq. 5.89, 5.91 and 5.92 in

$$i\hbar \frac{\partial \underline{\psi}}{\partial t} = \frac{-\hbar^2}{2m} \frac{\partial^2 \underline{\psi}}{\partial x^2} + V\underline{\psi}$$

$$i\hbar\left(\frac{-iE}{\hbar}\right)e^{-i/\hbar\ Et}.\psi = e^{-(iE/\hbar)t} \quad \left(\frac{-\hbar^2}{2m}\right)\frac{\partial^2 \psi}{\partial x^2} + ve^{-iE/\hbar\ t}.\psi$$

or $$E\psi = \frac{-\hbar^2}{2m}\frac{\partial^2\psi}{\partial x^2} + V\psi$$

$$\frac{\partial^2\psi}{\partial x^2} + \frac{2m}{\hbar^2}(E - V)\psi = 0 \qquad .....(5A.93)$$

Equation 5A.93 is the steady state form of schrodinger equation. In three dimension, we have

$$\frac{\partial^2\psi}{\partial x^2} + \frac{\partial^2\psi}{\partial y^2} + \frac{\partial^2\psi}{\partial z^2} + \frac{2m}{\hbar^2}(E - V)\psi = 0 \qquad .....(5A.94)$$

5A.12 Physical Significance of Wave Function

5A.12.1 Schrodinger's Interpretation

Schrodinger interpreted the wave function with the view that there is correspondence between wave-particle duality of light and matter waves.

The intensity of light, on the energy flowing per unit area per second is given by

$$I = C \cdot \frac{1}{2} \epsilon_o E^2 = N h v$$

(wave concept) (particle concept)

Thus, $N \alpha E^2$, N is the number of photons per unit volume.

Accordingly, Shcrodinger interpreted

$|\psi|^2 = \psi^* \psi$ as the number of particles per unit volume, the particle density i.e

$$|\psi|^2 = \psi^* \psi = \text{particle density} \qquad .....(5A.95)$$

If e is electronic charge, then

$$e|\psi|^2 = \text{charge density} \qquad .....(5A.96)$$

5A.12.2 Max Born's Interpretation

Max Born made a probabilistic interpretation to $|\psi|^2$. According to him

$$\psi^* \psi = |\psi|^2 = \text{probability density} \qquad .....(5A.97)$$

The probability of finding the particle described by wave function $\psi(\overline{r}, t)$ at position $\overline{r}$ and time it is proportional to $|\psi|^2$. Accordingly

$$e|\psi|^2 = \text{Probability of charge density} \qquad .....(5A.98)$$

The probability of finding a particle in a small volumle dv at $\bar{r}$ at time t is $|\psi(\bar{r},t)|^2$ dv.

Hence, number of particles in volume dv will be

$$dv = N_o |\psi(\bar{r},t)|^2 dv \qquad \text{......(5A.99)}$$

where N_o is the total number of particles. Since the total number of particles in entire space is N_o we have

$$\int_{\text{entire volume}} N_o |\psi(\bar{r},t)|^2 dv = N_o$$

or

$$\int_{\text{entire volume}} |\psi(r,t)|^2 dv = 1 \qquad \text{.....(5A.100)}$$

5A.13 Particle Trapped in One-Dimensional Box or Particle in Infinite Potential Well

Consider a particle of mass m trapped in a one-dimensional box extending from x = 0 to x = L as shown in Fig. 5A.9. The particle is bouncing back and forth between the walls of the box. Let L be the width of the box.

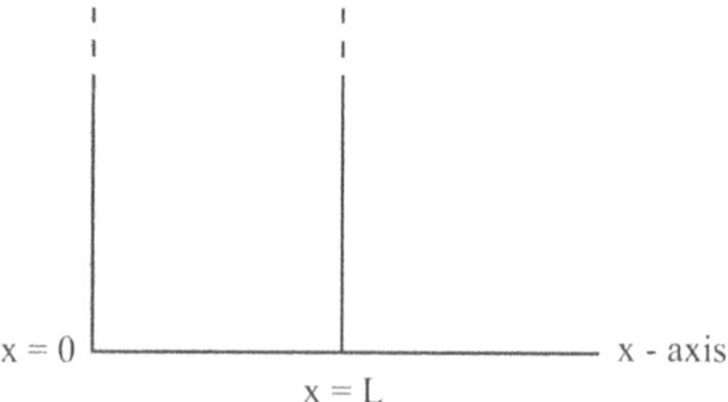

Fig. 5.9 Particle trapped in one-dimensional box of side L or in infinite potential well of width L

The potential energy V of the particle is infinite on both the sides of the box

$$V = 0 \text{ for } 0 < x < L$$

$$V = \infty \text{ for } x \leq o \text{ and } x \geq L$$

The particle cannot be found outside the box and so $\psi = 0$ for $x < 0$ and $x > L$. Within the box, $0 < x < L$, there is no force on the particle i.e $V = 0$.

The boundary condition for the particle are

$$\psi = 0 \text{ at } x = 0 \text{ and } x = L \qquad \text{.....(5A.101)}$$

$$\frac{d\psi}{dx} = 0 \text{ at } x = 0 \text{ and } x = L \qquad \text{......(5A.102)}$$

Within the box, the Schrodinger equation becomes

$$\frac{d^2\psi}{dx^2} + \frac{2m}{\hbar^2}(E - 0)\,\psi = 0$$

$$\frac{d^2\psi}{dx^2} + \frac{2m}{\hbar^2} E\,\psi = 0$$

$$\frac{d^2\psi}{dx^2} + K^2\psi = 0 \qquad \text{.....(5A.103)}$$

where

$$K^2 = \frac{2mE}{\hbar^2} \qquad \text{.....(5A.104)}$$

The general solution of eq. 5.103 is

$$\psi = A \sin Kx + B \cos Kx \qquad \text{.....(5A.105)}$$

Using Boundary conditions

1. $\psi = 0$ at $x = 0$

   we have $0 = A \sin 0 + B$

   $\Rightarrow \quad B = 0$

2. $\psi = 0$ at $x = L$

   $0 = A \sin KL$

   $\Rightarrow \quad A \neq 0$

   and

   $\sin KL = 0$

   $\sin KL = \sin n\pi$

$$K = \frac{n\pi}{L} \qquad \text{......(5A.106)}$$

Thus the wave function becomes

$$\psi = A \sin\frac{n\pi x}{L} \qquad \text{.....(5A.107)}$$

where n = 1, 2, 3, ...

Energy Eigen value

The energy is given by

$$K^2 = \frac{2mE}{\hbar^2}$$

$$E = \frac{K^2\hbar^2}{2m}$$

$$E = \frac{n^2\pi^2\hbar^2}{2mL^2} \qquad \text{.....(5A.108)}$$

Since $\hbar = \frac{h}{2\pi}$

Therefore

$$E = \frac{n^2\pi^2h^2}{2mL^2 \cdot 4\pi^2}$$

$$E = \frac{n^2h^2}{8mL^2} \qquad \text{.....(5A.109)}$$

where n = 1, 2, 3, ...

Hence it is clear form equation 5A.109 that inside an infinitely deep potential well or in an infinite square well, the particle cannot have any energy but can have only certain discrete energy corresponding to n = 1, 2, 3, ... each permitted energy is called is called eigen value of the particle.

Eigen Funciton

Applying normalisation condition to the wave function

$$\psi = A\sin\frac{n\pi x}{L} \quad \text{we get}$$

$$\int_0^L |\psi|^2\,dx = 1$$

$$\int_0^L |\psi|^2\,dx = \int_0^L A^2\sin^2\frac{n\pi x}{L}\,dx = 1$$

$$\int_0^L A^2\sin^2\frac{n\pi x}{L}\,dx = 1$$

$$\frac{A^2}{2}\int_0^L\left(1-\cos\frac{2n\pi x}{L}\right)dx = 1$$

$$\frac{A^2}{2}[L-0] = 1$$

$$A^2 = \frac{2}{L}$$

$$A = \sqrt{\frac{2}{L}}$$

∴ Therefore eigen function is given by

$$\psi_n = \sqrt{\frac{2}{L}} \sin \frac{n\pi x}{L} \qquad \text{......(5A.110)}$$

where n = 1, 2, 3, ...

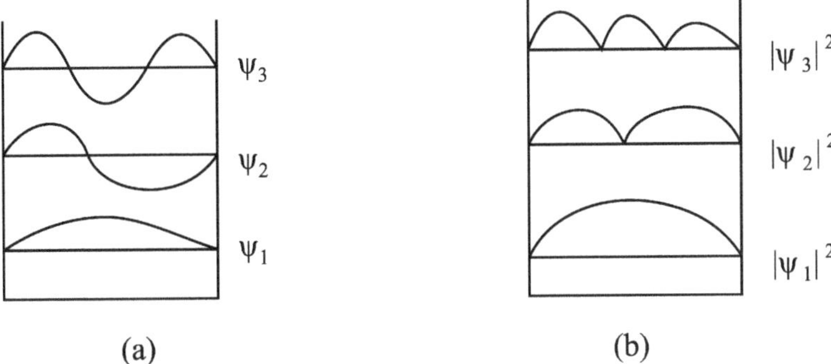

Fig. 5A.10 Parameters for one dimensional box with rigid walls
(a) Wave functions (b) Probability density

At a particular place in the box the probability of the particle being present may be very different for different quantum number. For instance $|\psi_1|^2$ has its maximum value of $\frac{2}{L}$ in the middle of the box, while $|\psi_1|^2 = 0$. A particle in the lowest energy level of n = 1 is most likely in the middle of the box, while a particle in the next higher state of n =2 is never there. Classical mechanics suggests the same probability for the particle being any where in the box.

5A.14 Harmonic Oscillator

In the case of simple harmonics motions, the restoring force F on a particle of mass m is proportional to

$$F = -Kx \qquad \text{.....(5A.111)}$$

The potential energy of the particle is given by

$$V(x) = -\int_0^x F(x)\, dx = K\int_0^x x\, dx = \frac{1}{2} Kx^2 \qquad .....(5A.112)$$

Substituting this value of the potential energy in the steady state form of Schrodinger equation,

$$\frac{\partial^2 \psi}{\partial x^2} + \frac{2m}{\hbar^2}(E - v)\,\psi = 0$$

$$\frac{\partial^2 \psi}{\partial x^2} + \frac{8\pi^2 m}{h^2}\left(E - \frac{1}{2} Kx^2\right)\psi = 0 \qquad .....(5A.113)$$

Let us assume that the solution is

$$\psi(x) = Ae^{-bx^2} \qquad .....(5A.114)$$

Differentiating we get

$$\frac{\partial \psi}{\partial x} = -2A\, bx\; e^{-bx^2} \qquad .....(5A.115)$$

Differentiating again

$$\frac{\partial^2 \psi}{\partial x^2} = -2A\, b\; e^{-bx^2} + 4\, Ab^2 x^2\, e^{-bx^2}$$

Substituting these values in eq. 5A.113, we get

$$-2\, Ab\, e^{-bx^2} + 4\, Ab^2 x^2\, e^{-bx^2} + \frac{8\pi^2 m}{h^2}\left(E - \frac{1}{2} Kx^2\right) Ae^{-bx^2} = 0$$

$$-2b + 4b^2 x^2 + \frac{8\pi^2 m}{h^2}\left(E - \frac{1}{2} Kx^2\right) = 0$$

$$\left[\frac{8\pi^2 m}{h^2} E - 2b\right] + \left[4b^2 - \frac{4\pi^2 mK}{h^2}\right] x^2 = 0 \qquad .....(5A.116)$$

The two sets of Quantities within brackets in equation (5.113) must be separately zero.

$$\therefore \qquad \frac{8\pi^2 m\, E}{h^2} - 2b = 0$$

$$\Rightarrow \qquad E = \frac{bh^2}{4\pi^2 m} \qquad .....(5A.117)$$

and $$4b^2 - \frac{4\pi^2 mK}{h^2} = 0$$

$$b = \frac{\pi\sqrt{mK}}{h} \qquad .....(5A.118)$$

Substituting for b from eq. 5A.118 in 5A.117, we get

$$E = \frac{\pi\sqrt{mK}}{h} \cdot \frac{h^2}{4\pi^2 m}$$

$$E = \frac{h}{4\pi}\sqrt{\frac{K}{m}} \qquad .....(5A.119)$$

The classical frequency ν is

$$\nu = \frac{1}{2\pi}\sqrt{\frac{K}{m}} \qquad .....(5A.120)$$

$$\therefore \quad E = \frac{1}{2} h\nu \qquad .....(5A.121)$$

The general solution of eq. 5A.113 is of the form

$$\psi(x) = A\,\psi_n(x)\, e^{-bx^2}$$

Where $\psi_n(x)$ is a polynomial in which the highest power of x is x^n

The solution leads to the energy values

$$E_n = \left(n + \frac{1}{2}\right) h\nu \qquad .....(5A.122)$$

where n is an integer

The energy of harmonic oscillator is quantised in the steps of hν, where ν is the classical frequency of oscillation. The energy is not zero, when n = 0. As the temperature approaches zero, the energy approaches to

$$E_o = \frac{1}{2} h\nu \qquad .....(5A.123)$$

and not zero and this energy is known as zero point energy.

Solved Numerical Problems

On de-Broglie wavelength

1. What is de-Broglie wavelength of an electron accelerated from rest through a potential difference of 100 volts?

Sol. Given that $V = 100$ volts

The de-Broglie wavelength of an electron accelerated through a potential difference of V volts is given by

$$\lambda = \frac{12.28}{\sqrt{V}} A^o$$

$$\therefore \quad \lambda = \frac{12.28}{\sqrt{100}} A^o$$

$$\lambda = 1.228 A^o$$

2. A particle of charge q and mass is accelerated from rest through a potential difference V. Calculate its de-Broglie wavelength, if particles is an electron and potential difference $V = 50$ volts?

Sol. Given that $V = 50$ volts

The de-Broglie wavelength of an electron accelerated through a potential difference of V volts is given by

$$\lambda = \frac{12.28}{\sqrt{V}} A^o$$

$$\therefore \quad \lambda = \frac{12.28}{\sqrt{50}} A^o$$

$$\lambda = 1.7366 A^o$$

3. Calculate the velocity and kinetic energy of an neutron having de-Broglie wavelength 1A°.

Sol. The de-Broglie wavelength is given by

$$\lambda = \frac{h}{mv}$$

$$\therefore \quad v = \frac{h}{m\lambda}$$

Given that $h = 6.62 \times 10^{-34}$ J.sec

$$m = 1.67 \times 10^{-27}\ \text{kg}$$

$$\lambda = 1\ \text{A}^\circ = 10^{-10}\ \text{m}$$

$$\therefore \qquad v = \frac{6.62 \times 10^{-34}\ \text{J.sec}}{1.67 \times 10^{-27}\ \text{kg} \times 10^{-10}\ \text{m}} = 3.96 \times 10^{3}\ \text{m/sec}$$

The kinetic energy of the neutron is given by

$$E = \frac{1}{2}\, m\, v^2$$

$$E = \frac{1}{2} \times 1.67 \times 10^{-27}\ \text{Kg} \times (3.96 \times 10^{3}\ \text{m/sec})^2$$

$$E = 1.309 \times 10^{-20}\ \text{Joule}$$

or

$$E = \frac{1.309 \times 10^{-20}}{1.6 \times 10^{-19}} = 0.082\ \text{ev}$$

4. A proton is moving with a speed of 2×10^8 m/sec. Find the wavelength of the matter wave associated with it?

Sol. The wavelength of the matter wave is given by

$$\lambda = \frac{h}{p}$$

since

$$p = m\, v$$

$$p = \frac{m_o v}{\sqrt{1 - \dfrac{v^2}{c^2}}}$$

Given that

$$v = 2.0 \times 10^8\ \text{m/sec}$$

$$m_o = 1.67 \times 10^{-27}\ \text{Kg}$$

$$c = 3 \times 10^8\ \text{m/sec}$$

$$p = \frac{1.67 \times 10^{-27}\ \text{Kg} \times 2.0 \times 10^8\ \text{m/s}}{\sqrt{1 - \left(\dfrac{2 \times 10^8}{3 \times 10^8}\right)^{2}}}$$

$$p = \frac{1.67 \times 10^{-19} \times 6}{\sqrt{5}}$$

$$\therefore \qquad \lambda = \frac{h}{p} = \frac{6.63\ 10^{-34} \times \sqrt{5}}{1.67 \times 10^{-19} \times 6}$$

$$\lambda = 1.47 \times 10^{-15}\ \text{m}$$

$$\lambda = 1.47 \times 10^{-5}\ \text{A}^{\circ}$$

5. Calculate the de-Broglie wavelength associated with a proton with a velocity equal to $\left(\frac{1}{20}\right)^{th}$ velocity of light.

Sol. The de-Broglie wavelength of a proton is given by

$$\lambda = \frac{h}{mv}$$

Given that $\qquad h = 6.62 \times 10^{-34}$ J-sec

$$m = 1.67 \times 10^{-27}\ \text{kg}$$

$$v = \frac{c}{20} = \frac{3 \times 10^{8}}{20}\ \text{m/sec}$$

$$\lambda = \frac{6.62 \times 10^{-34} \times 20}{1.67 \times 10^{-27} \times 3 \times 10^{8}}$$

$$\lambda = 2.643 \times 10^{-14}\ \text{m}$$

6. Calculate the de-Broglie wavelength of neutron of energy 28.8 ev. Given $h = 6.62 \times 10^{-34}$ J.sec, $m = 1.67 \times 10^{-27}$ kg.

Sol. The de-Broglie wavelength for non-relativistic particle of mass m is given by

$$\lambda = \frac{h}{\sqrt{2\,m\,E}}$$

here $\qquad m = 1.67 \times 10^{-27}$ kg

$$E = 28.8\ \text{ev}$$

$$= 28.8 \times 1.6 \times 10^{-19}\ \text{Joule}$$

$$\lambda = \frac{6.62 \times 10^{-34}\ \text{J.sec}}{\sqrt{2 \times 1.67 \times 10^{-27}\ \text{kg} \times 28.8 \times 1.6 \times 10^{-19}\ \text{Joule}}}$$

$$\lambda = 5.336 \times 10^{-12} \text{ m}$$

$$\lambda = 0.05336 \text{ A}^{\circ}$$

7. Calculate the de-Brglie wavelength of a neutron having kinetic energy of 1 ev.

Sol. The de-Broglie wavelength for a non-relativistic particle of mass m is given by

$$\lambda = \frac{h}{\sqrt{2\,m\,E}}$$

Given that

$$h = 6.62 \times 10^{-34} \text{ J.sec}$$

$$m = 1.67 \times 10^{-27} \text{ kg}$$

$$E = 1 \text{ ev} = 1.6 \times 10^{-19} \text{ Joule}$$

$$\lambda = \frac{6.62 \times 10^{-34} \text{ J.sec}}{\sqrt{2 \times 1.6 \times 10^{-19}\ 1.67 \times 10^{-27} \text{ kg}}}$$

$$\lambda = 2.87 \times 10^{-11} \text{ m} = 0.287 \text{ A}^{\circ}$$

8. Find the de-Broglie wavelength of a neutron of energy 12.8 Mev?

Sol. The de-Broglie wavelength for a non-relativistic particle of mass m is given by

$$\lambda = \frac{h}{\sqrt{2\,m\,E}}$$

Given that

$$h = 6.62 \times 10^{-34} \text{ J.Sec}$$

$$m = 1.67 \times 10^{-27} \text{ Kg}$$

$$E = 12.8 \text{ Mev}$$

$$= 12.8 \times 10^{6} \times 1.6 \times 10^{-19} \text{ Joule}$$

$$\therefore \quad \lambda = \frac{h}{\sqrt{2\,m\,E}}$$

$$\lambda = \frac{6.62 \times 10^{-34} \text{ J.Sec}}{\sqrt{2 \times 1.67 \times 10^{-27} \text{ Kg} \times 12.8 \times 10^{6} \times 1.6 \times 10^{-19} \text{ J}}}$$

$$\lambda = \frac{6.62 \times 10^{-34}}{8.28 \times 10^{-20}}$$

$$\lambda = 8.0 \times 10^{-15} \text{ m}$$

$$\lambda = 8.0 \times 10^{-5} \text{ A}^{\circ}$$

9. Calculate de-Broglie wavelength associated with nitrogen at 3.0 atmospheric pressure and 27 ° C (mass of N_2-atom = 4.65×10^{-26} kg).

Sol. The de-Broglie wavelength is given by

$$\lambda = \frac{h}{\sqrt{2\,m\,E}}$$

we know that

$$E = \frac{3}{2}KT$$

$$\therefore \qquad \lambda = \frac{h}{\sqrt{2m.\frac{3}{2}\ KT}}$$

$$\lambda = \frac{h}{\sqrt{2m\,.\,\frac{3}{2}\ KT}}$$

Given that $\qquad m = 4.65 \times 10^{-26}$ kg

$T = 27 + 273 = 300$ K

(Boltzmann constant) $K = 1.38 \times 10^{-23}$ J/K

$$\therefore \qquad \lambda = \frac{6.62 \times 10^{-34}\ Kg}{\sqrt{3 \times 4.65 \times 10^{-26} \times 1.38 \times 10^{-23} \times 300}}$$

$$\lambda = 2.754 \times 10^{-11}\ m$$

$$\lambda = 0.2754\ A^{o}$$

10. Show that de-Broglie wavelength a material particle of rest mass m_o and charge q accelerated from rest through a potential difference of V volts relativistically is given by

$$\lambda = \frac{h}{\sqrt{2m\,.\,qv\left(1 + \frac{qv}{2m_0\,c^2}\right)}}$$

Sol. The Kinetic energy is given by

$$E_K = qv$$

we know that

$$E^2 = p^2\,c^2 + m_o^{\,2}\,c^4$$

since $E = E_K + m_o c^2$

$$(E_K + m_o^2 c^2) = p^2 c^2 + m_o^2 c^4$$

$$(qv)^2 + m_o^2 c^4 + 2q\,v\,m_o c^2 = p^2 c^2 + m_o^2 c^4$$

$\therefore$ $$p^2 c^2 = q^2 v^2 + 2q\,v\,m_o c^2$$

$$p^2 = \frac{q^2 v^2}{c^2} + 2q\,v\,m_o$$

$$p^2 = 2\,m_o\,q\,v\left[1 + \frac{vq}{2\,m_o c^2}\right]$$

$\therefore$ $$P = \sqrt{2\,m_o\,q\,v\left[1 + \frac{vq}{2\,m_o c^2}\right]}$$

the wavelength $\lambda = \dfrac{h}{p}$

$\therefore$ $$\lambda = \frac{h}{\sqrt{2\,m_o\,qv\left(1 + \dfrac{vq}{2\,m_o\,c^2}\right)}}$$

11. A particle of rest mass m_o has a kinetic energy K. Show that its de-Broglie wavelength is given by

$$\lambda = \frac{hc}{\sqrt{K\left(K + 2\,m_o\,c^2\right)}}$$

Hence, calculate the wavelength of an electron of kinetic energy 1 Mev. What will be the value of λ is $K << m_o c^2$?

Sol. We know that

$$E^2 = p^2 c^2 + m_o^2 c^4 \qquad \text{.....(1)}$$

and E = Relativistic Kinetic energy (K) $+ m_o c^2$

or $$E = K + m_o c^2 \qquad \text{.....(2)}$$

from (1) and (2) $$(K + m_o c^2) = p^2 c^2 + m_o^2 c^4$$

$$K^2 + m_o{}^2 c^4 + 2K\ m_o c^2 = p^2 c^2 + m_o{}^2 c^4$$

$$p^2 c^2 = K^2 + 2K\ m_o c^2$$

$$p^2 = \frac{K}{c^2}(K + 2\ m_o c^2)$$

$$p = \frac{1}{c}\sqrt{K\left(K + 2\ m_o\ c^2\right)}$$

since $\lambda = \frac{h}{p}$

$$\therefore \quad \lambda = \frac{h}{\frac{1}{c}\sqrt{K\left(K + 2\ m_o\ c^2\right)}} \qquad .....(3)$$

If $K << m_o c^2$

then $K + 2\ m_o c^2 = 2\ m_o c^2$

$$\therefore \quad \lambda = \frac{hc}{\sqrt{2\ m_o\ K c^2}} = \frac{h}{\sqrt{2m_o K}}$$

from equation (3) we have

$$\lambda = \frac{hc}{\sqrt{K\left(K + 2\ m_o c^2\right)}} \qquad .....(4)$$

for an electron

$$m_o\ c^2 = 9.1 \times 10^{-31} \times (3 \times 10^8)^2 \text{ Joule}$$

$$= \frac{81.9 \times 10^{-15}}{1.6 \times 10^{-19}} = 0.51 \times 10^6 \text{ ev} = 0.51 \text{ Mev}$$

given K = 1 Mev

$\therefore$ from (4)

$$\lambda = \frac{hc}{\sqrt{1(1 + (2 \times 5.1))}} = \frac{hc}{\sqrt{2.02} \text{ Mev}}$$

$$\lambda = \frac{6.62 \times 10^{-34} \times 3 \times 10^8}{\sqrt{2.02} \times 1.6 \times 10^{-19} \times 10^6} \text{ m}$$

$$\lambda = 8.78 \times 10^{-13} \text{ m}$$

$$\lambda = 8.78 \times 10^{-3} A^o$$

12. Can a photon and an electron of the same momentum have the same wavelength? Compare their wavelength if the two have the same energy.

Sol. The momentum of the electron is given by

$$p_e = \frac{h}{\lambda_e} \qquad \text{.....(1)}$$

and the momentum of photon is

$$p_{ph} = \frac{hv}{c} = \frac{h}{\lambda_{ph}} \qquad \text{.....(2)}$$

from (1) and (2) we have

$$\lambda_e = \lambda_{ph}$$

hence, the photon and electron of the same momentum have the same wavelength.

The de-Broglie wavelength of the electron is

$$\lambda_e = \frac{h}{mv} = \frac{h}{\sqrt{2mE}} \qquad \text{.....(3)}$$

where E is the energy of an electron

the de-Broglie wavelength of photon is

$$\lambda_{ph} = \frac{h}{p}$$

but

$$E = h\nu = \frac{hc}{\lambda} = pc$$

$$\lambda_{ph} = \frac{h}{E/c} \qquad \text{.....(4)}$$

$$\therefore \quad \frac{\lambda_{ph}}{\lambda_e} = \frac{hc}{E} \cdot \frac{\sqrt{2mE}}{h}$$

$$\frac{\lambda_{ph}}{\lambda_e} = c\sqrt{\frac{2mE}{E^2}}$$

$$\frac{\lambda_{ph}}{\lambda_e} = \sqrt{\frac{2mc^2}{E}}$$

13. What would be the wavelength of quantum of radiant energy emitted, if an electron transmitted in to radiation and converted in to one Quantum?

Sol. We know that $E = mc^2$

and $E = h\nu = \dfrac{hc}{\lambda}$

$$\therefore \quad \lambda = \frac{h}{mc} = \frac{6.6 \times 10^{-34}}{9.1 \times 10^{-31} \times 3 \times 10^{8}} = 0.0242 \times 10^{-10} \text{ m}$$

$$= 0.0242 \text{ A}^{o}$$

On-Group and phase velocity

14. Prove that the de-Broglie wave velocity is a function of wavelength even in free space.

(or)

Show that the phase velocity of the de-Broglie waves associated with a moving particle having a rest mass m_o is given by

$$v_p = c\left[1 + \left(\frac{m_o c\, \lambda}{h}\right)\right]^{1/2}$$

Sol. We know that $\lambda = \dfrac{h}{mv} = \dfrac{h}{\dfrac{m_o\, v}{\sqrt{1 - v^2/c^2}}}$

or $\lambda = \dfrac{h\sqrt{1 - v^2/c^2}}{m_o\, v}$

$$\lambda = \frac{h}{m_o}\sqrt{\frac{1}{V^2} - \frac{1}{c^2}} \qquad \text{.....(1)}$$

we know that $V_p = \dfrac{c^2}{v}$

$$\therefore \quad v = \frac{c^2}{v_p} \qquad \text{.....(2)}$$

putting (2) in (1), we get

$$\lambda = \frac{h}{m_o}\sqrt{\frac{v_p^{\,2}}{c^4} - \frac{1}{c^2}}$$

$$\therefore \qquad \lambda = \frac{h}{m_o c}\sqrt{\frac{v_p^2}{c^2} - 1}$$

$$\frac{\lambda\, m_o c}{h} = \sqrt{\frac{v_p^2}{c^2} - 1}$$

$$\left(\frac{\lambda\, m_o c}{h}\right)^2 = \frac{v_p^2}{c^2} - 1$$

$$\frac{v_p^2}{c^2} = 1 + \frac{m_o c\, \lambda}{h}$$

$$v_p^2 = c^2\left[1 + \frac{m_o c\, \lambda}{h}\right]$$

$$v_p = c\sqrt{\left[1 + \frac{m_o c\, \lambda}{h}\right]}$$

15. The phase velocity of ripples on a liquid surface is $\sqrt{\frac{2\pi\, s}{\lambda\, \rho}}$, where s is surface tension and ρ the density of the liquid. Find the group velocity of the ripples?

Sol. We have

$$v_p = \sqrt{\frac{2\pi\, s}{\lambda\, \rho}}$$

$$v_p^2 = \frac{2\pi\, s}{\lambda\, \rho}$$

$$\lambda = \frac{2\pi\, s}{v_p^2\, \rho} \qquad \text{....(1)}$$

the wave number is given by

$$K = \frac{2\,\pi}{\lambda} = \frac{W}{v_p}$$

where w is the angular velocity

$$K = \frac{2\,\pi}{2\,\pi\, s}.v_p^2\, \rho$$

$$K = \frac{v_p^2 \rho}{s}$$

The group velocity is given by

$$v_g = \frac{dw}{dk}$$

as $$K = \frac{v_p^2 \rho}{s} = \frac{w^2}{K^2} \cdot \frac{\rho}{s}$$

$$K^3 = \frac{w^2 \rho}{s}$$

$$w^2 = \frac{K^3 s}{\rho}$$

$$w = \sqrt{\frac{s}{\rho}} K^{3/2}$$

$$v_g = \frac{dw}{dK} = \sqrt{\frac{s}{\rho}} \cdot \frac{3}{2} K^{1/2}$$

$$v_g = \frac{3}{2}\sqrt{\frac{sK}{\rho}} = \frac{3}{2}\sqrt{\frac{s}{\rho} \cdot \frac{v_p^2 \rho}{s}}$$

$$v_g = \frac{3}{2} v_p$$

16. An electron has de-Broglie wavelength 2.0×10^{-12} m. Find its Kinetic energy. Also find the phase and group velocities of its de-Broglie waves.

Sol. Total energy E of the electron is given by

$$E = \sqrt{(pc)^2 + (m_o c^2)^2}$$

we know that $$p = \frac{h}{\lambda}$$

$\therefore$ $$pc = \frac{hc}{\lambda} = \frac{6.63 \times 10^{-34} \times 3 \times 10^8}{2.0 \times 10^{-12} \times 1.6 \times 10^{-19}}$$

$$pc = 621.56 \text{ Kev}$$

and $$m_o c^2 = \frac{9.1 \times 10^{-31} \times (3 \times 10^8)^2}{1.6 \times 10^{-16}} = 511 \text{ Kev}$$

$$m_o c^2 = \frac{9.1 \times 10^{-31} \times (3 \times 10^8)^2}{1.6 \times 10^{-16}} = 511 \text{ Kev}$$

$$E = \sqrt{(621.56)^2 + (511)^2} = 804.64 \text{ Kev}$$

the Kinetic energy of the electron

$$K = E - m_o c^2$$

$$K = 804.64 - 511.00 = 293.64 \text{ Kev}$$

we know that $$m = \frac{m_o}{\sqrt{1 - v^2/c^2}}$$

$$\frac{m_o}{m} = \sqrt{1 - \frac{v^2}{c^2}}$$

$$\frac{m_o c^2}{m c^2} = \sqrt{1 - \frac{v^2}{c^2}}$$

$$1 - \frac{v^2}{c^2} = \frac{E_o^2}{E^2}$$

$$v = c\sqrt{1 - \left(\frac{E_o}{E}\right)^2}$$

$$v = c\sqrt{1 - \left(\frac{511}{804.64}\right)^2}$$

$$v = c\sqrt{1 - 0.4033}$$

$$v = 0.7724 \text{ c}$$

thus, the group velocity of the de-Broglie waves,

$$v_g = v = 0.7724c$$

phase velocity of de-Broglie waves

$$v_p = \frac{c^2}{v_g}$$

$$v_p = \frac{c^2}{0.7724\ c} = 1.29\ c$$

On Heisenberg Uncertainty principle

17. If the uncertainty in the location of a particle is equal to its de-Broglie wavelength, what is the uncertainty with velocity?

Sol. We know that $\Delta x \,.\, \Delta p \approx \dfrac{h}{2\pi}$

$$\Delta p \approx \frac{h}{2\pi\, \Delta x} \qquad .....(1)$$

But, according to the question

$$\Delta x = \lambda$$

$$\Delta x = \frac{h}{p} \qquad ....(2)$$

$$\therefore \qquad \Delta p = \frac{h\, p}{2\pi \,.\, h}$$

$$\Delta p \approx p$$

or $$\Delta (mv) \approx mv$$

$$m \Delta v \approx mv$$

$$\Delta v \approx v$$

18. Calculate the smallest possible uncertainty in the position of an electron moving with velocity $v = 3 \times 10^7$ m/sec.

Sol. According to the principle of uncertainty

$$(\Delta x)\ (\Delta p) = \frac{h}{2\ \pi}$$

$$\Delta x = \frac{h}{2\pi\ \Delta p} \qquad ...(1)$$

since $$\Delta p \approx p \approx mv$$

$$\Delta p \approx \frac{m_o v}{\sqrt{1 - v^2 / c^2}} \qquad \text{.....(2)}$$

putting (2) in (1) we get

$$\Delta x \approx \frac{h \sqrt{1 - v^2 / c^2}}{2\pi . m_o\ v}$$

given $h = 6.62 \times 10^{-34}$ J.sec

$v = 3.0 \times 10^7$ m/sec

$m_o = 9 \times 10^{-31}$ kg

$c = 3 \times 10^8$ m/sec

$$\Delta x \approx \frac{6.62 \times 10^{-34} \times \sqrt{1 - \left(3.0\times10^7 / 3\times10^8\right)}}{2 \times 3.14 \times 9.0 \times 10^{-31} \times 3.0 \times 10^7}$$

$$\Delta x = 3.88 \times 10^{-12} \text{ m}$$

$$\Delta x = 0.0388 \text{ A}^o$$

19. An electron is confined to a box of length 1.1×10^{-8} m. Calculate the minimum uncertainty in its velocity. Given $m = 9.1 \times 10^{-31}$ kg and $\hbar = 1.05 \times 10^{-34}$ J.sec

Sol. According to Heisenberg uncertainty principle

$$\Delta x . \Delta p \geq \hbar$$

or $$\Delta p \approx \frac{\hbar}{\Delta x}$$

since $$\Delta p = m \Delta v$$

where Δv is the uncertainty in the determination of velocity and m is the mass.

given $\Delta x = 1.1 \times 10^{-8}$ m

$\hbar = 1.05 \times 10^{-34}$ J.sec

$m = 9.0 \times 10^{-31}$ kg

$$m \Delta v = \frac{\hbar}{\Delta x}$$

$$\Delta v = \frac{\hbar}{m \Delta x} = \frac{1.05 \times 10^{-34} \text{ J.sec}}{9.0 \times 10^{-31} \times 1.1 \times 10^{-8}}$$

$$\Delta v = 1.06 \times 10^4 \text{ m/sec}$$

20. Calculate the uncertainty in the velocity of an electron which is confined in a 10 A° box.

Sol. According to Heisenberg uncertainty principle

$$\Delta x \ \Delta p \geq \frac{h}{2\pi}$$

$$\therefore \qquad \Delta p \approx \frac{h}{2\pi \ \Delta x}$$

$$m \Delta v \approx \frac{h}{2\pi \ \Delta x}$$

$$\Delta v = \frac{6.62 \times 10^{-34} \text{ J.sec}}{2 \times 3.14 \times 9.1 \times 10^{-31} \text{ Kg} \times 10 \times 10^{-10} \text{ m}}$$

$$\Delta v = 0.116 \times 10^{6} \text{ m/sec}$$

21. An electron has speed 600 m/sec with an accuracy of 0.005 %. Calculate the certainty with which we can locate the position of the electron.

Sol. According to Heisenberg uncertainty principle

$$\Delta x . \Delta p \geq \hbar$$

$$\Delta x . \Delta p \approx \frac{h}{2\pi}$$

$$\Delta x \approx \frac{h}{2\pi \ \Delta p}$$

$$\Delta x = \frac{h}{2\pi \ m \ \Delta v}$$

given $\qquad \Delta v = \frac{0.005}{100} \times 600 \text{ m/s}$

$$\Delta x = \frac{6.62 \times 10^{-34} \text{ J.sec} \times 100}{2 \times 3.14 \times 9.1 \times 10^{-31} \times 0.005 \times 600}$$

$$\Delta x = 0.39 \text{ m}$$

22. Hydrogen atom, say, has a radius of 0.5A°. Calculate the Kinetic energy needed by an electron to be confined to the atom.

Sol. The uncertainty in the position is

$$\Delta x = 0.5 \text{ A}^{\circ} = 0.5 \times 10^{-10} \text{ m}$$

∴ According to uncertainty principle

$$\Delta x . \Delta p_x \geq \hbar$$

or
$$\Delta x . \Delta p_x \approx \frac{h}{2\pi}$$

$$\Delta p_x \approx \frac{h}{2\pi . \Delta x}$$

$$\Delta p_x \approx \frac{6.63 \times 10^{-34}}{2 \times 3.14 \times 5 \times 10^{-11}}$$

$$\Delta p_x \approx 2.1 \times 10^{-24} \text{ Kg m/sec}$$

the momentum p of the electron must be at least equal to Δp_x .

therefore $p = 2.1 \times 10^{-24}$ Kg m/sec

∴ The Kinetic energy

$$K.E = \frac{p^2}{2m} = \frac{\left(2.1 \times 10^{-24}\right)^2}{2 \times 9.1 \times 10^{-31} \times 1.6 \times 10^{-19}}$$

$$K.E = 15.1 \text{ ev}$$

23. The speed of an electron is measured to be 5.00×10^3 m/sec to an accuracy of 0.003%. Find the uncertainty in determining the position of this electron?

Sol. According to Heisenberg uncertainty principle

$$\Delta x \ \Delta p \geq \hbar$$

$$\Delta x \ \Delta p \approx \frac{h}{2\pi}$$

since $p = mv = 9.0 \times 10^{-31} \times 5.00 \times 10^3$

$$= 4.5 \times 10^{-27} \text{ Kg- m/sec}$$

The uncertainty in the value of p is

$$\Delta p = \frac{0.003}{100} \times 4.5 \times 10^{-27}$$

$$\Delta p = 1.35 \times 10^{-31} \text{ Kg- m/sec}$$

Therefore the uncertainty in the position of this electron is

$$\Delta x = \frac{h}{2\pi \, \Delta p}$$

$$\Delta x = \frac{6.63 \times 10^{-34}}{2 \times 3.14 \times 1.35 \times 10^{-31}}$$

$$\Delta x = 7.82 \times 10^{-4} \text{ m}$$

24. A hydrogen atom is 0.53 A° in radius. Use uncertainty principle to estimate the minimum energy, an electron can have in this atom.

Sol. The uncertainty in the position is given by

$$\Delta x = 0.53 \text{ A}^\circ$$

$$\Delta x = 0.53 \times 10^{-10} \text{ m}$$

According to uncertainty principle

$$\Delta x \,.\, \Delta p_x \geq \hbar$$

$$\Delta p_x \geq \frac{h}{2\pi \, \Delta x}$$

or $$\Delta p_x \approx \frac{h}{2\pi \, \Delta x}$$

$$\Delta p_x = \frac{6.63 \times 10^{-34}}{2 \times 3.14 \times 0.53 \times 10^{-10}}$$

$$\Delta p_x \approx 19.8 \times 10^{-25} \text{ Kg-m/sec}$$

therefore $$\Delta p_x \approx p$$

$\therefore$ $$\text{K.E} = \frac{p^2}{2m} = \frac{\left(19.8 \times 10^{-25}\right)^2}{2 \times 9.1 \times 10^{-31}}$$

$$\text{K.E} = \frac{21.6 \times 10^{-19}}{1.6 \times 10^{-19}} = 13.5 \text{ ev}$$

25. If an excited state of hydrogen atom has a lifetime of 2.5×10^{-14} μs, what is the minimum error with which the energy of this state can be measured?

Sol. According to uncertainty principle

$$\Delta E \,.\, \Delta t \geq \hbar$$

or $$\Delta E . \Delta t \approx \frac{h}{2\pi}$$

$$\Delta E \approx \frac{h}{2\pi \Delta t}$$

given $$\Delta t = 2.5 \times 10^{-14} \mu \text{ sec}$$

$$\Delta t = 2.5 \times 10^{-14} \times 10^{-6} \text{ sec}$$

$$\Delta E = \frac{6.63 \times 10^{-34}}{2 \times 3.14 \times 2.5 \times 10^{-14} \times 10^{-6}}$$

$$\Delta E = \frac{6.63 \times 10^{-34}}{2 \times 3.14 \times 2.5 \times 10^{-14} \times 10^{-6}}$$

$$\Delta E = 4.22 \times 10^{-15} \text{ Joule}$$

$$\Delta E = \frac{4.22 \times 10^{-15}}{1.6 \times 10^{-19}} = 26.37 \text{ Kev}$$

26. Using the uncertainty relation $\Delta E . \Delta t = \frac{h}{2\pi}$, calculate the time required for the atomic system to retain rotation energy for a line of wavelength 6000 A° and width 10^{-4} A°.

Sol. We know that

$$E = h\nu$$

$$E = \frac{hc}{\lambda}$$

$\therefore$ $$\Delta E = \frac{-hc}{\lambda^2} \Delta\lambda$$

Given $$\Delta\lambda = 10^{-4} \text{ A}^\circ$$

$$\Delta\lambda = 10^{-4} \times 10^{-10} \text{ m}$$

from uncertainty relation

$$\Delta E . \Delta t = \frac{h}{2\pi}$$

$$\Delta t = \frac{h}{2\pi\,\Delta E} \approx \frac{h}{2\pi\left(\frac{hc}{\lambda^2}\right)\Delta\lambda}$$

$$\Delta t = \frac{\lambda^2}{2\pi c\,\Delta\lambda}$$

$$\Delta t = \frac{\left(6.0 \times 10^{-7}\right)^2}{2 \times 3.14 \times 3 \times 10^8 \times 10^{-14}} = 1.9 \times 10^{-8} \text{ second}$$

27. What is the minimum uncertainty in the frequency of a photon whose life time is about 10^{-8} second?

Sol. We know that

$$\Delta E.\,\Delta t \geq \frac{h}{2\pi} \qquad \text{.....(1)}$$

the energy of a photon is given by

$$E = h\nu$$

$$\Delta E = h\Delta\nu \qquad \text{.....(2)}$$

from (1) and (2)

$$h\Delta\nu\,.\,\Delta t \approx \frac{h}{2\pi}$$

$$\Delta\nu \approx \frac{1}{2\pi.\,\Delta t}$$

Given $\quad \Delta t = 10^{-8}$ sec

$$\Delta\nu \approx \frac{1}{2 \times 3.14 \times 10^{-8}}$$

$$\Delta\nu = 15.92 \times 10^6 \text{/sec}$$

On Schrodinger's equation

28. Find the energy of an electron moving in one dimension in an infinitely high potential box of width 1A° (mass of the electron is 9.1×10^{-31} kg and $h = 6.63 \times 10^{-34}$ J.sec)

Sol. The energy of a particle is given by

$$E = \frac{n^2h^2}{8mL^2}$$

Where n = 1,2,3.......

m is the mass of the particle; L is the width of the box

The minimum energy of the particle is obtained by substituting n = 1

$$\therefore \qquad E = \frac{h^2}{8mL^2}$$

Given,

$$m = 9.1 \times 10^{-31}\ \text{kg}$$

$$h = 6.62 \times 10^{-34}\ \text{J.sec}$$

$$L = 1\ A^o = 10^{-10}\ \text{m}$$

$$E = \frac{\left(6.62 \times 10^{-34}\ \text{J.sec}\right)^2}{8 \times 9.1 \times 10^{-31}\ \text{kg} \times \left(10^{-10}\text{m}\right)^2} = 6.63 \times 10^{-18}\ \text{Joule}$$

$$E = \frac{6.63 \times 10^{-18}}{1.6 \times 10^{-19}}\ \text{ev}$$

$$E = 37.69\ \text{ev}$$

29. Compute the lowest energy of a neutron confined to the nucleus where nucleus is considered a box with a size of 10^{-14} m

($h = 6.62 \times 10^{-34}$ J.sec, $m = 1.6 \times 10^{-24}$ g)

Sol. For the neutron, the energy is given by

$$E = \frac{\pi^2 h^2}{2m}\left[\frac{n_x^2}{l_x^2} + \frac{n_y^2}{l_y^2} + \frac{n_z^2}{l_z^2}\right]$$

for lowest energy state $n_x = n_y = n_z = 1$

for a cubical box of size 10^{-14} m

$$E = \frac{h^2}{8m}\left[\frac{1}{l^2} + \frac{1}{l^2} + \frac{1}{l^2}\right]$$

$$E = \frac{3h^2}{8ml^2} = \frac{3 \times \left(6.62 \times 10^{-34}\right)^2}{8 \times 1.6 \times 10^{-27}\ \text{kg} \times 10^{-28}\ \text{m}}$$

$$E = 10.29 \times 10^{-23}\ J$$

$$E = \frac{10.29 \times 10^{-23}}{1.6 \times 10^{-19}} = 6.43\ Mev$$

30. Find the expectation value $< x >$ of the position of a particle trapped in a box of length L.

Sol. The expectation value of the position of a particle with wave function ψ (x) is given by

$$< x > = \int_{-\infty}^{+\infty} x\, |\psi\ (x)|^2\, dx$$

we know that the wave function of a particle trapped in a box of a length L is given by

$$\psi\ (x) = \sqrt{\frac{2}{L}}\ \sin \frac{n\pi x}{L}$$

$$< x > = \frac{2}{L} \int_{o}^{L} x\ \sin^2 \left(\frac{n\pi x}{L}\right) dx$$

$$< x > = \frac{2}{L} \int_{o}^{L} x\ \frac{\left(1 - \cos \frac{2n\pi x}{L}\right)}{2}\ dx$$

$$< x > = \frac{1}{L} \left[\frac{x^2}{2}\right]_o^L$$

$$< x > = \frac{1}{L} \left[\frac{L^2}{2}\right] = \frac{L}{2}$$

31. A particle is in motion along a line between x = 0 and x = a with zero potential energy. At points for which x < 0 and x > a, the potential energy is infinite. The wave function for the particle in the nth state is given by

$$\psi_n = A \sin \frac{n\pi x}{a}$$

find the expression for the normalised wave function

Sol. The wave function of the particle is given by

$$\psi_n = A \sin \frac{n\pi x}{a}$$

The condition for normalisation is given by

$$\int_{-\infty}^{+\infty} |\psi(x)|^2 \, dx = 1$$

$$\int_{-\infty}^{+\infty} \left| A \sin\left(\frac{n\pi x}{a}\right) \right|^2 dx = 1$$

$$A^2 \int_{o}^{a} \sin^2 \frac{n\pi x}{a} \, dx = 1$$

$$\frac{A^2}{2} \left[1 - \cos 2\left(\frac{n\pi x}{a}\right)\right]_0^a dx = 1$$

$$\frac{A^2}{2} \left[x - \frac{\sin\,(2\,n\pi x/a)}{(2\,n\pi x/a)}\right]_0^a = 1$$

$$\frac{A^2}{2}(a) = 1$$

or $$A = \sqrt{\frac{2}{a}}$$

Therefore, the normalised wave function of the particle is

$$\psi_n = \sqrt{\frac{2}{a}} \sin\left(\frac{n\pi x}{a}\right)$$

32. An electron is bound in one dimensional potential box which has width 2.5×10^{-10}m. Assuming the height of the box to be infinite, calculate the lowest two permitted energy values of the electron.

Sol. The energy of particle of mass m is given by

$$E_n = \frac{n^2h^2}{8mL^2}$$

where n = 1, 2, 3,..........

Given

$$h = 6.62 \times 10^{-34} \text{ J.sec}$$

$$m = 9.1 \times 10^{-31} \text{ kg}$$

$$L = 2.5 \times 10^{-10}\ \text{m}$$

$$\therefore \qquad E_n = \frac{n^2\left(6.62 \times 10^{-34}\right)^2}{8 \times 9.1 \times 10^{-31} \times \left(2.5 \times 10^{-10}\right)^2}$$

$$E_n = 9.66 \times 10^{-19}\ n^2\ \text{Joule}$$

$$E_n = \frac{9.66 \times 10^{-19}}{6.6 \times 10^{-19}}\ n^2\text{ev}$$

$$E_n = 6.04\ n^2\ \text{ev}$$

The lowest two permitted energy values of the electron is obtained by putting

$n = 1, n = 2$

(i) For $n = 1$ we have

$$E_1 = 6.04 \times 1 = 6.04\ \text{ev}$$

(ii) For $n = 2$ we have

$$E_2 = 6.04 \times 2 \times 2 = 24.16\ \text{ev}$$

33. A particle is confined to move along x-axis has the wave function

$\psi = ax$ between $x = 0$ and $x = 1.0$

$\psi = 0$ else where.

Find the probability that particle can be found between $x = 0.35$ to $x = 0.45$. Also find the expectation value $\langle x \rangle$ of particle's position.

Sol. The probability of finding the particle between x_1 and x_2 when it is in n^{th} state is

$$p = \int_{x_1}^{x_2} |\psi_x|^2\ dx$$

Given $\psi_n = ax$ between $x = 0$ and $x = 1$

Given $x_2 = 0.45$ and $x_1 = 0.35$

$$\therefore \qquad p = \int_{0.35}^{0.45} (ax)^2\ dx$$

$$p = a^2 \int_{0.35}^{0.45} x^2\ dx$$

$$p = a^2 \left(\frac{x^3}{3}\right)_{0.35}^{0.45}$$

$$p = \frac{a^2}{3} = \left[(0.45)^3 - (0.35)^3\right]$$

$$p = \frac{a^2}{3}(0.0911 - 0.0428)$$

$$p = 0.0161\ a^2$$

The expectation value $< x >$ of the position of a particle described by a wave function ψ_n (x) is given by

$$< x > = \int_{-\infty}^{\infty} x\ |\psi_n(x)|^2\ dx$$

If the particle is trapped between x = 0 and x = 1

$$< x > = \int_0^1 x\ \left|(ax)^2\right|\ dx$$

$$< x > = a^2 \int_0^1 x^3\ dx$$

$$< x > = \left[\frac{a^2\ x^4}{4}\right]_0^1$$

$$< x > = \frac{a^2}{4}$$

$$< x > = 0.25\ a^2$$

34. Calculate the energy difference between the ground state and the first excited state for an electron in a one-dimesional rigid box length 10^{-10} cm (mass of electron = $9. \times 10^{-31}$kg and h = 6.62×10^{-34} Joule.sec)

Sol. The energy of a particle of mass m is given by

$$E_n = \frac{n^2 h^2}{8mL^2}$$

where, n = 1, 2, 3,.........

$$E_n = \frac{n^2(6.626 \times 10^{-34})^2}{8 \times 9.1 \times 10^{-31} \times (10^{-10})^2}$$

$$E_n = 6.03 \times 10^{-18} n^2 \text{ Joule}$$

or

$$E_n = \frac{6.03 \times 10^{-18}}{1.6 \times 10^{-19}} n^2$$

$$E_n = 38 n^2 \text{ ev}$$

(i) For the ground state, n = 1

$$\therefore \quad E_1 = 38 \times 1^2 = 38 \text{ ev}$$

(ii) For the first excited state, n = 2

$$E_2 = 38 \times 2 \times 2 = 152 \text{ ev}$$

The energy difference between the ground and first excited state

$$E_2 - E_1 = 152 \text{ ev} - 38 \text{ ev}$$

$$= 114 \text{ ev}$$

35. A particle is moving in one dimensional potential box of infinite height. What is the probability of finding the particle in a small interval Δ x, at the centre of the box when it is in the energy state, next to the least energy state.

Sol. The wave function of the particle is given by

$$\psi_n = \sqrt{\frac{2}{a}} \sin \frac{n\pi x}{a}$$

the state next to the least energy state in the second state

for n = 2

$$\psi_2 = \sqrt{\frac{2}{a}} \sin \frac{2\pi x}{a}$$

at the centre of the box

$$x = a/2$$

Therefore,

$$\psi_2 = \sqrt{\frac{2}{a}} \sin \frac{2\pi a}{2a} = 0$$

$\therefore$ probably of finding the particle in an interval Δ x at the centre of the box is

$$p = \psi_2^2 \Delta x = 0$$

36. Find the probability of finding a particle trapped in a box of length L in the region from 0.45 L to 0.55 L for the ground state and the first excited state

Sol. The eigen functions of a particle is given by

$$\psi_n = \sqrt{\frac{2}{L}} \sin \frac{n\pi a}{L}$$

the probability of finding the particle between x_1 and x_2 is given by

$$p = \int_{x_1}^{x_2} |\psi_n|^2 \, dx$$

$$p = \frac{2}{L}\int_{x_1}^{x_2} \sin^2\left(\frac{n\pi x}{L}\right) dx$$

$$p = \frac{2}{L}\int_{x_1}^{x_2} \frac{1}{2}\left(1 - \cos\frac{2n\pi x}{L}\right) dx$$

$$p = \frac{1}{L}\left[x - \frac{L}{2\pi n}\sin\frac{2n\pi x}{L}\right]_{x_1}^{x_2}$$

Given $x_1 = 0.45$ L and $x_2 = 0.55$ L; n =1

$$\therefore \quad p = \frac{1}{L}\left[\left(0.55 - \frac{L}{2\pi}\sin(1.10\pi)\right) - \left(0.45L - \frac{L}{2\pi}\sin .90\pi\right)\right]$$

$$p = \frac{1}{L}\left[\left(0.55 - \frac{1}{2\pi}\sin 198^\circ\right) - \left(0.45L - \frac{L}{2\pi}\sin .90\pi\right)\right]$$

$$p = (0.55 - 0.45) - \frac{1}{2\pi}(\sin 198^\circ - \sin 162^\circ)$$

$$p = 0.10 - \frac{1}{2\pi}(\sin 198^\circ - \sin 162^\circ)$$

applying $\quad \sin A - \sin B = 2\cos\frac{A+B}{2}\sin\frac{A-B}{2}$

we get $\quad p = 0.10 - \frac{1}{2\pi}[\cos 180^\circ \sin 18^\circ]$

$$p = 0.10 + \frac{0.3090}{3.14}$$

$$p = 19.8\%$$

similarly putting n = 2, we can get

$$p = 0.65\ \%$$

37. Substitute the wave function $\psi = xe^{-mwx^2/2\hbar}$ into Schrodinger equitation $\frac{\partial^2\psi}{\partial x^2} + \frac{2m}{\hbar^2}(E - V)\psi = 0$ where $V = \frac{1}{2}\ mw^2\ x^2$ show that $E = \frac{3}{2}w\hbar$

Sol. The given schrodinger equitation is

$$\frac{\partial^2\psi}{\partial x^2} + \frac{2m}{\hbar^2}\ (E - V)\ \psi = 0 \qquad .....(1)$$

where $\quad \psi = xe^{-mwx^2/2\hbar}$

differentiating wave function ψ with respect to x, we get

$$\frac{d\psi}{dx} = e^{-mwx^2/2\hbar} + x\left(\frac{-mw}{2\hbar}\ 2x\right)e^{-mwx^2/2\hbar}$$

$$\frac{d\psi}{dx} = e^{-mwx^2/2\hbar}\left(1 - \frac{mwx^2}{\hbar}\right)$$

again differentiating we get

$$\frac{d^2\psi}{dx^2} = e^{-mwx^2/2\hbar}\left(0 - \frac{mw}{\hbar}(2x)\right) + \left(1 - \frac{mwx^2}{\hbar}\right)\left(\frac{-mw}{2\hbar}2x\right)e^{-mwx^2/2\hbar}$$

$$\frac{d^2\psi}{dx^2} = e^{-mwx^2/2\hbar}\left(\frac{-2mwx}{\hbar} - \frac{mwx}{\hbar} + \frac{m^2w^2x^3}{\hbar^2}\right)$$

$$\frac{d^2\psi}{dx^2} = \frac{mwx}{\hbar}e^{-mwx^2/2\hbar}\left(-3 + \frac{mwx^2}{\hbar}\right) \qquad .....(2)$$

from (1) and (2)

$$\frac{mwx}{\hbar}e^{-mwx^2/2\hbar}\left(-3 + \frac{mwx^2}{\hbar}\right) + \frac{2m}{\hbar^2}\left(E - \frac{1}{2}\ mw^2x^2\right).\ xe^{-mwn^2/2\hbar} = 0$$

$$\frac{2m}{\hbar^2}\ x.\ e^{-mwx^2/2\hbar}\left(\frac{-3}{2}\ w\hbar + \frac{1}{2}\ mw^2x^2 + E - \frac{1}{2}mw^2x^2\right) = 0$$

$$\frac{2m}{\hbar^2} x.\, e^{-mwx^2/2\hbar} \left(\frac{-3}{2} w\hbar + \frac{1}{2} mw^2x^2 + E - \frac{1}{2}mw^2x^2 \right) = 0$$

From the above equation we can get

$$\left(\frac{-3}{2} \hbar w + E \right) = 0 \quad \text{or} \quad E = \frac{3}{2}\hbar w$$

38. Calculate the zero point energy of a liner harmonic oscillator of frequency 50 Hertz.

Sol. Given

$$\nu = 50\ H_z$$

the zero point energy is given by

$$E = \frac{1}{2} h\nu$$

$$E = \frac{1}{2} h\nu$$

$$E = \frac{1}{2} \times 6.62 \times 10^{-34}\ \text{J.sec} \times 50\ H_z$$

$$E = 1.66 \times 10^{-32}\ \text{Joule}$$

39. The energy of a linear harmonic oscillator in its third excited state is 0.1 ev. Calculate the frequency of vibration.

Sol. The energy of a linear harmonic oscillator

$$E_n = \left(n + \frac{1}{2} \right) h\nu$$

$$\nu = \frac{E_n}{\left(n + \frac{1}{2} \right) h}$$

Given, $n = 3$ $E_n = 0.1$ ev

$$\nu = \frac{0.1 \times 1.6 \times 10^{-19}}{\left(3 + \frac{1}{2} \right) \times 6.62 \times 10^{-34}}$$

$$\nu = 6.895 \times 10^{12}\ \text{Hertz}$$

40. The period of a linear harmonic oscillator is 1 second. Find its zero point energy.

Sol. Given T = 1 second

$$\nu = \frac{1}{T} = 1 \text{ sec}^{-1}$$

$$\text{zero point energy} = \frac{1}{2} h\nu$$

$$E = \frac{1}{2} \times 6.62 \times 10^{-34} \times 1$$

$$E = 3.315 \times 10^{-34} \text{ Joule}$$

41. An electron is trapped in an infinitely deep well of width a. If the electron is in its ground state, what fraction of time does it spend in the central one-third of the well?

Sol. The probability of finding the particle in given region is the fraction of time it spends. Therefore,

$$\text{fraction of time} = \int_{a/3}^{2a/3} \left(\sqrt{\frac{2}{a}} \sin \frac{\pi x}{a} \right)^2 dx$$

$$= \frac{2}{a} \int_{a/3}^{2a/3} \sin^2 \left(\frac{\pi x}{a} \right)$$

$$= \frac{2}{a} \int_{a/2}^{2a/3} \left(1 - \cos \frac{2\pi x}{a} \right) dx$$

$$= \frac{1}{a} \left(\frac{2a - a}{3} - \frac{a}{2\pi} \left\{ \sin \frac{2\pi}{a} \cdot \frac{2a}{3} - \sin \frac{2\pi}{a} \cdot \frac{a}{3} \right\} \right)$$

$$= \left(\frac{1}{3} - \frac{(\sin 4\pi/3 - \sin 2\pi/3)}{2\pi} \right)$$

$$= 0.61 \text{ or } 61\%$$

42. Consider a particle confined in one-dimensional box of width a. Find the probability that the particle is found between x = 0 and x = a/n, when it is in the nth state

Sol. The wave function ψ when the particle is in n^{th} state is

$$\psi = \sqrt{\frac{2}{a}} \sin \frac{n\pi x}{a}$$

Therefore, the probability

$$p = \int_{x=0}^{x=a/n} \left(\sqrt{\frac{2}{a}} \sin \frac{n\pi x}{a}\right)^2 dx$$

$$p = \frac{2}{a} \int_{x=0}^{x=a/n} \sin^2 \frac{n\pi x}{a} dx$$

$$p = \frac{2}{a} \int_{x=0}^{x=a/n} \frac{1}{2} \left(1 - \frac{a}{2n\pi} \cos \frac{2\pi nx}{a}\right) dx$$

$$p = \frac{1}{a} \left[\frac{a}{n} - \frac{a}{2nx} \sin \frac{2n\pi}{a} . \frac{a}{x}\right]$$

$$p = \frac{1}{a} \left(\frac{a}{n} - 0\right) = \frac{1}{n}$$

43. Prove that the eigen functions of a particle trapped in one-dimensional box are orthogonal.

Sol. If the wave functions associated with n^{th} and m^{th} state be ψ_n and ψ_m then they will be orthogonal provided

$$\int \psi_m^* \psi_n dv = 0$$

here

$$\Psi_m = \sqrt{\frac{2}{a}} \sin \frac{m\pi x}{a}$$

and

$$\Psi_n = \sqrt{\frac{2}{a}} \sin \frac{n\pi x}{a}$$

then

$$\int_0^a \psi_m^* \psi_n dx = \frac{2}{a} \int_0^a \left(\left(\sin \frac{m\pi x}{a}\right)\left(\sin \frac{n\pi x}{a}\right)\right) dx$$

$$= \frac{2}{a} \int_0^a \frac{1}{2} \left(\cos \frac{(m-n)\pi x}{a} - \cos \frac{(m+n)\pi x}{a}\right) dx$$

since m and n are integers, each integrand vanishes

$$\therefore \quad \int_0^a \psi_m^* \psi_n = 0$$

this proves that wave functions are orthogonal.

44. Find out the expectation value of

(i) position and (ii) momentum for a particle trapped in one-dimensional box of size a.

Sol. The expectation value of a quantity x is defined as

$$<x> = \int \psi^* x\ \psi\ dv = 0$$

here

(i)
$$<x> = \int_0^a \left(\sqrt{\frac{2}{a}} \sin\frac{n\pi x}{a}\right)(x)\left(\sqrt{\frac{2}{a}} \sin\frac{n\pi x}{a}\right) dx$$

$$<x> = \frac{2}{a}\int_0^a x \sin^2\frac{n\pi x}{a}\ dx$$

Integrating by parts and putting the limit, we get

$$<x> = \frac{a}{2}$$

(ii)
$$<p> = \int_0^a \left(\sqrt{\frac{2}{a}} \sin\frac{n\pi x}{a}\right)\left(\frac{-ih}{2\pi}\frac{\partial}{\partial x}\right)\left(\sqrt{\frac{2}{a}} \sin\frac{n\pi x}{a}\right) dx$$

$$<p> = \frac{2}{a}\int_0^a \left(\sin\frac{n\pi x}{a}\right)\left(\frac{-ih}{2\pi}\ \frac{a}{n\pi}\cos\frac{n\pi x}{a}\right) dx$$

$$<p> = \left(\frac{-ih}{n\pi^2}\right)\int_0^a \sin\frac{n\pi x}{a}\cos\frac{n\pi x}{a}\, dx$$

$$<p> = \frac{-ih}{2n\pi^2}\int_0^a \sin\frac{2n\pi x}{a}\ dx = 0$$

45. In a long chain molecule of length 5A°, electron may be treated to be free to move along the length. Calculate zero point energy, the energy gap between first two energy states of the electrons, and also the wavelength of absorption line arising from this transition.

Sol. The zero point energy is given by

$$E_1 = \frac{h^2}{8ma^2} = \frac{\left(6.62 \times 10^{-34}\right)^2}{8 \times 9.1 \times 10^{-31} \times \left(5 \times 10^{-10}\right)^2}$$

$$E_1 = 2.4 \times 10^{-19}\ J$$

$$E_1 = \frac{2.4\times10^{-19}}{1.6\times10^{-19}} = 1.5 \text{ ev}$$

the energy gap between first two energy states is

$$\Delta E = \frac{4h^2}{8ma^2} - \frac{h^2}{8ma^2}$$

$$\Delta E = 3.\ \frac{h^2}{8ma^2}$$

$$\Delta E = 3 \times 1.5 \text{ ev} = 4.5 \text{ ev}$$

the wavelength of absorption corresponding to the energy ΔE

$$\lambda = \frac{hc}{\Delta E} = \frac{hc}{4.5 \text{ ev}}$$

$$\lambda = \frac{6.62 \times 10^{-34} \times 3 \times 10^{8}}{4.5 \times 1.6 \times 10^{-19}}$$

$$\lambda = 2.76 \times 10^{-7}$$

$$\lambda = 2760 \text{ A}^{\circ}$$

Conceptual Questions

1. An electron, proton and an α-particle have the same de-Brogile wavelength, which one moves faster.

Ans. From de-Brogile, relation we have

$$\frac{h}{mv}$$

Since all the three particles have the same value of λ, the lightest moves fastest. Hence electron moves faster then the other two particles.

2. An electron, proton neutron and α-particle are moving with equal velocity. Which one will have smallest wavelength?

Ans. From de-Brogile relation, we have

$$\frac{h}{mv}$$

where v is the same for all the particles. Hence $\lambda \propto \frac{1}{m}$. Therefore, α-particle will have the smallest wavelength.

3. **An electron and a neutron have the same de-Brogile wavelength. Which particle has greater velocity?**

Ans. $\lambda_e = \lambda_n$ and $m_e < m_n$

$\therefore$ $v_e > v_n$

Hence electron has greater velocity then neutron.

4. **Two continuous wave trains of almost equal wavelength, λ and λ + dλ are super-imposed to produce wave packets. What would happen to the length of the packet of Δλ were increased?**

Ans. If Δλ were increased, then the difference between the two wavelength would be greater and the two wave train would get out of step move quickly this giving shorter wave packers. The lengths of packets would decrease.

5. **The atom in a solid possess a certain minimum zero-point energy even at 0 K, while no such restriction holds for the molecules in an ideal gas. Use the uncertainty principle to explain these statements.**

Ans. Each atom in a solid is limited to a certain definite region of space-otherwise the assembly of atoms would not be a solid. This uncertainty in position of each atom is therefore finite and its momentum and hence energy cannot be zero. The position of an ideal gas molecule is not restricted. So the uncertainty in its position is effectively infinite and its momentum and hence energy can be zero.

6. **What bearing would you think the uncertainty principle has on the existence of the zero-point energy of a harmonic oscillator?**

Ans. The oscillator cannot have zero energy because this would mean it is at rest in a definite position corresponds to an infinite momentum and hence energy uncertainty.

7. **Classically speaking, what force is exerted on a particle at the position of a step potential?**

Ans. The force is given by the gradient of the potential $f = \frac{V}{X}$. If V changes discontinuously, the force becomes indefinitely large. Such a force cannot be realized in particle; hence step potential are to be regarded (classically or quantum mechanically) as only limited cases of real physical situations.

8. Distinguish between a particle in a box and a free particle.

Ans. Motion and hence energy values of a particle in a box are restricted by the walls of the box. But a free particle has no boundary conditions.

9. What is significance of wavefunction?

Ans. Wave function (x, y, z, t), is a probability based function that is used to describe the wave like properties of a particle. It is obtained by solving the Schrodinger equation.

$| \quad x,y,z,t|^2$ is the probability per unit volume of finding the electron at (x, y, z) and at time t, $| \quad x,y,z,t|^2$ dxdydz is the probability of finding the electron in the small volume dxdydz at (x, y, z) at time t. The dimensions of the wave function are length $^{-3/2}$.

10. Define an operator, eigen function and eigen values.

Ans. Operator is the expression that acts on a function within some domain to produce new values occupying a given range. For example, the equation H E will be interpreted as operator H acting on function = operator E acting on function . The allowed solution $_n$ of the equation are often called eigen values and their corresponding energies E_n are called eigen values.

11. How does the results of quantum mechanics differs from those of classical mechanics in harmonic oscillator?

Ans. The results of quantum mechanics differs from those of classical mechanics in the following way.

1. The particle executing simple harmonic motion can have discrete energies i.e., $\frac{1}{2}hv, \frac{3}{2}hv, ...$
2. The minimum energy is not zero but it is $\frac{1}{2}$ hv, known as zero point energy.
3. The quantum mechanical harmonic oscillator has some probability outside the classical region and this probability is around 17%.

Objective Questions

1. What are the assumptions on which de Brogile forwarded the view that radiation exhibit dual nature.
   (a) Nature is symmetrical
   (b) Close parallelism between mechanical and geometrical optics
   (c) Bohr's theory of atomic structure
   (d) All the above

2. A particle possesses a kinetic energy E and mass m, the its de Brogile wavelength
   (a) $\frac{1}{h\sqrt{2mE}}$ (b) $h\sqrt{2mE}$
   (c) $\sqrt{\frac{2mE}{h}}$ (d) $\frac{h}{\sqrt{2mE}}$

3. The analogy between matter and radiation can be observed by
   (a) principle of least action of Hamilton
   (b) principle of least time of Fermat's
   (c) principle of action angle variables and with (a)
   (d) both (a) and (b)

4. An electron is accelerated through a potential difference of V volts. The de Brogile wavelength of the electron is
   (a) 12.27 V (b) $12.27\sqrt{V}$
   (c) $\frac{12.27}{\sqrt{V}}$ (d) $\frac{12.27}{V}$

5. In a dispersive medium, the relation between gropu velocity and phase velocity is given by
   (a) $V_g = V_{ph}$ (b) $V_g = V_{ph} - \lambda\frac{dV_{ph}}{d\lambda}$
   (c) $V_g = \frac{\omega}{K}$ (d) None of these

6. If the momentum of a particle is doubled then its de-Brogile wavelength
   (a) Halves (b) Doubles
   (c) Quadruples (d) Remains the same

7. The de-Broglie wavelength of an electron accelerated to a potential difference of V volts is given as

   (a) $\lambda = \frac{150}{V} \mathring{A}$ (b) $\lambda = \sqrt{\frac{150}{V}} \mathring{A}$

   (c) $\lambda = \frac{150}{V}$ cm (d) $\lambda = \sqrt{\frac{150}{V}}$ cm

8. The masses of neutron and electron are m_n and m_e respectively. If they have the same de-Brogile wavelength then their velocity should be in the ratio.

   (a) $\frac{m_e^2}{m_n^2}$ (b) $\frac{m_e}{m_n}$

   (c) $\frac{m_n}{m_e}$ (d) 1 : 1

9. The wavelength of a quantum of radiant energy emitted, if an electron transmitted in to radiation and went in to one quantum is

   (a) $\lambda = 0.0244 \mathring{A}$ (b) $\lambda = 0.0444 \mathring{A}$

   (c) $\lambda = 0.0541 \mathring{A}$ (d) None of these

10. de-Brogile wavelength of thermal neutron is given by

    (a) $\lambda = \frac{1}{\sqrt{T}} \mathring{A}$ (b) $\lambda = \frac{30.8}{\sqrt{T}} \mathring{A}$

    (c) $\lambda = \frac{40.8}{\sqrt{T}} \mathring{A}$ (d) $\lambda = 30.8 \sqrt{T} \mathring{A}$

11. An α-particle has a mass m_{4p} and a proton m_p. If they possers the same kinetic energy, then the ratio of the de-Brogile wavelength is

    (a) 1 : 2 (b) 2 : 1 (c) 4 : 1 (d) 1 : 4

12. If a charged particle of mass m is accelerated through a potential difference V volts, the de-Broglie wavelength is proportional to

    (a) V^2 (b) $V^{1/2}$ (c) V (d) $V^{-1/2}$

13. Momentum of a photon of energy hν is

    (a) hνc (b) hν (c) hν/c (d) $\frac{h\nu}{c^2}$

14. When the particle velocity is small compared with the velocity of light, then phase velocity of de-Brogile wave is

(a) ν (b) $\frac{\nu}{2}$ (c) $\frac{\nu}{3}$ (d) $\frac{\nu}{4}$

15. The de-Brogile hypothesis is associated with

(a) wave nature of all material particles (b) wave nature of α-particle only

(c) wave nature of radiations (d) wave nature of electron's only

16. In Davisson-Germer experiment, nickel crystal acts as

(a) Three dimensional diffraction grating (b) perfect reflector

(c) Perfect absorber (d) two dimensional grating

17. Heisenberg uncertainty relation is related to

(a) $(\Delta x)(\Delta P_x) \geq \hbar/2$ (b) $(\Delta E)(\Delta t) = \hbar$

(c) $\Delta J . \Delta\theta = \hbar$ (d) All are correct

18. The uncertainty principle cannot hold for the following pairs

(a) linear momentum and angle (b) angular momentum and angle

(c) position and momentum (d) energy and time

19. A nuclear is confined within the sphere of radius $r_o = 1.2 \times 10^{-13}$ cm. If P is the momentum then kinetic energy of the nuclear will be of the order of

(a) 1 mev (b) 2 Mev (c) 10 Mev (d) 15 Mev

20. To have bound state for an atom the total energy E of the system must be

(a) Negative (b) Positive (c) Zero (d) None of these

21. The value of the radius for which the total energy assumes its smallest value is

(a) $r_o = \frac{\hbar}{m_e}$ (b) $r_o = \frac{\hbar^2}{m_e}$ (c) $\frac{m_e}{\hbar}$ (d) $r_o = \frac{\hbar^2}{m_e^2}$

22. An electron is confined to a box of length 10^{-10} m. The minimum uncertainity in its velocity is

(a) 3.258×10^5 m/sec (b) 3.158×10^6 m/s

(c) 1.158×10^6 m/sec (d) 2.15×10^6 m/sec

23. The characteristic of wave function ψ are
    (a) complex single valued and infinite
    (b) complex, single valued, finite and continuous functions
    (c) real function, finite and discontinuous
    (d) complex, infinite and discontinuous function

24. The classical wave equation is written as

    (a) $\nabla^2\psi = \frac{\partial^2\psi}{\partial t^2}$ (b) $\nabla^2\psi = v^2\frac{\partial^2\psi}{\partial t^2}$

    (c) $\nabla^2\psi = \frac{1}{v^2}\frac{\partial^2\psi}{\partial t^2}$ (d) $\frac{\partial^2\psi}{\partial t^2} = V^2\nabla^2\psi$

25. The wave function for the motion of a particle in a potential well of width a is given as

$$\psi_n = B\sin\left(\frac{n\pi x}{a}\right), \text{ then B is}$$

    (a) $\frac{1}{a}$ (b) $\sqrt{\frac{1}{a}}$ (c) $\frac{2}{a}$ (d) $\sqrt{\frac{2}{a}}$

26. The energy operator is written as

    (a) E (b) $i\hbar\frac{\partial}{\partial t}$ (c) $\frac{\partial}{\partial t}$ (d) $-i\hbar\frac{\partial}{\partial x}$

27. When the wave function is normalized then

    (a) $\int_{-a}^{+a}\psi\psi^* dx = 1$ (b) $\psi\psi^* = 1$

    (c) $\int_{-1}^{+1}\psi\psi^* dx = 1$ (d) $\psi\psi^* dx = 1$

28. The Schrodinger time independent wave equation is written as

    (a) $\nabla^2\psi + \frac{\hbar^2}{2m}(E - V)\psi = 0$ (b) $\nabla^2\psi + \frac{2m}{\hbar^2}(V - E)\psi = 0$

    (c) $\nabla^2\psi + \frac{2m}{\hbar^2}(E - V)\psi = 0$ (d) $\left(-\frac{\hbar^2}{2m}\nabla^2 - V\right)\psi = -i\hbar\frac{\partial}{\partial t}\psi$

29. For a particle of mass m confined to a cubical box of side L, the allowed values of energy E are given as

(a) $\frac{n^2h^2}{2mL^2}$, n = 1,2,3 (b) $\frac{h^2L^2}{2mn^2}$, n = 1,2,3

(c) $\frac{n^2h^2}{mL}$, n = 1,2,3 (d) $\frac{n^2h^2}{8mL^2}$, n = 1,2,3

30. The equation $H\psi = E\psi$, which is eigen value

(a) H is the eigen value of E (b) E is the eigen value of H

(c) ψ is the eigen value of both E and H (d) None of these

31. In equation $H\psi = E\psi$, the value of H is

(a) $-\frac{\hbar^2}{2m}\nabla^2 + V$ (b) $\frac{\hbar^2}{2m}\nabla^2 - V$

(c) $-\frac{\hbar^2}{2m}\nabla^2 - V$ (d) $\frac{\hbar^2}{2m}\nabla^2 + V$

32. The Schrodinger equation is based on some assumptions. They are

(a) de-Brogile wavelength holds good for any particle moving in any particle moving in any field of force with potential energy V

(b) The wave function ψ is governed by the equation

$$\psi = \psi_o e^{-i\omega t}$$

(c) Only b

(d) Both (a) and (b)

33. The value of the potential V in the range $0 < x < a$ in case for a free particle is

(a) $V = \infty$ (b) V = one (c) V = 0 (d) V = two

34. Motion of a free particle in one dimensional box is a

(a) Circular (b) Straight line (c) Sinusoidal (d) Parabolic

35. The eigen values of a free particle in one dimensional box is given as

(a) $E_n = \frac{n^2}{2ma^2}$ (b) $E_n = \frac{n^2h^2}{2ma^2}$

(c) $E_n = \frac{n^2h^2}{8ma^2}$ (d) None of these

36. The Schrodinger equation for a free particle in one dimensional box is written as

(a) $\frac{\partial^2\psi}{\partial t^2}+k^2\psi=0$ (b) $\nabla^2\psi+k^2\psi=0$

(c) $\frac{\partial^2\psi}{\partial x^2}+k^2\psi=0$ (d) None of these

37. The general solution for the equation

$$\frac{\partial^2\psi}{\partial x^2}+k^2\psi=0 \text{ is given as}$$

(a) $\psi = A \sin Kx + B \sec Kx$ (b) $\psi = A \cos Kx + B \tan Kx$

(c) $\psi = A \sin Kx + B \cos Kx$ (d) $\psi = A \tan Kx + B \sec Kx$

38. When a particle is confined to a one dimensional box, its probability is given as

(a) $|\psi_n(n)|^2 = \frac{2}{a} \sin\left(\frac{n\pi x}{a}\right)$ (b) $|\psi_n(n)|^2 = \frac{2}{a} \sin^2\left(\frac{n\pi x}{a}\right)$

(c) $|\psi_n(n)|^2 = \frac{2}{a} \cos^2\left(\frac{n\pi x}{a}\right)$ (d) None of these

39. In the problem, of square potential step, the value of the potential is zero in the region

(a) $x < 0$ (b) $0 < x < 1$

(c) $0 < x < 3$ (d) None of these

40. In the problem of square potential step, the value of the potential is V_o in the region

(a) $x < 0$ (b) $x > 0$

(c) both (a) and (b) (d) None of these

Theoretical Questions

1. What is wave particle duality?
2. What are de-Broglie's matter wave?
3. Determine the de-Broglie wavelength of electron.
4. Discuss few important properties of matter wave.
5. Discuss the dual nature of matter and waves.

6. Find the de-Broglie wavelength of electron in ev.

7. A particle of charge q and mass m is accelerated from rest through a potential difference V. Find its de-Broglie wavelength.

8. Derive the formula for de-Broglie wavelength of a particle in terms of kinetic energy and its rest energy m_0c^2.

9. What are matter waves? Show that the wavelength λ associated with a particle of mass m and kinetic energy E is given by

$$\lambda = \frac{h}{\sqrt{2mE}}$$

10. Describe Davisson and Germer experiment to prove that electron possess wave nature.

11. Deduce an expression for de-Broglie wavelength of helium atom having energy at temperature T K.

12. What was de-Broglie waves? How do they help in the interpretation of Bohr's Quantisation rule?

13. Distinguish between the phase velocity (v_p) and group velocity (v_g) of a wave packet. Prove that

$$V_pV_g = c^2$$

14. Show that the phase velocity of de-Broglie wave is greater than the velocity of light, but the group velocity is equal to the velocity of particle with which the waves are associated.

15. Explain Heisenberg's uncertainty principle. Using this principle find the binding energy of an electron in an atom.

16. What is the physical significance of wave function ψ.

17. State and explain Heisenberg's uncertainty principle. Using this principle show that electron cannot reside in an atomic nucleus.

18. What is uncertainty principle. Apply this to find the radius of the Bhor's first orbit.

19. Which of the following wave function cannot be solutions to the Schrodinger equitation an and why ?

(a) $\psi = A \cos x$ (b) $\psi = A \sec x$ (c) $\psi = a \tan x$

(d) $\psi = A\ e^{x^2}$ (e) $\psi = A\ e^{-x^2}$ (f) $\psi = A\ e^{iKx}$

(g) $\psi = A\ e^{-\sin x}$

20. Which of the following wave functions shown in the Figure cannot have physical significance in the intervals shown and why?

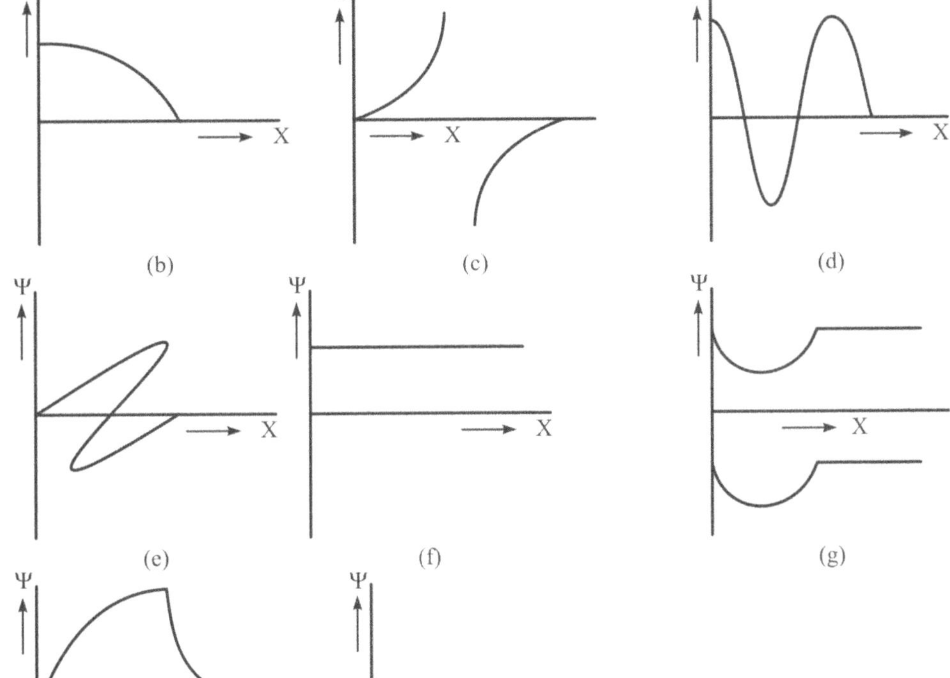

21. Show that $\psi = e^{i\alpha x}$, where α is some finite constant, is an acceptable wave function. Normalize it in the region $-a \leq x \leq a$.

22. Derive an expression for the wave function and energy of a particle confined in one dimensional potential box using Schrodinger wave equation.

23. Obtain time independent and time dependent Schrodinger wave equation. Explain ψ.

24. Derive Schrodinger's time independent wave equation. What is the physical significance of state function ψ used in this equation, what condition must it fulfil?

25. Find an expression for the energy state of a particle in one-dimensional box.

26. Calculate the normalize wave function of the particle in a box.

27. Solve Schordinger wave equation for a particle in one-dimesional rigid box of side L and having potential energy (V) as follows.

$$V(x) = \infty \quad \text{for } x < 0 \text{ and } x > L$$

$$V(x) = 0 \quad \text{for } 0 \le x \le L$$

28. A particle is in motion along a line between x = 0 and x = a with zero potential energy. At points for which x < 0 and x > a, the potential energy is infinite. Sovling Shcordinger's equation, obtain energy eigen values and normalized wave function for the particle.

Numerical Problems

1. Energy of a particle at absolute temperature T is of the order of K T. Calculate the wavelength of thermal neutron at 27 °C, given mass of neutron = 1.6×10^{-27} kg, K = 8.6×10^{-5} ev/deg

   [1.777 A°]

2. What is the energy of gamma ray photon having a wavelength of 1A° (h = 6.62×10^{-34} J.sec)

   [1.24×10^{4} ev]

3. An electron and a 150 gm baseball are travelling at 220 m/sec measured to an accuracy of 0.065%. Calculate and compare uncertainty in position of each of the bodies.

   [$\Delta x_e = 0.80 \times 10^{-3}$ m; $\Delta x_m = 4.9 \times 10^{-33}$ m]

4. An electron is confined to a box of length 10^{-9} m. Calculate the minimum uncertainty in its velocity

   [7.3×10^{5} m/sec]

5. Using uncertainty relation calculate the time required for the atomic system to retain rotation energy for a line of wavelength 6000 A° and width 10^{-4} A°.

   [1.9×10^{-4} sec]

6. A nucleon is confined to a nucleus of diameter 5×10^{-4} m. Calculate the minimum uncertainty in the momentum of the nucleon. Also calculate the minimum kinetic energy of the nucleon.

[0.33 meV]

7. Find the energy of an electron moving in one dimension in an infinitely high potential box of width 1 Å, given mass of the electron 9.11×10^{-31} kg and $h = 6.63 \times 10^{-34}$ J.s.

[5.68 ev]

8. An electron is bound by a potential which closely approaches an infinite square well width 2.5×10^{-10} m. Calculate the lowest three permissible Quantum energies the electron can have.

[54 ev]

9. A particle is moving in one dimensional potential box (of infinite height) of width 25A°. Calculate the probability of finding the particle within an interval of 5A° at the centres of the box when it is in its state of least energy.

[0.4]

10. Show that $\psi(x) = e^{ikx}$ is acceptable eigen-function, where K is some finite constant. For a region $-a \leq x \leq a$ normalize the given eigen-function.

11. The normalised state of a free particle is represented by a wave function.

$$\Psi(x) = Ne^{-\left(x^2/2a^2\right)+jKx}$$

Find the normalisation factor N and maximum probability of finding the particle.

$$[N = \frac{1}{a^{1/2}\pi^{1/4}};\ p(x) = \frac{1}{a\sqrt{\pi}}e^{-x^2/a^2}]$$

Answers to Objective Questions

1. (d)	2. (d)	3. (d)	4. (c)	5. (b)	6. (a)
7. (b)	8. (b)	9. (a)	10. (b)	11. (a)	12. (d)
13. (c)	14. (b)	15. (a)	16. (a)	17. (d)	18. (a)
19. (c)	20. (a)	21. (d)	22. (c)	23. (b)	24. (c)
25. (d)	26. (b)	27. (a)	28. (c)	29. (d)	30. (b)
31. (a)	32. (d)	33. (c)	34. (b)	35. (c)	36. (c)
37. (c)	38. (b)	39. (a)	40. (b)		

CHAPTER 5B

X-Ray Diffraction

If the hand be held between the discharge tube and the screen, the darker shadow of the bones is seen within the slightly dark shadow image of the hand itself…For brevity's sake I shall use the expression 'rays' and to distinguish them from the others of this name, I shall call them X-rays.

–Welhelm Conard Rontgen

5B.1 Introduction

X-rays being electromagnetic radiations, also undergo the phenomenon of diffraction as observed for visible light. However, unlike visible light, X-rays cannot be diffracted by ordinary optical grating because of their very short wavelengths. In 1912, a German physicist Max Van Laue suggested the use of a single crystal, to produce diffraction of X-rays. Since all the atoms in a single crystal are regularly arranged with interatomic spacing of the order of a few angstroms, a crystal can act as a three dimensional natural grating for X-rays. Friedrich and Knifing later successfully demonstrated the diffraction of X-rays from a thin single crystal of zinc blende (ZnS). The diffraction pattern obtained on a photographic film consisted of a series of dark spots arranged in a definite order. Such a pattern is called the 'Laue's pattern' and reflects the symmetry of the crystal. Apart from this, the phenomenon of X-ray diffraction has become an invaluable tool to determine the structures of single crystals and polycrystalline materials. It is also extensively used to determine the wavelength of X-rays.

5B.2 X-Ray Diffraction

X-rays have short wavelength of the order of 1 A^0 or 10^{-10} m. An ordinary grating which have 6000 lines per cm cannot produce diffraction with X-rays. The diffraction of X-rays was first studied by Max Laue by periodic array of atoms. The arranged atoms in the crystal grating correspond to the grating lines and the distance between two atoms, which is of the order of 10^{-8} cm, term a grating element, which can be used for X-ray diffraction. The difference between crystal grating and optical grating is that the atomic centres of diffraction in the crystal grating are not all in one plane but distributed in space. While in case of optical grating they are limited to one plane.

The crystal is thus a three-dimensional space grating rather than a two dimensional plane grating. It may be noted that the X-rays used for diffraction purposes should have wavelength which is the most appropriate for producing diffraction effects. Since $\sin\theta$ should be less than unity.

Equation $$2d \sin\theta = \lambda \quad .....(5B.1)$$

Gives $$\lambda < 2d \quad .....(5B.2)$$

Normally

$$d \sim 3A^{o}$$

$$\therefore \quad \lambda < 6\,A^{o} \quad .....(5B.3)$$

Longer wavelength X-rays are unable to resolve the details of the structure on the atomic scale whereas shorter wavelength X-rays are diffracted through angles which are too small to be measured experimentally.

Note

In X-ray diffraction studies, the probability that the atomic planes with right orientations are exposed to X-rays is increased by adopting one of the following methods.

1. A single crystal is held stationary and beam of white radiation is inclined on it at a fixed glancing angle θ i.e. θ is fixed while λ varies.

   Different wavelengths presents in the white radiations select the appropriate reflecting planes out of the numerous present in the crystal, such that the Bragg's condition is satisfied. This technique is called the Laue's technique.
2. A single crystal is held in the path of monochromatic radiations and is rotated about an axis i.e λ is fixed while θ varies. Different sets of parallel atomic planes are exposed to incident radiation for different values of θ and reflection take place from those atomic planes for which d and θ satisfy the Bragg's law. This method is known as rotating crystal method.
3. The sample in the powdered form is placed in the path of monochromatic X-ray i.e. λ fixed while both θ and d varies. Thus a number of small crystallites with different orientations are exposed to X-rays. The reflections take place for these values of d, θ and λ which satisfy the Bragg's law. This method is called the powder method.

5B.3 Bragg's Law

Bragg's has given a very simple interpretation of the diffraction pattern obtained by Laue. A crystal may be regarded as a stack of parallel planes of atoms. There are several sets of planes oriented in different directions. Bragg has shown that any diffracted ray can be regarded as if it were reflected from one of these systems of planes as though it were reflected from a mirror parallel to planes.

The diffracted beams are found only when reflections from parallel planes of atoms interfere constructively. As X-rays penetrate the crystal, they are scattered by the atoms in the successive planes as shown in Fig. 5B.1.

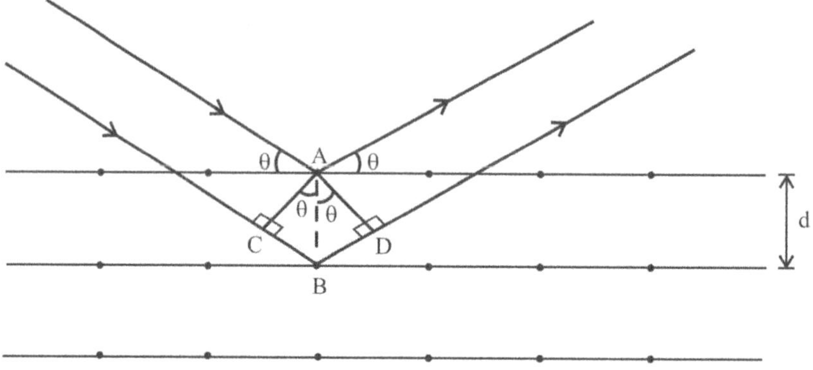

Fig. 5B.1 Scattering of X-rays.

5B.3.1 Derivation of Bragg's law

Fig. 5B.1 shows a particular set of atomic planes in a crystal, d being the interplanar spacing. Suppose an X-ray beam is incident at a glancing angle θ, it is scattered by the atoms like A and B in random directions. Constructive interference takes place only between those scattered waves which are reflected specularly and have a path difference $n\lambda$, where λ in the wavelength of X-rays and n is an integer.

The path difference for the wave reflected form adjacent planes is

$$CB + BD = d \sin \theta + d \sin \theta$$

$$CB + BD = 2\, d \sin \theta \qquad \text{.....(5B.4)}$$

For constructive interference

$$\text{path difference} = n\lambda$$

i.e. $$2d \sin \theta = n\lambda \qquad \text{.....(5B.5)}$$

where $n=1, 2, 3$........

Equation (5B.5) is called Bragg's equation.

It should be noted that:

1. Bragg's law is the consequence of periodicity of atoms in crystals.
2. The law is quite general and is applicable to all types of lattices.
3. The maximum value of θ is 1. Therefore Bragg reflection can occur only for wavelength $\lambda \leq 2d$ and d is of the order of 1 A°. This is why we cannot use visible light.
4. For given order n, θ decreases with λ, for lower λ, θ is small.
5. To use Bragg's law, continuous values of θ has to be provided and then equation $2 d \sin \theta = n\lambda$ will be satisfied for particular value of θ.

5B.4 Bragg's Spectrometer

Bragg devised an apparatus to investigate the structure of single crystal by using X-rays. It is used to measure glancing angle θ. He used crystal as reflection grating and not as transmission grating.

The experiment arrangement is shown is Fig. 5B.2

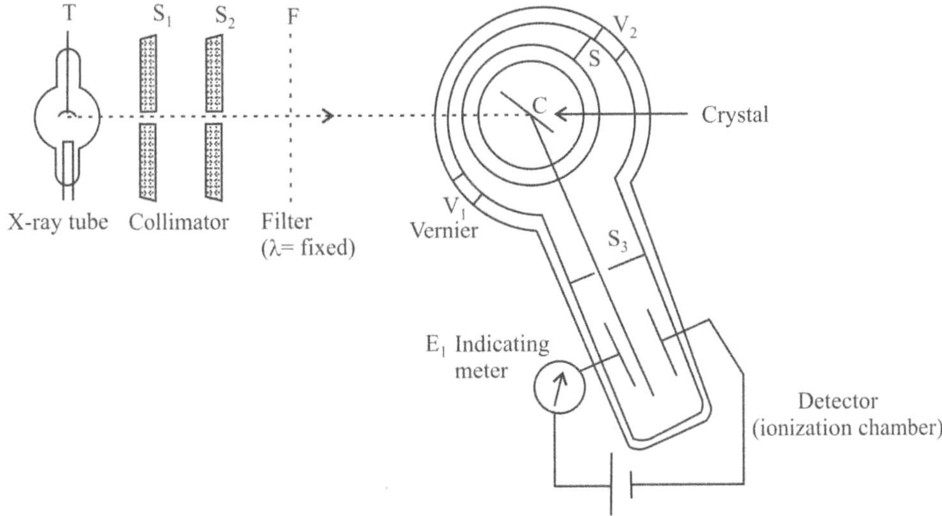

Fig. 5B.2

X-rays from X-ray tube is collimated using two adjustable slits, S_1 and S_2 are made to fall on a crystal C which is mounted in wax on the spectrometer table. The venire scale V_2 is attached to the table on which the crystal is placed. The venire scale V_2 is attached to the table on which the crystal is placed. The vernier scale V_2 is capable to move on the circular scale S and determine the position of the crystal. A strong monochromatic X-ray beam is incident on the crystal face. The reflected beam after passing through the slit S_3 enters an ionisation chamber mounted on an arm which can be rotated about the same axis as the crystal. Its position can be read by the venier V_1. The gas in the chamber is ionised by the X-rays. The resulting ionisation current, measured by the indicating meter E_1 is a measure of the intensity of X-rays reflected by the crystal.

The crystal is rotated through small angles, while the arm carrying the ionisation current is rotated through double the angles, and the ionisation current is measured each time.

The curve of intensity which is measure of ionisation current against glancing angle θ is plotted. For certain values of glancing angle, θ, the intensity of ionisation current increases abruptly as shown is Fig. 5B.3.

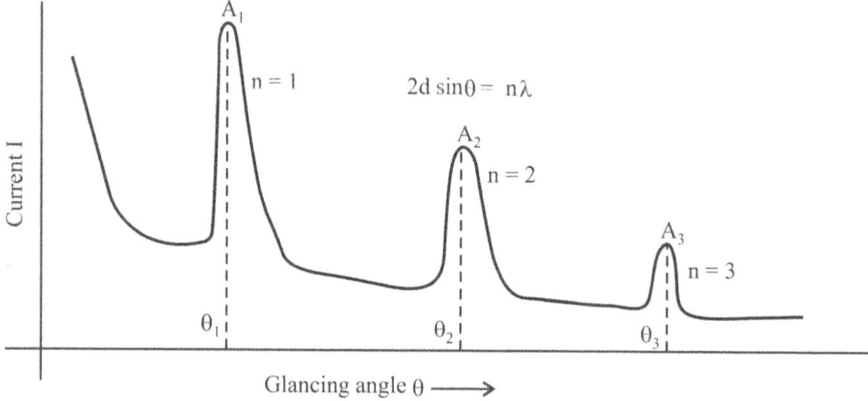

Fig. 5B.3 Variation of current I with glancing angle θ.

5B.4.1 Application of Bragg's spectrometer

According to Bragg's equation

$$2 \, d \sin \theta = n \, \lambda$$

For 1st, 2nd and 3rd order, the values of n are 1, 2, and 3 respectively and values of θ are θ_1, θ_2 and θ_3 respectively.

$$\text{Sin } \theta_1 : \ \sin \theta_2 : \sin \theta_3 : = 1 : 2 : 3$$

From the figure 5B.3, the glancing angles θ_1, θ_2 and θ_3 can be measured and it can be seen that

$$\text{Sin } \theta_1 \ : \sin \theta_2 : \sin \theta_3 = 1 : 2 : 3$$

From the observed value of θ and known value of d and n, the wavelength of X-ray can be calculated.

Suppose the wavelength of incident X-ray is known then the ratios of the interplanar spacing can be determined.

Suppose for a particular crystal used on a Bragg's spectrometer, strong reflections from the sets of planes (100), (110) and (111) are obtained for angles θ_1, θ_2 and θ_3 respectively in the first order.

Then from Bragg's equation, we have

$$2d_{100} \sin \theta_1 = \lambda$$

$$2d_{110} \sin \theta_2 = \lambda$$

$$2d_{111} \sin \theta_3 = \lambda$$

Hence $$d_{100} : d_{110} : d_{111} = \frac{1}{\sin\theta_1} : \frac{1}{\sin\theta_2} : \frac{1}{\sin\theta_3}$$

Braggs found that kCl crystal

$$\theta_1 = 5^\circ\ 23'; \theta_2 = 7^\circ\ 36'; \theta_3 = 9^\circ\ 25'$$

Hence for kCl

$$d_{100} : d_{110} : d_{111} = \frac{1}{\sin 5^\circ\ 23'} : \frac{1}{\sin 7^\circ\ 36'} : \frac{1}{\sin 9^\circ\ 25'}$$

$$= \frac{1}{1} : \frac{1}{1.414} : \frac{1}{1.744}$$

$$= 1 : \frac{1}{\sqrt{2}} : \frac{1}{\sqrt{3}}$$

Theoretically, this ratio is found to hold a simple cubic lattice structures. Hence it is concluded that the kCl crystal has a simple cubic structure.

When the first order reflections form the three faces (100), (110) and (111) of NaCl are compared, the ratios between the interplaner spacing are found as

$$d_{100} : d_{110} : d_{111} = 1 : \frac{1}{\sqrt{2}} : \frac{1}{\sqrt{3}}$$

Which agree with the theoretical values for a face centred cubic structure. He concluded that NaCl has fcc structure.

5B.5 Powder Crystal Method

This method for the analysis of crystal structure developed by Debye and Scherrer in 1916. In this method, the specimen, instead of being a single crystal, is a finely powdered crystal. It is prepared by crushing the commonly available polycrystalline material, thus eliminating the tedious process of growing the single crystal. The minute crystals called crystallites of this powder are all randomly oriented so that they make all possible angles with the incident radiation. Thus all orders of reflection from all possible atomic planes are recorded at the same time.

The experimental arrangement is shown in Fig. 5B.4.

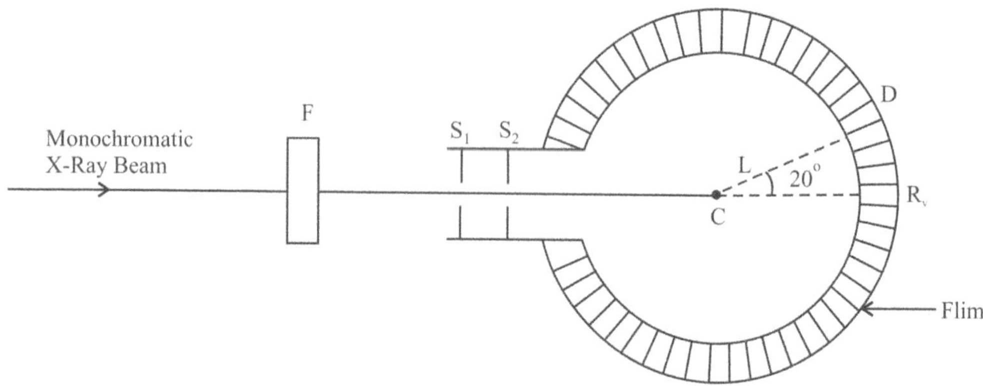

Fig. 5B.4 Powder Crystal Method.

It consists of a cylindrical camera called the Debye-Scherrer camera, whose length is smaller as compared with the diameter. The finely powered sample is filled in a thin capillary tube or is simply pasted on a wire by means of a binder and mounted at the centre of the camera. A collimated beam of monochromatic X-ray is allowed to fall on thin walled capillary tube containing powdered crystal. Since the specimen contain a large number of small crystallites with random orientations almost all the possible θ and d values are available. The diffraction takes place for these values of d and θ which satisfy Bragg's condition i.e. $2\,d \sin\theta = n\,\lambda$.

Also, since for a particular value of the angle of incidence, numerous orientations of a particular set of planes are possible, the diffracted rays corresponding to fixed values of θ and d lie on the surface of a cone with its apex at the sample and the semi vertical angle 2θ (Fig. 5B.5).

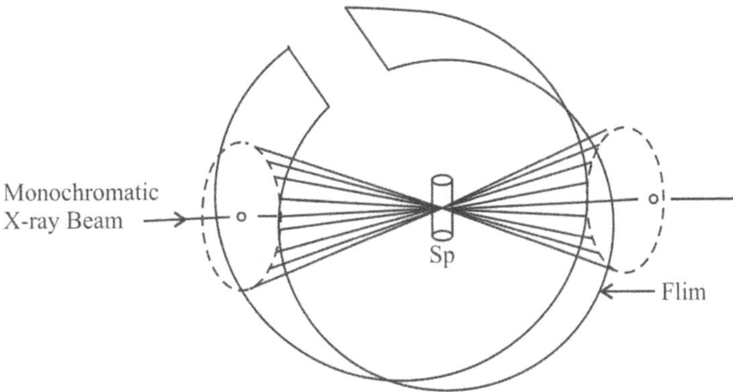

Fig. 5B.5

Different cones are observed for different sets of d and θ for a particular values of n, and for different combination of θ and n for a particular value of d. The transmitted X-rays move out of the camera through an exit hole located diametrically opposite to the entrance of the camera. Each cone of the reflected beam leaves two impressions on the film (Fig. 5B.6), which are in the form of arc on either side of the exit hole with their centres coinciding with the hole.

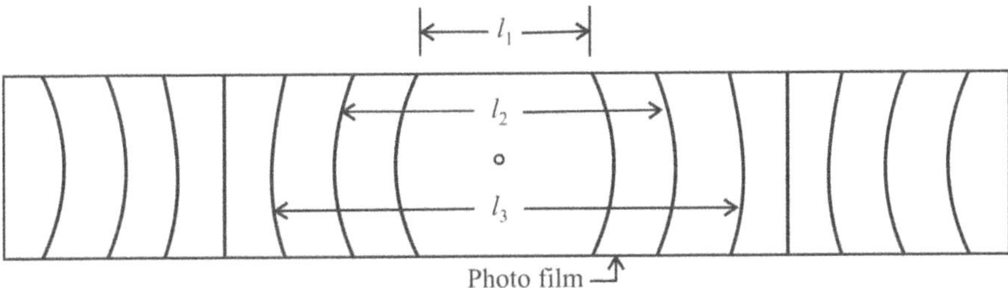

Fig. 5B.6

The film is exposed for a few hours in order to obtain reflected lines of sufficiently high intensity. It is then removed from the camera and developed. The arc produced by reflected rays appear dark on the developed film. The angle θ corresponding to a particular pair of arcs is related to the distance S between the arcs on

$$4\,\theta\ (\text{radian}) = S/R$$

where R is the radius of the camera. If θ is measured in degrees,

$$4\theta\ (\text{degrees}) = \frac{57.296\ S}{R} \qquad .....(5B.6)$$

The calculation can be made simpler by taking the radius of the camera in the multiples of 57.296. For example. taking R= 57.296 mm, We get θ (degrees)

$$\theta\text{ (degrees)} = \frac{S}{4}\text{ (mm)} \qquad .....(5B.7)$$

Knowing all the possible θ's and considering only first order reflection from all possible planes, we can calculate the interplanar spacing for various sets of parallel planes as

$$d = \frac{\lambda}{2\sin\theta} \qquad ...(5B.8)$$

These d values are used to determine the space lattice of the crystal structure.

5B.6 The Laue's Method

An experimental arrangement used to produce Lane's pattern is shown in Fig. 5B.7.

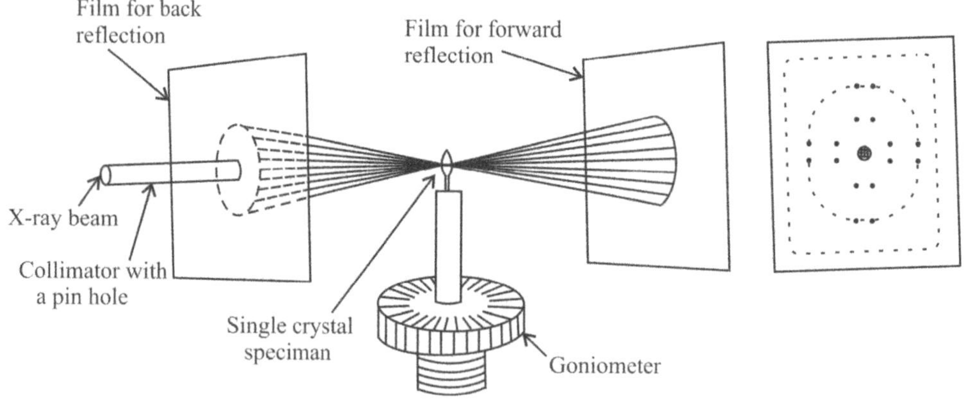

Fig. 5B.7

It consists of a flat plate camera which contains a collimator with a fine hole to obtain a very fine beam of X-rays. The sample is placed on a goniometer which can be rotated to change the orientation of the single crystal. Two flat photographic films are used, one for receiving the transmitted diffracted beam and the other for receiving the reflected diffracted beam for back reflection experiment. Such experiments are performed particularly where there is an excessive absorption of X- rays in the crystal.

Initially a single crystal specimen having dimensions of the order of 1mm × 1mm × 1mm is held stationary in the path of white X-ray having wavelength ranging from 0.2 to 2 A^o. Since the crystal contains a number of sets of parallel atomic planes with different interplanar spacing, diffraction is possible for certain values of a λ and d which satisfy the Bragg's condition. Thus diffraction spots are produced on the photographic films as shown in Fig. 5B.7. The crystal can be rotated with the help of goniometer to change its orientation with respect to the incident beam. By doing so, the diffraction condition may be satisfied for a new set of atomic planes and hence a diffraction pattern of different types may be obtained on the photographic film. The symmetry of the crystal is, however, reflected in each pattern.

The Laue's method is mostly used to determine the crystal symmetry. For example, if a crystal having four-fold symmetry is oriented so that its axis is parallel to the beam, the resulting Laue's pattern also exhibits the four-fold symmetry. The symmetry of the pattern helps to determine the shape of the unit cell. It is however, not practicable to determine the structure of the crystal by this method. It is because a number of wavelengths may be reflected from a single plane in different orders and may superpose at a single point resulting in the loss of a number of reflections. The symmetry of the Laue's pattern also helps to orient the crystal for various solid state experiments

Another application of Laue's methods is the determination of imperfections or stains in the crystal. An imperfect or strained crystal has atomic planes which are not exactly plane but are slightly curved. Thus instead of sharp diffraction spots one gets streaks in the Laue's pattern. This type of streaking on Laue's photographs is called asterism.

5B.7 Rotating Crystal Method

In this method, a monochromatic beam of X-rays is incident on a single crystal mounted on a rotating spindle such that one of its crystallographic axes coincides with the axis of rotation, which is kept perpendicular to the direction of the incident beam.

The single crystal having dimension of the order of 1 mm is positioned at the centre of cylindrical holder concentric with the rotating spindle as shown in Fig. 5B.8

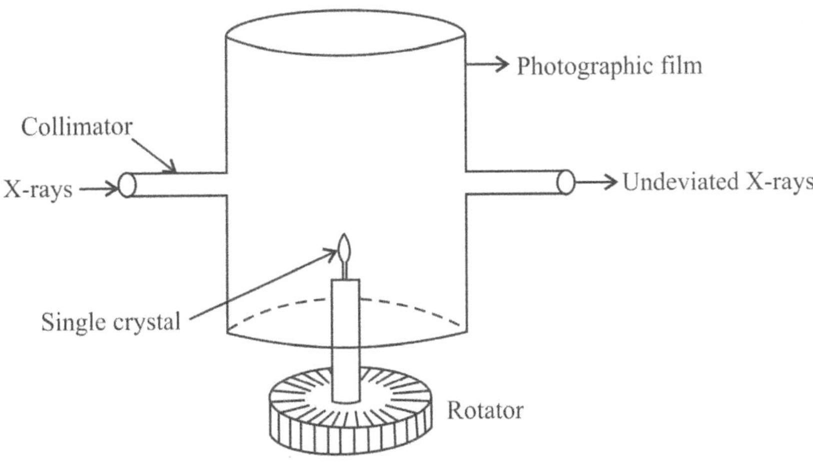

Fig. 5B.8

The diffraction takes place from those planes which satisfy the Bragg's law for a particular angle of rotation. The planes parallel to the axis of rotation diffract the incident rays in a horizontal plane. However, reflection cannot be observed for those planes which always contain the incident beam. The planes inclined to the rotation axis produce reflection above or below the horizontal plane depending upon the angle of inclination.

The horizontal lines produced by diffraction spots on the photographic film are called layer lines. If the crystal is positioned such that its C-axis coincides with the axis of rotation, all the planes with miller indices of the type (hko) will produce the central line.

Likewise the planes having miller indices of the type (hkl) and $\left(hk\overline{l}\right)$ will produce the layer lines above and below the central line respectively and so on. These layer lines are shown in Fig. 5B.9. The vertical spacing between the layer lines depends on the distance between the lattice points along the C-axis. Hence the distance C can be measured from the photographic film. Similarly one can determine the translation vector $\bar{a}$ and $\bar{b}$ on mounting the crystal along $\bar{a}$ and $\bar{b}$ axes respectively. Thus the dimensions of the unit cell can be easily determined

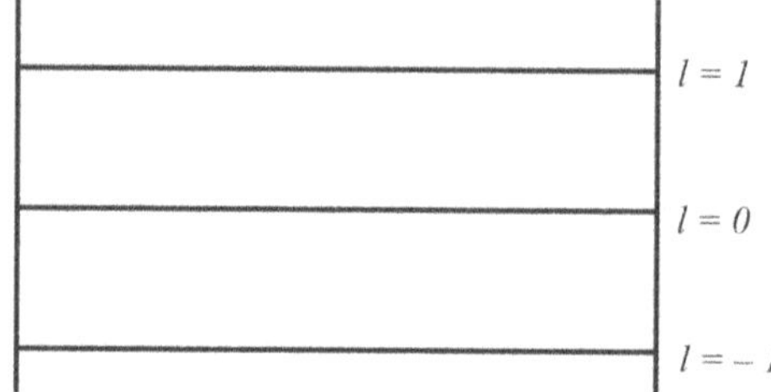

Fig. 5B.9 Layer lines produced after flattening the film.

5B.8 Compton Effect

When a beam of monochromatic radiation (X-rays) of sharply defined frequency were incident on a material of low atomic number like carbon, the rays suffered a change of frequency on scattering. The scattered beam contains two wavelengths. In addition to the expected incident wavelength, there exists a line of larger wavelength. The change of wavelength is due to loss of energy of the incident X-rays. This phenomenon is known as Compton Effect.

5B.8.1 Derivation

Compton considered that an X-ray photon strike an electron, assumed to be initially at rest and is scattered away from its original direction of motion.

If the initial photon has the frequency λ associated with it, the scattered photon has the lower frequency λ' as shown in Fig. 5B.10.

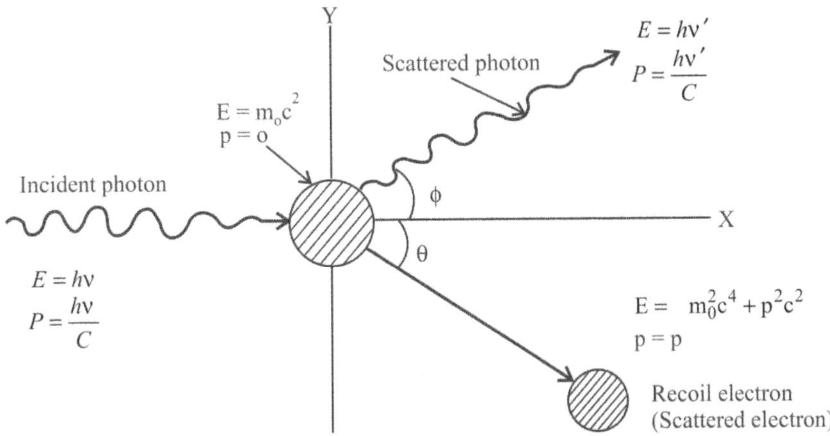

Fig. 5B.10

The momentum of scattered photon $= \frac{h\nu'}{c}$

In the original photon direction

Initial momentum = Final momentum

$$\frac{h\nu}{c} + 0 = \frac{h\nu'}{c}\cos\phi - P\cos\theta \qquad \text{.....(5B.9)}$$

In the perpendicular direction

Initial momentum = Final momentum

$$0 = \frac{h\nu'}{c}\sin\phi - P\sin\theta \qquad \text{.....(5B.10)}$$

$$P\sin\theta = \frac{h\nu'}{c}\sin\phi \qquad \text{.....(5B.11)}$$

Solving equation 5B.9 and 5B.10 we can write

$$Pc\cos\theta = h\nu - h\nu'\cos\phi \qquad \text{.....(5B.11)}$$

$$Pc\sin\theta = h\nu - h\nu'\sin\phi \qquad \text{.....(5B.12)}$$

Squaring and adding equation (5B.11) and (5B.12), we get

$$P^2 c^2 = (h\nu)^2 - 2(h\nu)(h\nu')\cos\phi + (h\nu')^2 \qquad \text{.....(5B.13)}$$

According to the principle of conservation of energy

$$h\nu + m_o c^2 = h\nu' + E$$

$$E = h(\nu - \nu') + m_o c^2 \qquad .....(5B.14)$$

Also from relativistic mechanics, we have

$$E = \sqrt{P^2c^2 + m_0^2 c^4} \qquad .....(5B.15)$$

Comparing equation (5B.14) and (5B.15) we get

$$\left[h(\nu - \nu') + m_o c^2\right] = P^2 c^2 + m_0^2 c^4$$

$$h^2(\nu - \nu^1)^2 + m_0^2 c^4 + 2 m_o c^2 h(\nu - \nu') = P^2C^2 + m_0^2 c^4$$

$$P^2c^2 = h^2(\nu - \nu') + 2m_o c^2 h(\nu - \nu')$$

$$P^2c^2 = h^2\nu^2 - 2(h\nu)(h\nu') + h^2\nu'^2 + 2m_o c^2(h\nu) - 2m_o c^2(h\nu') \qquad .....(5B.16)$$

Comparing equation (5B. 13) and (5B. 16), we get

$$h^2\nu^2 - 2(h\nu)(h\nu')\cos\phi + (h^2\nu'^2)$$

$$= h^2\nu^2 - 2(h\nu)(h\nu') + (h^2\nu'^2) + 2m_o c^2 h\nu - 2m_o c^2 h\nu'$$

$$(h\nu)(h\nu')\cos\phi = (h\nu)(h\nu') - m_o c^2 h\nu + m_o c^2 h\nu'$$

Dividing both the sides by h^2c^2, we get

$$\frac{\nu\nu'}{c^2}\cos\phi = \frac{\nu\nu'}{c^2} - \frac{m_o\nu}{h} + \frac{m_o\nu'}{h}$$

$$\frac{m_o}{h}(\nu - \nu) = \frac{\nu\nu'}{c^2} - \frac{\nu\nu'}{c^2}\cos\phi$$

$$\frac{\nu}{c} = \frac{1}{\lambda} \text{ and } \frac{\nu'}{c} = \frac{1}{\lambda'}$$

$$\frac{m_o}{h}\left(\frac{c}{\lambda} - \frac{c}{\lambda'}\right) = \frac{1}{\lambda\lambda'} - \frac{1}{\lambda\lambda'}\cos\phi$$

$$\frac{m_o c}{h}\left(\frac{\lambda' - \lambda}{\lambda\lambda'}\right) = \frac{1}{\lambda\lambda'}(1 - \cos\phi) \qquad .....(5B.17)$$

$$\lambda' - \lambda = \frac{h}{m_o c}(1 - \cos\phi)$$

Compton shift

$$\Delta\lambda = \lambda' - \lambda = \frac{\lambda}{m_o c}(1 - \cos\phi) \quad .....(5B.18)$$

$$\Delta\lambda = 0.0242(1 - \cos\phi) \ \ A^o \quad .....(5B.19)$$

From equation 5B.18, we conclude that:

1. The wavelength λ' of the scattered photon in greater than the wavelength λ of the incident photon
2. $\Delta\lambda$ is independent of the incident wavelength
3. $\Delta\lambda$ has the same value for all substances containing free electrons
4. $\Delta\lambda$ only depends upon the angle of scattering
5. When $\phi = 90^o$ then $\cos\phi = 0$

$$\therefore \Delta\lambda = \left(\frac{h}{m_o c}\right) = \lambda_c \text{ is called Compton wavelength} \quad .....(5B.20)$$

$$\Delta\lambda = \lambda_c = o.0242 \ \ A^o \quad ....(5B.21)$$

$$\Delta\lambda = \lambda_c = 2.42 \text{ p m} \quad .....(5B.22)$$

6. $\Delta\lambda$ has the maximum value at $\phi = \pi$, which is

$$\Delta\lambda = \frac{h}{m_o c}(1 \text{ - } \cos\phi)$$

$$\Delta\lambda = \frac{h}{m_o c}\left[(1 \text{ - } (-1))\right]$$

$$\Delta\lambda = \frac{2h}{m_o c} = 2 \times 0.0242 \ \ A^o \quad ....(5B.23)$$

$$\Delta\lambda = 4.84 \ \text{p.m} \quad .....(5B.24)$$

7. The Compton shift $\Delta\lambda$ is found to vary with the angle at which scattered rays are observed.

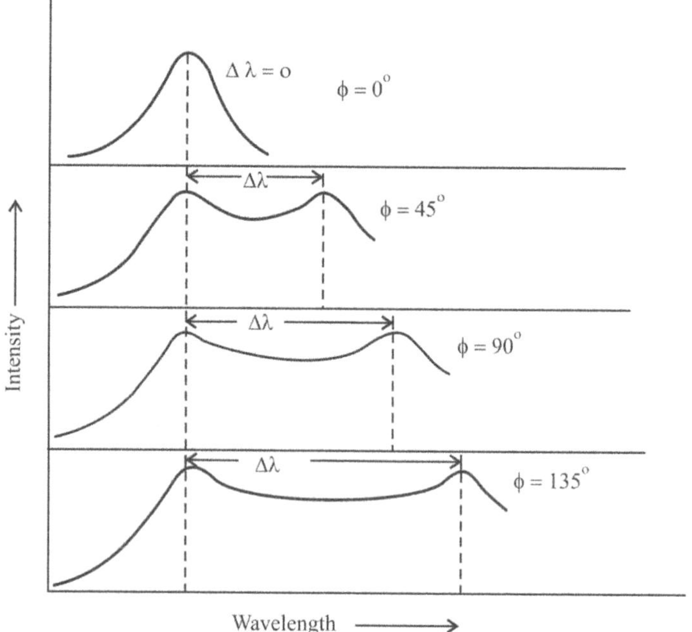

Fig. 5B.11 Variation of intensity with wavelength.

5B.8.2 Direction of Recoiled Compton Electron

We know that

$$PC\cos\theta = h\nu - h\nu'\cos\phi$$

$$Pc\cos\theta = h\nu'\cos\phi$$

$$\therefore \quad \tan\theta = \frac{h\nu'\sin\phi}{h\nu - h\nu'\cos\phi} = \frac{\nu'\sin\phi}{\nu - \nu'\cos\phi} \qquad .....(5B.25)$$

We have

$$\frac{1}{\nu'} = \frac{1}{\nu} + \frac{h}{m_o c^2}(1-\cos\phi)$$

$$\therefore \quad \frac{1}{\nu'} = \frac{1}{\nu} + \frac{h}{m_o c^2}.2\sin^2\frac{\phi}{2}$$

$$\nu' = \frac{\nu}{1+\left(\frac{h}{m_o c^2}\right)2\sin^2\frac{\phi}{2}} \qquad .....(5B.26)$$

Putting (5B.26) in (5B.25), we get

$$\tan\theta = \frac{\dfrac{\nu\sin\phi}{1+\left(\dfrac{h\nu}{m_o c^2}\right)2\,\sin^2\dfrac{\phi}{2}}}{\nu-\dfrac{\nu\cos\phi}{1+\left(\dfrac{h\nu}{m_o c^2}\right)2\,\sin^2\dfrac{\phi}{2}}}$$

$$\therefore\quad \tan\theta = \frac{\sin\phi}{1+\left(\dfrac{h\nu}{m_o c^2}\right)2\,\sin^2\dfrac{\phi}{2}-\cos\phi}$$

$$\tan\theta = \frac{2\,\sin^2\dfrac{\phi}{2}\cos\dfrac{\phi}{2}}{1+\left(\dfrac{h\nu}{m_o c^2}\right)2\,\sin^2\dfrac{\phi}{2}-\;1+2\sin\dfrac{\phi}{2}}$$

$$\tan\theta = \frac{2\,\sin^2\dfrac{\phi}{2}\cos\dfrac{\phi}{2}}{2\,\sin^2\dfrac{\phi}{2}\left[\dfrac{h\nu}{m_o c^2}+1\right]}$$

$$\tan\theta = \frac{2\cos\dfrac{\phi}{2}}{\sin\dfrac{\phi}{2}\left[1+\dfrac{h\nu}{m_o c^2}\right]}$$

$$\tan\theta = \frac{2\cos\dfrac{\phi}{2}}{\sin\dfrac{\phi}{2}\left[1+\dfrac{h\nu}{m_o c^2}\right]} \qquad \text{......(5B.27)}$$

$$\cot\frac{\phi}{2} = \left(1+\frac{h\nu}{m_o c^2}\right)\tan\theta \qquad \text{.....(5B.28)}$$

$$\tan\theta = \frac{\cot\frac{\phi}{2}}{\left(1+\frac{h\nu}{m_oc^2}\right)} \quad .....(5B.29)$$

This equation reveals that the angle of recoil electron θ depends on the scattering angleϕ. If $\phi = 0$, $\theta = 90^o$ and if $\phi = 180$, $\theta = 0$. It shows that the electron can recoil only in the onward direction at angles less than 90°; while a photon can scattered in any direction.

5B.8.3 Kinetic energy of Recoiled electron

The maximum kinetic energy of the recoiled electron is given by

$$K.E_{max} = h(\nu - \nu')$$

$$K.E_{max} = h\left(\frac{c}{\lambda} - \frac{c}{\lambda'}\right) \quad .....(5B.30)$$

$$K.E_{max} = hc\left[\frac{\lambda' - \lambda}{\lambda\lambda'}\right] \quad .....(5B.31)$$

We know that the Compton shift is given by

$$\lambda' - \lambda = \frac{h}{m_oc}(1-\cos\phi) \quad .....(5B.32)$$

For maximum shift $\phi = 180^o$

$$\Delta\lambda = \lambda' - \lambda = \frac{h}{m_oc}\left[(1-(-1))\right]$$

$$\lambda' - \lambda = \frac{2h}{m_oc} \quad .....(5B.33)$$

$$\therefore \quad K.E_{max} = \frac{hc}{\lambda\lambda'}\left(\frac{2h}{m_oc}\right)$$

$$K.E_{max} = \frac{2h^2}{m_o\lambda\left(\lambda + \frac{2h}{m_oc}\right)}$$

Since $$\lambda = \frac{c}{\lambda}$$

$$\therefore \quad K.E_{max} = \frac{2h^2\nu}{m_oc\left[\frac{c}{\lambda} + \frac{2h}{m_oc}\right]}$$

$$K.E_{max} = \frac{2h^2\nu.m_oc.\nu}{m_oc\left[m_oc^2 + 2h\lambda\right]}$$

$$K.E_{max} = \frac{2h^2\nu^2 / m_0c^2}{1 + 2h\nu / m_0c^2} \qquad .....(5B.34)$$

Equation (5B.34) is the required expression for maximum kinetic energy of the recoil electron.

5B.8.4 Compton Effect is not observed with visible light

The Compton shift is given by

$$\Delta\lambda = \frac{h}{m_oc}(1 - \cos\phi)$$

$$\Delta\lambda = 0.0242(1 - \cos\phi)\ A^o$$

Maximum value of $\Delta\lambda$ is for $\phi = 180^o$

$$\therefore \quad \cos\phi = -1$$

$$(\Delta\lambda)_{max} = 0.0484\ A^o$$

This means that Compton effect can be detected only for those radiations whose wavelength in not grater than a few A^o. For visible light

$$\lambda = 4000 - 8000\ A^o$$

Then $(\Delta\lambda)_{max}$ = *0*.001% of the initial wavelength which cannot be detected.

5B.9 Practical Applications of X-rays

1. *X-ray crystallography*

   X-ray analysis enables us to know about the internal structure, symmetry and hence properties of the crystal.

For example,

For a kCl crystal

$$\theta_1 = 5^\circ\ 23', \theta_2 = 7^\circ\ 37' \text{ and } \theta_3 = 9^\circ\ 25'$$

$$As \quad d_{100} : d_{110} : d_{111} = \frac{1}{\sin\theta_1} : \frac{1}{\sin\theta_2} : \frac{1}{\sin\theta_3}$$

$$d_{100} : d_{110} : d_{111} = \frac{1}{\sin 5^\circ\ 23'} : \frac{1}{\sin\ 7^\circ\ 37'} : \frac{1}{\sin\ 9^\circ\ 25'}$$

$$d_{100} : d_{110} : d_{111} = \frac{1}{0.0938} : \frac{1}{0.1326} : \frac{1}{0.1636}$$

$$d_{100} : d_{110} : d_{111} = \frac{1}{1} : \frac{1}{1.414} : \frac{1}{1.744}$$

$$d_{100} : d_{110} : d_{111} = 1 : \frac{1}{\sqrt{2}} : \frac{1}{\sqrt{3}}$$

Theoretically, for the planes (100), (110), (111) for a cubic structure, the distance between successive planes will be given by

$$d_{100} = \frac{a}{\sqrt{1^2 + 0 + 0}} = a$$

$$d_{110} = \frac{a}{\sqrt{1^2 + 1^2 + 0}} = \frac{a}{\sqrt{2}}$$

$$d_{111} = \frac{a}{\sqrt{1^2 + 1^2 + 1^2}} = \frac{a}{\sqrt{3}}$$

$$d_{100} : d_{110} : d_{111} = a : \frac{a}{\sqrt{2}} : \frac{a}{\sqrt{3}}$$

$$= 1 : \frac{1}{\sqrt{2}} : \frac{1}{\sqrt{3}}$$

Thus it becomes clear from X-ray analysis that KCl belongs to the simple cubic system.

Now the X-ray analysis is not only used for simple crystal but also used for examining the properties of alloy systems.

2. X-rays are widely used for testing materials in industries. X-ray photography techniques are used to study the structure of cellulose, Rubber fibres, plastics etc, to distinguish between real and artificial gems and to defect pearls in oysters.

3. X-rays are playing an important role in medical science. X-rays are utilized in the examination and in the diagnosis of the diseases or fracture in the body. X-rays are not only used for medical diagnosis, but also used for cure i.e radiography and radiotherapy.

Solved Numerical Problems

1. Calculate the longest wavelength that can be analysed by rock salt crystal of spacing d = 2.82 A°

(a) in the first order (b) in the second order

Sol. According to Bragg's equation

$$2d \sin\theta = n\lambda$$

$$\therefore \quad \lambda = \frac{2d\sin\theta}{n}$$

(a) In the first order

$n = 1$, d=2.82 A°, θ=90° (for the longest wavelength)

Therefore

$$\lambda = \frac{2\times\ 2.82\ A^o \times \sin 90^o}{1}$$

$$\lambda = 5.64\ A^o$$

(b) In the second order

$$\lambda = \frac{2\times 2.82A^o \times \sin 90^o}{2}$$

$$\lambda = 2.82\ A^o$$

2. What wavelengths in a beam containing the range 0.2 A° to 1 A° will be reflected when incident at 9° upon the cube face of a rock salt crystal.(Given d = 2.814 A°)

Sol. Given d = 2.814A°

$$\theta = 9^o$$

We know that

$$2d \sin\theta = n\lambda.$$

$$\therefore \quad \lambda = \frac{2d\sin\theta}{n}$$

For n = 1

$$\lambda_1 = \frac{2 \times 2.814\text{A}^\circ \times \sin 9^\circ}{1}$$

$$\lambda_1 = 0.880\ \text{A}^\circ$$

For n = 2 $$\lambda_2 = \frac{2 \times 2.814\text{A}^\circ \times \sin 9^\circ}{2} = 0.4402\ \text{A}^\circ$$

For n = 3 $$\lambda_3 = \frac{2 \times 2.814\ \text{A}^\circ \times \sin 9^\circ}{3} = 0.2936\ \text{A}^\circ$$

For n = 4 $$\lambda_4 = \frac{2 \times 2.814\ \text{A}^\circ \times \sin 9^\circ}{4} = 0.2201\ \text{A}^\circ$$

For n = 5 $$\lambda_5 = \frac{2 \times 2.814\ \text{A}^\circ \times \sin 9^\circ}{5} = 0.1760\ \text{A}^\circ$$

The reflected beam contains four wavelengths

$$0.880\ \text{A}^\circ,\ 0.440\ \text{A}^\circ,\ 0.293\ \text{A}^\circ \text{ and } 0.220\ \text{A}^\circ$$

3. Calculate the smallest glancing angle at which the K_α line of copper (λ = 1.449 A°) will be reflected from a Quartz crystal which has an atomic spacing d = 4.225 A°. What is the highest reflecting order that could be observed with this radiation?

Sol. The smallest angle of reflection will correspond to n=1 in the Bragg relation

$$2d \sin\theta = n\lambda$$

$$\theta_{min} = \sin^{-1}\left(\frac{\lambda}{2d}\right)$$

$$\theta_{min} = \sin^{-1}\left(\frac{1.449}{2 \times 4.225}\right)$$

$$\theta_{min} = 10^\circ 30'$$

Since $\sin\theta < 1$, the highest possible value of n is such that

$$\frac{n\lambda}{2d} < 1$$

$$\frac{\lambda}{2d} = 0.01821$$

or

$$n \times 0.01821 < 1$$

4. The first order Bragg maximum of electron diffraction in a nickel crystal (d= 0.4086 A°) occurred at a glancing angle of 65°. Calculate the Broglie wavelength of the electrons and their velocity.

Sol. **According to Bragg's equation for the diffraction of X-rays**

$$2d \sin\theta = n\lambda$$

For first order,

$$n=1$$

$$\therefore 2d \sin\theta = \lambda$$

or

$$\lambda = 2 \times 0.4086 \times 10^{10} \times \sin 65^{o}$$

$$\lambda = 2 \times 0.4086 \times 10^{10} \times 0.9063$$

$$\lambda = 0.7406 A^{o}$$

$$\lambda = 0.7406 \times 10^{-10} m$$

The velocity of electrons

$$v = \frac{h}{m\lambda}$$

$$v = \frac{6.6 \times 10^{-34}}{9.1 \times 10^{-31} \times 0.7406 \times 10^{-10}}$$

$$v = 9.793 x 10^{6} \ m/sec$$

5. 10 keV electrons are passed through a thin film of a metal for which atomic spacing is 5.5 × 10^{-11} m. What is the angle of deviation of the first order diffraction minimum?

Sol. Given

$$E = 10 \text{ keV}$$

$$E = 10^{4} \text{e V} = 10^{4} \times 1.6 \times 10^{-19} \text{ J} = 1.6 \times 10^{-15} \text{ J}$$

$$m = 9.1 \times 10^{-31} \text{Kg}$$

$$d = 5.5 \times 10^{-11} \text{ m}$$

$$n = 1$$

$$h = 6.63 \times 10^{-34}$$

The wavelength is given by

$$\lambda = \frac{h}{\sqrt{2mE}}$$

$$\therefore \lambda = \frac{6.63\times10^{-34}}{\sqrt{2\times9.1\times10^{-31}\times1.6\times10^{-15}}}$$

$$\lambda = 1.23\times10^{-11}\text{m}$$

According to Bragg's law

$$2d\ \sin\theta = n\ \lambda$$

$$2\times 5\times 10^{-11}\times \sin\theta = 1\times1.23\times10^{-11}$$

$$\sin\theta = 0.1118$$

6. A beam of X-rays is incident on a NaCl crystal with lattice spacing 0.282 nm. Calculate the wavelength of X-rays of the first order Bragg's reflection takes place at a glancing angle of $8^\circ\ 35'$. Also calculate the maximum order of diffraction possible.

Sol. Given

$$d = 0.282\text{ nm}$$

$$n = 1$$

$$\theta = 8^\circ\ 35'$$

$$2d\ \sin\theta = n\ \lambda$$

$$\lambda = \frac{2\times0.282\times10^{-9}\times\sin(8^\circ\ 35')}{1}$$

$$\lambda = 0.819\times10^{-9}\text{ m}$$

$$\lambda = 0.819\times10^{-10}\text{ m} = 0.819\text{ A}^\circ$$

For maximum order of diffraction, θ should be 90°

$$n_{max}\lambda = 2d\ \sin 90^\circ$$

$$n_{max} = \frac{2d}{\lambda} = \frac{2\times0.282\times10^{-19}}{0.819\times10^{-10}} = 6.8$$

$$n_{max} = 7$$

7. In Bragg's reflection of X-ray, a reflection was found at 30° glancing angle with lattice planes of spacing 1.87 A°. If this is a second order reflection, calculate the wavelength of X-rays?

Sol. According to Bragg's equation

$$2d \sin \theta = n \lambda$$

Given, $\theta = 30^\circ$; $n = 2$ and $d = 1.87A^\circ$

$$\lambda = \frac{2d\sin\theta}{n}$$

$$\lambda = \frac{2\times 1.87 \times \sin 30^\circ}{2}$$

$$\lambda = 1.87 \times 0.5 = 0.935 \text{ A}^\circ$$

8. X-rays of the wavelength 0.36×10^{-8} cm are diffracted by a Bragg crystal spectrograph at a glancing angle of 4.8°. Find the interplaner separation of atomic planar in the crystal?

Sol. From Bragg's law

$$2d \sin \theta = n \lambda$$

$$\therefore d = \frac{n\lambda}{2\sin\theta}$$

where d is the inter planar separation of atomic planes in the crystal

Given $n = 1, \lambda = 0.36 \times 10^{-8}$ cm, $\theta = 4.8^\circ$

$$\therefore d = \frac{0.36\times 10^{-8}}{2\times \sin(4.8^\circ)}$$

$$\therefore \quad d = \frac{0.36\times 10^{-8}}{0.1674}$$

$$d = 2.15 \times 10^{-8} \text{cm}$$

9. Calculate the longest wavelength that can be analysed by a rock salt crystal of spacing 2.82 A° in the second order.

Sol. From Bragg's

$$2d \sin \theta = n \lambda$$

$$\therefore \quad \lambda = \frac{2d\sin\theta}{n}$$

For the longest wavelength $(\sin\theta)_{max}=1$

$$\lambda_{max} = \frac{2d}{n}$$

$$\lambda_{max} = \frac{2\times 2.82\ A^o}{2}$$

$$\lambda_{max} = 2.82A^o$$

10. An X-ray beam of wavelength 0.97A° is obtained in the third order reflection at 60° from the crystal plane. Another beam is obtained in the first order after reflection at 30° form the same crystal plane. Find the wavelength of the second X-ray beam.

Sol. According to Bragg's equation

$$2d\sin\theta = n\lambda$$

For first beam, $\theta = 60^o$; $n = 3$ and $\lambda = 0.97^o$

$$\therefore\ 2d\sin 60^o = 3\times 0.97\ A^o \quad (1)$$

Let us consider that λ' be the wavelength of the second beam

$$2d\sin 60^o = n'\lambda'$$

Given

$$\theta' = 30^o \text{ and } n = 1$$

$$\therefore\ 2d\sin 30^o = \lambda' \quad (2)$$

From 1 and 2, we have

$$\frac{\lambda'}{3\times 0.97\ A^o} = \frac{2d\sin 30^o}{2d\sin 60^o}$$

$$\frac{\lambda'}{3\times 0.97\ A^o} = \frac{\left(\frac{1}{2}\right)}{\left(\frac{\sqrt{3}}{2}\right)}$$

$$\therefore\ \frac{\lambda'}{3\times 0.97\ A^o} = \frac{1}{\sqrt{3}}$$

$$\lambda' = \frac{3\times 0.97\ A^o}{\sqrt{3}}$$

$$\lambda' = 1.68\ A^o$$

11. Calculate the fractional change in wavelength of an X-ray of wavelength 0.400 A° that undergoes a 90° Compton scattering from an electron.

Sol. We know that, the Compton shift is given by

$$\lambda' - \lambda = \frac{h}{m_o c}(1 - \cos\phi)$$

$$\lambda' - \lambda = 0.0242\ (1 - \cos 90^\circ)\ A^\circ$$

$$\lambda' - \lambda = 0.0242\ A^\circ$$

$$\frac{\lambda' - \lambda}{\lambda} = \frac{0.0242\ A^\circ}{0.400\ A^\circ} = 0.0605$$

12. X-rays of wavelength 10.0 pm are scattered from a target

a. Find the wavelength of the X-rays scattered through 45°

b. Find the maximum wavelength present in the scattered X-rays

c. Find the maximum kinetic energy of the recoil electrons?

Sol. **a.**

$$\lambda' - \lambda = \frac{h}{m_o c}(1 - \cos\phi)$$

$$\lambda' - \lambda = 0.0242(1 - \cos 45^\circ)\ A^\circ$$

$$\lambda' - \lambda = 0.0242 \times (0.292)\ A^\circ$$

$$\therefore \quad \lambda' = \lambda + 0.0242 \times (0.292)\ A^\circ$$

$$\lambda' = 0.10\ A^\circ + 0.0242 \times (0.292)\ A^\circ$$

$$\lambda' = (0.1 + 0.00708)\ A^\circ$$

$$\lambda' = 0.10708\ A^\circ$$

b. $\lambda' - \lambda$ is maximum when $\phi = 180^\circ$

$$\lambda' = \lambda + 0.0484\ A^\circ = 0.10\ A^\circ + 0.0484\ A^\circ$$

$$\lambda' = 0.1484\ A^\circ$$

c. $$KE_{max} = h(\nu - \nu') hc \left(\frac{1}{\lambda} - \frac{1}{\lambda'}\right)$$

$$= \frac{\left(6.62 \times 10^{-34}\ J - sec\right)\left(3 \times 10^{8}\ m/_{sec}\right)}{10^{-10}} \left(\frac{1}{0.107} - \frac{1}{0.148}\right)$$

$$KE_{max} = 6.54 \times 10^{-15}\ J$$

13. Calculate the wavelength of X-rays scattered at 18° from a carbon block of the frequency of the incident rays is $1.8 \times 10^{-10}\ sec^{-1}$.

Sol. Given

$$\nu = 1.8 \times 10^{-10}\ sec^{-1}$$

$$\nu = \frac{C}{\lambda} = \frac{3.0 \times 10^{8}}{1.8 \times 10^{18}}\ m = 1.67 \times 10^{-10}\ m$$

$$\Delta\lambda = \frac{h}{m_o c}(1 - \cos\phi)$$

$$\Delta\lambda = 0.0242\left(1 - \cos\ 180^{o}\right)\ A^{o}$$

$$\Delta\lambda = 0.0484\ A^{o}$$

$$\lambda' = \lambda + 0.0484\ A^{o}$$

$$\lambda' = 1.67\ A^{o} + 0.0484\ A^{o}$$

$$\lambda' = 1.72\ A^{o}$$

14. An X-ray photon is found to have its wavelength doubled on being scattered through 90°. Find the wavelength and energy of the incident photon

(m_o of electron $= 9.0 \times 10^{-31}$ Kg)

Sol. If λ had λ' be the wavelength of incident X-ray photon and scattered photon respectively and ϕ be the angle of scattering, then Compton shift in given by

$$\Delta\lambda = \lambda - \lambda' = \frac{h}{m_0 c}(1 - \cos\phi)$$

Hence

$$\phi = 90^{o} \text{ and } \lambda' = 2\lambda$$

$$\therefore \quad \Delta\lambda = 2\lambda - \lambda = \lambda$$

$$\lambda = \frac{h}{m_0 c}(1 - \cos\phi)$$

$$\lambda = \frac{h}{m_0 c} = \frac{6.62 \times 10^{-34}}{4.0 \times 10^{-31} \times 3.0 \times 10^{8}}$$

$\lambda = 0.0245\ A^{o}$

The energy of incident photon

$$E = hv = \frac{hc}{\lambda} = \frac{6.62 \times 10^{-34} \times 3.0 \times 10^{8}}{0.0245 \times 10^{-10}}$$

$$E = 8.106 \times 10^{-14} \text{ Joule}$$

15. A photon of frequency v is scattered by an electron initially at rest. Verify that the maximum kinetic energy of the recoil electron is

$$K.E_{max} = \frac{2h^2v^2/m_0c^2}{1 + 2hv/m_oc^2}$$

Sol. We know that

$$K.E_{max} = h(\nu - \nu') = hc\left(\frac{\lambda - \lambda'}{\lambda\lambda'}\right)$$

Also

$$\lambda - \lambda' = \frac{h}{m_o c}(1 - \cos\phi)$$

$$\lambda - \lambda' = \frac{2h}{m_o c} \qquad \text{for maximum K.E}$$

$$K.E_{max} = \frac{hc}{\lambda\lambda'}\left(\frac{2h}{m_o c}\right)$$

$$= \frac{2h^2}{m_o\ \lambda\lambda'}$$

$$= \frac{2h^2}{m_0\lambda\left(\lambda + \frac{2h}{m_0 c}\right)} \qquad \text{since } \nu = \frac{c}{\lambda}$$

$$= \frac{2h^2\nu}{m_o c\left(\frac{c}{\nu} + \frac{2h}{m_o c}\right)}$$

$$= \frac{2h^2\gamma}{(m_o c)\left(\frac{c}{\nu}\right)\left(1 + \frac{2h\nu}{m_o c^2}\right)}$$

$$= \frac{2h^2\nu^2}{\left(m_o c^2\right)\left(1 + \frac{2h\nu}{m_o c^2}\right)}$$

$$K.E_{max} = \frac{\frac{2h^2\nu^2}{m_o c^2}}{\left(1 + \frac{2h\nu}{m_o c^2}\right)} \qquad \text{hence proved.}$$

16. For what wavelength photon does Compton scattering result in a photon whose energy is one half of the original photon, at scattering angle of 45°? In which region of electromagnetic spectrum does such a photon lie?

(Compton wavelength of the electron = 0.0241 A°)

Sol. we know that the Compton shift in given by

$$\Delta\lambda = \lambda' - \lambda = \frac{h}{m_o c}(1 - \cos\phi)$$

where $\frac{h}{m_o c}$ is Compton wavelength

Given $\frac{h}{m_o c} = 0.0242\ A^o$ and $\phi = 45^o$

$$\therefore\ \Delta\lambda = 0.0242\ A^o\left(1 - \cos 45^o\right)$$

$$\Delta\lambda = 0.0242\ A^o \left(1 - \frac{1}{\sqrt{2}}\right)$$

$$\Delta\lambda = 0.007\ A^o$$

If $\frac{hc}{\lambda'}$ is the energy of the incident photon, then the energy of the scattered photon could be

$$\frac{1}{2}\left(\frac{hc}{\lambda}\right)$$

$$\therefore \quad \frac{hc}{\lambda^1} = \frac{hc}{2\lambda}$$

or $\quad \lambda' = 2\lambda$

Since $\quad \lambda' - \lambda = \Delta\lambda$

$\therefore \quad 2\lambda - \lambda = \Delta\lambda$

$$2\lambda = \Delta\lambda + \lambda$$

or $\quad \lambda = \Delta\lambda$

$\therefore \quad \lambda = 0.07A^o$

17. Show that the minimum value of energy which a photon must have so that it may transfer half of its energy to an electron at rest is about 256 keV.

Sol. The energy of the scattered photon is given by

$$\frac{hc}{\lambda'} = \frac{hc}{2\lambda}$$

$$\therefore \quad \lambda' = 2\lambda$$

We know that the Compton shift in given by

$$\Delta\lambda = \lambda' - \lambda = 2\lambda - \lambda = \lambda$$

$$\lambda = \frac{h}{m_0 c}(1 - \cos\phi)$$

For minimum energy of incident photon, λ should be maximum.

That is $\theta = 180^o$

$$\lambda_{max} = \frac{h}{m_o c}\left(1 - \cos 180^o\right) = \frac{2h}{m_o c}$$

$$= \frac{2 \times 6.62 \times 10^{-34}}{9.1 \times 10^{-31} \times 3 \times 10^8}$$

$$\lambda_{max} = 0.0484 \times 10^{-10}\,\text{m}$$

$\therefore$ Minimum energy of the incident photon

$$E_{min} = \frac{hc}{\lambda_{max}} = \frac{6.62 \times 10^{-34} \times 3.0 \times 10^{8}}{0.0484 \times 10^{-10}}$$

$$= 256 \times 10^{3}\ \text{eV}$$

$$E_{min} = 256\ \text{keV}$$

18. when X-ray of energy 100 keV strikes a target, they are scattered at an angle 30°. Find the energy of the recoiled electrons?

Sol. we now that the Compton shift is given by

$$\Delta\lambda = \lambda' - \lambda = \frac{h}{m_o c}(1 - \cos\phi) \qquad \text{.....(1)}$$

Kinetic energy of the recoiled electron

= Decrease in energy of the photon

$= h\nu - h\nu'$

Form equ. 1, we can write

$$\frac{\lambda'}{hc} - \frac{\lambda}{hc} = \frac{1}{m_o c^2}(1 - \cos\phi)$$

$$\frac{1}{h\nu'} - \frac{1}{h\nu} = \frac{1}{m_o c^2}(1 - \cos\phi)$$

$$\frac{1}{h\nu'} = \frac{1}{h\nu} + \frac{1}{m_o c^2}(1 - \cos\phi)$$

Given $E = h\nu = 100$ keV; $m_o = 9.1 \times 10^{-}31$ Kg; $c = 3.10^{8}$m/sec; $\theta = 30°$

$$\therefore \quad m_o c^2 = \frac{9.1 \times 10^{-31} \times (3 \times 10^{8})^2}{1.6 \times 10^{-19}} = 5.11\ keV$$

$$\therefore \quad \frac{1}{h\nu'} = \frac{1}{100} + \frac{1}{511}(1 - \cos 30°)$$

$$= \frac{1}{100} + \frac{1}{511}(1 - 0.866)$$

$$= \frac{1}{100} + \frac{0134}{511}$$

$$\frac{1}{h\nu'} = 0.0102 \quad \text{or} \quad h\nu' = 98.04 \text{ keV}$$

∴ The energy of the recoeil electron

$$= h\nu - h\nu' = (100 - 98.04) \text{ keV}$$

$$= 1.96 \text{ keV}$$

19. Calculate Compton shift if X-rays of λ = 1.00 A° are scattered from a carbon block. The scattered radiation is viewed at 90° to the incident beam.

Sol. If λ and λ′ be the wavelength of incident and scattered X-ray photons, then Compton shift (Δ λ) is given by

$$\Delta\lambda = \lambda' - \lambda = \frac{h}{m_o c}(1 - \cos\phi)$$

Given $\lambda = 1.00 \text{ A}^\circ$, $\phi = 90^\circ$, $h = 6.63\times10^{-34}$ J-sec;

$C = 3.0\times10^8$ m/sec and $m_o = 9.1\times10^{-31}$ Kg

$$\therefore \quad \Delta\lambda = \frac{6.63\times10^{-34}}{9.1\times10^{-31}\times3\times10^8}(1 - \cos 90^\circ)$$

$$= 0.0243\times1 \text{ A}^\circ$$

$$\Delta\lambda = 0.0243 \text{ A}^\circ$$

20. In a Compton scattering experiment, X-ray of wavelength 0.015 A° is scattered at 60°. Find the wavelength of the scattered X-rays.

Sol. We know that the Compton shift is given by

$$\Delta\lambda = \lambda' - \lambda = \frac{h}{m_o c}(1 - \cos\phi)$$

$$\therefore \quad \lambda' - \lambda = \frac{2h}{m_o c}\left(\sin^2\frac{\theta}{2}\right)$$

Given $\lambda = 0.015 \text{ A}^\circ = 0.015\times10^{-10}$ m and $\phi = 60^\circ$

$$\lambda' = \lambda + \frac{2h}{m_o c}\left(\sin^2\frac{\theta}{2}\right)$$

$$= \lambda + \frac{2\times 6.63\times 10^{-34}}{9.1\times 10^{-31}\times 3\times 10^8}\left(\sin^2 30^o\right)$$

$$= 0.015\times 10^{-10} + \frac{2\times 6.63\times 10^{-34}}{9.1\times 10^{-31}\times 3\times 10^8}\left(\frac{1}{2}\right)^2$$

$$= 0.015\times 10^{-10} + 0.012\times 10^{-10}$$

$$= 0.027\times 10^{-10} \text{ m}$$

$$\lambda' = 0.027 \text{ A}^o$$

21. X-ray of wavelength λ_o=2.00 A° are scattered from a block of carbon. The scattered X-rays are observed at an angle of 45° to the incident beam. Find the fraction of energy lost by the photon in this collision.

Sol. From Compton shift we have

$$\Delta\lambda = \lambda' - \lambda = \frac{h}{m_o c}(1-\cos\phi)$$

Given $\phi = 45^o$

$$\lambda' = \lambda_o + \frac{2h}{m_o c}\left(1-\cos 45^o\right)$$

$$= \lambda_o + 0.0243 \text{ A}^o\left(1-\frac{1}{\sqrt{2}}\right)$$

$$= 2.00 \text{ A}^o + 0.0243 \text{ A}^o\left(\frac{\sqrt{2}-1}{\sqrt{2}}\right)$$

$$\lambda' = 2.00 \text{ A}^o + 0.007 \text{ A}^o = 2.007 \text{ A}^o$$

$$\lambda' = \lambda_o + \Delta\lambda = 2.007 \text{ A}^o$$

∴ The fraction of energy lost by the photon in this collision

$$\frac{\Delta E}{E} = \frac{\Delta\lambda}{\lambda'}$$

$$\frac{\Delta E}{E} = \frac{0.007\ A^o}{2.007\ A^o} = 0.0035$$

22. A beam of gamma radiation having photon energy 510 keV is incident on a foil of aluminium. Calculate the wavelength of scattered radiation at 90°.

Sol. Energy of each photon $E = h\nu = \frac{hc}{\lambda} = 510\ keV$

$$\lambda = \frac{hc}{510 \times 10^3 \times 1.6 \times 10^{-19}}$$

$$= \frac{6.63 \times 10^{-34} \times 3.0 \times 10^8}{510 \times 10^3 \times 1.6 \times 10^{-19}}$$

$$\lambda = 0.0244 \times 10^{-10}\ m$$

According to the Compton scattering formula

$$\lambda' - \lambda = \frac{h}{m_o c}(1 - \cos\phi)$$

$$\lambda' = \lambda + \frac{h}{m_o c}(1 - \cos\phi)$$

Given $\theta = 90^o$

$$\lambda' = \lambda + \frac{h}{m_o c}$$

$$= 0.0244 \times 10^{-10} + \frac{6.63 \times 10^{-34}}{9.1 \times 10^{-31} \times 3 \times 10^8}$$

$$= 0.0244 \times 10^{-10} + 0.0243 \times 10^{-10}$$

$$\lambda' = 0.0487 \times 10^{-10}\ m = 0.0487\ A^o$$

23. The Bragg angle corresponding to the first order reflection from (111) plane in a cubic crystal is 30° when X-rays of wavelength 1.75 A° are used. Find the inter atomic spacing.

Sol. We know that

$$d_{hkl} = \frac{a}{\sqrt{h^2 + k^2 + l^2}}$$

$$\therefore d_{111} = \frac{a}{\sqrt{3}}$$

We know that $2d\sin\theta = n\lambda$

$$2 \times \frac{a}{\sqrt{3}} \sin 30^{\circ} = 1.75 \text{ A}^{\circ}$$

$$\therefore \quad a = \frac{1.75 \times \sqrt{3} \text{ A}^{\circ}}{2 \times \sin 30^{o}} = 3.031 \text{ A}^{\circ}$$

24. Calculate the density of lattice points in lattice plane (100) in a simple cubic lattice of constant 2.5A°

Sol. we know that

$$d_{hkl} = \frac{a}{\sqrt{h^2 + k^2 + l^2}}$$

$$d_{hkl} = \frac{2.5 \times 10^{-10}}{\sqrt{1+0+0}} = 2.5\text{A}^{\circ}$$

Volume of unit cell

$$a^3 = (2.5 \times 2.5 \times 2.5) \times 10^{-30} \text{ m}^3$$

$$V = 15.625 \times 10^{-30} \text{ m}^3$$

Therefore,

$$\text{Density of lattice points} = \frac{nd_{100}}{V} \qquad \text{(For n = 1 primitive lattice)}$$

$$\therefore \quad \text{Density of lattice points} = \frac{d_{100}}{V} = \frac{2.5 \times 10^{-10}}{15.625 \times 10^{-30}}$$

$$\therefore \quad \text{Density of lattice points} = 1.6 \times 10^{19} \text{ atoms/m}^3$$

25. The Bragg angle corresponding to the first order reflection from (111) plane in a simple cubic crystal in 30° when X-rays of wavelength 1.57 A° are used. Calculate

(i) Inter atomic spacing

(ii) Ratio of intercepts on the three axes by (123) planes

Sol. (i)

$$d = d_{hkl} = \frac{a}{\sqrt{h^2 + k^2 + l^2}} = \frac{a}{\sqrt{1+1+1}} = \frac{a}{\sqrt{3}}$$

Now $2d \sin \theta = n\lambda$

$$d = \frac{a}{\sqrt{3}},\ n = 1,\ \theta = 30^o,\ \lambda = 1.57\ A^o$$

$$\therefore \quad \frac{2a}{\sqrt{3}} \sin 30^o = 1.57 \times 10^{-10}$$

$$\therefore \quad a = 2.72\ A^o$$

Suppose the lattice plane cuts intercepts on the axes l_1, l_2 and l_3 then

$$l_1 : l_2 : l_3 = pa : qb : rc$$

Since $a = b = c$

Therefore

$$p : q : r = \frac{1}{h} : \frac{1}{k} : \frac{1}{l} = \frac{1}{1} : \frac{1}{1} : \frac{1}{3}$$

Therefore $p : q : r = 6 : 3 : 2$

Conceptual Questions and Answers

1. **Distinguish between X-rays and γ -rays**

Ans. 1. γ -rays have much shorter wavelength than X-rays

2. γ - rays are more penetrating than X-rays

3. γ -rays originate from energy changes in the nuclei of atoms. But X-rays originate from energy changes associated with the extra-nuclear electronic structure of the atom.

2. **Can X-ray of wavelength greater than 2d be diffracted from the crystal?**

Ans. No; since $2d \sin \theta = nd$ and $\lambda_{max} = 2d$

3. **How is X-ray reflection different from optical reflection?**

Ans. (i) In optical reflection, reflection is possible for all angles of incident, But, in X-ray reflection, reflection is governed by Bragg's law. When a narrow parallel beam of X-rays in incident on a crystal, it will clearly make some angle, θ, with any chosen set of (hkl) planes. Bragg showed that a diffraction maximum will occur corresponding to a specular reflection from these planes only if $n\lambda = 2d \sin\theta$, where λ is the wavelength of the monochromatic X-rays and d is spacing between successive hkl planes

(ii) In the case of optical reflection, reflection occurs at a single layer. But in the case of X-ray reflection, reflection occurs at several layers in the crystal.

4. **Explain why it is not possible to deduce the position of hydrogen atom from X-ray diffraction. How can this difficulty be overcome by using the neutron diffraction method.**

Ans. If more than one kind of atoms are present in the crystal, the atom with the greater number of electrons has the greater scattering power. For light elements, the scattering power is directly proportional to the number of electrons around the atom. It is because of this reason that the scattering ability of hydrogen atom is very small and thus whatever effect it produces is overshadowed by the effect of neighbouring atoms which contain larger number of electrons. Thus the position of hydrogen atoms cannot be deduced from X-ray diffraction. However their position can be determined by neutron diffraction since the latter involves the interaction with the nuclei and the interaction is about the same with all nuclei.

5. **Both photoelectric and Compton Effect arise due to the action of photons on electrons, but the two effects are not the same. Explain this?**

Ans. In the photoelectric effect, the photon completely disappears and all of its energy is given to the photoelectron. In Compton Effect, only a part of the energy of the incident photon is given to the Compton electron. The energy of the scattered photon is less than the energy of the incident photon.

6. **Explain why Compton effect is experimentally not observed for visible light rays.**

Ans. We know that $\Delta\lambda = \frac{h}{m_o c}(1-\cos\phi) = 0.024(1-\cos\phi)\ A^o$

$\Delta\lambda$ depends only on the scattering angle ϕ. Maximum value of $\cos\phi = -1$ when $\phi = 180^o$. Hence the maximum value of $\Delta\lambda = 0.048\ A^o$. This means that Compton effect can be detected only for those radiations whose wavelength is not greater than a few A^o. For visible light $\left(\lambda \approx 5000\ A^o\right)$, $(\Delta\lambda)_{max}$ is only about 0.001% of the initial wavelength which cannot be detected.

7. **Explain the presence of unmodified line in Compton scattering?**

Ans. When a photon collides with a bound electron, its wavelength does not change.

8. What is the reason for the spread in the wavelength of the Compton-shifted X-ray?

Ans. The main reason for the spread in wavelength of the Compton-shifted X-rays is that the initial electron is in general not stationary but has a momentum spread even inside the atom. The correction due to the binding of the electron can be taken in to account in terms of its momentum distribution.

9. In Compton scattering, one neglect the effect of the nucleus on the X-ray. Why?

Ans. The cross section of Compton scattering is proportional to the inverse of the square of the scattering particle. Hence

$$\frac{\text{cross section of photon–electron scattering}}{\text{cross section of photon – nucleus scattering}} = \left(\frac{m_N}{m_e}\right) \gg 1$$

Therefore, the effect of the nucleus on the X-ray is negligible.

10. Explain why we cannot use visible light for studying the crystal structure.

Ans. From Bragg's law,

$$2d \sin\theta = n\lambda$$

Let us consider that n = 1 (first order)

As the maximum value of sin θ is 1 2d = λ

As d is of the order of few A°, hence wavelength λ must also be of the same order. For visible light the λ varies from 5000 A° to 8000 A°, which is very high. Hence Bragg's condition is not satisfied. That is why we cannot use visible light for studying crystal structure. For this X-rays are used because X-ray wavelengths are of few A° (1A°)

Objective Questions

1. Bragg's equation $2d \sin\theta = n\lambda$ will have no solution if

(a) $\lambda<2d$ (b) $\lambda<d$

(c) $\lambda>d$ (d) $\lambda>2d$

2. When X-rays fall on a crystal of lattice constant d and the glancing angle θ, then of the following equations, the one which gives diffraction maximum of nth order in accordance with Bragg's formula is

(a) $2d \sin\theta = n\lambda$ (b) $2d \sin\theta = \lambda$

(c) $2d \sin\theta = \lambda/n$ (d) $2n \sin\theta = d$

3. X-ray can be diffracted from crystals as the diffraction of light form a grating. The crystal grating for X-rays has a grating element
   (a) about 1/1000 of wavelength of X-ray
   (b) equal to the distance between any two nearest atom
   (c) equal to the size of the crystal
   (d) that is, an integral multiple of the wavelength of X-rays.
4. A patient is asked to drink Barium Sulphate ($BaSo_4$) solution for examining the stomach by X-rays, because X-rays are
   (a) replaced by heavy atoms (b) move absorbed by heavy atoms
   (c) less absorbed by heavy atoms (d) diffracted by heavy atoms
5. The Compton effect in X-ray proves that
   (a) X-rays have wave nature
   (b) electrons have wave nature
   (c) X-rays have corpuscular (particle) nature
   (d) X-rays have quantum nature
6. The Compton effect is associated with
   (a) positive rays (b) visible light
   (c) β-rays (d) γ-rays
7. The Compton shift depends on
   (a) material of target (b) angle of scattering
   (c) energy of incident light (d) frequency of incident light
8. In Compton effect the electron that takes part is
   (a) free
   (b) emitted from nucleus by decay of neutrons
   (c) may be free or bound
   (d) bound
9. The Compton shift is maximum, along the direction of scatted photon which makes an angle
   (a) 180° with incident direction (b) 45° with incident direction
   (c) 0° with incident direction (d) 90° with incident direction

10. In an experiment of Compton scattering, the X-ray of wavelength λ are scattered from a carbon block through an angle 90° to the incident X-ray photon beam. The wavelength of scattered radiation is λ^1, then the direction of recoil electrons makes an angle

(a) $\theta = \tan^{-1}\left(\frac{\lambda}{\lambda^1}\right)$

(b) $\theta = \tan^{-1}\left(\frac{\lambda \sin\theta}{\lambda^1 - \lambda}\right)$

(c) $\theta = \tan^{-1}\left(\frac{\lambda^1}{\lambda}\right)$

(d) $\theta = \tan^{-1}\left(\frac{\lambda}{\lambda^1 - \lambda\cos\theta}\right)$

Theoretical Questions

1. What do you mean by diffraction of X-rays?
2. Derive Bragg's equation for reflection of X-rays by crystal planes.
3. Describe Bragg's spectrometer and derive the necessary formula and explain how it is used to study the structure of crystals?
4. Write short note on powder crystal method for X-ray diffraction.
5. Discuss Laue's principle of X-ray diffraction and obtain the diffraction condition for a simple cubic lattice.
6. What is Compton Effect? Derive appropriate expression for the change in wavelength of a photon when it is scattered by an electron. Discuss the conclusion drawn based on this effect.
7. What is Compton effect? Derive the expression for the direction of recoiled Compton electron. Show that the Compton electron can recoil only in onward direction at angles less than 90°?
8. A photon of frequency ν is scattered by an electron initially at rest. Prove that the maximum kinetic energy of the recoil electron is given by

$$K.E_{max} = \frac{2h^2\nu^2}{m_oc^2\left[1 + \frac{2h\nu}{m_oc^2}\right]}$$

9. What is Compton Effect? Explain the physical significance of Compton shift given by

$\lambda' - \lambda = \frac{h}{m_oc}(1 - \cos\phi)$, where the symbols have their usual meaning.

10. Derive an expression for Compton shift. What is Compton wavelength? Explain why Compton shift is not observed with visible light?

Numerical Problems

1. X-rays having wavelength of 2.6 A° are diffracted at an angle of 20° for the second order in Bragg's spectrometer. Find the spacing constant of the crystal?

[7.60 A°]

2. X-rays of wavelength 1.6 A° are diffracted by X-ray spectrograph at an angle of 30° in the second order. Calculate inter atomic spacing?

[3.2 A°]

3. The first order Bragg's maximum of electron diffraction in a nickel crystal d=0.4086 A° occurred at a glancing angle of 65°. Calculate the de-Broglie wavelength of the electrons and the velocity?

[0.7406 A°, 9.793×10^6 m/s]

4 In Bragg's reflection for X-rays, a reflection was found at 30° glancing angle with lattice planes of spacing 1.87 A°. If this is second order reflection, calculate the wavelength of X-rays?

[9.35 A°]

5. X-rays of wavelength 1.0 A° are scattered from a carbon block. Find:

(a) the wavelength of the scattered beam in a direction making 90° with the incident beam

(b) Kinetic energy imparted to the recoil electron

(c) direction of the recoil electron

Given, h = 6.63×10^{-34} J-sec, c = 3×10^8 m/s and 1 eV = 1.6×10^{-19} Joule and m_o of electron = 9.1×10^{-31} Kg.

[1.0243 A°; 295 eV; 44.4°]

6. A photon collides with a free electron and is scattered at right angles to its initial direction. Find the percentage change in its energy as result of this Compton collision, when the photon is a

(i) microwave photon of λ =3.0 cm [8.1×10^{-9} %]

(ii) visible light photon of λ = 5000 A° [4.86×10^{-4} %]

(iii) X-ray photon of λ= 1.0 A° [2.4 %]

(iv) γ-ray photon of λ = 0.0124 A° [66 %]

In view of your results discuss the relative importance of the Compton Effect in the different regions of electromagnetic spectrum.

[2.46×10^{-12}m; $4.844\times^{-14}10^{-12}$m,$4.074\times10^{-14}$J]

7. A beam of γ–radiation having photon energy 510 keV is incidents on a foil of aluminium. Calculate the wavelength of the radiation at 90° and also the energy and direction of the emission of the corresponding electron.

[2.46×10^{-12} m; 4.844×10^{-12} m, 4.074×10^{-14} J]

8. Gamma photon of 1.2 A° wavelength suffers Compton scattering at $\frac{\pi}{3}$. Find percentage increase in wavelength?

[1%]

9. If the maximum energy imparted to an electron in Compton scattering is 45 keV, what is the wavelength of the incident photon?

[9.39×10^{-2} A°]

10. Determine the maximum scattering angle in a Compton experiment for which the scattered photon can produce a positron –electron pair?

[$\theta = 60^\circ$]

Answers to Objective Questions

1. (d) 2. (a) 3. (b) 4. (d) 5. (d) 6. (d)
7. (b) 8. (a) 9. (a) 10. (a)

CHAPTER 6A

Dielectric Properties of Materials

'O tell me, when along the line
From my full heart the message flow
What currents are induced in thine?
One click from the will end my woes'
Through many an ohm the weber flow
And clicked the answer back to me.
'I am thy Farad, Staunch and true,
Charged to a volt with love for thee'

- James Clerk Maxwell

6A.1 Introduction

Insulating materials are also termed as dielectrics. The difference in the name between dielectric and insulator lies in the application to which these materials are used. When these materials are used to prevent the flow of electricity through them on the application of potential difference, then they are called as insulator or passive dielectrics. On the other hand, if they are used for charge storage then they are called as dielectrics or active dielectrics.

Materials such as glass, rubber, wood and porcelain provides electrical insulation between two conductors at certain potential difference, and also serve as an electrical charge storage under certain other circumstances.

In an ideal dielectrics materials, all the electrons are tightly bound to the nucleus of the atom. As a result, there are no free electrons available for conduction. This is the reason for them to be non-conducting. Generally the dielectrics are non-metallic materials of high specific resistance and have negative temperature coefficient of resistance.

6A.2 Dipole

A pair of equal and opposite charges separated by a small distance constitutes an electric dipole. The product of magnitude of one of the charges q and the distance of their separation d is called the dipole moment.

$$\mu = q \times d$$

Therefore, polarisation is defined as the process of creating or inducing dipoles in a dielectric material by an external electric field. The polarisation increases with the field up to the critical value.

On the basis of polarisation concept, the dielectrics are the materials that have either permanent dipoles or induced dipoles in the presence of an applied electric field. They are classified into two categories, namely polar and non-polar dielectrics.

6A.3 Non-Polar Dielectrics

A dielectric material in which there is no permanent dipole existence in the absence of an external field is called non-polar dielectrics.

This is because in these types of dielectrics, the centroid of positive and negative charges of the molecules constituting the dielectric material coincide.

Example:

H_2, N_2, O_2, CO_2 and CH_4. These molecules are called non-polar molecules.

6A.4 Polar Molecules

A dielectric material in which there is an existence of permanent dipole even in the absence of an external field is called a polar dielectric

This is because, in these types of dielectrics, the centre of the positive charges and negative charges of the molecules constituting the dielectric material do not coincide even in the absence of an external field.

Example: H_2O, HCl and CO. These molecules are called polar molecules.

6A.5 Behaviour of Dielectrics in External Electric Field: Electric Polarisation of Matter

When a dielectric material is placed in electric field, its electrical properties are modified. In case of non-polar dielectric (zero permanent dipole moment of molecules), the positive and negative charges of molecules experience electrostatic forces in opposite directions. Therefore, the centres of gravity of two charges are separated from each other (The separation is limited to a small fraction of molecular diameter by strong restoring forces set up by the changing charge configuration in the molecules). The molecules thus acquire an induced electric-dipole movement in the direction of the filed.

Fig. 6A.1 [No field]

Let us now consider a dielectric whose molecules are polar having permanent electric dipole moments. In the absence of any external electric field, the individual dipoles (molecules) are oriented at random and no net dipole moment is observed in the dielectric (Fig. 6A.1)

When the dielectric is placed in an electric field, the molecules become sufficiently aligned in the direction of the field (Figure 6A.2) (The alignment is however, incomplete due to the thermal agitation of the molecules. It becomes more and more perfect as the electric field is increased or the temperature is decreased). Thus a net dipole moment is produced in the dielectric

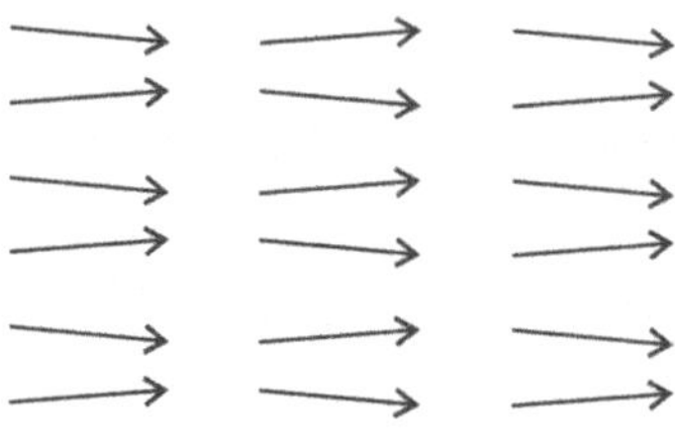

Fig. 6A.2 [Field $\vec{E}$]

6A.6 Dielectric in an Electric Field

Suppose a slab of dielectric material is placed in the uniform electric field $\vec{E}_0$ set up between the parallel plates of a charged capacitor, Figure 6A.3. The slab becomes electrically polarised. That is, its molecules become electric dipoles oriented in the direction of the filed. The net effect is the appearance of negative charge on one face of the slab in front of the positively charged plate and equal positive charges on the face in front of the negatively charged plate. The net charge in the interior of the slab remains zero, because the positive side of one polarised molecule is near the negative side of its neighbour.

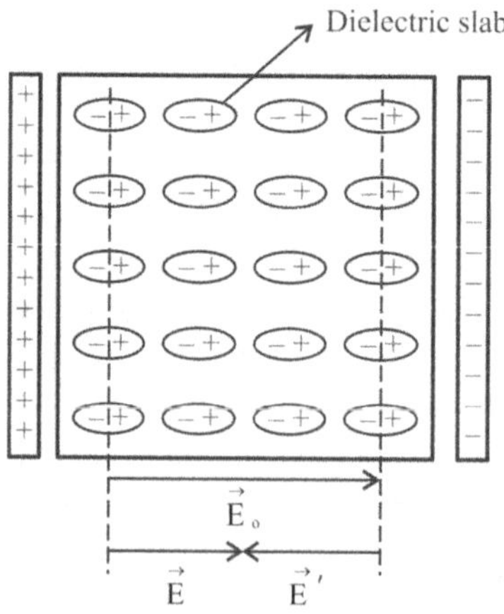

Fig. 6A.3

The polarisation charges induced on the two faces of the slab produce their own electric field $\vec{E}'$ (The induced electric field $\vec{E}'$ is smaller than the applied electric field $\vec{E}_0$ because the charges induced on the faces of the dielectric slab are 'bound' to the molecules lying in the surface of the slab. They cannot move to an extent that the induced field may be equal to the applied field) which opposes the external field $\vec{E}_0$. Consequently the resultant field $\vec{E}$, within the dielectric is less than the field $\vec{E}_0$ in the free space, but points in the same direction as $\vec{E}_0$. The field in the rest of the free space is still $\vec{E}_0$. Hence we conclude that when a dielectric is placed in an electric field, the field 'within' the dielectric is weakened (but not reduced to zero).

The weakening of electric field within the dielectric is illustrated in Fig.6 A.4 and 6 A.5. The Fig.6 A.4 shows the original field. In Fig.6 A.5, not all the lines of force leaving the positive plate of the capacitor penetrate the dielectric, some of them terminate on the charge induced on the dielectric.

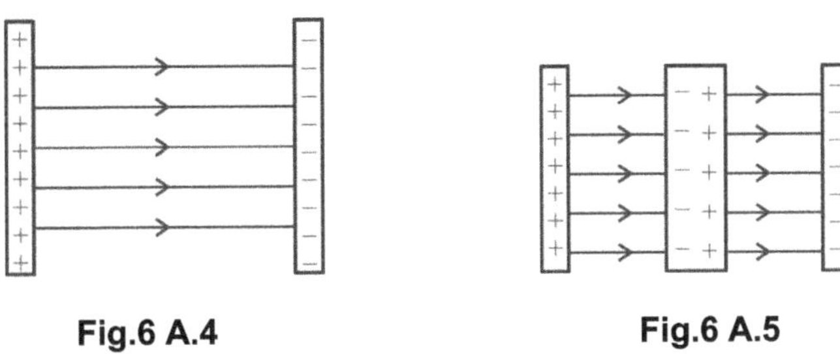

Fig.6 A.4 Fig.6 A.5

6A.7 Relative Permittivity or Dielectric Constant ϵ_r

The dielectric characteristics are determined by the dielectric constant.

The dielectric constant or relative permittivity ϵ_r is defined as the ratio between the permittivity of the medium ϵ and the permittivity of free space ϵ_0

i.e.
$$\epsilon_r = \frac{\epsilon}{\epsilon_0}$$

ϵ_r is a dimensionless Quantity and varies widely from material to material. ϵ_r has a value of unity for vacuum and for all other dielectrics ϵ_r is always greater than 1. Since $\epsilon_r > 1$, we can write

$$\epsilon_r = 1 + \chi_e$$

Where χ_e is called electric susceptibility. Permittivity of a medium or material indicates the polarisation nature of a material.

(1) ϵ_r is a constant for a given isotropic material when the applied field is static (d,c) and is referred to as static dielectric constant. But, when the material is subjected to an alternating field, it becomes a frequency dependent complex Quantity.

(2) ϵ_r values varies with direction in the case of anisotropic materials.

6A.8 Polarisation Vector P

Polarisation vector measures the extent of polarisation in a unit volume of dielectric matter.

Polarisation is defined as the induced dipole moment per unit volume of the dielectric. P is a vector Quantity, and its direction is along the direction of the applied field. If μ is the average induced dipole moment per unit molecule and N is the number of molecules per unit volume, then the polarisation is given by

$$P = N\mu \quad \text{.....(6A.1)}$$

The polarisation P can also be defined as the induced charge per unit area, which can be explained as follows:

Let the polarisation of the dielectric slab as a whole, Fig.6 A.6 give rise to +q and –q induced charge on the respective faces of the dielectric slab of thickness t and volume V kept between the two plates of a capacitor as shown in Fig.6 A.6.

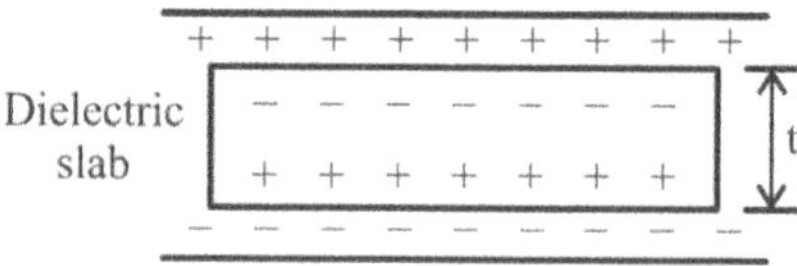

Fig. 6A.6

Therefore, the dipole moment of the slab is given by qt. Thus the dipole moment per unit volume of the dielectric or the polarisation is given by

$$P = \frac{qt}{V}$$

$$P = \frac{qt}{At}$$

where V=At, and A is the area of the slab

$$\therefore \quad P = \frac{q}{A}$$

Thus polarisation can be defined as the induced surface charge per unit area.

But $\frac{q}{A}$ is the induced charge density. Therefore magnitude of polarisation is equal to the induced charge density.

The unit of polarisation is Cm^{-2}

6A.9 Electric Susceptibility χ_e

In a large number of dielectrics it is found that the polarisation is directly proportional to the external field applied E. for such cases

$$P \propto E$$

Thus the relation between P and E can be given as

$$P = \varepsilon_o \chi_e E$$

or
$$\chi_e = \frac{P}{\varepsilon_o E}$$

where χ_e is a constant, called the dielectric susceptibility of the medium, which is defined as:

The ratio of polarisation to the net electric field $\varepsilon_o E$ as modified by the induced charge on the surface of the dielectric is called susceptibility.

6A.10 Relation between Polarisation P Susceptibility χ_e the Dielectric Constant $\in_r$

Let us consider a parallel plate capacitor as shown is Fig.6 A.7 between which an electric field E_o exists.

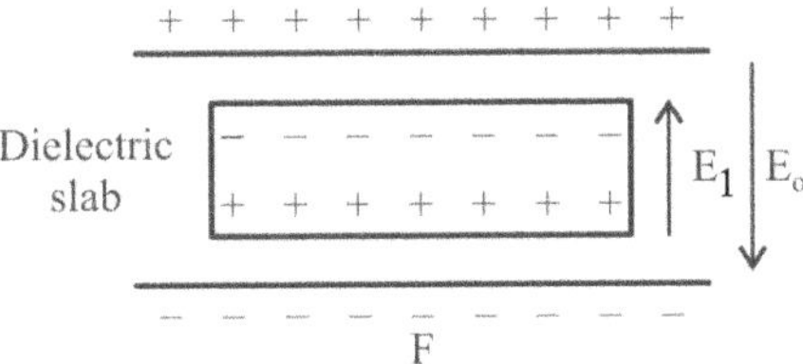

Fig. 6A.7

If σ is the surface charge density (i.e. charge per unit area), then from Gauss Law

$$E_o = \frac{\sigma}{\in_o} \qquad \text{......(6 A.2)}$$

If a dielectric slab is placed between the plates of the capacitor then, due to polarisation charges appear on the two faces of the slab and establish another field E_1 within the dielectric. This field will be in a direction opposite to that of E_o.

∴ The resultant field E can be given as

$$E = E_o - E_1 \qquad \text{.....(6 A.3)}$$

If σ_s is the surface charge density on the slab then, by equation (6 A.2), we can write

$$E_1 = \frac{\sigma_s}{\epsilon_o} \qquad \text{.....(6 A.4)}$$

Therefore from (6 A.2), (6A.3) and (6A.4)

$$E = \frac{\sigma}{\epsilon_o} - \frac{\sigma_s}{\epsilon_o}$$

$$\epsilon_o E = \sigma - \sigma_s \qquad \text{.....(6 A.5)}$$

Since $\sigma_s = P$ (Polarisation is charge/ unit area) .....(6 A.6)

Also by gauss law, the electric flux density or electric displacement density D is given by

$$D = \sigma \text{ (D is charge/ Unit area)} \qquad \text{.....(6 A.7)}$$

∴ Equation (6 A.5) can be written as

$$\epsilon_o E = D - P \qquad \text{......(6 A.8)}$$

Or

$$D = \epsilon_o E + P \qquad \text{......(6A.9)}$$

But from electrostatics we know that

$$D = \epsilon E$$

$$D = \epsilon_o \epsilon_r E \qquad \text{...(6A.10)}$$

Therefore e.g (6 A.9) can be written as

$$\epsilon_o \epsilon_r E = \epsilon_o E + P$$

or

$$\epsilon_o E(\epsilon_r - 1) = P$$

$$\epsilon_r - 1 = \frac{P}{\epsilon_o E} \qquad \text{.....(6 A.11)}$$

we know that $\frac{P}{\epsilon_o E} = \chi_e$ .....(6 A.12)

$$\therefore \epsilon_r - 1 = \chi_e$$

$$\chi_e = \epsilon_r - 1 \qquad \text{.....(6 A.13)}$$

6A.11 Polarisability α

Let us consider an individual atom in a dielectric material. Let the material be subjected to an electric field $\vec{E}$. It is found that the induced dipole moment μ is acquired by the atom is proportional to the strength of the electric field $\vec{E}$.

i.e $$\mu \propto E$$

$$\mu = \alpha E \quad \text{.....(6 A.14)}$$

Where α is the proportionality constant called Polarisability. Its unit is F m^2

Note: Polarisability is not a bulk property of the material, but it is the property of an individual atom or molecule.

6A.12 Electrical Polarisation Mechanism

There are four different types of mechanism through which electrical polarisation can occur in a dielectric material when they are subjected to an external electric field.

They are:

(a) Electronic polarisation

(b) Ionic polarisation

(c) Orientation polarisation

(d) Space charge or interfacial polarisation.

6A.12.1 Electronic Polarisation

The electronic polarisation occurs, due to the displacement of the positively charged nucleus and the negatively charged electron cloud in opposite directions within a dielectric material upon applying an external electric field E.

Thus the speration created between the charges induces a dipole. This process occurs throughout the material and the material as a whole will be polarised. Electronic polarisation is more in liquid and solid dielectrics than in gases.

Therefore, the induced dipole moment

$$\mu = \alpha_e E \quad \text{.....(6 A.15)}$$

Where α_e is the electronic polarisability. The electronic polarisability for a rare gas atom is given by

$$\alpha_e = \frac{\epsilon_o (\epsilon_r - 1)}{N} \quad \text{.....(6 A.16)}$$

Where N is the number of atoms per unit volume.

6A.12.2 Expression for electronic Polarisability (α_e)

Let us consider one of the constituent atom of a dielectric material in the absence of an electric field E. Let the radius of the atom be R and its atomic number be Z as shown in Fig.6 A.8.

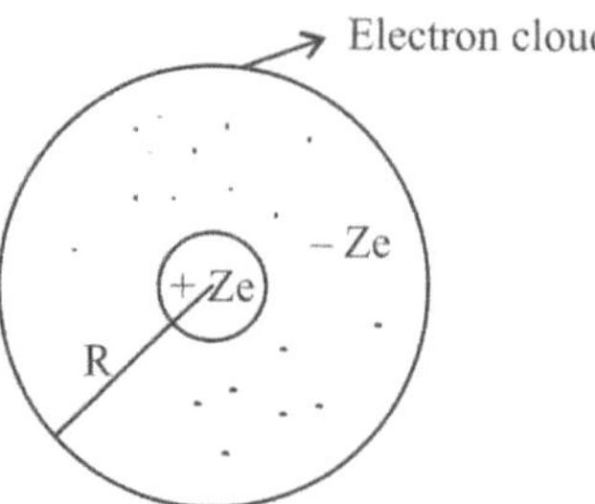

Fig.6 A.8

Here the nucleus of the charge +Ze is surrounded by an electronic cloud of charge –Ze. In the atom, the negative charge distribution due to its electron is spherically symmetric. Therefore, the charge density for electron cloud is given by

$$\rho = \frac{-Ze}{\frac{4}{3}\pi R^3} \qquad \text{.....(6 A.17)}$$

$$\rho = \frac{-3}{4}\left(\frac{Ze}{\pi R^3}\right)$$

But when an electric field is applied, the nucleus and the electron cloud experience a Lorentz force of magnitude Ze E in opposite direction. Hence the nucleus and the electron cloud are pulled by a distance as shown in Fig 6 A.9

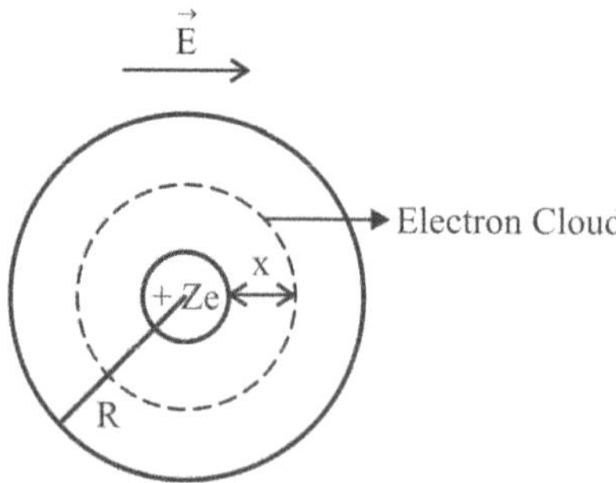

Fig.6 A.9

Since the nucleus is much heavier than the electron cloud it is assumed that only the electron cloud is displaced upon applying an electric field. Let the electron cloud displacement be x with respect to the centre of the nucleus.

Hence the Lorentz force acting over the electron cloud is

$$= -Ze\,E \qquad \text{.....(6 A.18)}$$

But according to Gauss theorem a coulomb attractive force is said to act over the nucleus due to the electron cloud in a sphere of radius x and this force tends to oppose the displacement.

Then, the coulomb force $= Ze \times E$ .....(6 A.19)

According to Coulomb's, the electric field at any point in a charged conductor is $\frac{1}{\epsilon_o}$ times the surface charge density i.e charge per unit area.

$$\therefore \qquad E = \frac{\text{total charge enclosed in a sphere of radius x } (q)}{4\,\pi\,\epsilon_o\,x^2} \qquad \text{.....(6A.20)}$$

But the total charge enclosed in a sphere of radius x is

$$q = \frac{4}{3}\pi x^3 \rho \qquad \text{.....(6 A.21)}$$

where ρ is negative charge density and $\frac{4}{3}\pi x^3$ is volume of the sphere of radius x.

Substituting for ρ from equation (6 A.17), we get

$$q = \frac{-3}{4}\left(\frac{Ze}{\pi R^3}\right) \times \frac{4}{3}\pi x^3$$

$$q = \frac{-ZeX^3}{R^3} \qquad \text{.....(6 A.22)}$$

Substituting for q is equation (6 A.20), we get

$$E = \frac{-Ze\,x^3}{R^3} \times \frac{1}{4\pi\,\epsilon_o\,x^2} \qquad \text{.....(6 A.23)}$$

Hence

Coulomb force = Ze E

$$= Ze\left(\frac{-Ze\,x^3}{R^3} \times \frac{1}{4\pi\,\epsilon_o\,x^2}\right)$$

Coulomb force $= \dfrac{-Z^2e^2x}{4\pi \epsilon_o R^3}$ .....(6 A.24)

These two forces i.e the Lorentz force and the coulomb force are equal in magnitude but opposite in direction. As a result, a equilibrium is reached. Hence at equilibrium.

Lorentz force = Coulomb force

$$-Ze\,E = \frac{-Ze^2x}{4\pi \epsilon_o R^3} \qquad .....(6\ A.25)$$

or

$$E = \frac{Zex}{4\pi \epsilon_o R^3} \qquad .....(6\ A.26)$$

Hence $x = \dfrac{4\pi \epsilon_o R^3}{Ze} E$ .....(6 A27)

Thus, the displacement of the electron cloud is proportional to the applied field E. Due to this displacement the atom acts as a dipole.

∴ The induced dipole moment

$$\mu = Ze \times x$$

$$\mu = (Ze)\left(\frac{4\pi \epsilon_o R^3}{Ze} E\right)$$

or

$$\mu = 4\pi \epsilon_o R^3 E \qquad .....(6\ A.28)$$

But $\mu = \alpha_e E$ .....(6 A.29)

Comparing equation (6 A.28) and (6 A.29), we get

$$\alpha_e = 4\pi \epsilon_o R^3 \qquad .....(6\ A.30)$$

Thus the electronic Polarisability is proportional to the volume of the atom and is independent of temperature.

The polarisation vector $P = N\mu$

$$\therefore P = N \alpha_e E \qquad .....(6\ A.31)$$

But we know that

$$P = \epsilon_o E(\varepsilon_r - 1) \qquad ......(6\ A.32)$$

∴ Equating equation (6 A.31) and (6 A.32) we get

$$N\alpha_e E = \epsilon_o E(\varepsilon_r - 1)$$

$$\alpha_e = \frac{\epsilon_o(\varepsilon_r - 1)}{N} \quad \text{.......(6 A.33)}$$

6A.12.3 Ionic Polarisation

Ionic polarisation occur only in ionic solids such as NaC*l* which posses ionic bonds. It does not occur in covalent crystals such as diamond, silicon and germanium.

When ionic solids are subjected to an external electric field, the adjacent ion of opposite sign undergo displacement as shown in Fig. 6 A.10.

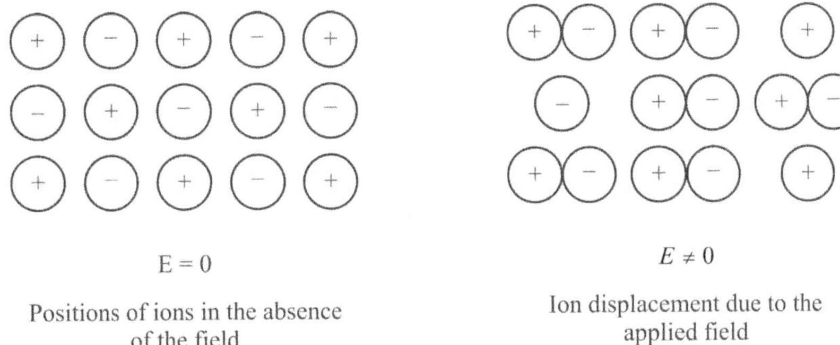

Fig .6 A.10

6A.12.4 Expression for Ionic Polarisability ($\propto_i$)

Let an electric field E is applied to an ionic solid consisting of one cation and one anion per unit cell, This, applied field causes the positive ions and negative ions to get displaced through a distance x_1 and x_2 respectively from their equilibrium position thereby inducing a dipole moment. Thus the resultant displacement per unit cell is

$$\mu = e\ (x_1 + x_2) \quad \text{.....(6 A.34)}$$

Due to the application of electric field a restoring force is said to act over the cation and anion.

Thus restoring force

$$F = \beta_1\ x_1 = \beta_2\ x_2 \quad \text{.....(6 A.35)}$$

Hence

$$x_1 = \frac{F}{\beta_1} \text{ and } x_2 = \frac{F}{\beta_2} \quad \text{.....(6 A.36)}$$

where β_1 and β_2 are restoring force constants of cation and anion. This restoring force constant depends upon the mass of the ion and angular frequency of the molecule in which the ions are present.

Thus

$$x_1 = \frac{e\,E}{m w_o^2} \text{ and } x_2 = \frac{e\,E}{M w_o^2} \qquad \text{.....(6 A.37)}$$

where m is the mass of cation and M the mass of the anion.

Substituting equation (6 A.37) and (6 A.34), we get

$$\mu = e\left[\frac{e\,E}{m w_o^2} + \frac{e\,E}{M w_o^2}\right]$$

$$\mu = \frac{e^2\,E}{w_o^2}\left[\frac{1}{m} + \frac{1}{M}\right] \qquad \text{.....(6 A.38)}$$

But $\mu = \alpha_i\, E$ .....(6 A.39)

Comparing equation (6 A.38) and (6 A.39), we get

$$\alpha_i\, E = \frac{e^2\,E}{w_o^2}\left[\frac{1}{m} + \frac{1}{M}\right]$$

$$\alpha_i = \frac{e^2}{w_o^2}\left[\frac{1}{m} + \frac{1}{M}\right] \qquad \text{.....(6 A.39)}$$

Thus ionic polarisability is inversely proportional to the square of the natural frequency of the ionic molecule and directly proportional to it reduced mass. The ionic polarisability is also independent of temperature.

6A.12.5 Orientation polarisation (Dipolar polarisation)

Orientation polarisation occurs in dielectric material which possess molecules with permanent dipole moment (i.e. in polar molecules).

In the absence of an external electric field, because of random orientation of the dipoles due to thermal agitation the material has net zero dipole moment. But under the influence of an external applied electric field, each of the dipoles undergoes rotation so as to reorient along the direction of the field as shown in Fig.6 A.11. Thus the material itself develops electrical polarisation.

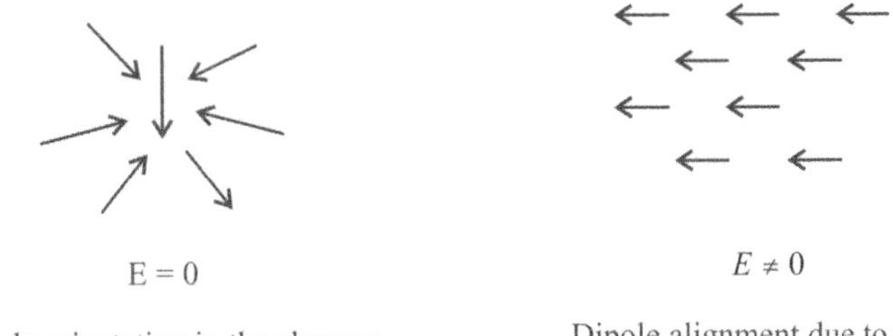

Fig.6 A.11

This polarisation occur in ferroelectric material such as $BaTiO_3$ and $PbTiO_3$ and produces a very high dielectric constant in these materials.

In the case of polar dielectrics, the orientation polarisability α_o is given by

$$\alpha_o = \frac{\mu^2}{3K_BT} \quad .....(6\ A.40)$$

Where K_B in the Boltzmann constant, T is the Temperature and μ is the permanent dipole moment. Thus, the orientational polarisabililty is inversely proportional to absolute temperature.

6A.13 Space Charge (Interfacial Polarisation)

The space charge polarisation occur in multiphase dielectric materials. When such materials are subjected to an external electric field, especially at high temperature the charges get accumulated at the interface or at the electrodes because of sudden change in conductivity as shown in Fig.6 A.12. Since, the accumulation of charges with opposite polarity occurs at opposite parts in the low resistive phase, it leads to the development of dipole moment within the low resistivity.

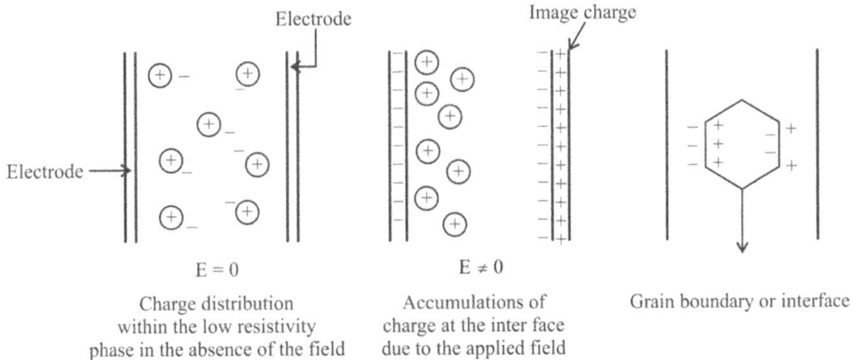

Fig.6 A.12

Grain boundaries often lead to interfacial polarisation as they can trap charges migrating under the influence of an applied field as shown in Fig. 6 A.12. This type of polarisation occur in same materials like $LiTaO_3$, Li NbO_3 and in certain glasses containing Li_2 O or Na_2O. This polarisation in not an important factor in most common dielectrics used in capacitor and other applications.

6A.14 Total Polarisability (α_T)

The total polarisability α_T of a material is thus given by the sum of the electric, ionic and orientational plarisation.

$$\alpha_T = \alpha_e + \alpha_i + \alpha_o \qquad \text{.....(6 A.41)}$$

Since, the space charge polarisability is very small when compared to other types of polarisabilities it is neglected.

Thus, the total polarisability of a material is given by

$$\alpha_T = 4\pi \in_o R^3 + \frac{e^2}{\omega_o^2}\left(\frac{1}{m} + \frac{1}{M}\right) + \frac{\mu^2}{3K_BT}$$

(Substituting for α_e, α_i andα_e)

Hence, the total polarisability ρ is given by

$$P = N\alpha_T E$$

$$P = NE\left[4\pi \in_o R^3 + \frac{e^2}{\omega_o^2}\left(\frac{1}{m} + \frac{1}{M}\right) + \frac{\mu^2}{3K_BT}\right] \qquad \text{.....(6 A.42)}$$

The above equation is known as Langevin-Debye equation.

6A.15 Internal Field (Local Field)

The electric field acting at an atom in a dielectric is known as the internal field or local field E_{int}. is different from the applied external field.

The internal field, or the local field, in the electric field acting at an atom of a solid or liquid dielectric subjected to an external field. It is the result of the applied field and the field due to all the surrounding dipoles.

6A.16 Calculation of the Internal Field (Einstein's model for one dimensional atomic array)

To calculate local field, we use Einstein's model for the determination of the local field in the case of one dimensional atomic array shown in Fig.6 A.13. Let X, A, B, C, D, E and

F are few equi-spaced atomic dipoles created by an external electric field $\vec{E}$ applied across a dielectric from left to right. Let equi-spaced atomic dipoles are separated by a distance'd' and P_i is the dipole moment induced in each atom due to the applied field E.

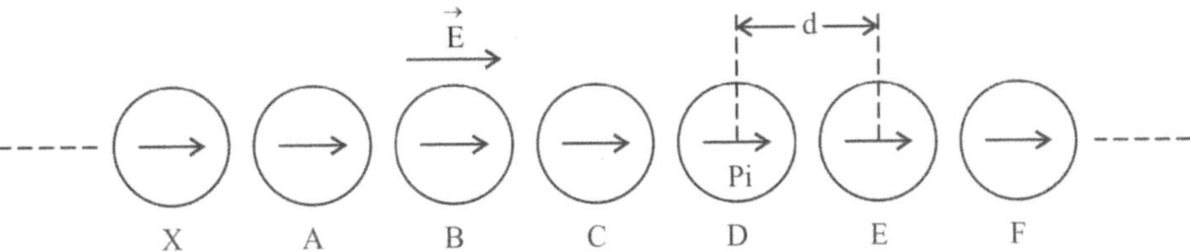

Fig.6 A.13

The internal field at the atom site C is the sum of the applied field and the fields produced by induced dipoles of all other atoms B, A........lying on the left of C and D, E......lying on the right of C.

The radial and transverse component of the electric field produced by the atomic dipole of induced dipole moment P at a point in terms of radial co-ordinates (r, θ) are

$$E_r = \frac{1}{4\pi \epsilon_o} \frac{2P \cos\theta}{r^3} \quad .....(6\text{ A.43})$$

$$E_\theta = \frac{1}{4\pi \epsilon_o} \frac{\sin\theta}{r^3} \quad .....(6\text{ A.44})$$

The field produced at the atom site C due to the atomic dipole at B is obtained by putting $r = d$ and $\theta = 0$ in equation 6 A.43

$$E_{CB} = \frac{2P_i}{4\pi \epsilon_o d^3} \quad .....(6\text{ A.45})$$

Similarly, the field produced at C due to the atomic dipole at D is obtained by putting $r = -d$ and $\theta = 180^\circ$ in equation 6 A.43

$$E_{CD} = -\frac{1}{4\pi \epsilon_o} \frac{2P_i}{(-d)^3} = \frac{2P_i}{4\pi \epsilon_o d^3} \quad .....(6\text{ A.46})$$

Therefore the induced electric field at C would be

$$E_{CB} + E_{CD} = \frac{P_i}{\pi \epsilon_o d^3} \quad ......(6\text{ A.47})$$

and similarly the combined induced electric field due to atomic dipoles A and E at C would be

$$E_{CA} + E_{CE} = \frac{2\,P_i}{4\pi\,\epsilon_o\,(2d)^3} - \frac{2\,P_i}{4\pi\,\epsilon_o\,(-2d)^3}$$

Since for E_{CA}, r = 2d and θ = 0; and for E_{CE}, r = – 2d and θ = 180°

$$\therefore E_{CA} + E_{CE} = \frac{P_i}{\pi\,\epsilon_o\,(2d)^3} \qquad \text{.....(6 A.48)}$$

Similarly the contribution of the combined field due to the atomic dipoles X and F at C would be

$$E_{CX} + E_{CF} = \frac{P_i}{\pi\,\epsilon_o\,(3d)^3} \qquad \text{.....(6 A.49)}$$

and so on

Thus, the internal field or local field at atom site C

$$E_i = E + \frac{P_i}{\pi\,\epsilon_o\,d^3} + \frac{P_i}{\pi\,\epsilon_o\,(2d)^3} + \frac{P_i}{\pi\,\epsilon_o\,(3d)^3}$$

Hence, the internal field or local field at atomic site due to all atoms of atomic array is

$$E_i = E + \frac{P_i}{\pi\,\epsilon_o\,d^3} + \frac{P_i}{\pi\,\epsilon_o\,(2d)^3} + \frac{P_i}{\pi\,\epsilon_o\,(3d)^3} + \ldots\ldots\ldots\ldots$$

$$E_i = E + \frac{P_i}{\pi\,\epsilon_o\,d^3}\left[1 + \frac{1}{2^3} + \frac{1}{3^3} + \ldots\ldots\ldots\ldots\right]$$

$$E_i = E + \frac{P_i}{\pi\,\epsilon_o\,d^3}\left[\sum_{i=1}^{\infty} \frac{1}{n^3}\right] \qquad \text{.....(6 A.50)}$$

where n is an integer

Since $\displaystyle\sum_{i=1}^{\infty} \frac{1}{n^3} \approx 1.2$

$$\therefore \quad E_i = E + \frac{1.2\,P_i}{\pi\,\epsilon_o\,d^3} \qquad \text{.....(6 A.51)}$$

The internal field calculation for three dimensional picture of the dielectric is quite complicated and depends upon the crystal structure.

6A.17 Calculation of the Internal Field (Local Field) : Lorentz method for Cubic structure

The following method suggested by Lorentz is used to establish the relation between the applied field E and the internal field E_{int}.

The given dielectric is placed in between a parallel plate capacitor and is uniformly polarised by applying an external field E.

To evaluate E_{int}, we must calculate the total field acting on an atom A (dipole) within the dielectric. For this to calculate, imagine a spherical cavity (Lorentz sphere) around the atom A (dipole) inside the dielectric as shown in Fig.6A.14. Further it is assumed that, the radius r of the cavity is large compared to the radius of the atom. The medium lying external to the cavity is treated as a continuous medium as far as the atom A is concerned.

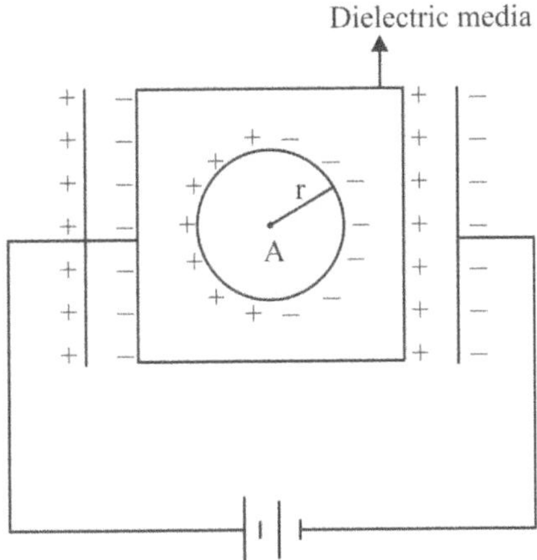

Fig.6 A.14 [Calculation of Internal Field for Cubic Structure]

The interaction of the atom A (dipole) with other dipoles within cavity is treated microscopically, since, the medium very close to the atom A (dipole) within the cavity is considered to be discrete.

The internal field or local field, acting on the central atom (dipole) is thus given by

$$E_{int} = E_1 + E_2 + E_3 + E_4 \quad \text{.....(6A.52)}$$

where

E_1 is the field due to the charges on the plates

E_2 is the field due to the polarising charges lying at the external surface of the dielectric medium. This is known as depolarisation field.

E_3 is the field due to the polarisation charges lying on the surface of the sphere which is known as the Lorentz field.

E_4 is the field due to other dipoles lying within the sphere.

Field E_1

When a dielectric medium is polarised due to an electric field E, the displacement vector D is given by

$$D = \epsilon_o E + P \quad .....(6\ A.53)$$

From field theory, the field at A due to the charge density on the plates is given by

$$D = \epsilon_o E_1 \quad .....(6\ A.54)$$

Therefore equating equation (6 A.53) and (6 A.54)

we get

$$\epsilon_o E_1 = \epsilon_o E + P$$

$$E_1 = E + \frac{P}{\epsilon_o} \quad .....(6\ A.55)$$

Field E_2

E_2 is the field due to the polarisation charges on the external surface of the dielectric. This field acts in a direction opposite to the external field. Therefore, it is given by

$$E_2 = -\frac{P}{\epsilon_o}$$

Field E_3

E_3 is the field due to polarisation charges on the surface of the cavity. This field value is calculated as follows.

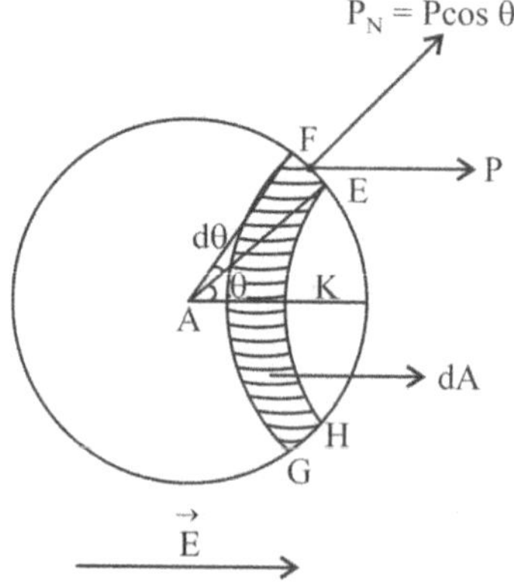

Fig.6 A.15

$$d\,A = 2\,\pi \times EK \times EF$$

Consider an enlarged view of the cavity as shown in Fig.6 A.15. In order to calculate this field acting on the atom A in the direction of the field consider an elemental ring E F G H perpendicular to the field direction. The area of the ring shaped element along the surface of the sphere in given by

Where EF = r d θ is the arc length. From the Fig.6 A.15; $\sin\theta = \frac{Ek}{r}$,

Therefore $\quad EK = r \sin\theta$

Therefore $\quad d\,A = 2\,\pi \times r \sin\theta \times r\,d\,\theta \qquad$ .....(6 A.57)

Hence $\quad d\,A = 2\,\pi\,r^2 \sin\theta\,d\,\theta \qquad$ .....(6 A.58)

From the definition of polarisation, defined as the induced surface charge per unit area i.e $P = \frac{q}{A}$, the charge q on dA is equal to the induced surface area normal component of polarisation and surface area.

i.e. $\quad q = p_N\,.\,d\,A \qquad$ .....(6 A.59)

The polarisation r is parallel to E. Its component normal to dA is $p_N = P\cos\theta$

$$\therefore\ q = P\cos\theta\,dA \qquad .....(6\ A.60)$$

Substituting for d A from equation (6,58)we get

$$q = P\cos\theta\,.\,2\pi\,r^2 \sin\theta\,d\,\theta \qquad .....(6\ A.61)$$

Thus, the electric field dE_3 at A due to the charge q is given by

$$dE_3 = \frac{q}{4\pi \in_o r^2} \qquad .....(6A.\ 62)$$

Substituting for q from equation (6 A.61) in (6 A.62) we get

$$dE_3 = \frac{P\cos\theta\,.\,2\pi\,r^2\,.\sin\theta\,d\theta}{4\pi\ \in_o\ r^2}$$

$$dE_3 = \frac{P\cos\theta\ .\sin\theta\,d\theta}{2\in_o} \qquad .....(6A.\ 63)$$

The above field is having two components, one component (cosθ) along the applied field and the other component (sin θ) perpendicular to the field E.

The field at A due to the perpendicular component is Zero. Since, the components are symmetrically distributed around the axis, they will cancel out each other

$$dE_3 = \frac{P \cos\theta \sin\theta \, d\theta}{2 \epsilon_o} \times \cos\theta$$

$$dE_3 = \frac{P \cos^2\theta \sin\theta \, d\theta}{2 \epsilon_o} \quad \text{.....(6 A.64)}$$

Therefore, the field at A due to the surface charge on the cavity is obtained by integrating over the whole surface of the sphere.

i.e

$$E_3 = \frac{P}{2 \epsilon_o} \int_o^{\pi} \cos^2\theta \sin\theta \, d\theta$$

$$E_3 = \frac{P}{2 \epsilon_o} \times \left(\frac{2}{3}\right) \qquad \text{since } \int_o^{\pi} \cos^2\theta \sin\theta \, d\theta = \frac{2}{3}$$

$$E_3 = \frac{2P}{6 \epsilon_o}$$

$$E_3 = \frac{P}{3 \epsilon_o} \quad \text{.....(6 A.65)}$$

Field E_4

This is the field due to the dipoles within the cavity which depends on the crystal structure.

This field E_4 is zero for a spherically symmetric system i.e. for a cubic structure. Hence, assuming the dielectric material to be of a cubic structure E_4 is taken to be zero in this case i.e for a cubic system $E_4 = 0$ .....(6 A.66)

Elements like, carbon, silicon, and germanium have cubic structure.

Thus the internal field is given by

$$E_{int} = E_1 + E_2 + E_3 + E_4$$

$$E_{int} = E + \frac{P}{\epsilon_o} - \frac{P}{\epsilon_o} + \frac{P}{3 \epsilon_o} + 0$$

$$E_{int} = E + \frac{P}{3 \epsilon_o} \quad \text{.....(6 A.67)}$$

The above equation is the Lorentz relation. Thus, it is seen that, the local field is larger than the macroscopic field E by an additional factor $\frac{P}{3\epsilon_o}$

6A.18 Claussius – Mossotti Equation

The polarisation for a single atom is given by

$$p = N\alpha_T\ E_{int} \qquad \text{.....(6 A.68)}$$

Where α_T is the total polarisability and E_{int} the internal field.

we know that

$$\alpha_T = \alpha_e + \alpha_i + \alpha_o \qquad \text{.....(6 A.69)}$$

For a cubic structure element, since there are no ions and permanent dipoles

$$\alpha_i = \alpha_o = 0$$

$$\therefore\ p = N\alpha_e\ E_{int} \qquad \text{.....(6 A.70)}$$

Substituting for E_{int}

$$p = N\alpha_e\left(E + \frac{p}{3\epsilon_o}\right)\quad \text{since } \alpha_T = \alpha_e \qquad \text{.....(6 A.71)}$$

we also know that

$$P = \epsilon_o\ E(\epsilon_r - 1) \qquad \text{.....(6 A.72)}$$

Substituting for P in equation (6 A.71) we get

$$N\ \alpha_e\left(E + \frac{P}{3\ \epsilon_o}\right) = \epsilon_o\ E(\epsilon_r - 1) \qquad \text{.....(6 A.73)}$$

$$N\ \alpha_e\left(E + \frac{\epsilon_o\ E}{3\ \epsilon_o}(\varepsilon_r - 1)\right) = \epsilon_o\ E(\epsilon_r - 1)$$

$$N\ \alpha_e E\left(1 + \frac{\epsilon_o}{3\ \epsilon_o}(\epsilon_r - 1)\right) = \epsilon_o\ E(\epsilon_r - 1)$$

$$N\ \alpha_e = \frac{\epsilon_o\ (\epsilon_r - 1)}{\dfrac{3\ \epsilon_o + \epsilon_o\ (\epsilon_r - 1)}{3\ \epsilon_o}} = \frac{3\epsilon_o^{\,2}\ (\epsilon_r - 1)}{\epsilon_o\ \ (3 + \epsilon_r - 1)}$$

$$N\alpha_e = \frac{3\in_o^2(\in_r - 1)}{\in_o(\in_r + 2)}$$

$$\therefore N\alpha_e = \frac{3\in_o(\in_r - 1)}{(\in_r + 2)} \quad .....(6\ A.74)$$

$$\frac{N\alpha_e}{3\in_o} = \frac{\in_r - 1}{\in_r + 2} \quad .....(6\ A.75)$$

The above relation is known as the claussius – Mossotti equation. This relates the microscopic quantity α_e to the macroscopic quantity $\in_r$

Claussius- Mossotti equation in terms of

(a) Susceptibility

$$\frac{N\alpha_e}{3\in_o} = \frac{\chi_e}{\chi_e + 3} \quad .....(6\ A.76)$$

Since $\in_r = 1 + \chi_e$

(b) Refractive index

At higher frequencies i.e. at optical frequencies, only electronic polarisation occurs. Hence $\in_r = n^2$ (assuming that the material has no magnetic properties), where n is the refractive index of the material

Therefore $$\frac{N\alpha_e}{3\in_o} = \frac{n^2 - 1}{n^2 + 2} \quad .....(6\ A.77)$$

6A.19 Dielectric Loss

If a dielectric material is subjected to an alternating voltage, the electrical energy is absorbed by the material and is dissipated in the form of heat.

This dissipation of heat energy is called as dielectric loss.

Whenever a dielectric material is subjected to an alternating voltage the molecules will be in a state of electrostatic stress due to polarisation. For each polarity reversal of the applied voltage, the stress is also reversed; and this inform makes the dipoles to switch their orientation. In order to do so, the dipoles are required to overcome a sort of internal friction which involves a loss of energy. Thus the energy loss due to friction is always dissipated as heat by the dielectric material to the surroundings. This is termed as dielectric loss.

Dielectric loss is a major problem involving heat generation and in dominating in high voltage application. When an a.c voltage is applied to a pure dielectric material like vacuum or purified gasses, there is no consumption of energy. As the charging current leads the applied voltage by an angle 90° as shown in Fig.6 A.16

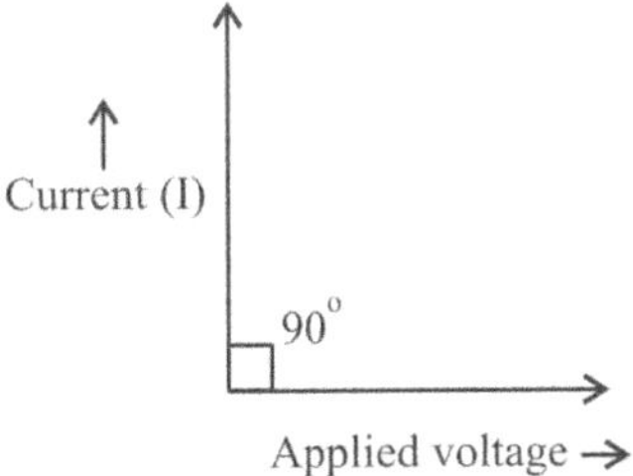

Fig.6 A.16

and the polarisation of the dielectric is in phase with the voltage. But when an a.c. voltage is applied to commercial dielectrics, the charging current does not lead the applied voltage by 90°; but is always less than 90° as shown in (Fig.6 A.17)

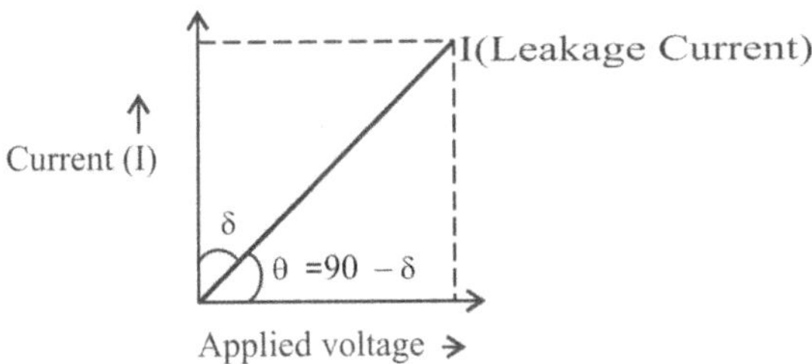

Fig. 6 A.17

The complementary angle $\delta = 90 - \theta$ is called the dielectric loss angle. This angle is a measure of the power or heat dissipated in each cycle. In most of the dielectrics this angle δ is negligible.

The dielectric power loss for a dielectric material used in a capacitor with capacitance C and applied voltage V is given by

$$P = V\, i \cos\theta \qquad \text{.....(6 A.78)}$$

Where V is the applied voltage in volts

i is the current through the capacitor in amp. The value of I is given by the relation

$$I = \frac{V}{X_C},$$

where X_C is the capacitative reactance given by

$$X_C = \frac{1}{2\pi\, f\, c}$$

$$\therefore\ p = V.\frac{V}{X_C}\ \cos\,(90 - \delta)$$

Since $\theta = 90 - \delta$ .....(6 A.79)

$$p = \frac{V^2}{X_C}\ \cos\,(90 - \delta) \qquad .....(6\ A.80)$$

Since $\cos\,(90 - \delta) = \sin\,\delta$

But $\sin\,\delta$ is negligiblly small in most of the dielectrics.

$\text{Sin}\,\delta\ = \tan\,\delta$

$$\therefore\ p = 2\pi\, V^2\ f\, c\, \tan\,\delta \qquad .....(6\ A.81)$$

Here, $\tan\,\delta$ is called the power factor. It is also referred to as tangent loss. When the factors like voltage, frequency and capacitance are constant, the dielectric loss is dependent only on δ. The dielectric loss should be low in a dielectric material.

6A.19.1 Factors Affecting Dielectric Loss

The dielectric loss increases with;

1. High frequency of the applied voltage
2. High applied voltage
3. High temperature and humidity

6A.20 Dielectric Breakdown

Whenever the voltage applied to a dielectric material exceeds a critical value, a large amount of current flows thorough it. As a result of this, the dielectric loses its insulating property. This phenomenon is known as dielectric breakdown.

Thus breakdown indicates the failure of the dielectric material.

The electric field at which this breakdown occur is called the dielectric strength. Also,

$$\text{Dielectric strenght} = \frac{\text{Breakdown voltage}}{\text{Thickness of the dielectric}}$$

It is expressed in KV/cm. Dielectric strength decreases with increase of temperature and humidity.

6A.20.1 Important Types of Breakdown

1. Intrinsic breakdown
2. Thermal breakdown
3. Electro chemical breakdown
4. Discharge Breakdown
5. Defect Breakdown

6A.21 Specific Applications of Dielectrics

The dielectric material based on their nature of state has a variety for applications in electrical engineering field.

The applications of different dielectric materials in capacitors with their functions are listed below. The major role of dielectric materials in capacitor is to store electrical energy. Based on the nature of the dielectric material used, the capacitors falls into different groups.

1. Capacitors with vacuum, air or other gases as dielectrics are used in radio frequency and low frequency measuring circuits.
2. Capacitors with mineral oil as dielectrics are used in high voltage applications, where a large value of capacitance is required.
3. Capacitors with a combination of solid and liquid dielectric are used in the applications where precision is not so important but a high value capacitance is required. In electric power distribution system for power correction, these type of capacitors are used. Examples; Glass, mica, oil Impregnated paper dielectric, mineral oil, caster oil etc.
4. Capacitors with only solid dielectric like Sodium, glass, titanium oxide etc, are used in laboratory as standard capacitor.

6A.22 Frequency Dependence of Dielectric Constant

The total polarization P, the total polarizability α, and the dielectric constant ϵ_r of a dielectric in an alternating field all depend on the case with which the dipoles can reverse alignment with each reversal of the field. Some polarizability mechanism do not permit sufficiently rapid reversal of the dipole alignment. In such a process the time required to reach the equilibrium orientation is called the relaxation times, and its reciprocal, the relaxation frequency. When the frequency of a particular field exceeds that of the relaxation frequency of a particular polarization process, the dipoles cannot reorient fast enough and operation of the process ceases. Since the relaxation frequencies of all four polarization process are different, it is possible to separate the different contributions experimentally as shown in Fig.6 A.18.

Four frequency ranges are found over which the separate polarization mechanism operate.

In such a highly covalent solid as diamond, only electronic polarization is present. The $\in_r$ can be measured optically from the index of refraction.

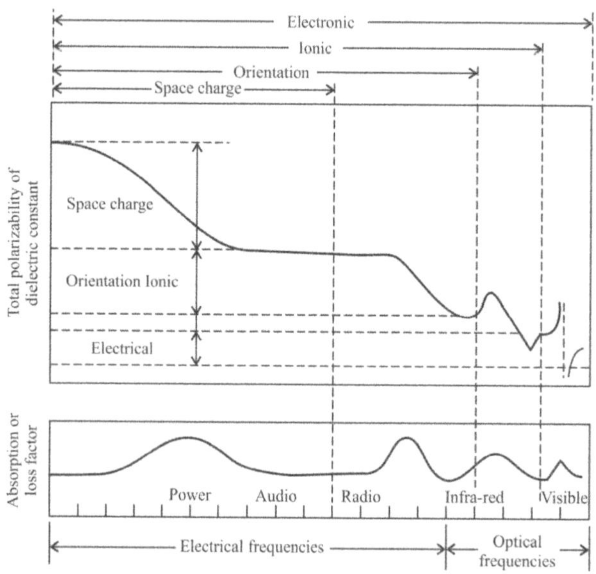

Fig. 6A.18 [Variation of the total polarizability and dielectric absorption as a function of frequency. Each contribution to the polarizability decays as its characteristic resonant frequency exceeds]

Ionic materials cease to contribute to the total polarization at infrared frequencies. With such materials it is possible to measure separate ionic and electronic contributions by making both optical and electrical measurement of $\in_r$. The orientation and space charge polarization only function at lower frequencies.

6A.23 Ferroelectricity

If the centre of gravity of the positive and negative charge in a substance do not coincide in the absence of the applied electric field, the substance had an electric dipole moment and is said to be spontaneously polarized. Such a substance is called ferroelectric. Ferroelectric materials contain small regions known as ferroelectric domains, which are polarized in different directions even in the absence of an electric field, when temperature exceeds a certain value called the ferroelectric curie temperature, the substance losses its ferroelectric properties. The phenomenon of ferroelectricity is exhibited by following materials: Rochelle salt, barium titanate ($BaTiO_3$), lead titanate ($PbTiO_3$), lead Zirconate titanate (PZT), lead lanthanum Zirconate titanate (PLZT), potassium niobate ($KNbO_3$), Potassium dihydrogen phosphate (KH_2PO_4) etc.

6A.23.1 Characteristics of Ferroelectrics

The ferroelectric material possesses following typical properties:

1. They lack centre of symmetry i.e. centre of positive charge does not coincide with centre of negative charge.
2. Their static dielectric constant changes with temperature (Fig 6A.19) according to the relation

$$\in \ = \frac{C}{T - Tc} \text{ provided } T > T_C$$

where C is a constant independent of temperature, usually called curie constant and T_C is ferroelectric curie temperature.

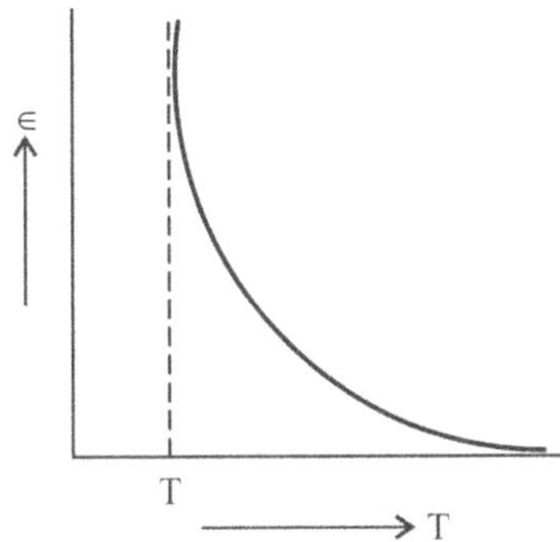

Fig. 6A.19 [Dielectric Constant against temperature]

3. They posses spontaneous electric polarization below T_C i.e polarization without the help of an external electric field. The variation of spontaneous polarization with temperature is shown in Fig.6 A.20.

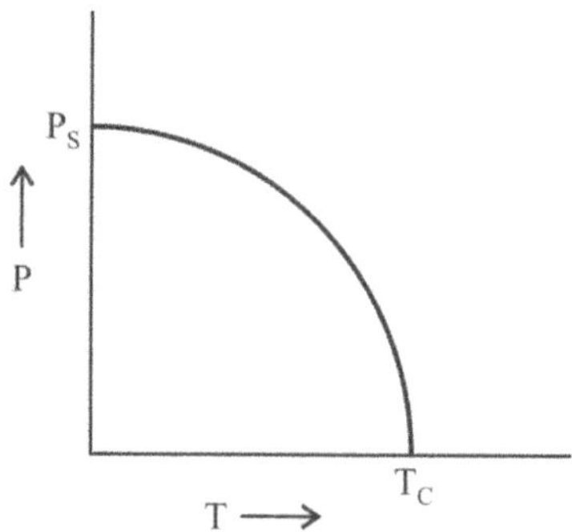

Fig.6 A.20 [Spontaneous Polarisation against temperature]

4. They exhibit hysteresis similar to B-H curve under the action of an alternating voltage. Such a curve is known as P-E curve and is shown in fig.6A.21.

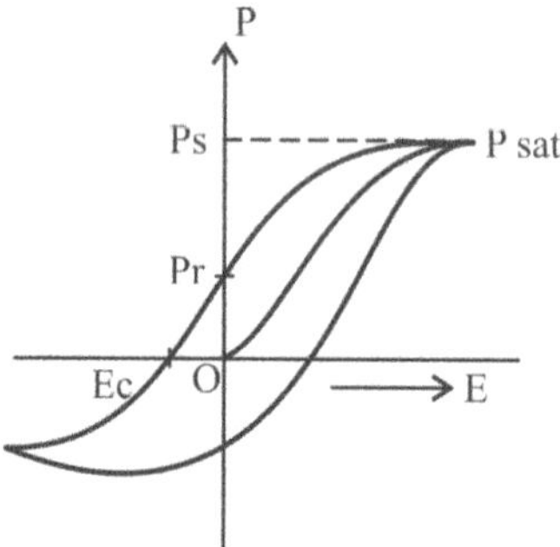

Fig.6 A.21 [P-E Hysteresis loop]

6A.23.2 Applications of Ferroelectric Material

Ferroelectric materials are widely used in

(a) Capacitors

(b) Ferroelectric memories

(c) Pyroelectric detectors

(d) Ferroelectric thin film waveguides

(e) Ferroelectric thin film memory displays

6A.24 Piezoelectricity

A piezoelectric material is a dielectric material in which an electric potential appears across a crystal if the dimensions of the crystals are changed by the application of mechanical force. Thus, potential is produced by the displacement of external charges.

The effect is reversible i.e if a varying potential is applied to the proper axis of the crystal, it will change the dimensions of the crystal thereby deforming it. This phenomenon is known as piezoelectricity materials showing this effect are Rochelle salt (K Na C_4 H_4 O_6 $4H_2$ O), tourmaline, lead zirconate (pb ZrO_3) lead titanate (pb TiO_3), barium titanate ($BaTiO_3$) etc.

6A.24.1 Applications of Piezoelectric Materials

Piezoelectric material have wide spread applications in:

(a) Crystal oscillators

(b) Delay lines

(c) Medical ultrasound application

(d) Gas ignitors

(e) Displacement tranducers

(f) Accelerometers

(g) Piezoelectric transformers

(h) Impact printer head

6A.25 Pyroelectricity

Pyroelectric effect is the change in spontaneous polarisation when the temperature of the specimen is changed. The pyroelectric coefficient λ is defined as the change in polarization per unit temperature change of the specimen.

$$\lambda = \frac{dP}{dt}$$

Change in polarisation results in change in external field and hence charge on the surface. As it is possible to detect a charge of 10^{-16} coulomb with suitable electrometer, temperature change as small as 10^{-6} °C can be measured using the pyroelectric effect.

6A.26 Applications of Pyroelectric Material

1. The pyroelectric materials such as $BaTiO_3$, $LiNibO_3$ etc are used to make very good infra-red detectors which can operate at room temperature.
2. Materials such as TGS, $NaNO_2$ and PZT ceramics are used in the construction of pyroelectric image tubes.

Solved Numerical Problems

1. Copper is a Fcc crystal with a lattice constant 3.61A° and atomic number 29. If the average displacement of electrons relative to the nucleus is 1×10^{-18} m, on applying an electric field calculate the electronic polarisation.

Sol. Given that

$$x = 1 \times 10^{-18}\,\text{m}$$

$$a = 3.61 \times 10^{-10}\,\text{m}$$

$$Z = 29$$

$$P = ?$$

For Fcc crystal, number of atoms per unit cell is 4

$$\therefore \text{Number of electrons per m}^3 = \frac{(\text{number of atoms/ unit cell})(\text{no.of electrons / atom})}{a^3}$$

$$\therefore \text{Number of electrons per m}^3 = \frac{4\times29}{(3.61\times10^{-10})^3}$$

$$= 2.465 \times 10^{30}\ \text{electrons/m}^3$$

Since the electron density in copper is equal to the number of atoms per unit volume $N = 2.465 \times 10^{30}$ atom/m^3

$$P = N\,\mu$$

Since $$\mu = e \times x$$

$$\therefore P = N\,e\,x$$

$$P = 2.465 \times 10^{30} \times 16 \times 10^{-19} \times 1 \times 10^{-18}$$

$$P = 3.944 \times 10^{-7}\ \text{c/m}^2$$

2. If a NaC*l* crystal is subjected to an electric field of 1000 V/m and the resulting polarisation is 4.3×10^{-8} c/m^2. Calculate the relative permittivity of NaC*l* ?

Sol. We know that

$$P = \epsilon_o\,(\epsilon_r - 1)E$$

$$\therefore \frac{P}{\epsilon_o\,E} = \epsilon_r - 1$$

$$\in_r = 1 + \frac{p}{\in_o E}$$

Given $P = 4.33 \times 10^{-8}$ c/m^2

$E = 1000$ V/m

$$\in_r = 1 + \frac{4.3 \times 10^{-8}}{8.86 \times 10^{-12} \times 1000}$$

$$\in_r = 5.86$$

3. Calculate the polarsiability and relative permittivity in hydrogen gas with a density of 9.8×10^{26} atoms/m^3. Given the radius of the hydrogen atom to be 0.50 A°.

Sol. Given that

$N = 9.8 \times 10^{26}$ atoms/m^3

$R = 0.50 \times 10^{-10}$ m

$\alpha_e = ?$

$\in_r = ?$

We know that

$$\alpha_e = 4\pi \in_o R^3$$

$$\alpha_e = 4 \times 3.14 \times 8.85 \times 10^{-12} \times \left(0.50 \times 10^{-10}\right)^3$$

$$\alpha_e = 1.3890 \times 10^{-41} \text{ Fm}^2$$

We know that

$$\alpha_e = \frac{\in_o \times (\in_r - 1)}{N}$$

$$N\alpha_e = \in_o (\in_r - 1)$$

$$\in_r = 1 + \frac{N \alpha_e}{\in_o}$$

$$\in_r = 1 + \frac{9.8 \times 10^{26} \times 1.389 \times 10^{-41}}{8.85 \times 10^{-12}}$$

$$\in_r = 1.0015$$

4. Find polrasiation (P) in a dielectric material with dielectric constant, $\epsilon_r = 2.8$, when $D = 3 \times 10^{-8}$ c/m^2

Sol. We know that

$$\vec{P} = (\epsilon_r - 1)\epsilon_o \vec{E}$$

and

$$\vec{D} = \epsilon E = \epsilon_o \epsilon_r \vec{E}$$

$$\therefore P = \frac{(\epsilon_r - 1) D}{\epsilon_r}$$

Give that $\epsilon_r = 2.8$ and $D = 3 \times 10^{-8}$ c/m^2

$$P = \frac{(2.8 - 1) \times 3 \times 10^{-8}}{2.8}$$

$$P = 1.93 \times 10^{-8} \text{ c/m}^2$$

5. The electronic polarisability of argon atom is 1.75×10^{-40} Fm^2. What is the static dielectric constant of solid Argon, of its density is 1.8×10^3 Kg/m^3 (given atomic weight of Argon = 39.95 and $N_A = 6.025 \times 10^{26}$/K Mole)

Sol. Given that

$$\alpha_e = 1.75 \times 10^{-40} \text{ Fm}^2$$

$$\rho = 1.8 \times 10^3 \text{ Kg/m}^3$$

$$\text{At.wt} = 39.95$$

$$N_A = 6.025 \times 10^{26}\text{/K mole}$$

and $\epsilon_r = ?$

Number of atoms/ unit volume

$$N = \frac{N_A \rho}{\text{At. wt}} = \frac{6.025 \times 10^{26} \times 1.8 \times 10^3}{39.95}$$

$$\therefore \quad N = 2.7146 \times 10^{28} \text{ atoms/m}^3$$

we know that

$$\epsilon_r = \frac{N \alpha_e}{\epsilon_o} + 1$$

$$\epsilon_r = \frac{N\,\alpha_e}{\epsilon_o} + 1$$

$$\epsilon_r = \frac{2.7146 \times 10^{28} \times 1.75 \times 10^{-40}}{8.85 \times 10^{-12}} + 1$$

$$\epsilon_r = 0.53679 + 1$$

$$\epsilon_r = 1.53679$$

6. The electrical susceptibility of a material is 35.4×10^{-12} $C^2/N.m^2$. What are the values of dielectric constant and the permittivity of the material ?

Sol. We know that the dielectric constant ϵ_r in related to the electrical susephibility χ_e as

$$\epsilon_r = 1 + \frac{\chi_e}{\epsilon_o}$$

$$\epsilon_r = 1 + \frac{35.4 \times 10^{-12}}{8.85 \times 10^{-12}}$$

or $$\epsilon_r = 1 + 4 = 5$$

The permittivity, ϵ is given by

$$\frac{\epsilon}{\epsilon_o} = \epsilon_r$$

or $$\epsilon = \epsilon_o \epsilon_r$$

$$\epsilon = 5 \times 8.85 \times 10^{-12}\ c^2 / N - m^2$$

$$\epsilon = 44.25 \times 10^{-12}\ c^2 / N - m^2$$

7. Calculate the electric polarizability of an argon atom, given $\epsilon_r = 1.0024$ at NTP and $N = 2.7 \times 10^{25}$ atom / m^3

Sol. We know that

$$P = \epsilon_o (\epsilon_r - 1) E$$

and $$P = \alpha_e N\,E$$

$$\therefore\ \alpha_e N\,E = \epsilon_o (\epsilon_r - 1)\,E$$

$$\alpha_e = \frac{\epsilon_o (\epsilon_r - 1)}{N}$$

Given

$$\epsilon_r = 1.0024$$

$$\epsilon_o = 8.85 \times 10^{-12}\ C^2 / N - m^2$$

$$N = 2.7 \times 10^{25}\ \text{atoms/m}^3$$

$$\alpha_e = \frac{8.85 \times 10^{-12}(1.0024 - 1)}{2.7 \times 10^{25}}$$

$$\alpha_e = 7.9 \times 10^{40}\ Fm^2$$

8. Consider an electron of charge –e moving in a circular orbit radius a_o about a charge +e in a field directed at right angles to the plane of the orbit. Show that the polarizability, α is approximately $4\pi \epsilon_o a_o^3$

Sol. The electric field responsible for separating positive and negative charges is

$$E_{int} = \frac{P}{3\epsilon_o}$$

where P is the polarisation vector or density of polarisation.

The dipole moment of the polarized system will be

$$\mu = \frac{4}{3}\pi a_o^3 P$$

We know that

$\mu = \alpha E_{int}$, where α is the polarizability

$$\alpha = \frac{\mu}{E_{int}}$$

$$\alpha = \frac{\mu}{P/3\epsilon_o} = \frac{\mu\, 3\epsilon_o}{P}$$

$$\alpha = \frac{\frac{4}{3}\pi a_o^3 P . 3\epsilon_o}{P}$$

$$\alpha = 4\pi \epsilon_o a_o^3$$

9. The atomic weight and density of sulphur are 32 and 2.08 gm $(cm)^{-3}$ respectively. The electronic polarisability of the atom is 3.28×10^{-40} F.m^2. If sulphur solid has solid symmetry, what will be its relative dielectric constant?

Sol. According to Claussius-Mossotti

$$\frac{\epsilon_r - 1}{\epsilon_r + 2} = \frac{N\,\alpha_e}{3\,\epsilon_o}$$

If N_A is the Avagadro's number, ρ the density and M the molecular weight, then

$$N = \frac{N_A \rho}{M}$$

where $N_A = 6.023 \times 10^{26}$/K mole

$$\frac{\epsilon_r - 1}{\epsilon_r + 2} = \frac{N_A\,\alpha_e \rho}{3M\,\epsilon_o}$$

$$\frac{\epsilon_r - 1}{\epsilon_r + 2} = \frac{6.023 \times 10^{26} \times \left(2.08 \times 10^3\right) \times 3.28 \times 10^{-40}}{3 \times 32 \times 8.85 \times 10^{-12}}$$

$$\frac{\epsilon_r - 1}{\epsilon_r + 2} = 0.483$$

$$\epsilon_r - 1 = 0.483\,(\epsilon_r + 2)$$

$$\epsilon_r - 1 = 0.483\,\epsilon_r + 2 \times 0.483$$

$$\epsilon_r - 0.483\,\epsilon_r = 1 + 0.966$$

$$\epsilon_r = \frac{1.966}{0.517}$$

$$\epsilon_r = 3.8$$

10. The following data refers to a dielectric material $\epsilon_r = 4.94$ and $n^2 = 2.69$ where n is the index of refraction. Calculate the ratio between electronic and ionic polarisability of this material?

Sol. Given that

$$\epsilon_r = 4.94$$

$$N^2 = 2.69$$

$$\frac{\alpha_e}{\alpha_i} = ?$$

The Claussius-Mossotti equation is

$$\frac{\epsilon_r - 1}{\epsilon_i + 2} = \frac{N\alpha}{3\epsilon_o} \qquad \text{......(1)}$$

At optical frequency $\epsilon_r = n^2$ and at this frequency only electronic polarisation occur. Hence, the claussius-Mossott equation can be written as

$$\frac{N\alpha_e}{3\epsilon_o} = \frac{n^2 - 1}{n^2 + 2} \qquad \text{.....(2)}$$

We know that

$$\alpha_T = \alpha_e + \alpha_i + \alpha_o$$

Neglecting α_o, we get

$$\alpha_T = \alpha_e + \alpha_i$$

$$\therefore \quad \frac{N\alpha_T}{3\epsilon_o} = \frac{\epsilon_r - 1}{\epsilon_r + 2}$$

or

$$\frac{N(\alpha_e + \alpha_i)}{3\epsilon_o} = \frac{\epsilon_r - 1}{\epsilon_r + 2} \qquad \text{.....(3)}$$

Dividing equation (3) by equation (2), we get

$$\frac{\alpha_e + \alpha_i}{\alpha_e} = \frac{(\epsilon_r - 1)(n^2 + 2)}{(\epsilon_r + 2)(n^2 - 1)}$$

$$1 + \frac{\alpha_i}{\alpha_e} = \frac{(4.94 - 1)(2.69 + 2)}{(4.94 + 2)(2.69 - 1)}$$

$$1 + \frac{\alpha_i}{\alpha_e} = \frac{(3.94)(4.69)}{(6.94)(1.69)}$$

$$\frac{\alpha_i}{\alpha_e} = 1.5755 - 1$$

$$\frac{\alpha_i}{\alpha_e} = 0.5755$$

Hence $\frac{\alpha_e}{\alpha_i} = 1.7376$

11. A parallel plate capacitor of area 6.5 cm^2 and a plate separation d of 4mm has a charge of 200 p.c calculate the potential difference between the plates if a dielectric of $\epsilon_r = 3.5$ is introduced between the plates.

Sol. Given that

Area of plate of capacitor = 6.50×10^{-4} m^2

Distance between the plates d = 4×10^{-3} m

The capacitance of a capacitor is given by

$$C = \frac{\epsilon_o \epsilon_r A}{d} = \frac{Q}{V}$$

$$V = \frac{Qd}{\epsilon_o \epsilon_r A}$$

$$V = \frac{200 \times 10^{-12} \times 4 \times 10^{-3}}{\left(8.86 \times 10^{-12}\right) \times \left(3.5 \times 6.5 \times 10^{-4}\right)}$$

$$V = 39.8 \text{ Volt}$$

Conceptual Questions and Answers

1. **What are dielectrics?**

Ans. Dielectrics are the materials used for charge storage, by polarisation of the molecules. They are also called as active dielectrics. The dielectric material increases the capacitance or charge storage ability of a capacitor.

2. **Mention few properties of dielectric materials.**

Ans. The properties of dielectric materials are as follows:

1. They are non-metallic materials
2. They have high specific resistance
3. They have negative temperature co-efficient of resistance

3. **Mention the types of polarisation mechanism in a dielectric material?**

Ans. There are four different types of polarisation mechanisms in the dielectric material

(i) Electronic polarisation (ii) Ionic polarisation

(iii) Orientation polarisation (iv) Space charge polarisation

4. Define local field?

Ans. It is electric field acting at an atom of a solid or liquid dielectric when subjected to an external electric field. It is the resultant of the applied field and the field due to the surrounding dipoles.

5. What is claussius-Mossotti equation? Give the expression?

Ans. The equation which relates the microscopic quantity α_e to the macroscopic quantity ϵ_r is called the Claussius-Mossotti equation.

Its expression is

$$\frac{N\alpha_e}{3\epsilon_o} = \frac{\epsilon_r - 1}{\epsilon_r + 2}$$

6. What is relaxation time? How does it vary for different types of polarisation?

Ans. The time required by dipoles to reorient in response to an electric field known as relaxation time. The relaxation time varies for each of the various contributions to the polarization. For electronic polarization response is fast and relaxation time is small, for Ionic polarisation response is slower, for dipolar polarisation response is still slower and finally for space charge polarisation response is quite slow and relaxation time is large.

7. Explain dielectric loss?

Ans. It is the loss of electrical energy in the form of heat by a dielectric material during polarisation process when subjected to an ac voltage.

8. Give the factors affecting dielectric loss?

Ans. The factor affecting dielectric loss are high frequency of the applied voltage, high applied voltage, increase of temperature and humidity.

9. Define power factor in a dielectric material?

Ans. Power factor is a measure of the power loss in a dielectric material occurring during polarisation process. It should be low.

10. Define dielectric loss angle?

An. The complementary angle $\delta = 90 - \theta$ is called the dielectric loss angle. This angle is a measure of the power of heat dissipated in each cycle when a dielectric material is subjected to ac voltage.

11. Mention the name of different dielectric breakdown mechanism?

Ans. The different dielectric breakdown mechanism is as follows:

1. Intrinsic breakdown
2. Thermal breakdown
3. Electro chemical breakdown
4. Discharge breakdown
5. Defect breakdown

12. What is electrostriction?

Ans. When any material undergoes polarisation, its ions and electronic clouds are displaced, causing the development of a mechanical strain in the material. This effect is seen in all materials subjected to an electric field and is known as electrostriction.

13. What are piezoelectric materials?

Ans. Materials that develops voltage upon the application of stress and develop strain on an application of electric field are known as piezoelectric materials.

14. What is piezoelectricity?

Ans. The phenomenon of conversion of mechanical (strain) energy in the electrical energy and vice-versa is known as piezoelectricity.

15. What are ferroelectric materials?

Ans. A ferroelectric material develops a spontaneous polarization in response to an external field. The polarisation does not go away when the external field is removed. The direction of the polarisation is reversible.

Objective Questions

1. The correct relation for a dielectric medium is

(a) $D = \epsilon_0 (\epsilon_r - 1)E$ (b) $D = \epsilon_0 E + P$

(c) $E = \epsilon D$ (d) $D = \epsilon_0 P$

2. When a monoatomic gas atom is placed in a uniform electric field E, the displacement of nucleus is proportional to

(a) Independent of E (b) E^2

(c) E^3 (d) E

3. The relation between dipole moment, poarsiability and electric field is

(a) $\mu = \alpha E^2$ (b) $E = \mu \alpha$

(c) $\mu = \alpha / E$ (d) $\mu = \alpha E$

4. Dielectrcis are

(a) Hole conductors (b) Electric conductors

(c) Electric insulators (d) materials that works under low voltages

5. A dielectric material is one which has

(a) Free charges (b) No molecules

(c) No free charges or electrons (d) None of these

6. Electric field E, the polarisation ρ produced is expressed as

(a) $P = \in_o E$ (b) $E = \in_o (\in_e - 1) P$

(c) $P = \in_r (\in_o - 1) E$ (d) $P = \in_o (\in_r - 1) E$

7. For a given dielectric, as the temperature increases, the ionic polarisability?

(a) First increases then decreases

(b) Decreases

(c) Increases

(d) Remains unchanged

8. If $\in_r$ is the relative permittivity of an isotropic medium, then the electric susceptibility, χ_e is

(a) $\in_o (\in_r - 1) E$ (b) $\in_r / \in_o$

(c) $\in_r \in_o$ (d) $\in_o (\in_r - 1)$

9. When an electric field is applied to a capacitor containing dielectric medium

(a) the positive and negative charges are pushed to one side

(b) the electric dipoles becomes randomly oriented

(c) all the dipoles will disappear

(d) the positive and negative charges are pushed in opposite direction.

10. The orientation polarizability per molecule in a polyatomic gas is proportional to

(a) $\frac{1}{T^2}$ (b) T^2

(c) $\frac{1}{T}$ (d) T

11. Choose the correct relation for a dielectric

(a) $D = P \times E$ (b) $D = \epsilon_o E + P$

(c) $D = \epsilon_o E$ (d) $D = \epsilon_o P$

12. The clausius-Mossotti relation is

(a) $\frac{\epsilon_r - 1}{\epsilon_r + 2} = \frac{N\alpha}{3\epsilon_o}$ (b) $\frac{\epsilon_r + 1}{\epsilon_r - 1} = \frac{3N\alpha}{\epsilon_o}$

(c) $\frac{\epsilon_r + 1}{\epsilon_r - 2} = \frac{N\alpha}{3\epsilon_o}$ (d) $\frac{\epsilon_r - 1}{\epsilon_r + 1} = \frac{N\alpha}{3\epsilon_o}$

13. The temperature dependent polarization is

(a) Ionic polarisation (b) Deformation polarisation

(c) Electronic polarisation (d) Orientational Polarisation

14. When an electric field is applied to capacitor filled with a dielectric medium, then

(a) No electric field is induced within the dielectric

(b) the field within the dielectric is weakened

(c) the electric field within the dielectric is increased by a factor K

(d) the induced electric field is greater than the applied field

15. For a given dielectric, the electronic polarisability α_e

(a) Independent of temperature (b) Decreases with temperature

(c) Increases with temperature (d) None of the above

16. The electronic polarisability α_e of mono atomic gas atom is

(a) $\frac{4}{3}\pi \epsilon_o R^3$ (b) $4\pi \epsilon_o R^3$

(c) $4\pi \epsilon_o$ (d) $4\pi \epsilon_o R$

17. If α_d is the deformation polarisability then total polarisability α of polar dielectric is expressed as

(a) $\alpha_d + \frac{P_o^2}{3\,KT^2}$ (b) $\alpha_d + \frac{P_o^2}{3\,KT}$

(c) $\alpha_d + \frac{P_o}{3\,KT}$ (d) $\alpha_d + \frac{P_o}{3\,KT^2}$

18. In a dielectric, the polarisation is

(a) Square function of applied field

(b) linear function of applied field

(c) exponential function

(d) None of the above

19. The spontaneous polarisation is related with

(a) atoms (b) non-polar molecules

(c) permanent dipole moment (d) free electrons

20. The Lorentz field equation for local field E_L is

(a) $E + \frac{P}{\epsilon_o}$ (b) $E + \frac{3\epsilon_o E}{P}$

(c) $E + \frac{P}{3\epsilon_o}$ (d) $\frac{P}{3\epsilon_o}$

21. Piezoelectric effect is the production of electricity by

(a) Pressure (b) Varying field

(c) Chemical effect (d) Temperature

22. The loss factor in a dielectric depends on

(a) its conductivity only

(b) the initial polarisation of the dielectric

(c) The Thickness of the dielectric

(d) the real part of the dielectric constant

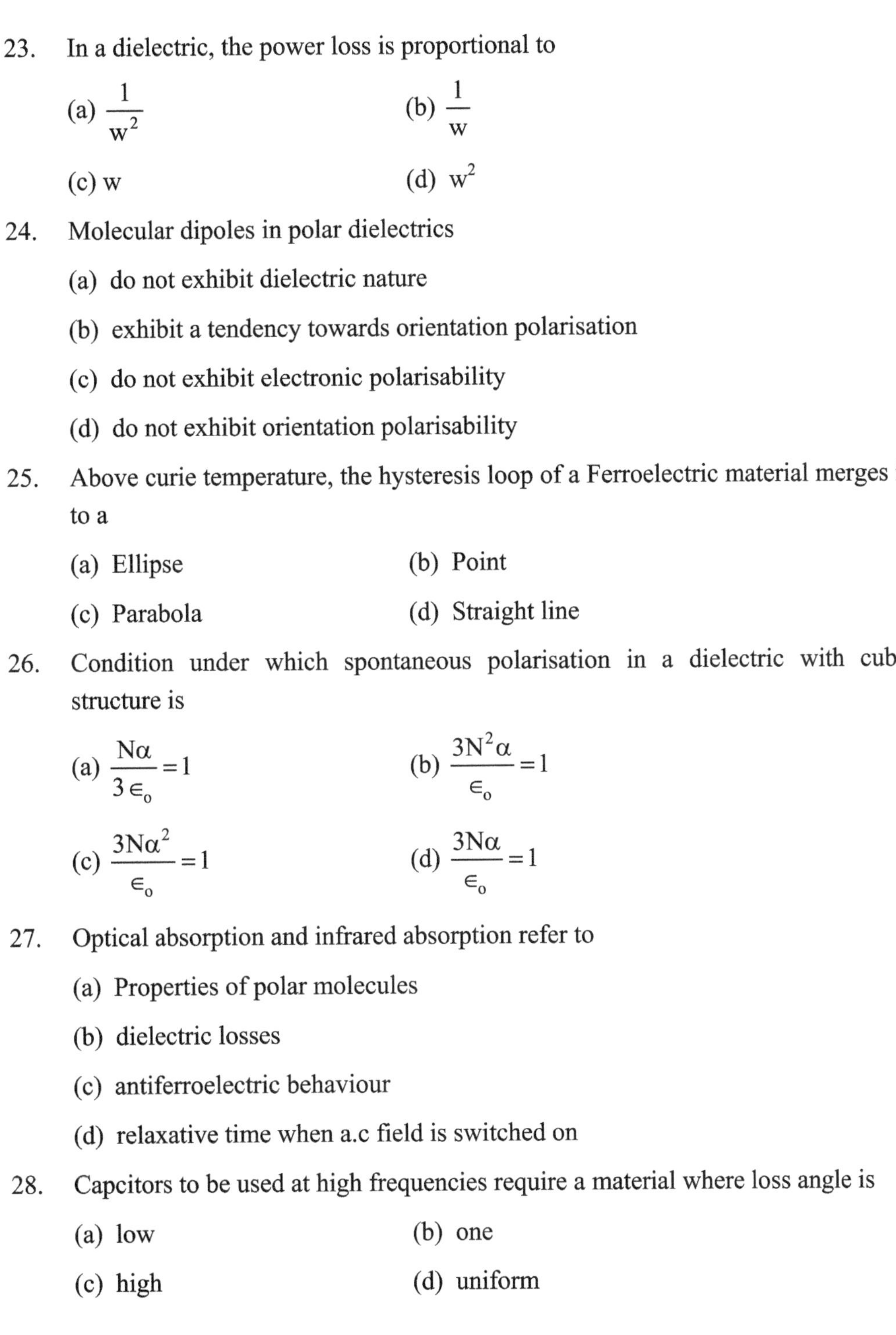

23. In a dielectric, the power loss is proportional to

(a) $\frac{1}{w^2}$ (b) $\frac{1}{w}$

(c) w (d) w^2

24. Molecular dipoles in polar dielectrics

(a) do not exhibit dielectric nature

(b) exhibit a tendency towards orientation polarisation

(c) do not exhibit electronic polarisability

(d) do not exhibit orientation polarisability

25. Above curie temperature, the hysteresis loop of a Ferroelectric material merges in to a

(a) Ellipse (b) Point

(c) Parabola (d) Straight line

26. Condition under which spontaneous polarisation in a dielectric with cubic structure is

(a) $\frac{N\alpha}{3\epsilon_o}=1$ (b) $\frac{3N^2\alpha}{\epsilon_o}=1$

(c) $\frac{3N\alpha^2}{\epsilon_o}=1$ (d) $\frac{3N\alpha}{\epsilon_o}=1$

27. Optical absorption and infrared absorption refer to

(a) Properties of polar molecules

(b) dielectric losses

(c) antiferroelectric behaviour

(d) relaxative time when a.c field is switched on

28. Capcitors to be used at high frequencies require a material where loss angle is

(a) low (b) one

(c) high (d) uniform

29. In an isotropic dielectric medium
    (a) D and E are perpendicular to each other
    (b) D and E are inclined at 45° to each other
    (c) D and P are perpendicular to each other
    (d) D and E have the same direction.

Theoretical Questions

1. Define dipole?
2. Define dipole moment?
3. Explain the phenomenon of polarisation in dielectrics?
4. What are polar dielectrics? Give examples.
5. What are non-polar dielectrics? Give examples.
6. Define dielectric constant or relative permittivity?
7. Define polarization vector?
8. Define electric susceptibility.
9. Give the relation between susceptibility and the dielectric constant.
10. Discuss in detail the various polarisation mechanism in a dielectric material.
11. What are the effects of temperature on the polarisation mechanism in a dielectric material.
12. Derive an expression for the internal field in a dielectric material.
13. Derive an expression for Claussius-Mossotti relation.
14. Define loss tangent or tangent loss in a dielectric material.
15. What is meant by local field? How it is calculated for a cubic structure? Deduce Clausius-Mossotti equation?
16. What is dielectric break down.
17. Derive an expression for dielectric loss in a dielectric material.
18. Explain the phenomenon of spontaneous polarisation in ferroelectric materials.
19. Write a short note on the application of dielectric materials.
20. In what way do ferroelectric differ from ordinary dielectric.

Numerical Problems

1. A parallel plate capacitor is charged by connecting it to a 60 volt battery. The battery is then disconnected and a dielectric liquid is filled between the plates to fill the air gap? The potential difference is now reduced to 30V? What is the dielectric constant of the liquid?

[2]

2. A parallel plate capcatior consists of 2 plates each of area 4×10^{-4} m^2.They are separated by a distance 1.5×10^{-3} m and filled with a dielectric of relative permittivity 6. Calculate the charge on the capacitor if it is connected to a 100 volt d.c battery.

[1.41×10^{-9} Cloumb]

3. The dielectric constant of He gas at N.T.P is 1.0000684. Calculate the electronic polorizability of the atom if the gas contains 2.7×10^{25} atoms per m^3 and hence evaluate the radius of the helium atoms?

$$\left[\alpha_e = 2.242\times10^{-41}\,Fm^2 \;\; R = 0.59\ A^{o}\right]$$

4. Silicon has a relative permittivity of 11.7 at frequencies high enough to ignore all but its electric (optical) polarisability. It contains 4.82×10^{28} atom/m^3. Calculate the dipole moment of each atom in a field of 10^4 V/m and also find the effective distance at this field strength between the centre of the electron cloud in each atom and the nucleus?

[8.77×10^{-8} A°]

5. An electric field of 6×10^4 Volt/meter is applied on a sample of Neon gas at N.T.P. Calculate the dipole moment induced if the dielectric constant of Neon at N.T.P is 1.000134. Also find out the atomic polarizability of Neon?

[$m = 2.65 \times 10^{-36}$ cm; $\alpha = 4.9 \times 10^{-23}$ m^3]

6. Determine the percentage of ionic polarizability in the sodium chloride crystal has the optical index of refraction and the static dielectric constant 1.5 and 5.6 respectively?

[51.4%]

7. The value of $\in_r$ for glass is 6.75 at frequencies 10^9 Hz. What percentage may be attributed to ionic polarisability? Given, refractive index of glass as 1.5?

[78.3%]

Answers to Objective Questions

1. (b)	2. (d)	3. (d)	4. (c)	5. (c)	6. (d)
7. (d)	8. (d)	9. (d)	10. (c)	11. (b)	12. (a)
13. (d)	14. (b)	15. (a)	16. (b)	17. (b)	18. (b)
19. (c)	20. (c)	21. (a)	22. (b)	23. (c)	24. (b)
25. (d)	26. (a)	27. (d)	28. (a)	29. (d)	

CHAPTER 6B

Magnetic Properties of Materials

I happen to have discovered a direct relation between magnetism and light, also electricity and light, and the field it opens is so large and I think rich.
-Michael Farady

6B.1 Introduction

Magnetic materials play a prominent role in modern technology. These materials are widely used in industrial electronics, entertainment electronics and computer industry. The traditional methods of data storage and retrieval are rapidly being replaced by magnetic storage. Based on the response of materials to external magnetic field, they are divided into different groups depending on the alignment of magnetic moments within the materials, such as diamagnetic, paramagnetic, ferromagnetic, anti ferromagnetic and ferrimagnetic. The magnetic materials are classified as hard and soft materials.

6B.2 Magnetic Field

The space surrounding a magnet where its magnetic influence is experienced is known as the magnetic field.

Magnetic induction $B = \frac{F}{m}$ is the force per unit north pole placed at that point.

where m is the pole strength

S.I unit of magnetic induction is teasla (T) or wb/m^2

C.G.S unit of B is Gauss (G)

$$1\ G = 10^{-4}\ T$$

The magnetic field strength is denoted by H. The unit of H are amper-turns per meter in S.I.system and oersted (O_e) in gaussion system.

6B.3 Magnetic Dipole

Magnetic dipole consists of two equal and opposite magnetic poles separated by a small distance. When it is placed in an external magnetic field, it experiences a torque. Due to this torque, dipole sets itself in a particular direction.

A compass needle, bar magnet, current carrying loop, coil and solenoid experience a torque in an external magnetic field and hence, all are magnetic dipoles. Every magnetic dipole has a magnetic dipole moment denoted by $\vec{\mu}_m$. If a magnetic dipole of magnetic dipole moment $\vec{\mu}_m$ is kept in an external magnetic field $\vec{B}$, the torque acting on the dipole is given by

$$\vec{\tau} = \vec{\mu}_m \times \vec{B} \qquad .....(6B.1)$$

The magnitude of this torque is

$$\vec{\tau} = \vec{\mu}_m \vec{B} \sin\theta \qquad .....(6B.2)$$

where θ is the angle between $\vec{\mu}_m$ and $\vec{B}$. The direction of magnetic moment is taken as that from south pole to north pole.

When a coil of N turns, each of area of cross section A is placed in a magnetic field $\vec{B}$ in such a way that the plane of the coil makes an angle θ with the field, a torque acts on the coil whose magnitude is given by

$$\tau = \text{Ni A B Sin}\ \theta$$

$$\tau = \text{NiA}\ (\text{B}\sin\theta) \qquad .....(6B.3)$$

Comparing equation (6 B.2) and (6 B.3), we have the magnetic dipole moment of a current carrying coil

$$\mu_m = \text{NiA} \qquad .....(6B.4)$$

Thus, the S.I unit of magnetic dipole moment = Amper – m^2

If τ is the torque experienced by a magnetic dipole in an external magnetic field $\vec{B}$, then the magnetic moment $\vec{\mu}_m$ of the dipole is defined as

$$\vec{\tau} = \vec{\mu}_m \times \vec{B} \qquad .....(6B.5)$$

Unit of $\vec{\mu}_m$ = ampere – metre ^{2}

6B.4 Magnetic Intensity $(\vec{H})$

A material medium is magnetized when placed in a magnetic field. The degree to which the magnetic field can magnetize a material medium, may be represented by a magnetic vector $\vec{H}$, called the magnetic intensity of the field

$$\text{Unit of } (\vec{H}) = \frac{\text{ampere} - \text{turn}}{\text{metre}}$$

6B.5 Magnetisation $(\vec{M})$

The total magnetic moment per unit volume of the material is know as magnetisation

$$\vec{M} = \frac{\mu_m}{V} \quad \text{.....(6B.6)}$$

If a material having area A and length l possesses pole strength P then

$$\vec{M} = \frac{\mu_m}{V} = \frac{P \times l}{A \times l} = \frac{P}{A} \quad \text{.....(6B.7)}$$

Therefore magnetisation can also be defined as the pole strength per unit area

$$\text{Unit of } \vec{M} = \frac{\text{ergs}}{\text{oersted cm}^3} \quad \text{C.G.S unit}$$

$$= \frac{\text{Ampere} - \text{turn}}{\text{metre}} \quad \text{S.I unit}$$

6B.6 Magnetic Permeability

Magnetic permeability is defined as the ratio of the magnetic induction $\vec{B}$ to the intensity $\vec{H}$. It is denoted as μ

$$\text{Hence } \mu = \frac{\vec{B}}{\vec{H}} \quad \text{.....(6B.8)}$$

$$\text{Unit of } \mu = \text{weber / ampere} - \text{m}$$

$$= \frac{\text{Henry}}{\text{metre}}$$

$$= \frac{\text{H}}{\text{m}}$$

Magnetic permeability of free space is given by

$$\mu_o = 4\pi \times 10^{-7} \text{ H/m}$$

$$\mu_o = \frac{\vec{B}_o}{\vec{H}}$$

Hence in term of μ_o

$$\vec{B} = \mu_o\left(\vec{H} + \vec{M}\right) \quad .....(6B.9)$$

The ratio of magnetic permeability of a material to the magnetic permeability if the material is free space in termed as relative permeability (μ_r)

or $$\frac{B}{B_o} = \frac{\mu}{\mu_o} = \mu_r \quad \text{(dimension less Quantity)} \quad .....(6B.10)$$

Relative permeability of a material is a measure of the degree to which the material can be magnetized.

6B.7 Magnetic Susceptibility χ

Magnetic susceptibility is defined as the ratio of magentisation vector $\vec{M}$ of a magnetic material to the magnetic field intensity $\vec{H}$. It is denoted by χ.

$$\chi = \frac{\vec{M}}{\vec{H}} \quad .....(6B.11)$$

Unit of χ = dimensionless

For vacuum $\chi = 0$

6B.8 Relation between $\vec{B}, \vec{H}$ and $\vec{M}$

Let us consider the material in the form of a Rowland ring having a toroidal winding of N turns around it. When a current i_o is sent throught the winding, the material ring is magnetized along its circumferential length. The current i_o is the real current which magnetizes the ring.

Fig. 6 B.1 shows the section of the magnetized ring

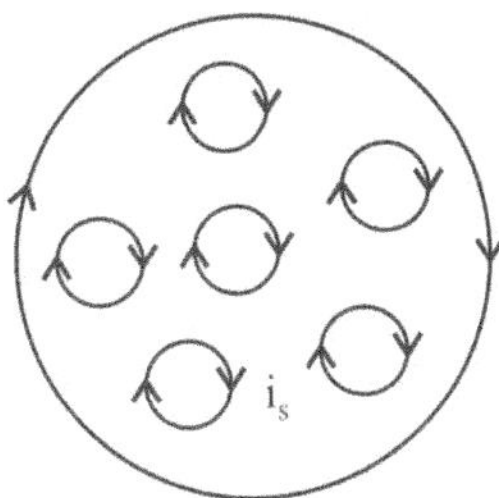

Fig. 6B.1

The small circles represent the current-loops. Now except at the periphery (outer circle), every portion of each loop is adjacent to another loop in which the current at the point of contact is in opposite direction. Hence net current inside is zero.

The current in the outer loop remains uncancelled. Thus the whole network of electronic currents, within the material can be replaced by a current i_s circulating around the surface of the ring. This current is know as 'Amperian Surface Current'.

Let A be the cross-sectional area and l the circumferential length of the ring

$$\therefore \text{ Volume} = A\,l$$

The ring behaves as a large dipole of magnetic moment = i_s A

$$\therefore \text{ Magnetisation} = \frac{\text{Magnetic moment}}{\text{Volume}}$$

$$\text{Magnetisation} = \frac{i_s A}{Al} = \frac{i_s}{l} \text{ amperes / m} \qquad \text{.....(6B.12)}$$

The magnetisation thus equals the Amperian surface current per unit length. This is also known as 'Magnetising current'.

Now, the magnetic thus density B with the material of the ring arises due to free current i_o and due to Amperian surface current i_s.

$$B = \mu_o \left(\frac{Ni_o}{l} + \frac{i_s}{l} \right) \qquad \text{.....(6B.13)}$$

But $\qquad \dfrac{i_s}{l} = M$

$$\therefore\ B = \mu_o \left(\frac{Ni_o}{l} + M \right)$$

or

$$\frac{B}{\mu_o} - M = \frac{Ni_o}{l} \qquad .....(6B.14)$$

The quantity $\frac{B}{\mu_o} - M$ is of significance in magnetism and is known as 'magnetic field intensity' H.

$$H = \frac{B}{\mu_o} - M \qquad .....(6B.15)$$

or

$$B = \mu_o (H + M) \qquad .....(6B.16)$$

and

$$\vec{H} = \frac{Ni_o}{l} = ni_o \qquad .....(6B.17)$$

where n is the number of turns per unit length. Thus the value of H depends only on the free current i_o and is independent of the material.

If the Rowland ring is empty, then $\vec{M} = 0$.

The flux density in vacuum is

$$\vec{B}_o = \mu_o \vec{H} \qquad .....(6B.18)$$

6B.9 Atomic Origin of Magnetisation

The nuclear model of atomic structure proposed by Rutherford helps to explain the magnetic properties of the material. Magnetic properties arise as a consequence of magnetic moments associated with the atoms. The magnetic moment of the atom originates from three sources.

1. The Orbital motion of electrons

The motion of an electron constitutes a current and the circular path of electron is identical to a current loop. Such a current loop behaves as an elementary magnet having a magnetic moment. The total orbital magnetic moment of an atom is the sum of orbital magnetic moment of individual electrons. The total orbital magnetic moment of an atom is the sum of orbital magnetic moment of individual electrons.

2. **The electron spin**

   The spinning motion of an electron also has a magnetic moment associated with it.

3. **The nuclear spin**

   The magnetic moment of the nucleus is about 1/2000 as much as of an electron. So it is neglected while studying magnetic properties of material.

   Thus orbital and spin motion of electron contribute to magnetic moments. In many cases, the magnetic moments of electrons in an atom cancel each other and an atom may not possess a permanent magnetic moment. When such atoms are placed in a magnetic field, the orbital motion of electrons undergoes changes and the atoms acquire an induced magnetic moment.

6B.10 Magnetic Moment of Atom: Bohr Magneton

The magnetic moment associated with orbital motion of the electron is related to orbital angular momentum L by relation.

$$\mu_{m\,(orb)} = \frac{-e}{2m} L \qquad \text{.....(6B.19)}$$

The orbital magnetic momentum is space quantized and its projection L_x in a given direction takes only $(2l + 1)$ specific values

$$L_x = m l \,\hbar$$

$$\therefore \; \mu_{m\,(orb)} = \frac{-e\hbar}{2m} . ml$$

$$\therefore \; \mu_{m\,(orb)} = -m_l \cdot \mu_B \qquad \text{.....(6B.20)}$$

where $\mu_B = \frac{e\hbar}{2m} = 9.274 \times 10^{-24}\ \text{Am}^2$ is called Bohr Magneton.

Thus

$$\mu_B = \frac{e\hbar}{2m} = \frac{eh}{4\pi m} = 9.27 \times 10^{-24}\ \text{A–m}^2 \quad \text{(in S.I unit)} \qquad \text{.....(6B.21)}$$

And $$\mu_B = \frac{e\hbar}{2m} = \frac{eh}{4\pi m} = 0.927 \times 10^{-20}\ \text{erg / oersted} \quad \text{(in C.G.S unit)} \qquad \text{.....(6B.22)}$$

No electron can have a magnetic moment below μ_B or this is the minimum value of projection of magnetic moment of the electron in an arbitrary direction.

The spin angular momentum and spin magnetic dipole moment of an electron are related through

$$\mu_B \text{ (spin)} = \frac{-e}{m} S_Z \qquad \text{.....(6B.23)}$$

where, S_Z is $\pm \dfrac{\hbar}{2}$

$$\therefore \mu_m \text{ spin} = \pm \frac{e\hbar}{2m} = \pm \mu_B \qquad \text{.....(6B.24)}$$

Each orbital in an atom accommodates two electrons which orient themselves in opposite directions and revolve in opposite directions. Because of this the spin and orbital magnet moments of the pair cancel each other. Consequently, an atom having completely filled electron sub shells or shells does not possess permanent magnetic moment. Hence, magnetic moment of an atom is determined only by the electrons which occupy partially filled shells.

The magnetic moment of a free atom is mainly determined by uncompensated spins of the outer electrons.

The magnetic moment of a molecule is not equal to the sum of the magnetic moments of the atoms constituting it. It is because a chemical bond formed between the atoms requires a certain rearrangement of the valence electrons

6B.11 Classification of Magnetic Materials

On the basis of the behaviour of magnetic material when placed in a magnetic field, all material can be classified as:

1. Diamagnetic materials
2. Paramagnetic material
3. Ferromagnetic material
   a. Anti-ferromagnetic materials
   b. Ferrimagnetic materials

6B.11.1 Diamagnetic Materials

The atoms in diamagnetic material do not possess permanent magnetic moment. When such material is placed in a magnetic field, there is change in the orbital motion of electron. Therefore, the magnetic moment is induced in an atom. As a result the material becomes magnetized. The magnitude of induced magnetic moment is proportional to the applied field and direction of induced magnetic moment will be opposite to that of external magnetic field (Fig.6 B.2)

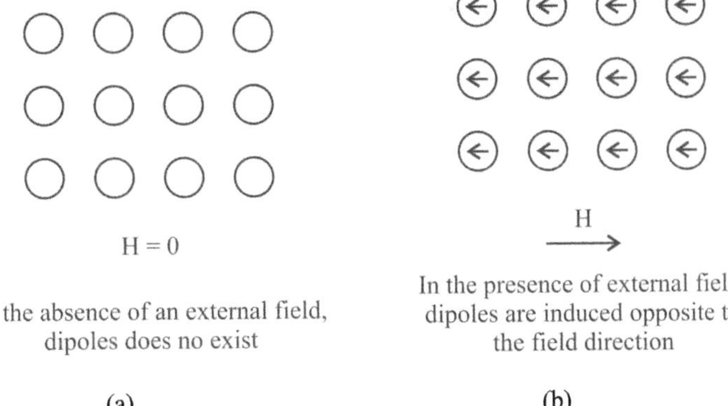

Fig. 6B.2 The atomic dipole configuration for diamagnetic material
(a) In the absence of an external field, dipole does not exist
(b) In the presence of external field, dipoles are induced opposite to the field direction

Examples of diamagnetic material are, Bismuth, antimony, gold, mercury etc.

Properties

1. The induced dipoles and magnetisation vanish as soon as the applied field is switched off.
2. In a diamagnetic materials, the magnetic lines of forces due to an applied field are repelled (Fig. 6B.3)

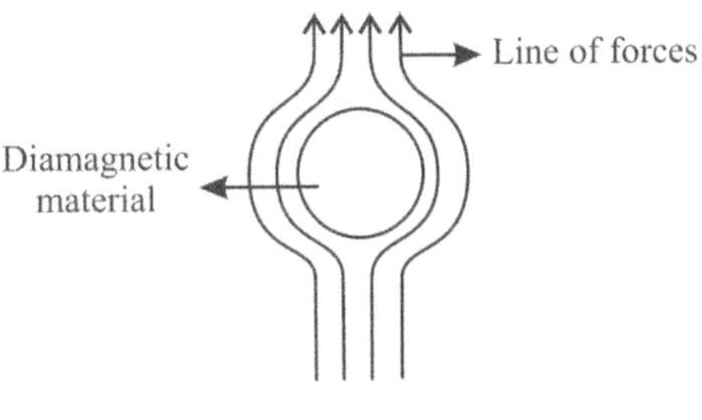

Fig. 6B.3

3. Relative permeability μ_r is less than unity ($\mu_r < 1$)
4. Magnetic susceptibility is independent of temperature (do not obey curie's law)
5. λ_{Dia} is negative and of order fo 10^{-5}
6. Diamagnetic materials have the tendency to move from the stronger to the weaker parts in a non-uniform magnetic field.

6B.11.2 Paramagnetic Materials

The materials which when placed in a magnetic field, becomes weakly magnetized, in the direction of applied external field, are called as paramagnetic materials.

When the materials is subjected to external magnetic field magnetic dipoles tend to align in the direction of field and material becomes magnetised Fig. 6B.4

Fig. 6B.4 [The atomic dipole configuration for a paramagnetic material (a) In the absence of an external field (b) In the presence of external field, dipole tend to align along the field direction]

Example of paramagnetic materials are, platinum, aluminium, chromium, manganese etc.

Properties

1. In paramagnetic materials, the magnetic lines of forces due to applied field are attracted towards the materials, Fig. 6B.5

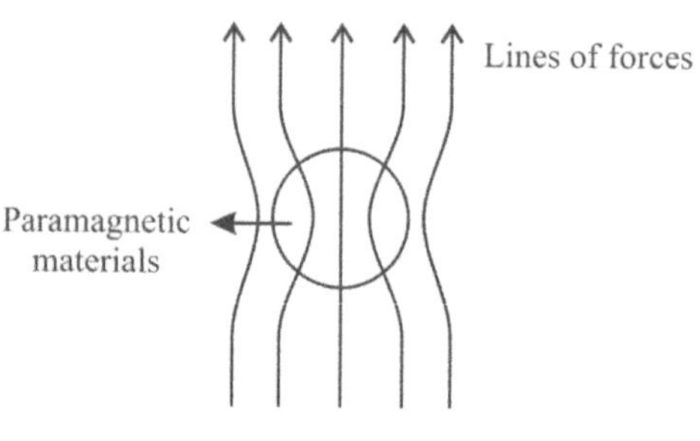

Fig. 6B.5

2. Relative permeability μ_r is greater than unity ($\mu_r > 1$)
3. χ_{Para} is small and positive 10^{-3}

4. Magnetic susceptibility decreases with increasing temperature i.e. they obey the curie's law (Fig. 6B.6)

$$\chi_{para} = \frac{C}{T} \qquad \text{where C = constant} \qquad .....(6B.25)$$

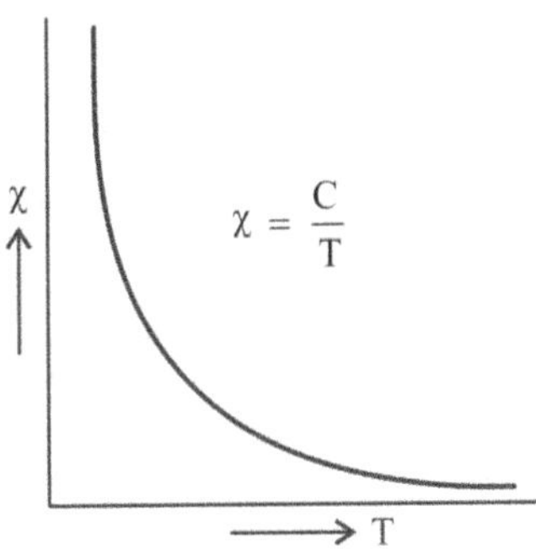

Fig. 6B.6 (Paramagnetic material)

5. The aligning force on the permanent moments of the atoms with ordinary magnetic fields is rather small, so that paramagnetic effect is weak.

6. Paramagnetic materials have the tendency to move from weaker to the stronger parts in a non-uniform magnetic field.

7. Langevin's expression for paramagnetic susceptibility is

$$\chi = \frac{\vec{M}}{\vec{H}} = \frac{\vec{M}_s\,\alpha}{3\,\vec{H}} \qquad .....(6\text{ B}.26)$$

with $\vec{M}_s = \mu_m\,N$

and $\alpha = \dfrac{\mu_m\,\vec{B}}{K_B\,T}$

Thus $$\chi = \frac{\mu_m^2\,\mu_o N}{3K_B\,T} \qquad .....(6\text{ B}.27)$$

where K_B is Boltmann's constant

6B.11.3 Ferromagnetic Materials

The materials which are strongly magnetised even in the absence of external magnetic field are called as ferromagnetic materials

Examples of ferromagnetic materials are - Iron, nickel, cobolt, gadolinium and their alloys.

Properties

1. Relative permeability μ_r is much higher than unit ($\mu_r >> 1$) (100 to 1000)
2. All ferromagnetic material becomes paramagnetic above a temperature called curie temperature T_C.
3. In ferromagnetic materials, the magnetic lines of forces due to applied field are strongly attracted towards the materials (Fig. 6B.7).

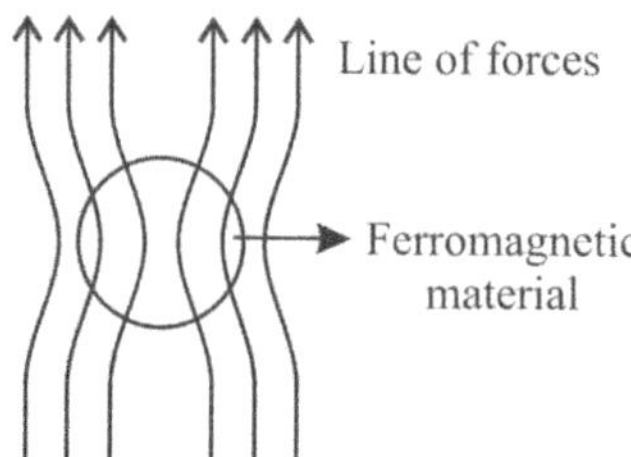

Fig. 6B.7

4. Magnetic susceptibility decreases with rise in temperature according to curie - weiss law (Fig. 6B.8).

$$\chi_{ferro} = \frac{C}{T - T_c} \quad \text{......(6B.28)}$$

where C is constant and T_c is curie temperature

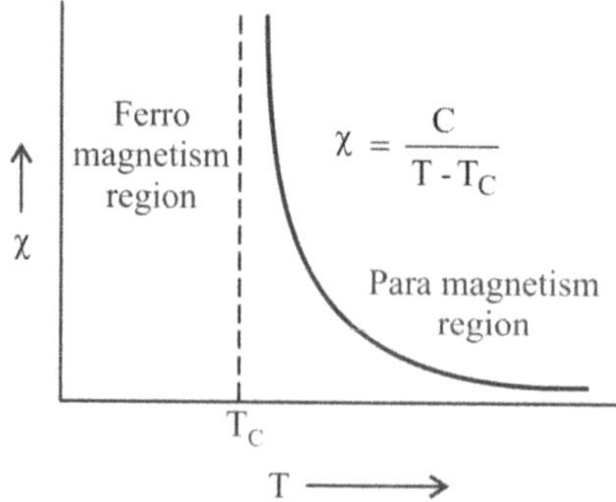

Fig. 6B.8

5. χ_{ferro} is large and positive.
6. The source of ferromagnetism is the spin of the electrons.
7. Ferromagnetism is the property of a material to be strongly attracted to a magnetic field and to become a powerful magnet.
8. Ferromagnetic materials like Fe, Co, Ni, Gd have incomplete inner shells. These shells can be completed by using Hund's rule.

   The difference between dia, para and ferromagnetic materials is shown in Fig. 6B.9.

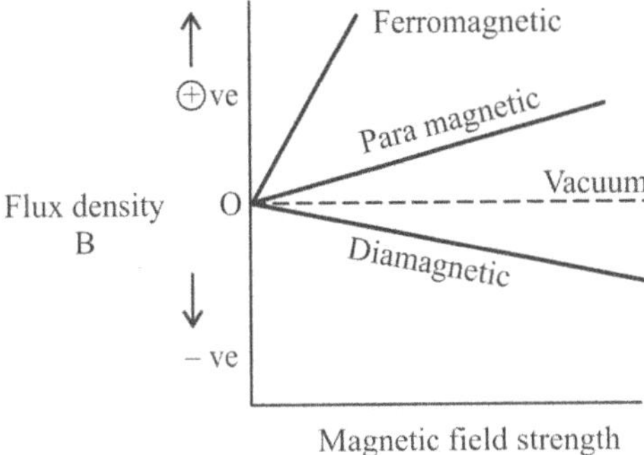

Fig. (6B.9)

Ferromagnetic materials can be subdivided as

(a) Anti ferromagnetic materials and

(b) Ferrimagnetic materials

6.11.3.1 (a) Antiferromagentic Materials

In anti ferromagnetic materials, the spin magnetic moments of neighbouring atoms are aligned in an anti parallel order as shown in Fig. 6B.10 and are equal in magnitude.

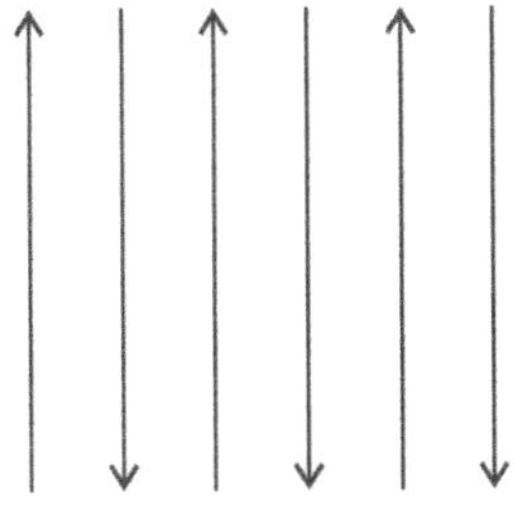

Fig. 6B.10

Properties of Anti-ferromagnetic materials

1. Anti-ferromagnetic materials exhibit a spontaneous alignment of magnetic moments below a critical temperature.
2. In anti-ferromagnetic materials the neighbouring atoms are aligned in the antiparallel direction and their magnitudes are equal. So the net magnetic moment = 0

   Example of anti-ferromagnetic materials are - Mno, Mns, Cr_2O_3 , NiCr, etc.

3. The magnetic susceptibility is positive and of the order of 10^{-3} to 10^{-5}.
4. The susceptibility χ increases with increase in temperature and reaches at maximum value at a certain temperature called Neel temperature T_N.
5. Above Neel temperature the material becomes paramagnetic. Fig. 6 B.11 shows the variation of susceptibility with temperature.

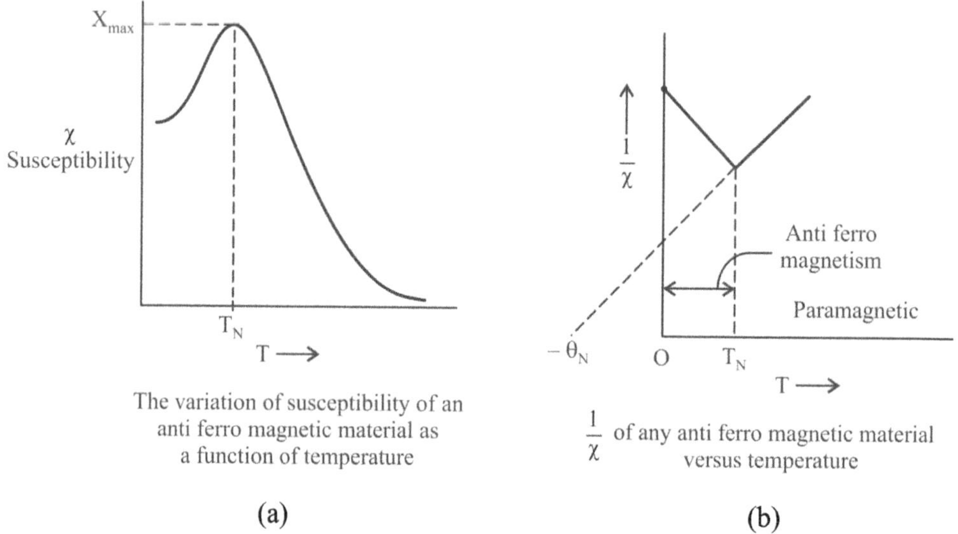

Fig. 6B.11 (a) The variation of susceptibility of an anti-ferromagnetic material as a function of temperature (b) $\frac{1}{\lambda}$ of an anti-ferromagnetic material versus temperature

In the paramagnetic state, the variation of inverse susceptibility $\frac{1}{\chi}$ with temperature is linear

The extrapolation of the paramagnetic line to $\frac{1}{\chi} = 0$

yields a negative temperature, known as extrapolated Neel temperature θ_N.

The variation of χ_A with temperature obeys a modified Weiss law

$$\chi_A = \frac{C}{T-(-\theta_N)} = \frac{C}{T+\theta_N}$$

$$\chi_A = \frac{C}{T+\theta_N} \qquad T > T_N \qquad \text{.....(6B.29)}$$

The Neel temperature is often below room temperature.

6.11.3.2 Ferrimagnetic Materials

In ferrimagnetic materials, the magnetic moment of unequal magnitude are aligned in opposite directions. (Fig. 6B.12).

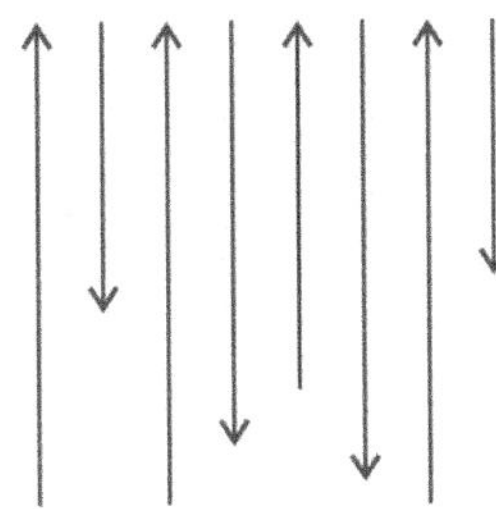

Fig.6B.12

Properties of Ferrimagnetic Materials

1. It result form antiparallel spin coupling and incomplete magnetic moment cancellation.
2. In this case, the moments of the two sublattices are unequal and there is a resultant magnetic moment. The material is called a ferrimagnet.
3. This moment disappears above a curie temperature T_c, thermal energy randomizes the individual magnetic moments and the materials becomes paramagnetic.
4. Ferrimagentic materials, generally known as ferrites are represented by the formula XO Fe_3 O_4, where X stands for Mn, Co, Ni, Mg, Zn, Cu etc.

6B.11.4 Applications of Ferrimagentic Materials or Ferrites

They are especially suitable for the following applications:

1. Magnetically hard ferrites for permanent magnets
2. Magnetically soft ferrites for transformers and induction

3. Rectangular loop ferrites for data storage.
4. Ferrities and garnets for microwave applications

6B.12 Langevin's theory of Diamagnetism

Let us consider an electron of charge 'e' moving in the circular orbit 'r' then current through the circle is given as

$$I = \frac{\text{Charge}}{\text{Time taken}} = \frac{e}{t} = \frac{e}{\frac{s}{v}} = \frac{ev}{2\pi r} \quad .....(6B.30)$$

Due to this current, a magnetic moment is created which is given by

$$\mu_m = I.A = \frac{ev}{2\pi r} . \pi r^2$$

$$\mu_m = \frac{e\,v\,r}{2} \quad .....(6B.31)$$

We know that an external magnetic field accelerates or decelerates the orbiting electrons, which results in a change in magnetic moment μ_m.

The external magnetic field induces an electric field, which in turn, exerts an electrostatic force F, on the orbiting electron which is

$$F = ma = e\,E \quad .....(6B.32)$$

where E is the electric fields strength and m is the mass of the electron. Now acceleration of the electron is given by

$$a = \frac{dv}{dt} = \frac{eE}{m} \quad .....(6B.33)$$

We can write the electric field as the ratio of the induced voltage or emf V_e per unit length L

$$E = \frac{V_e}{L} \quad .....(6\ B.34)$$

According to Lenz's law, a change in an external magnetic flux ϕ induces in a loop-shaped wire an emf which opposes, the change in flux is given by

$$V_e = -\frac{d\phi}{dt} = -\frac{d(BA)}{dt} \quad .....(6B.35)$$

Hence

$$a = \frac{dv}{dt} = \frac{eE}{m} = \frac{V_e e}{mL}$$

$$\frac{dv}{dt} = -\frac{eA}{mL}\frac{dB}{dt}$$

$$\frac{dv}{dt} = -\frac{e\pi r^2}{2\pi r m}\frac{dB}{dt} \qquad \text{......(6B.36)}$$

$$\therefore \frac{dv}{dt} = -\frac{e\ r}{2\ m}\frac{dB}{dt}$$

or

$$\therefore \frac{dv}{dt} = -\frac{e\ r}{2\ m}\frac{dB}{dt} \qquad \text{.....(6B.37)}$$

A change in the magnetic field strength from O to B yields a change in velocity of the electrons

$$\int_{v1}^{v2} dv = -\frac{e\,r}{2m}\int_{0}^{B} dB \qquad \text{.....(6B.38)}$$

$$\Delta v = -\frac{e\,r\,B}{2m} \qquad \text{.....(6B.39)}$$

This change in electron velocity yields in turn a change in magnetic moment,

Therefore, the change in magnetic moment is given by

$$\Delta\,\mu_m = \frac{e\,\Delta v\,r}{2} = -\frac{e^2\,r^2\,B}{4\,m} \qquad \text{......(6B.40)}$$

The change in magnetic moment has a direction opposite to the applied magnetic field B.

Atom can have more than one electron, and orbits of the electrons need not necessarily be circular

∴ The average radius of the orbit $\overline{r}$ is related to the radius of the atom as

$$\overline{r}^2 = \frac{2}{3}r_o^2 \qquad \text{.....(6B.41)}$$

If N is the number of atoms per unit volume and each atom has Z electrons, the magnetic moment per kilogram atom.

$$\Delta \bar{\mu}_m = \frac{-N e^2}{4 m} B Z \left(\frac{2}{3}\right) r_o^2$$

$$\Delta \bar{\mu}_m = -N Z \left[\frac{e^2}{6 m} r_o^2\right] B$$

$$\Delta \bar{\mu}_m = -N Z \mu_o \left[\frac{e^2}{6 m} r_o^2\right] H \qquad \text{.....(6B.42)}$$

The atomic susceptibility is given by

$$\chi = \frac{\Delta \bar{\mu}_m}{H} = -\frac{\mu_o Z N e^2}{6 m} r_o^2 \qquad \text{.....(6B.43)}$$

For,

$m = 9.1 \times 10^{-13}$ Kg

$r_o = 0.1$ nm

$N = 5 \times 10^{28} / m^3$

and $e = 1.6 \times 10^{-19}$ Coulomb

$$\chi = -3 \times 10^{-6} \qquad \text{....(6B.44)}$$

For most diamagnetic material χ is of the order of 10^{-6}. Thus an external field induces in all atoms and ions a diamagnetic moment equivalent to a negative susceptibility. Since the electronic structure does not vary with temperature, the diamagnetic susceptibility is independent of temperature.

6B.13 Langevin's theory of Paramagentism

It is considered that the magnetic moments arising from the orbital motion of electrons causes paramagnetic behaviour and the interaction between magnetic moment is neglected. In the absence of an external magnetic field, the magnetic axes of molecules are uniformly distributed in all directions and the sum of projections of magnetic moments in any direction is zero. Therefore, there is no net magnetisation.

The potential energy corresponding to a magnetic dipole in a magnetic field of induction B is given by

$$U = -\mu_m \cdot B \qquad \text{......(6 B.45)}$$

$$U = -\mu_m \, B \cos\theta \qquad .....(6\,B.46)$$

where θ is the angle between the magnetic field and the magnetic dipole. The potential energy attains its minimum value when direction of dipole orientation coincides with the direction of the magnetic field.

When paramagnetic material is subjected to a magnetic field, the action of the field turns the dipoles so that they are aligned with the magnetic field. Thermal agitation opposes the ordering action of the magnetic field. So the degree of orientation of magnetic dipoles in a magnetic field is determined both by the field strength and temperature.

Let us consider a system of N dipoles each carrying a dipole moment μ_m . They are randomly oriented in all directions in the absence of external magnetic field. Therefore, all directions of orientation are equally probable as a result of which the net magnetic moment of unit volume of material is zero. Let the system be in thermal equilibrium at temperature T. At any instant the number of magnetic moments N (θ) d θ that lie within a solid angel $d\Omega$ is proportional to the solid angle which is equal to $2\pi \sin\theta \, d\theta$

Thus

$$N(\theta)\, d\theta \quad \propto \quad 2\pi \sin\theta \, d\theta \qquad ......(6\,B.47)$$

Now suppose that the gas of magnetic moments is subjected to external magnetic field of induction B. The magnetic field causes orientation of dipoles along its direction.

According to Maxwell - Boltzman statistics, the number of dipoles whose potential energy is U, is proportions to Boltzman factor $e^{-U/KT}$

∴ The number of dipoles dN which have potential energy U and lying within the solid angle $d\Omega$

$$= A \, d\Omega \, e^{-U/KT} \qquad .....(6\,B.48)$$

Substituting the value of $d\Omega$ and U from equation (6 B.46) and (6 B.47) we get

$$d N = A(2\pi \sin\theta \, d\theta) \; e^{\frac{\mu_m \, B \cos\theta}{KT}} \qquad .....(6\,B.49)$$

substitute

$$A = 2\pi A \qquad .....(6\,B.\,51)$$

and

$$\beta = \frac{\mu_m \, B}{K\,T} \qquad .....(6\,B.51)$$

$$\therefore \qquad d N = A \sin\theta \, d\theta \, . \, e^{\beta \cos\theta} \qquad ......(6\,B.52)$$

$$d\,N = A\ e^{\beta \cos\theta} \sin\theta\, d\theta \qquad \text{.....(6 B.53)}$$

The total number of dipoles N within the unit volume having potential energy U is given by

$$N = \int_0^{\pi} d\,N = A\int_0^{\pi} e^{\beta \cos\theta}\,.\,\sin\theta\, d\theta \qquad \text{......(6 B.54)}$$

Each dipole carries a magnetic momen μ_m, the net magnetic moment contributed by dN dipoles is given by

$$P = d\,N\,\mu_m \cos\theta \qquad \text{.......(6 B.55)}$$

The resultant magnetisation is d N μ_m cos θ

The total magnetisation

$$M = \mu_m \int d\,N\ \cos\theta$$

$$M = A\ \mu_m \int_0^{\pi} e^{B\cos\theta} \cos\theta\ \sin\theta\ d\theta \qquad \text{......(6 B.56)}$$

From equation (6 B.54)

$$A = \frac{N}{\int_0^{\pi} e^{\beta\ \cos\theta}\ \sin\theta\, d\theta}$$

$$M = N\,\mu_m = \frac{\int_0^{\pi} e^{\beta\ \cos\theta}\ \cos\theta\,\sin\theta\, d\theta}{\int_0^{\pi} e^{\beta\ \cos\theta}\ \sin\theta\, d\theta} \qquad \text{.....(6 B.57)}$$

Let us put $y = \cos\theta$

$$dy = -\sin\theta\, d\theta$$

$$M = \frac{N\,\mu_m \int_{+1}^{-1} e^{\beta y}\ y\,(-\,dy)}{\int_{+1}^{-1} e^{\beta y}\,(-\,dy)}$$

$$M = N\mu_m \frac{\left[\frac{e^{\beta y}}{\beta} y\right]_{+1}^{-1} - \left[\frac{e^{\beta y}}{\beta^2}\right]_{+1}^{-1}}{\left[\frac{e^{\beta y}}{\beta}\right]_{+1}^{-1}}$$

$$M = N\mu_m \frac{\left[\frac{-e^{-\beta}}{\beta} \frac{-e^{\beta}}{\beta}\right] - \left[\frac{e^{-\beta}}{\beta^2} - \frac{e^{\beta}}{\beta^2}\right]}{\left[\frac{e^{-\beta}}{\beta} - \frac{e^{\beta}}{\beta}\right]}$$

$$M = N \,.\, \mu_m \frac{\frac{-2}{\beta}\left\{\left[\frac{e^{-\beta} + e^{\beta}}{2}\right] - \left[\frac{e^{\beta} - e^{-\beta}}{2\beta}\right]\right\}}{\frac{-2}{\beta}\left[\frac{e^{\beta} - e^{-\beta}}{2}\right]}$$

$$= N \,.\, \mu_m \frac{\cos h\beta - \frac{1}{\beta}\sin h\beta}{\sin h\beta} \qquad \text{.....(6B.58)}$$

$$M = N \,.\, \mu_m \left(\cot h\beta - \frac{1}{\beta}\right) \qquad \text{.....(6B.59)}$$

or

$$M = N\,\mu_m\,L(\beta) \qquad \text{.....(6B.60)}$$

where $L(\beta) = \left(\cot h\beta - \frac{1}{\beta}\right)$ is called the Lagevin function .....(6B.61)

Case I

When $\mu_m \beta >> KT$ and $\beta >> 1$

$$\left|L(\beta)\right|_{\beta \to \infty} = \left|\left(\cot h\beta - \frac{1}{\beta}\right)\right|_{\beta \to \infty}$$

$$= \left|\frac{e^{\beta} + e^{-\beta}}{e^{\beta} - e^{-\beta}} - \frac{1}{\beta}\right|_{\beta \to \infty}$$

$$= \left| \frac{1+e^{-2\beta}}{1-e^{-2\beta}} - \frac{1}{\beta} \right|_{\beta \to \infty}$$

$$= 1$$

$$\therefore \qquad M_s = N\,\mu_m \qquad \text{.....(6B.62)}$$

where M_s denotes the saturation value of magnetisation. It will be attained either when the applied field B is very large and the temperature is very low. It corresponds to the condition where all the dipoles are completely aligned in the field direction.

Case II $\mu_m\,\beta << KT$ and $\beta << 1$

i.e magetnic field is not very strong For values $\beta << 1$, the Langevin function is given by

$$L(\beta) = \left[\frac{1}{\beta} + \frac{\beta}{3} - \frac{\beta^2}{45} + \frac{2\beta}{245} + \dots\dots\dots\right] - \frac{1}{\beta}$$

$$L(\beta) \approx \frac{1}{\beta} + \frac{\beta}{3} - \frac{1}{\beta} = \frac{\beta}{3}$$

$$L(\beta) = \frac{\beta}{3} = \frac{\mu_m\,B}{3\,K\,T} \qquad \text{.....(6B.63)}$$

Form e.g (6 B.60) magnetisation may be written as

$$M = \frac{N\,\mu_m\,\beta}{3}$$

$$M = \frac{N\,\mu^2_m\,B}{3\,K\,T} \qquad \text{since } B = \mu_o\,H \qquad \text{.....(6B.64)}$$

$$\therefore \qquad M = \frac{N\,\mu_o\,\mu^2_m\,H}{3\,K\,T} \qquad \text{.....(6B.65)}$$

Thus susceptibility

$$\chi = \frac{M}{H} = \frac{N\,\mu_o\,\mu^2_m}{3\,K\,T} \qquad \text{......(6B.66)}$$

$$\chi = \frac{C}{T} \qquad .....(6\ B.67)$$

where the curie constant C is given by

$$C = \frac{N\,\mu_o\,\mu^2_m}{3\,K} \qquad .....(6\ B.68)$$

Equation (6 B.68) is the curie law.

From equation (6 B.64) it is seen that

$$M \propto \frac{B}{T} \qquad .....(6\ B.69)$$

Which implied the increasing B tend to aling the magnetic moment in the material so that M increases whereas increasing T tends to disorient the moments as a result of which M decreases.

6B.14 Ferromagnetic Domains

In order to explain the fact that a piece of ferromagnetic material may exist in an magnitised state and may be magnetised to saturation by application of a field of a few oersted only. Weiss introduced the concept of ferromagnetic domains. He postulated that any specimen consists of a large number of small regions or domains which are spontaneously magnetised, the magnetisation being appropriate to the temperature T. In the demagnetised state the magnetisation vectors of these domains are oriented in such a way that the specimen doe not possess a resultant magnetic moment. In other words, the overall magnetisation is given by the sum of domain vectors. Magnetisation occurs only when some of these vectors are caused to align themselves more or less parallel to an applied field.

In the case of a single crystal, the crystal is to be pictured as made up of a very large number of regions each magnetised to the degree of saturation in one of the directions parallel to one of the cube edges.

The increase in the value of the resultant magnetic moment of the specimen under the action of applied magnetic field may be imagined to take place according to domain theory, by two independent processes as suggested by R. Becker by an increase in the volume of domains which are favourably oriented with respect to the field at the expense of unfavourable oriented domains or by rotation of the directions of magnetisation towards the direction of the field i.e from easy to hard direction. These two methods by which the resultant magnetisation may change are shown in Fig. 6 B.13. Closer examination reveals that in weak field the magnetisation changes usually by means of domain boundary displacement and so the domain change in size. In strong fields the magnetisation usually changes by means of rotation of direction of magentisation.

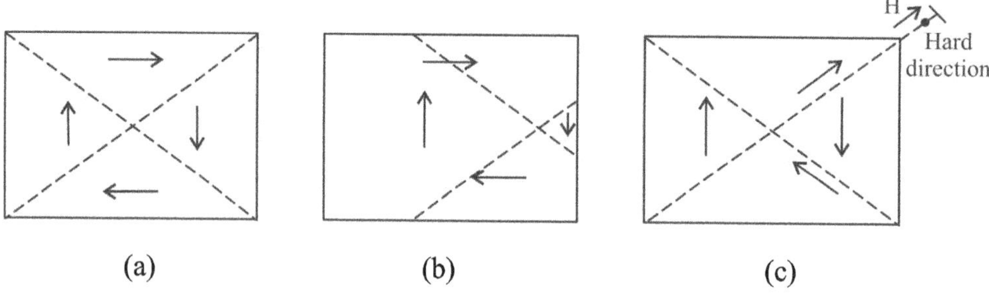

Fig. 6B.13

6B.15 Origin of Domains

The most direct experimental evidence for the existence of domains furnished by microphotographs of domains boundaries by the technique of 'magnetic powder pattern'. This method applied originally by Bitter has, in the hands of W.C. Elmore and H.J. Williams and his collaborates provided, convincing proof that domains exists and behave as expected theoretically. The powder pattern method consists in placing a drop of colloidal suspension of finely divided ferromagnetic material such as magnetite, on the carefully prepared surface of the ferromagnetic crystal under study. It is found on observation through microscope that the colloidal particles in the suspension becomes strongly concentrated about certain well-defined line; which represent the boundaries between domains magnetised in different directions. The reason why the colloidal particles concentrate near these boundaries is that in their vicinity there exists a very strong magnetic field which attracts the magnetic particles.

We may understand the origin of domains by considering the structure shown in Fig.6B.14, each representing a cross-section through ferromagnetic single crystal. In Fig. 6B.14 (a) we have saturated configuration consisting of a single domain. Because of the free magnetic poles formed on the surface of crystal this configuration will have high magnetic energy $\frac{1}{8\pi}\int H^2\, dv$. The magnetic energy for square cross- section will be of the order of $M_s^2 \approx 10^6$ erg/ cm^2, where M_s is saturated magnetisation.

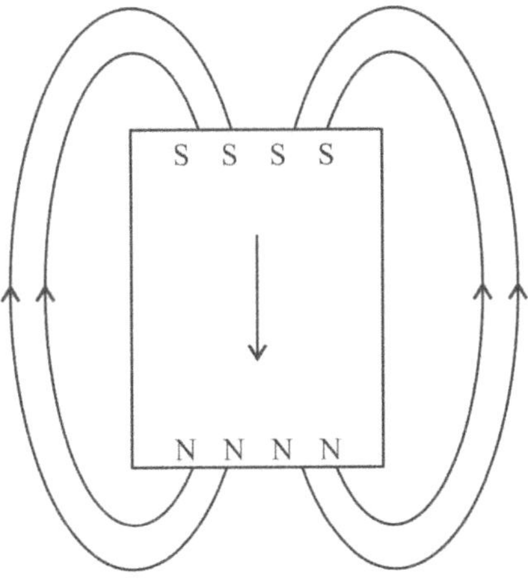

Fig. 6B.14 (a)

In Fig 6B.14 (b) the magnetic energy has been reduced by one-half as a resultant of dividing the crystal into two domains magnetised in opposite directions.

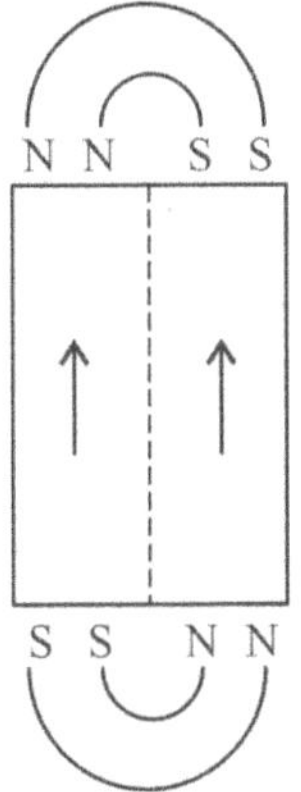

Fig.6B.14 (b)

The sub-division process may be carried further as in Fig. 6B.14(c) with N domains. It turns out that the magnetic energy is reduced to approximately $\frac{1}{N}$ of the magnetic energy of saturated configuration (6B.14 (a)).

The sub-division process continues until the energy required to establish an additional boundary layer or interface separating two domains magnetised oppositely is greater than the reduction in the magnetic field energy. A domain structure as shown in Fig.6 B.14(d) and 6 B.14(e) has the Zero magnetic energy.

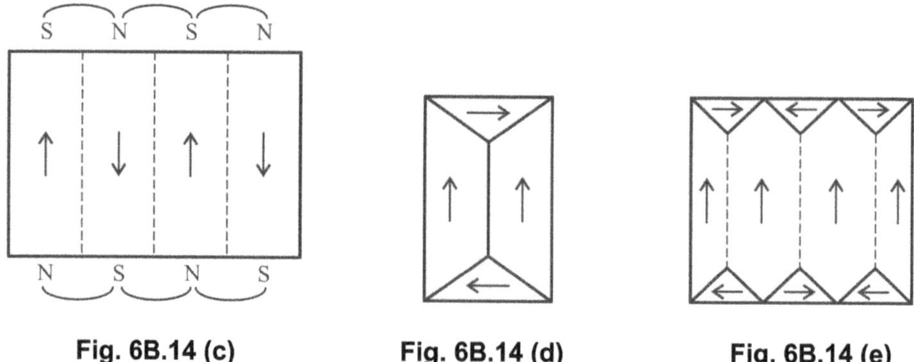

Fig. 6B.14 (c) Fig. 6B.14 (d) Fig. 6B.14 (e)

Here, the boundaries of triangular prism domains termed 'Closure domains' near the end faces of crystal make equal angles of 45° with the magnetisation directions. Therefore, the component of magnetisation normal to the boundary is continuous on crossing the boundary and no poles are formed any where in the crystal. As there are no poles there is no magnetic field associated with the magnetisation and we speak of the flux circuit being completed within the crystal, thus giving rise to the phrase 'domains of closure' for domains near the surface of the crystal which act to complete the flux circuit.

The energy required to form a domain of closure is essentially determined by crystalline 'anisotropy energy'. The anisotropy energy tends to make the magnetisation of domain line up along a certain crystallographic axis. The axis thus favoured is known as preferred axis.

The number of domains and domain structure will depend, to a large extent, on the shape and size of the crystal under consideration. The size of the domain for a particular domain structure can also be obtained from the principle of minimum energy. The volume of domains may vary from 10^{-2} to 10^{-6} cm^3.

6B.16 Retentivity, Coercivity and Hysteresis

When a specimen of ferromagnetic material is placed in a magnetic field, the specimen is magnetised by induction. As the magnetic intensity H is varied, the flux density B in the ferromagnetic material does not vary linearly with H. In other words, the permeability μ is not a constant but varies with H and also depends upon the past history of the material. The variation in B with variation in H is shown in Fig. 6B.15.

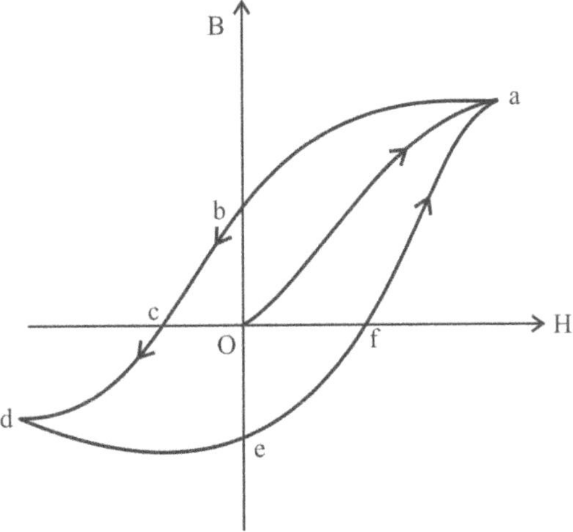

Fig. 6B.15

The point O represents an initially unmagnetised specimen and a zero magnetic intensity. As H is increased B also increases, but not uniformly, and a point such as 'a' is reached.

If H is now decreased, B also decreases but following a path ab (and not the original path oa). Thus, B lags behind H. When H becomes zero, B still has a value equal to ob. This magnetic flux density remaining in the specimen in the absence of any external field is called 'residual magnetism'

The power of retaining this magnetism is called the 'retentivity' or 'remanence' of the specimen. Thus, the retentivity of a specimen is a measure of the magnetic flux remaining in the specimen when the magnetising force (magnetic intensity) is removed.

If the magnetic intensity H is now increased in the reverse direction, the value of B further decreases, still lagging behind H, and becomes zero when H has a value equal to oc. This value of the magnetic intensity is called the 'coercive force' or 'coercivity' of the specimen. Thus, coercivity is a measure of the magnetic intensity required to destroy the residual magnetism of the specimen.

As H is increased beyond oc, the speciment is increasingly magnetised in the opposite direction, and a point such as d is reached.

By taking H back from its negative maximum value, through zero, to its original positive maximum value, a similar curve defa is obtained. At points b and e where the specimen is magnetised in the absence of any external magnetic intensity, it is said to be a'permanent magnet'. It is thus found the flux density B always lags behind the magnetic intensity H. The lagging of B behind H is called 'hystersis'. The closed curve abcdefa which represent a cycle of magnetisation of the specimen is known as the 'hysteresis curve' (or loop) of the specimen.

6B.17 Loss of Energy due to Hysteresis

In the process of magnetisation of a ferromagnetic material through a cycle, a loss of energy is always involved. For aligning the molecular magnets in the direction of applied magnetic filed, the work is done by the magnetising field against the mutual attractive force acting between them. This energy spent in magnetising the specimen is not completely recoverable on reversing the magnetising field. Due to the group force, some of the molecular magnets remain aligned even when H = O. Hence, a coersive force in the reversed direction is applied to demagnetise the sample completely. Thus, there in a loss of energy is taking a sample of ferromagnetic substance through a cycle of magnetisation.

6B.18 Hysteresis Loss from B – H Curve

Let us consider a ring of ferromagnetic material having a coil of N-turns wound uniformly over it. The permeability of the medium is μ, circumferential length of coil is *l* and cross-sectional area is A. When a current *i* ampere is passed through the winding, it produces a flux of magnetic flux density B. The total magnetic flux linked with the coil is

$$\phi = NBA \quad \text{wb-turns} \qquad \text{.....(6 B.70)}$$

The magnetising field H in the material due to current following through N-turns is given by

$$H = \frac{Ni}{l} \quad \text{amp- turns / m}$$

Hence,
$$i = \frac{Hl}{N} \quad \text{amp} \qquad \text{.....(6 B.71)}$$

As the current i is increased, the flux linked with the coil changes and hence an e.m.f.e is induced in the coil. The e.m.f. so induced is given by

$$e = \frac{-d\phi}{dt} = \frac{-d}{dt}(N\,B\,A) \qquad \text{.....(6 B.72)}$$

$$e = -NA\frac{dB}{dt} \qquad \text{.....(6 B.73)}$$

The magnetism indicates that induced emf opposes the increasing current. The work done by the current i against the induced e. m. f. e. in time dt is given by

$$d\,W_1 = e\,I\ dt$$

$$d\,W_1 = \left(NA\ \frac{dB}{dt}\right)\left(\frac{Hl}{N}\right) dt$$

$$d\,W_1 = H\,A\,l\,d\,B \qquad \text{.....(6 B.74)}$$

Hence, the work done per unit volume of the material is

$$dW = \frac{HAldB}{Al} = HdB \qquad \text{.....(6 B.75)}$$

Let us consider the BH curve shown in Fig 6B.16

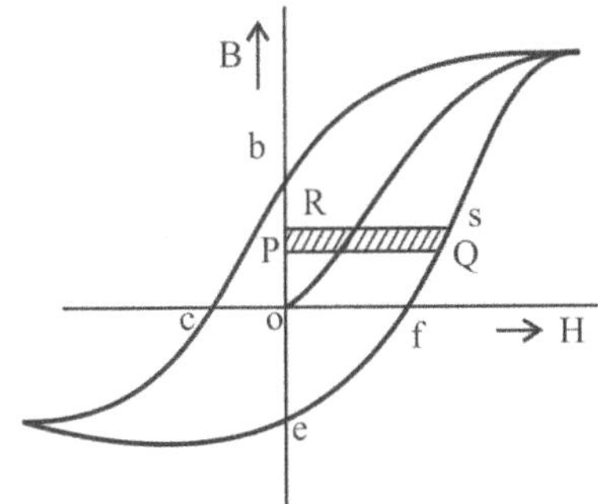

Fig. 6B.16

From the Fig. 6B.16

$$PQ = H \text{ and } PR = dB$$

$\therefore$ Area of P Q R S $= HdB$ ......(6 B.73)

The amount of work done given by equation (6 B.75) is the loss of energy per unit volume in taking the material through a complete cycle of magnetisation.

The loss of energy per unit volume per cycle is

$$\oint HdB = \text{area of BH curve} \qquad \text{.....(6 B.76)}$$

6B.19 Hysteresis Loss from I – H Curve

The magnetic flux density B is given by

$$B = \mu_o\,(I + H) \qquad \text{......(6 B.77)}$$

Differentiating (6B.77) we get

$$dB = \mu_o \left(dI + dH\right) \qquad .....(6\ B.78)$$

Substituting the value of dB from equation (6 B.78) in equation

$$\oint HdB = \text{area of BH curve, we get}$$

$$\oint HdB = \mu_o \oint HdH + \mu_o \oint HdH$$

The term $\oint HdH$ is zero because curve between H and dH is a straight line. Hence hysteresis loss

$$\oint HdB = \mu_o \oint HdI$$

$$\oint HdB = \mu_o \times \text{area of IH curve} \qquad .....(6\ B.79)$$

This is Warburg's law.

6B.20 Uses of Hysteresis Curve

By using the hysteresis curve of various ferromagnetic materials we can select the material which gives minimum hysteresis curve when put to cycle of magnetisation. From hysteresis curve, an idea of magnetic properties like susceptibility, permeability, retentivity, coercivity of a ferromagnetic material can be made.

6B.21 Selection of Magnetic Materials

(a) Permanent Magnets

The materials used for permanent magnet must have the following characteristics.

(i) High retentivity so that the magnet may cause strong magnetic field

(ii) High coercivity so that magnetisation should not come out by strong external fields, mechanical ill-treatment and temperature changes.

Example of permanent magnets are 'cobalt steel', ticonal, alnico, vicalloy etc.

(b) Electromagnets

The magnetic materials that are used for preparing the electromagnets should have maximum flux density with comparatively small magnetizing field and low hysteresis loss. Electrical steels (soft magnetic materials) is used for electromagnets.

(c) Transformer, cores, electromoters, Generators and Telephone Diaphragms.

The magnetic materials which have the following characteristics can be used for the above purposes.

(i) The highest possible permeability

(ii) Magnetic core material should be capable of being easily magnetised or demagnetised.

(iii) The core loss should be low.

Examples of materials are permalloy, ferrites, garnets etc.

(d) Magnetic shielding

The magnetic materials used for magnetic shield must have high saturation induction and very low coercivity. Examples are super mendur and perm alloy.

6B.22 Soft and Hard Magnetic Materials

Based on the area within a hystersis loop, the magnetic materials are broadly categorised into two types, namely soft magnetic material and hard magnetic materials.

6B.22.1 Soft Magnetic Materials

1. Materials having narrow hysteris loop are known as soft magnetic materaisl shown in Fig. 6B.17

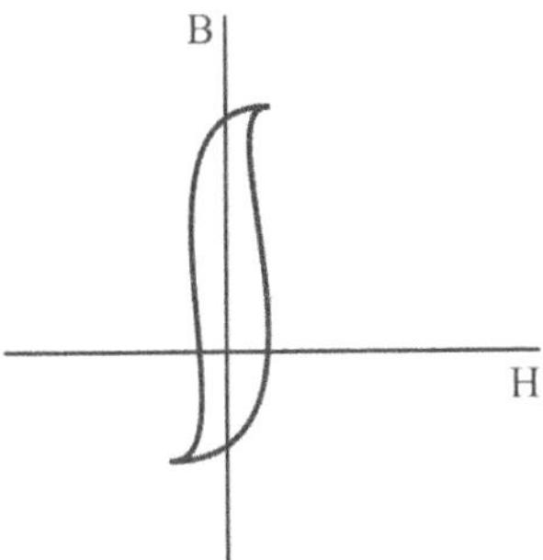

Fig. 6B.17

2. They can be magnetised and demagnetised easily.
3. Because of small area hysteresis loop it exhibits lower power losses. Therefore, they are used in high frequency applications.
4. When the magnetic flux through a conducting material changes, voltages are induced which causes eddy current. Eddy current increases the energy loss. So it is desirable to

minimise eddy current losses in soft magnetic materials. Eddy current losses can be reduced by increasing the electrical resistivity of the material. The resistivity of the material can be increased by using thin laminations of ferromagnetic material, which are insulated from each other.

The soft magnetic materials are categorised into four groups.

1. Heavy duty Flux multiplier:They are the cores of transformers, motors and generators. Electrical steel is used for manufacturing cores.
2. Light duty flux multiplier:These are cores of small special purpose transformer inductors etc, used in communication equipment. Ni-Fe-alloys and soft ferrites are used in these application.
3. Square loop materials: They are used in magnetic amplifier, saturable core devises, computers etc. Here also Ni –Fe alloys and soft ferrites are used.
4. Microwave system components: Soft ferrites and garnets are used in these applications.

6B.22.2 Hard Magnetic Materials

These materials have:

1. High resistance to demagnetisation
2. High resonance
3. High permeability
4. High coercive field
5. Large hysteresis loop as shown in Fig. 6B.18

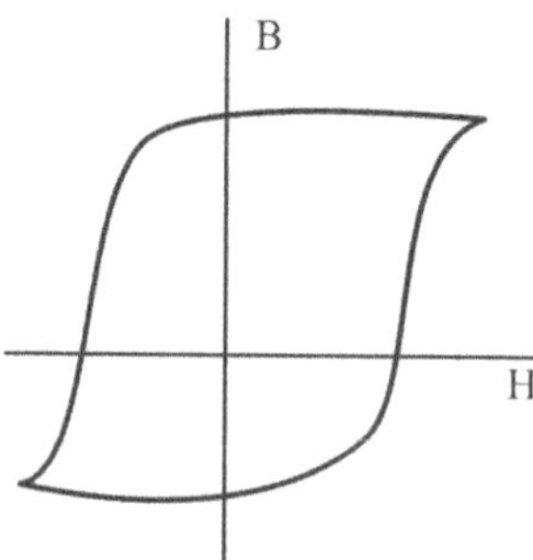

Fig. 6B.18

The hard magnetic materials are magnetised in a magnetic field strong enough to orient the magnetic moments of their domains in the direction of the applied field. Part of energy of the applied field is converted into potential energy which is store in the permanent magnet produced. A permanent magnet in the fully magnetised condition is thus in a relatively high energy state as compared to a demagnetised state.

The power or external energy of hard material is directly related to the size of hystresis loop.

Comparison between Hard and soft magnetic materials is given in Table 6 B.1

Table 6B.1 Comparison between Hard and Soft Magnetic materials

Soft Magnetic Materials	Hard Magnetic Materials
1. These are the magnetic materials having narrow hysteresis loop.	1. These are the magnetic materials with large area hyseresis loop.
2. They are magnetised and demagnetised easily.	2. They have high resistance to demagnetization.
3. They have a high initial permeability and a lower coercivity.	3. They have high permeability and, a high coercive field and high remanence
4. These materials reach their saturation magnetisation with a relatively low applied field.	4. There materials reach to saturation magnetisation with a strong magnetic field.
5. It retains magnetism only in a magnetic field.	5. It retains magnetism even in the absence of magnetic field.

Solved Numerical Problems

1. Magnetic susceptibility of a medium is 948×10^{-11}. Calculate the permeability and relative permeability

Sol. Given that

$$\chi = 948 \times 10^{-11}$$

Formla used

$$\mu_r = 1 + \chi$$

$$\mu_r = 1 + 948 \times 10^{-11}$$

$\mu_r \approx 1$ or slightly more than 1

$\mu_r = \mu / \mu_o$

$\mu = \mu_r \ \mu_o$

$\mu = 4\pi \times 10^{-7}$ H/m

2. A solenoid consisting of 100 turns and carrying 10 amp current is 0.5 m long. calculate

(a) Magenetomotive force

(b) Total flux if area of cross section is 0.0004 m^2

Sol. Given that

N = 100

i = 10 amp: l = 0.5 m ; A = 0.0004 m^2

Formula used

(a) Magnetomotvie force = N i

$= 100 \times 10$

$= 1000$ amp.turns

(b) flux ϕ = B.A

$$= \mu_r \cdot \mu_o HA = \mu_r \cdot \mu_o \frac{Ni}{l} A$$

$$= 1 \times 4\pi \times 10^{-7} \times \frac{1000}{0.5} \times 0.0004$$

$$\phi = \frac{16 \times 3.14}{5} \times 10^{-7}$$

$$\phi = 10.048 \times 10^{-7}$$

$$\phi = 1.048 \times 10^{-6} \text{ weber.}$$

3. An iron rod of volume 10^{-3} meter^3 and relative permeability 1200 is placed inside a long solenoid wound with 5 turns per cm. If a current of 0.5 amp is passed through the solenoid, find the magnetic moment of the rod.

Sol. We Know that

$$B = \mu_o (H + I)$$

or $$I = \frac{B - \mu_o H}{\mu_o}$$

If μ is the permeability of material then

$$B = \mu H$$

$$\therefore \quad I = \frac{(\mu - \mu_o)H}{\mu_o}$$

$$I = \left(\frac{\mu}{\mu_o} - 1\right) H$$

But we know that

$$\frac{\mu}{\mu_o} = \mu_r$$

$$\therefore \; I = (\mu_r - 1) H$$

We know that

$$H = ni$$

$$\therefore \quad I = (\mu_r - 1)\, ni$$

Given that

$$\mu_r = 1200$$

$$n = 5 \text{ turns / cm} = 500 \text{ turns / m}$$

and $$i = 0.5 \text{ amp}$$

$$\therefore \quad I = (1200 - 1)\, 500 \times 0.5$$

$$I = 2.997 \times 10^5 \text{ A/m}$$

$$I = 3.0 \times 10^5 \text{ A/m}$$

we know that the magnetisation I is the magnetic moment per unit volume

$\therefore$ Magnetic moment

$$M = I \times V$$

$$M = 3.0 \times 10^5 \times 10^{-3}$$

$$M = 3 \times 10^2 \text{ A} - \text{m}^2$$

4. The magnetic susceptibility of silicon is $- 0.5 \times 10^{-5}$. What is the intensity of magnetisation and magnetic flux density in a magnetic field of intensity 9.9×10^4 amp/ m.

Sol. Given that

$$\chi = -0.5 \times 10^{-5}$$

$$H = 9.9 \times 10^{4} \text{ amp/m}$$

Formula used

The intensity of magnetisation is given by

$$M = \chi \ H$$

$$M = -0.5 \times 10^{-5} \times 9.9 \times 10^{4} \text{ amp/m}$$

$$M = -0.495 \text{ amp/m}$$

Magnetic flux density B is given by

$$B = \mu_o H (1 + \chi)$$

$$B = 4\pi \times 10^{-7} \text{ H/m} \times 9.9 \times 10^{-4} \text{ amp/m } (1 - 0.5 \times 10^{-5})$$

$$B = 0.124 \text{ Wb/m}^2$$

5. A bar magnet has a coercivity 5×10^{3} amp/m. It is desired to demagnetise it by inserting it inside a solenoid 10 cm long and having 50 turns. What current should be sent through the solenoid.

Sol. We know that the coercivity H is given by

$$H = \frac{ni}{l}$$

where i is the current flowing through the solenoid.

$$i = \frac{Hl}{n}$$

Given that

$$H = 5 \times 10^{3} \text{ amp/m}$$

$$l = 10 \text{ cm} = 10 \times 10^{-2} \text{ m}$$

$$n = 50$$

$$i = \frac{5 \times 10^{3} \times 10 \times 10^{-2}}{50} = 10 \text{ amp}$$

6. Prove that a super conductor (B = 0) is a perfect diamagnet.

Sol. We know that

$$B = \mu_o (H + I)$$

since $B = 0$

$$0 = \mu_o (H + I)$$

$$\therefore \quad H + I = 0$$

$$I = -H$$

Susceptibility is given by

$$\chi = I/H = -1$$

It is true for diamagnet. Hence, super conductor is a perfect diamagnet.

7. Find the relative permeability of a ferromagnetic material if the field of strength 220 amp/m produces a magnetisation of 3300 amp/m in it?

Sol. Given that

$$B = 220 \text{ amp /m}$$

$$M = 3300 \text{ amp /m}$$

Formula used

$$\mu_r = 1 + \frac{I}{H}$$

$$\mu_r = 1 + \frac{3300}{220}$$

$$\mu_r = 16$$

8. Diamagnetic $Al_2\ O_4$ is subjected to an external magnetic field of 10^5 A/m. Evaluate magnetisation of magnetic flux density in $Al_2\ O_3$

(susceptibility of $Al_2\ O_3 = 5 \times 10^{-5}$)

Sol. Given that

$$B = 10^5 \text{ A/m}$$

$$\chi = -5 \times 10^{-5}$$

$$\mu_o = 12.57 \times 10^{-7} \text{ H/m}$$

Formula used

$$M = \chi\ H$$

$$M = -5 \times 10^{-5} \times 10^5 = -5 \text{ A/m}$$

$$B = \mu_o (H+M) = (12.57 \times 10^{-7} \text{ H/m}) (105 \text{ A/m} - 5 \text{ A/m})$$

$$B = 0.126 \text{ Wb/m}^2$$

9. The susceptibility of paramagnetic $FeCl_3$ is 3.7×10^{-3} at $27\,°C$. what will be the value of its suceptiblity μ_r at $200°\,K$ and $500°\,K$?

Sol. Given that

$$\chi = 3.7 \times 10^{-3}$$

$$T = 27\,°C = 300\,°K$$

Curie constant

$$C = \chi\, T$$

$$C = 3.7 \times 10^{-3} \times 300$$

$$C = 1.11°\,K$$

$$\chi_{200°K} = \frac{C}{T} = \frac{1.11}{200} = 5.55 \times 10^{-3}$$

$$\chi_{500°K} = \frac{C}{T} = \frac{1.11}{500} = 2.22 \times 10^{-3}$$

10. The area of a B – H loop is 0.5 cm^2, whereas on X-axis 1 cm represents $H = 10^3$ A/m and on Y-axis 1 cm represent B = 1 Tesla. If the volume of the specimen is $10^{-3}\,m^3$ and the frequency of a.c is 50 Hz. calcualte the hystresis power loss.

Sol. Area of B H loop $= 0.5 \times 10^{-3} \times 1\ J/m^3$

Hysstresis power loss = area of B. H loop

$$p = nv$$

$$p = 50 \times 10^{-3} \times 0.5 \times 10^3$$

$$p = 25 \text{ watt}$$

11. The area of hystersis loop drawn between B and H is 100 m^2. Each unit space along the vertical axis represent 0.01 wb/ m^2 and each unit space along the horizontal axis represents 40 amp/m. Determine the hystersis loss per cycle.

Sol Given that

Area = 100 m^2

Unit space along Y axis = 0.01 wb/ m^2

Unit space along X axis = 40 amp/m

Formula used

Hystresis loss/ cycle = area × unit space along Y axis × Unit space along X axis

Hystresis loss / cycle = 100 × 0.01 × 40

$= 40\ J/m^3$

12. An iron ring of mean circumferential length 30 cm and cross-section area of 1 cm^2 is wound uniformly with 300 turns of wire. When a current of 0.032 amp flows in the winding, the flux in the ring is 2×10^{-6} weber. Find the flux density in the ring, the magnetic intensity and the permeability of iron.

Sol. Given that

$$l = 30 \text{ cm} = 0.3 \text{ m}$$

$$A = 1 \text{ cm}^2 = 10^{-4} \text{ m}^2$$

$$i = 0.32 \text{ amp}$$

$$\phi_B = 2 \times 10^{-6} \text{ weber}$$

Formula used,

$$B = \frac{\phi}{A} = \frac{2\times10^{-6}}{10^{-4}} = 2 \times 10^{-2} \text{ wb/m}^2$$

Magnetic intensity $H = \frac{Ni}{l} = \frac{300 \times 0.032}{0.3} = 32$ amp turns / m

Permeability $\mu = \frac{B}{H} = \frac{2 \times 10^{-2}}{32} = 6.25 \times 10^{-4}$ wb/ amp.m

Relative permeability

$$\mu_r = \frac{\mu}{\mu_o} = \frac{6.25 \times 10^{-4}}{4\pi \times 10^{-7}} = 500 \text{ unit}$$

13. Prove that $\mu_r = 1 + \chi$

Sol. We know that

$$\vec{B} = \mu_o \left(\vec{H} + \vec{M}\right)$$

Dividing the expression by $\vec{H}$

$$\frac{\vec{B}}{\vec{H}} = \mu_o \left(1 + \frac{\vec{M}}{\vec{H}}\right)$$

since $\frac{\vec{M}}{\vec{H}} = \chi$

and $\frac{\vec{B}}{\vec{H}} = \mu$

$$\therefore \qquad \mu = \mu_o (1 + \chi)$$

$$\frac{\mu}{\mu_o} = (1 + \chi)$$

$$\mu_r = (1 + \chi)$$

Conceptual Questions and Answers

1. **Define intensity of magnetisation of a magnetic material. How does it vary with temperature of paramagnetic material**

Ans. The intensity of magnetisation of a substance is defined as the magnetic moment develop per unit volume, when it is subjected to a magnetising field.

The intensity of magnetisation of paramagnetic substance is inversely proportional to its absolute temperature.

2. **Define the term magnetic flux**

Ans. Magnetic flux through a surface is defined as the number of magnetic field lines passing normally through a surface.

3. **Define magnetic permeability**

Ans. It is defined as the ratio of the magnetic induction (B) to the magnetic intensity H

$$\mu = \frac{B}{H}$$

4. **What is a diamagnetic substance**

Ans. A substance which when placed in a magnetic field gets feebly magnetised in a direction opposite to that of the magnetising fields, is called a diamagnetic substances.

5. **Name any two diamagnetic substance**

Ans. Zinc and Bismuth.

6. **Why is diamagnetism impendent of temperature**

Ans. The atoms of a diamagnetic do not have intrinsic magnetic dipole moment. On placing a diamagnetic sample in a magnetic field, the magnetic moment of the sample is always opposite to the direction of the field, it is not affected by the thermal motion of the dipoles.

7. What is paramagnetic substance

Ans. A substance which when placed in a magnetic field gets feebly magnetised in the direction of the magnetising field, is called a paramagnetic substance.

8. Name any two paramagnetic substances

Ans. Alumnimum and Manganese.

9. Why does a paramagnetic sample display greater magnetisation (for the same magnetising field) when cooled.

Ans. When cooled, the tendency of the thermal agitation to disrupt the alignement of magnetic dipoles decreases in case of paramagnetic materials. Hence, they display greater magnetisation.

10. How does the magnetic induction of a paramagnetic material vary with temperature

Ans. It decreases with the increase of temperature.

11. Why do magnetic field llines prefer to pass through ferrimagnetic substance (e.g. iron) than through air

Ans. The ferromagnetic substances have relative permeability quite greater than 1. hence the magnetic field line prefer to pass through them.

12. What is hystersis

Ans. The lag of intensity of magnetisation behind the magnetising field during the process of magnetisation and demagnetisation of a ferromagnetic material is called hysteresis.

13. What type of magnetic material is used in making permanent magnets

Ans. For making permanent magnet, a material having high coercivity is used.

14. Soft iron is used to make electromagnets, why?

Ans. It is because of the fact that the hystresis loop for soft iron is narrow. Due to this, the loss of energy per unit volume is small, when soft iron is taken over complete cycle of magnetisation.

15. Which is the material used in making the core of a transformer or a moving coil galvanometer.

Ans. Soft iron is used in making the core of the galvanometer. It makes the magnetic field strong as magnetic field lines tend to cross through a ferromagnetic.

16. Why soft iron is used in making the core of a transformer

Ans. The area of the hystresis loop for soft iron is very small. Since energy dissipated during a complete cycle of magnetisation and demagnetisation is proportional to the area of the hystresis loop, a small amount of energy will be wasted, when the core of the transformer is made of soft iron.

17. What do we infer from the large value of the susceptibility of the material? For example, susceptibility of iron is more than that of copper. Explain?

Ans. The susceptibility of a material is defined as the ratio of the intensity of magnetisation (I) to the magnetic intensity (H).

Therefore $$\chi_m = \frac{I}{H}$$

Thus, greater the value of the susceptibility of a material, greater will be the value of intensity of magnetisation for it i.e., more easily it can be magnetised. Thus, greater value of susceptibility for iron means that iron can be magnetised more easily than copper.

Objective Questions

1. Metals which are magnetised along the direction of external magnetising field by orientation of their atomic magnets are called.

   (a) Diamagnetic (b) Para magnetic

   (c) Anti-ferromagnetic (d) ferromagnetic

2. The unit of magnetic induction B is

   (a) Newton / amp-m (b) Newton-weber

   (c) weber/m^2 (d) weber

3. The energy loss per unit of the material per cycle is equal to μ_o times the area of

   (a) B-H loop (b) M-H loop

   (c) χ-H loop (d) I-H loop

4. The change in angular velocity of an electron of an atom when placed in a magnetic field density B is

   (a) e B (b) $\pm \frac{e}{2m} B$

   (c) $\frac{B}{2me}$ (d) $\frac{-e}{2 m B}$

5. A ferromagnetic material, when placed in an external magnetic field. The magnetic domains

(a) have no relation with the field (b) decrease in size

(c) increase in size (d) may increase or decrease

6. The three vector B,H and I in a magnetic material are related as

(a) $B = \mu_o (H + I)$ (b) $B = \dfrac{I}{H}$

(c) $B = \dfrac{H}{I}$ (d) $B = H + I$

7. The relative permeability of platinum is greater than unity and susceptibility has a small positive value. It must be

(a) non-magnetic (b) ferromagnetic

(c) diamagnetic (d) paramagnetic

8. The Hysteresis is

(a) lagging of the magnetising force, H with respect to the flux density B

(b) lagging of the flux density B with respect to the magnetising force H

(c) the ratio of B and H

(d) None of the above

9. The susceptibility for a diamagnetic substance is

(a) Zero (b) negative

(c) low but positive (d) positive and high

10. When a ferromagnetic material goes through a hystresis loop, the magnetic susceptibility

(a) has a fixed value (b) may be zero

(c) may be infinity (d) may be negative

11. Magnetic susceptibility χ is defined as

(a) I/H (b) H/I

(c) $I \times H$ (d) I^2/H^2

12. The core of an electromagnet is made of soft iron because soft iron has
    (a) small susceptibility and small retentivity
    (b) hystresis
    (c) directional property
    (d) susceptibility independent of temepratur.
13. Soft iron is used to manufacture electromagnet because of their
    (a) magnetic saturation limit is high and retentivity and coercive force are small
    (b) area of hysteresis curve is large
    (c) retentivity is high
    (d) coercive force is high
14. The relative permeability is represented by μ_r and the susceptibility by χ for a magnetic substance. Then for a paramagnetic substance.
    (a) $\mu_r > 1, \chi < 0$ (b) $\mu_r > 1, \chi > 0$
    (c) $\mu_r < 1, \chi < 0$ (d) $\mu_r < 1, \chi > 0$
15. when a ferromagnetic material goes through a hystresis loop the magnetic susceptibility
    (a) has a fixed value (b) may be zero
    (c) may be infinity (d) may be negative
16. The hystresis loop for the material of permanent magnet is
    (a) short and wide (b) tall and narrow
    (c) short and narrow (d) tall and wide
17. The most suitable material for making permanent magnet is
    (a) soft iron (b) steel
    (c) copper (d) aluminium
18. χ_1 and χ_2 are the susceptibilities of a diamagnetic substance at temperature T_1 K and T_2 K respectively. Then
    (a) $\chi_1 T_1 = \chi_2 T_2$ (b) $\chi_1 = \chi_2$
    (c) $\chi_1 T_2 = \chi_2 T_1$ (d) $\chi_1 \sqrt{T_1} = \chi_2 \sqrt{T_2}$
19. The material for making permanent magnets should have
    (a) high retentivity, high coercivity (b) high retentivity, low coercivity
    (c) low retentivity, low coercivity (d) low retentivity, high coercivity

20. According to curie law

 (a) Magnetic susceptibility of paramagnetic salt is directly proportional to absolute temperature.

 (b) The magnetic susceptibility is inversely proportional to the absolute temperature.

 (c) for strong fields and low temperatures, the magnetisation attains a minimum value.

 (d) the saturation magnetisation in a paramagnetic salt is dependent only on the magnetic field and not an temperature.

21. Above curie temperature

 (a) a diamagnetic substance becomes paramagnetic

 (b) a ferromagnetic substance becomes paramagnetic

 (c) a paramagnetic substance becomes anti ferromagnetic

 (d) a paramagnetic substance becomes ferromagnetic

22. Magnetic materials which can readily be magnetized in either direction, are called

 (a) high hystresis loss materials (b) hard magnetic materials

 (c) soft magnetic materials (d) low hysteresis loss materials

23. The area of the hystresis loop of a ferromagnetic or ferrimagnetic crystal gives

 (a) the volume of the domains that are present in the crystal

 (b) the value of spontaneous magnetisation of the crystal

 (c) the energy that is consumed in traversing the loop

 (d) the number of domains present in the crystal.

24. The most suitable materials for making electromagnet should have

 (a) low retentivity, low coercivity (b) high retentivity, high coercivity

 (c) high retentivity, low coercivity (d) low retentivity, high coercivity

25. Susceptiblity of a magnetic substance is found to depend on temperature and the strength of the magnetising field. The material is a

 (a) Diamagent (b) superconductor

 (c) paramagnet (d) ferromagent

26. If a magnetic field is applied to a diamagnetic substance, then
    (a) a magnetic moment will be induced whose direction is opposite to the field
    (b) the specimen will develop magnetic domains
    (c) it is attracted towards higher field-region
    (d) the magnetic dipoles tend to point in the same direction as the applied field
27. Property possessed by ferromagnetic substance only is
    (a) directional property (b) attracting magnetic substances
    (c) hysteresis (d) magnetic field on the loop.

Theoretical Questions

1. Define permeability and magnetic susceptibility
2. Prove that $\mu = \mu_o(1 + \chi_m)$
3. Prove that $B = \mu_o (H + I)$
4. Distinguish diamagnetic, paramagnetic and ferromagnetic materials
5. What are diamagnetic, paramagnetic and ferromagnetic
6. Explain diamagnetism, Para magnetism and ferromagnetism on the basis of magnetic dipoles of the atoms.
7. Prove that the diamagnetic susceptibility is negative and is independent of temperature.
8. Describe Langevin's theory of diamagnetism
9. What is magnetisation curve? Explain residual magnetism, coercive force and hystersis.
10. Describe Langevin's theory of Paramagnesium.
11. What does hysteresis and cycle of magnetisation mean? Prove that the energy dissipated per cycle of magnetisation is μ_o times the are of I-H curve.
12. What is hysteresis curve? Explain residual magnetism, coercive force and hystresis loss.
13. Explain the formation of domains in the ferromagnetic material and discuss how the hysteresis curve is explained on the basis of domain theory.
14. Describe the importance of hysteresis curve. How would you use the hysteresis curve for selecting the material for the construction of a permanent magnet.
15. Distinguish between soft magnetic material and Hard magnetic materials.

Numerical Problems

1. A magnetic material has a magnetisation of 3000 amp/m and flux density of 0.005 wb/m^2. Calculate the magnetic force and the relative permeability of the material

   [4.058]

2. A magnetising field of 1600 A/m produces a magnetic flux of 2.4×10^{-5} weber in an iron bar of cross-sectional area 0.2 cm^2. Calculate permeability and susceptibility of the bar.

   [7.5×10^{-4} N/A^2 ; 596.13]

3. Calculate the hourly loss of energy in the iron core of a transformer, the hysteresis loop of which is equivalent in area to 3000 ergs / cm^2. Given, frequency 50 cycles/sec, density of iron 7.5 gm/cc and weight of the core 12 Kg.

   [8.64×10^{11} erg]

4. A sample of paramagnetic salt contains 2.0×10^{24} atomic dipoles each of dipole moment 1.5×10^{-23} J/T. The sample is placed under a homogeneous magnetic field by 0.84 T and cooled to a temperature of 4.2 K. The degree of magnetic saturation achieved is equal to 15%. What is the total dipole moment of the sample for a magnetic field of 0.98 T and a temperature of 2.8 K (assume curie's law)

   [7.875 J/T]

5. An electron in the ground state of hydrogen atom is revolving in anti clock wise direction in a circular orbit. The atom is placed in uniform magnetic induction $\vec{B}$ such that the normal to the electron orbit makes an angle of 30 ° with the magnetic field. Calculate the torque experienced by the orbiting electron.

   $$\left[\frac{e\,h\,B}{8\,\pi\,m}\right]$$

6. A bar magnet has a coercivity of 4×10^3 A/m. It is desired to demagnetise it by inserting it inside a solenoid 12 cm long and having 60 turns what current should be sent through the solenoid.

   [n = 500 turns / metre i = 8.0 amp]

7. An iron ring has a cross-section area 10 cm^2 and air gap of width 2 mm and mean length 60 cm including the air gap. Find the number of turns required to produce the flux density of 1 weber / $merter^2$. Given $\mu_{iron} = 10^{-3}$ H/m.

   [600 ampt – turns]

8. Estimate the number of ampere - turns necessary to produce a flux of 0.001 weber around an iron ring of 6 cm^2 cross – section and 20 cm mean diameter, having an air gap of 2 mm. The permeability of iron is 1200 μ_o.

   [694 amp – turns]

9. A magnetic field of 100 gauss generates a flux of 2600 Maxwells in a long bar of iron. Calculate the permeability and intensity of magnetisation of the bar which has a uniform crosssection of the bar which has a uniform crosssection of 2 cm^2.

   [m = 13, M = 10.332 amp /m]

10. An iron rod of volume 10^{-3} meter ^{3} and relative permeability 1200 is placed inside a long solenoid wound with 5 truns per cm. If a current of 0.5 amp is passed through the solenoid, find the magnetic moment of the rod.

   [300 A - m^2]

Answers to Objective Questions

1. (b)	2. (c)	3. (d)	4. (b)	5. (d)	6. (a)
7. (d)	8. (b)	9. (b)	10. (c)	11. (a)	12. (a)
13. (a)	14. (b)	15. (c)	16. (c)	17. (b)	18. (b)
19. (a)	20. (b)	21. (b)	22. (c)	23. (b)	24. (a)
25. (d)	26. (a)	27. (c)			

CHAPTER 6C

Ultrasonics

As soon as we get rid of the backroom attitude and brought our apparatus fully into the Department with an inexhaustible supply of living patients with fascinating clinical problems, we were able to get ahead really fast. Any new technique becomes more attractive if its clinical usefulness can be demonstrated without harm, indignity or discomfort to the patient.

- Ian Donald

6C.1 Introduction

Ultrasonic waves are those sound waves whose frequency lies above the audible limit of about 20 kHz. Ultrasound must be differentiated from supersonic waves which is the study of the effects produced by objects that travel through a medium at a faster speed than the wave they generate.

6C.2 Classification of Sound Waves

Sound waves are longitudinal-mechanical waves. They can be propagated in solids, liquid and gases. The particles of the medium transmitting such a wave oscillate in the direction of propagation of the wave itself. There is a large range of frequencies within which the longitudinal-mechanical waves can be generated.

The three basic different sound waves can be classified as audible, infrasonic wave and ultrasonic waves and their properties are given in Table 6C.1.

Table 6C.1 Properties audible waves, infrasonic and ultrasonic waves.

S.No.	Audible waves	Infrasonic waves	Ultrasonic waves
1.	Their frequency ranges from 20 Hz to 20 KHz	Their frequency lies below 20 Hz	Their frequencies lies above 20 KHz
2.	Source of producing audible waves are (violin, human vocal cords, vibrating air column etc)	They are generated by sources like earthquake waves.	They can be produced mainly by either magnetostriction effect or piezoelectric effect.

6C.3 Basic Properties of Ultrasound Waves

1. The speed of propagation of ultrasonic waves increases with increase in frequency.
2. The wavelength of the waves is very small and the waves exhibit negligible diffraction effects.
3. They can travel over a long distances as highly directional beam and without appreciable loss of energy.
4. They are highly energetic owing to the high frequencies involved.
5. They produce cavitation effects in liquids.

6C.4 Production of Ultrasonics

Ultrasonic can be generated by number of ways, depending upon the range of frequencies.

For lower frequencies, mechanical type generators like galton whistle are used, and the high frequency ultrasonic is produced either by magnetostriction transducer or crystal transducer.

By magnetostriction method frequencies of the order of 300 KHz can be generated while crystal transducer can generate frequencies upto 15 MHz. The two phenomenon are

(i) Magnetostriction generators

(ii) Piezo-electric generators

6C.4.1 Magnetistriction Method

(a) ***Principle:*** When a magnetic field is applied parallel to the length of a ferromagnetic rod made of material such as iron or nickel, a small elongation or contraction occurs in its length. This phenomenon is known as magnetostriction.

The change in length depends on the intensity of the applied magnetic field and nature of the ferromagnetic material. It does not depend on the direction of the field. When the rod is placed inside a magnetic coil carrying alternating current, the rod suffers a change in length for each half cycle of alternating current. This means that the rod vibrates with a frequency twice that of the frequency of the alternating current. The amplitude of vibration is usually small; but if the rod is made to vibrate with its natural frequency i.e., if the frequency of the alternating current coincides with the natural frequency of the rod, due to resonance, the amplitude of vibration increases considerably. The natural frequency of vibration of the rod is given by

$$f = \frac{1}{2l}\sqrt{\frac{E}{\rho}} \qquad .....(6C.1)$$

where l is the length of the rod, E is the Young's Modulus of the material of the rod and ρ is the density. If the frequency of the alternating current is sufficiently high, ultrasonic waves are produced from the ends of the rod. Frequencies ranging from a few hundred hertz to about 300 KHz can be produced

(b) ***Construction:*** The magnetostriction transducer is as shown in Fig. 6C.1.

It consists of a premagnetised rod of nickel or iron magnetised initially by passing direct current in the coil which is wound over it. The rod is clamped at the centre. The two sections of the rod are surrounded by the coils L_1 and L_2 in the plate and grid circuits of a triode valve as shown in figure 6C.1. The length of the rod depends upon its material, desired frequency and the velocity of ultrasonics in the rod.

The diameter of the rod varies from 3 mm to 6 mm. The frequency of alternating current can be adjusted with the variable condenser C.

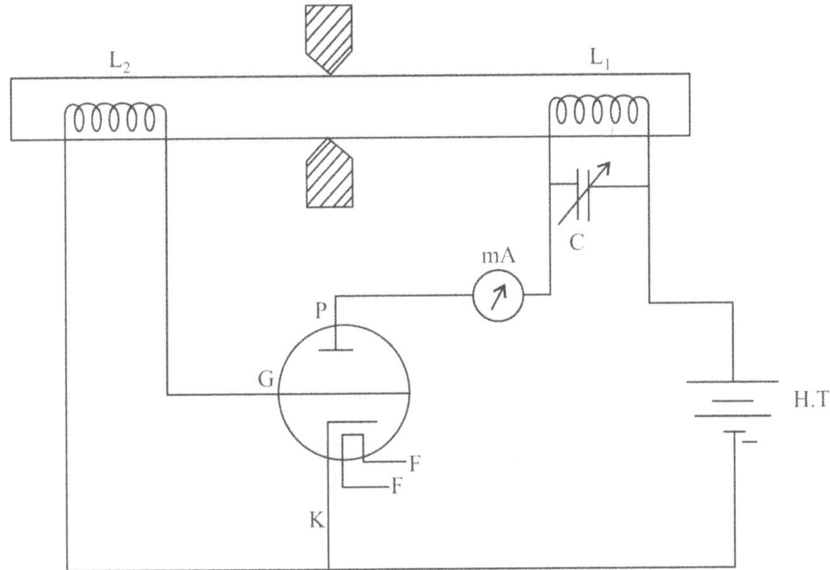

Fig. 6C.1 Magnetostriction transducer.

(c) ***Working:*** By adjusting the variable condenser C, high frequency oscillation current are set up in the plate circuit. This high frequency current flowing through the coil L_1 produces changes in the length of the rod. Due to this the rod alternatively extends and contracts; and the rod starts vibrating. Due to the vibrations of the rod, the magnetic flux linked with L_2 changes and hence an e.m.f is induced in it.

This induced emf is fed to the grid; amplified by the triode valve and feedback to the coil L_2. Thus the oscillations of the circuit are maintained. The rod vibrates at the frequency of the applied voltage. When the frequency of applied voltage matches the natural frequency of the rod; resonance occurs. The rod vibrates with maximum amplitude and sets the surrounding medium into longitudinal vibrations; thus producing the ultrasonic waves.

By adjusting the length of the rod and the capacity of condenser ultrasonic waves of different frequencies can be produced.

The frequency of oscillation of the L-C circuit is given by the expression

$$f = \frac{1}{2\pi}\sqrt{\frac{1}{LC}} \quad \text{.....(6C.2)}$$

By varying C, this frequency can be adjusted to be in tune with the natural frequency of the rod. Under this resonance condition, sustained oscillations and hence ultrasonic waves are produced by thc rod.

A rod of 0.1 m length can give out frequency of 20 KHz. Nickel, alloys of nickel and cobalt ferrites are the widely used as magnetostriction materials. Generally a bundle of fine insulated wire of such materials are employed just to avoid eddy currents.

(d) ***Merits:***

1. Large output power can be generated.
2. Magnetostrictive materials are easily available and inexpensive.

(e) ***Demerits:***

1. Generation of frequencies beyond 300 KHz is not possible.
2. When a single rod ferromagnetic material is used eddy current losses appear.
3. Since the elastic constant varies with magnetisation, one cannot get a single frequency oscillations.

6C.4.2 Piezoelectric Method

The Curie brothers discovered that certain asymmetric crystals exhibit piezoelectric effects.

In piezoelectric effect, when one pair of opposite faces of certain crystals such as quartz is subjected to pressure, in the other pair of opposite faces, opposite electric charges are developed. When the faces are subjected to tension instead of pressure, the sign of the developed charges reverses.

The converse of piezoelectric effect called inverse piezoelectric effect is also true. According to this if alternating voltage is applied to one pair of faces, along the direction of the other pair of opposite faces the dimension varies i.e., it expands and contracts periodically thereby generating elastic waves. The natural frequency of the crystal is given by

$$f = \frac{1}{2l}\sqrt{\frac{E}{\rho}} = \frac{v}{2l} \qquad \text{.....(6C.3)}$$

where E is Young's modulus of the material of the crystal

ρ is its density and

v is the velocity of longitudinal waves in the crystal

When the frequency of alternating voltage matches with the natural frequency of the crystal, resonant vibrations are produced resulting in generation of ultrasonic waves.

Commonly used piezoelectric transducers include ones construction from crystals of quartz. Rochelle salt, and ammonium dihydrogen phosphate (ADP). Crystalline quartz is the widely used material for the preparation of piezoelectric transducer.

6C.5 Quartz Crystal Transducers

Consider a natural quartz crystal plate of thickness t and length l in a shape of a hexagonal prism with a pyramid attached to each end. The line joining the end points of these pyramids is called the optic axis or Z-axis.

The three lines which pass through the opposite corners of the crystal constitutes its three X-axes called electrical axes. And line perpendicular to side of hexagon are called Y-axes. Thin plates cut perpendicular to X-axes are called X cut plates and those cut perpendicular to Y-axes are called Y cut axes.

In a quartz crystal, if an electric voltages is applied in the direction of the electrical axis, mechanical stress is produced in direction of Y-axis. The converse is also true i.e., mechanical stress applied along Y-axis yields electrical charges along X-axis.

Suppose there is an X-cut crystal plate of thickness t and length l. When an alternating voltage is applied across the faces of these plates along the electrical axes, then alternating stresses and strains are set up both in its thickness and length.

If the frequency of the alternating voltage happens to be equal to the natural frequency of vibration of the plate, then resonance will occur resulting in large amplitude of vibration.

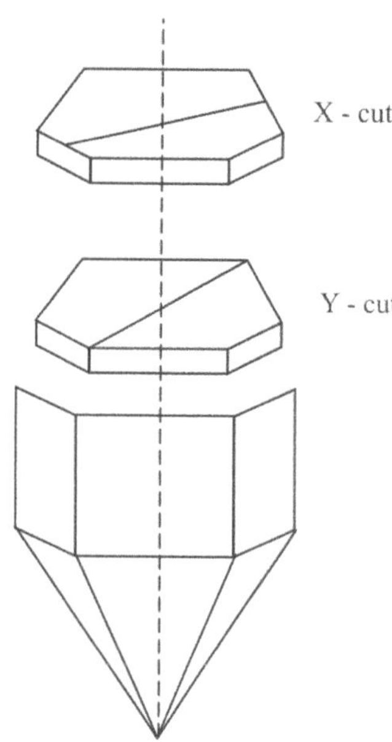

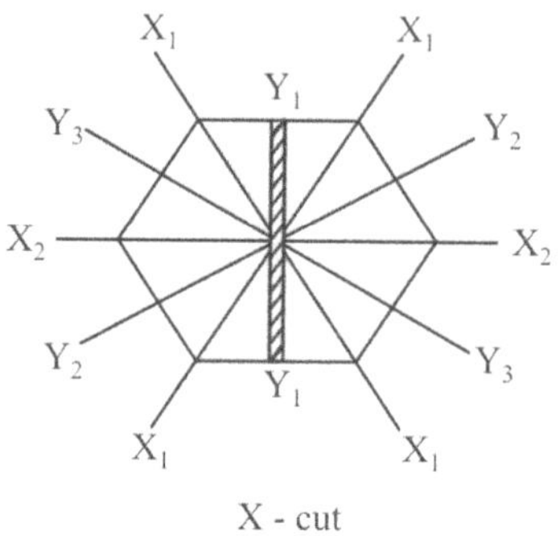

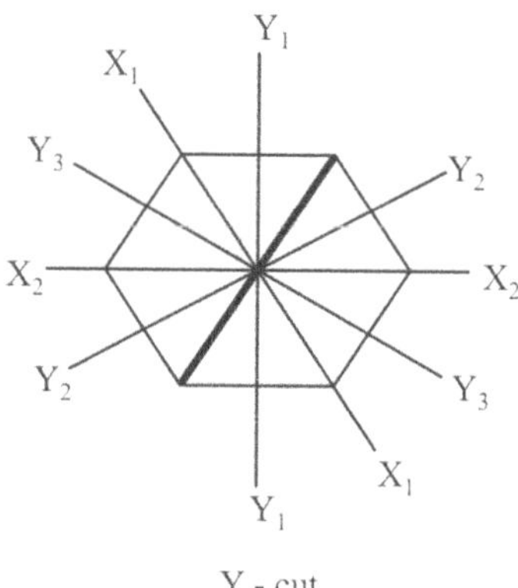

Fig. 6C.2

Then the frequency of the thickness vibration (along X axis) is

$$f = \frac{P}{2t}\sqrt{\frac{Y}{\rho}} \qquad .....(6C.4)$$

and the frequency of the length vibration is

$$f = \frac{P}{2l}\sqrt{\frac{Y}{\rho}} \qquad ....(6C.5)$$

where P = 1, 2, 3, ... for fundamental first overtone, second overtone, etc.

Y is the Young's Modulus along appropriate direction

ρ is the density of the crystal.

6C.6 Piezoelectric Oscillator

In 1917 Langevin utilised the piezoelectric effect for producing ultrasonic waves. The experimental set up is shown in Fig. 6C.3.

A thin slice of quartz crystal was subjected to a very high frequency alternating voltage provided by a triode valve oscillator. One end of the oscillatory circuit is connected to the plate of the valve and the other end to grid. The quartz crystal is placed in between two metal plates A and B to form a parallel plate capacitor with the crystal as directed, this is connected in parallel to the variable condenser C_1, by adjusting the variable condenser the frequency of the oscillatory plate is tuned to the natural frequency of the crystal. The tuning circuit is coupled to the amplifier T through capacitor C_g which provides the positive feedback to the amplifier. The oscillations generated by the tuning circuit are sustained and the output is applied to the electrodes of the piezoelectric crystal. When the frequency of oscillations of oscillator matches with the natural frequency of the piezoelectric crystal, at this stage, the crystal is set into mechanical vibrations and ultrasonic waves are generated. Frequencies as high as 15×10^6 Hz can be obtained by this method as

$$f = \frac{1}{2\pi}\sqrt{\frac{1}{LC}} \qquad \text{.....(6C.6)}$$

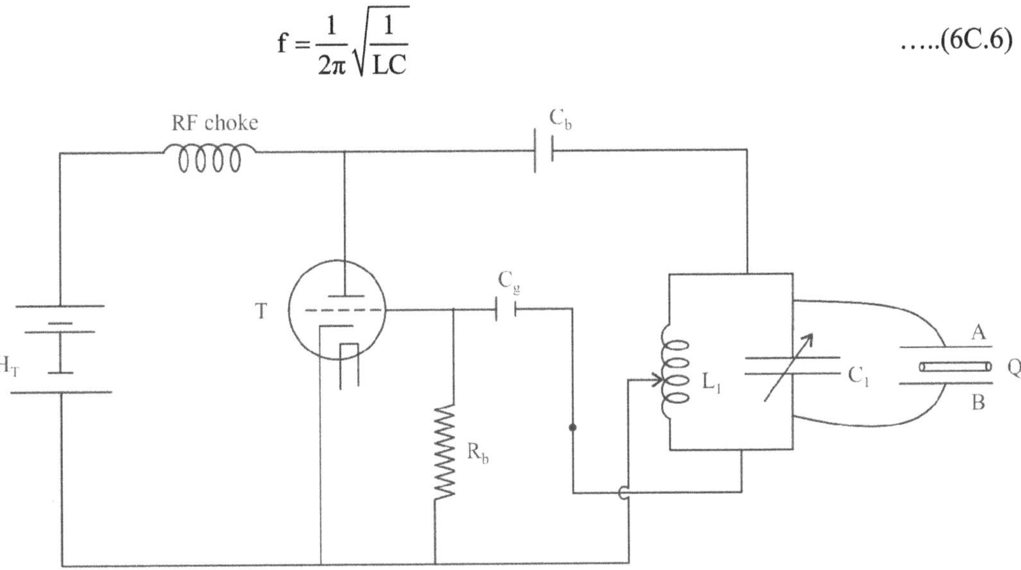

Fig. 6C.3 Piezoelectric oscillator.

6C.7 Methods to Determine Ultrasonic Waves

6C.7.1 Kundt's Tube Method

Kundt's tube is a long glass tube supported horizontally with an air column in it when the ultrasonic waves are passed in the Kundt's tube, the lycopodium powder sprinkled in the tube collects in the form of heaps at the nodal points and is blown off at the antinodal points. This method is used provided the wavelength is not very small.

6C.7.2 Flame Method

When a sensitive flame is moved in a medium where the ultrasonic waves are present, the flame remains stationary at antinodes and flickers at the nodes.

6C.7.3 Thermal Detector Method

A thin platinum wire is used whose resistance varies proportional to air velocity going to pass it. The resistance of the wire is minimum at nodes and maximum at antinodes. Exposure to the ultrasonic wave result in cooling of the wire causing the resistance fluctuation due to the amplitude of the ultrasonic waves. Therefore the ultrasonic waves amplitude can be calculated.

6C.7.4 Piezoelectric Detector

A quartz crystal can be used for the detection of the ultrasonic waves based on piezoelectric effect. One pair of opposite faces of the crystal is subjected to ultrasonic waves which exert mechanical pressure on the crystal and as a result, opposite charges are developed on the other pair of opposite face perpendicular to the first. These fluctuating charges are detected after amplification with the help of the amplifier.

6C.7.5 Acoustic Diffraction Method

This method is based on the fact that ultrasonic waves which consists of alternate compressions and rarefactions change the density of the medium through which they pass. This leads to a periodic variation of refractive index of the liquid, such a liquids column subjected to ultrasonic waves constitutes an acoustical grating. If the monochromatic light is passed through the waves the liquid causes diffraction of light.

Figure 6C.4 shows the experimental arrangement stationary ultrasonic waves are produced in a liquid contained in a glass tube. The density and so the refractive index of the liquid is maximum at the nodal point and minimum at the antinodal points. Hence the nodal area acts as opaque region while antinodal area acts as transparent region for light. The liquid column thus resembles the ruled grating.

The grating period d equal to $\lambda/2$ and is given by

$$d \sin \theta = n\lambda$$

$\lambda =$ wavelength of monochromatic light beam

N = order of minima

An acoustic diffraction grating produced by a liquid column subjected to ultrasonic waves.

If λ_a is the wavelength of ultrasonic waves then $d = \lambda_a/2$

$$\therefore \qquad \frac{\lambda_a}{2} \sin\theta = n\lambda$$

or
$$\lambda_a = \frac{2n\lambda}{\sin\theta} \qquad \text{.....(6C.7)}$$

If f is the frequency of ultrasonic wave then

$$\text{velocity } v = f\lambda_a = \frac{2fn\lambda}{\sin\theta} \qquad \text{.....(6C.8)}$$

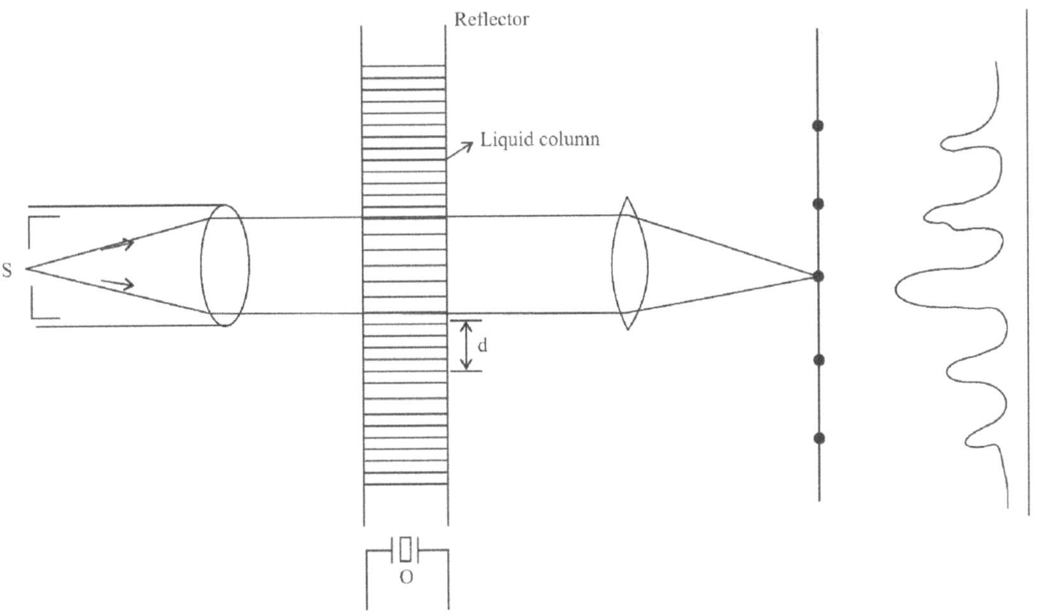

Fig. 6C.4 Acoustic grating.

6C.8 Applications of Ultrasonic Waves

Ultrasonic waves have found many applications mainly due to reasons:

1. At sufficiently high frequency, an almost parallel beam of plane wave can be propagated which is highly directional and focussed.
2. As the wavelengths are small the measurements can be made in a small sample without affecting the physical condition such as temperature, density etc.
3. The wavelengths of high frequency waves are of the order of molecular and electronic free path and the frequencies are of the same order as the relaxation frequencies of the certain internal molecular energy exchange.

6C.8.1 Sound Signalling

Using this technique, the ultrasonic waves of high frequency are used for sound signalling. These waves are focussed as narrow directional beam, are very suitable for directional signalling with negligible diffraction effects, the large amount of energy can be directed along the desired direction.

6C.8.2 Detection of Submarines, Icebergs, etc in Sea

The presence of submerged object in sea like icebergs, submarines, a shoal of fish in sea etc., can be easily detected using the pulses of ultrasonic signals that are sent out at short intervals. The reflected echo is received in the ultrasonic receiver and the time interval between the transmitted and received signal is noted. Knowing the velocity of sound in water, the distance between the source and the target can be calculated. It is also known as SONAR (Sound Navigation and Ranging).

6C.8.3 Thickness Measurement

The advantage of using ultrasonic waves for thickness measurement using echo principle is that there is no need for drilling holes of otherwise inflicting damage to the piece.

A piezoelectric transducer is used to send pulse and to collect the return pulse, which when connected to electrical signal. The time of flight of sound waves in material is found, assuming the velocity of sound wave be constant,

$$\therefore \qquad \text{Thickness} = \frac{v \times t}{2}$$

where v = velocity of sound in material

t = time

6C.8.4 Depth Sounding

The depth of sea or the depth of water below a ship can be found out by using the echo sounding principle. Ultrasonic waves generated by a crystal transducer in the range of 40-50 kHz are directed towards the bottom of the sea. The time lag between the sending of ultrasonic signals and its return back as an echo to the ship is recorded. Knowing the velocity and with measurement of the time lag, the depth of sea water below the ship can be calculated

$$\text{Depth of sea} = \frac{\text{Velocity of sound in sea} \times \text{time}}{2} \qquad \text{.....(6C.9)}$$

6C.8.5 Cavitation Effect

Microscopic bubbles of about 10^{-9} to 10^{-8} in size are always present in a liquid. A decrease in pressure above the liquid causes an intense evaporation in the bubbles and lead to their growth. The growth of bubbles led to their collapse. The entire process of growth and collapse of a bubble occur within one millisecond (10^{-3} sec).

During the collapse of a bubble, a shock wave in formed causing an abrupt increase in the temperature of the gas within the bubble. This effect is called cavitation effect. The formation and explosion of bubbles accounts for erosion and pitting of an ultrasonic transducer kept in the liquid. The bubbles have two effects although:

(a) They produce a clouds in front of the transducer and hinder the propagation of ultrasonic waves

(b) Frequent explosion of bubbles destroys the surface of the transducer causing pits. Nevertheless it has successful industrial applications like ultrasonic cleaning, ultrasonic emulsification etc.

6C.8.6 Non-Destructive Testing

These are characterized by the low intensity of the sound wave used. No damage is caused to the specimen material either physically or chemically.

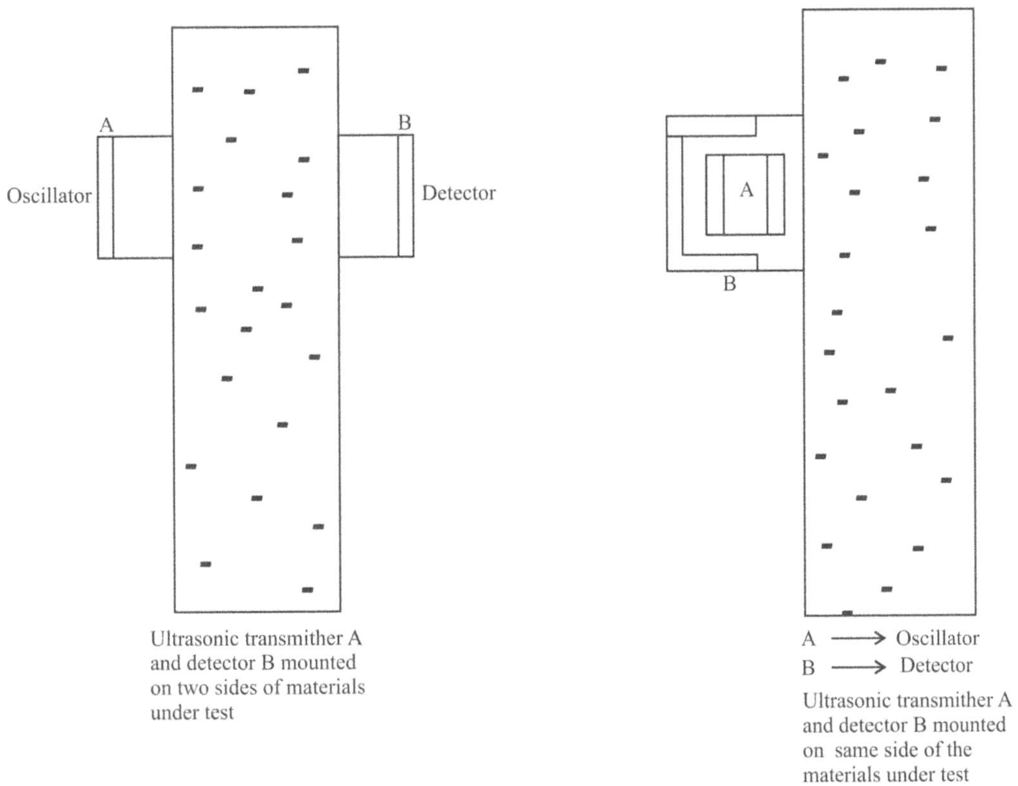

Fig. 6C.5

The instrument that is used to detect the defects such as cracks, flaws, holes in metal casting using non-destructive testing technique with the help of ultrasonic is called inspectoscope ultrasonic signals of frequencies ranging from 500 KHz to 15 MHz are generated by transmitter A and the detector B is mounted on the opposite faces of the material under test with the standard thickness material (T) under study, the time t_1, is noted when put to text. The return signal time t_2 is noted. The material is considered to be defective if $t_1 \neq t_2$. If there is no access to the other side of the material for mounting the detector, the transmitter, detector assembly is used as in Fig. 6C.5.

Transmitter A sends the signal and receiver B receives after the signal either reflected from boundary of material or from flaw, crack, hole etc. If transmission time is greater than receiver time, the material is considered to be defective.

Such a method is called Non-Destructive Testing (NDT)

6C.9 Applications of Ultrasound in Medicine

1. Pulsed Doppler ultrasound can be used to measure the velocity gradient across a blood vessel as well as the velocity of the heart wall or specific valves in the heart.
2. In the echocardiogram which utilizes a M scan technique, movement of the valves and other structures of the hearts are displayed as a function of time and usually in conjunction with an electrocardiogram.
3. A well established clinical application of ultrasonic imaging using the A scan mode of display is the encephalogram, which is used in determining the location of the medicine of the brain.
4. Important application involves diagnostic scanning of the eye for the diagnosis of retinal detachment, intraocular tumours, lens dislocation etc., using echo ophthalmoscope.
5. In obstetrics and gynecology ultrasonic imaging permits visualization of very small structures such as the ovaries and tubes and permits examination of a foetus as early as the 4 week stage.
6. Ultrasonic tomography techniques in which information from scans taken from many different vantage points are combined mathematically, provide increased detail in the visualization of certain parts of the body.

Solved Numerical Problems

1. A quartz crystals has thickness 0.001 m is vibrating at resonance, calculate the fundamental frequency. Given that Y for quartz = 7.9×10^{10} N/m^2 and ρ for quartz = 2650 Kg/m^3

Sol. We know that

$$v = \sqrt{\frac{Y}{\rho}}$$

$$v = \sqrt{\frac{7.9 \times 10^{10}}{2650}} = 4561 \text{ m/sec}$$

For the fundamental mode of vibration the thickness should be equal to $\lambda/2$.

Hence,

$$\lambda = 2t = 2 \times 0.001 = 0.002 \text{ m}$$

Now Velocity $\quad v = \nu\lambda$

$$\therefore \quad \nu = \frac{v}{\lambda} = \frac{5461}{0.002} = 2.730 \times 10^6 \text{ Hz}$$

$$\nu = 2.730 \text{ MHz}$$

2. A piezoelectric crystal with vibrating length 3×10^{-3} m has density 3.5×10^3 Kg/m^3. If it is made of a material of Young's Modulus 8×10^{10} N/m^2, what is its fundamental frequency.

Sol. The fundamental frequency is given by

$$\nu = \frac{1}{2t}\sqrt{\frac{Y}{\rho}}$$

$$\nu = \frac{1}{2 \times (3 \times 10^{-3})}\sqrt{\frac{8 \times 10^{10}}{3.5 \times 10^3}}$$

$$\nu = \frac{2\sqrt{2}}{6 \times 10^{-3}} \times \sqrt{\frac{10^8}{35}}$$

$$\nu = \frac{2\sqrt{2} \times 10^4}{6 \times 10^{-3} \times 5.916}$$

$$\nu = \frac{20 \times 1.414 \times 10^6}{6 \times 5.916}$$

$$\nu = 0.7967 \times 10^6 \text{ Hz}$$

$$\nu = 0.7967 \text{ MHz}$$

3. **A piezoelectric crystal has a thickness of 0.002 m. If the velocity of sound wave in crystal is 5750 m/s, calculate the fundamental frequency of the crystal.**

Sol. We know that

Frequency $$\nu = \frac{v}{\lambda} = \frac{v}{2t}$$

$$\nu = \frac{5720}{2 \times 0.002} = \frac{5720}{0.004}$$

$$\nu = 1.4375 \times 10^6 \text{ Hz}$$

$$\nu = 1.4375 \text{ MHz}$$

4. **Calculate the capacitance required to produce ultrasonic waves of 10^6 Hz with an inductance of 1 henry**

Sol. The frequency of LC circuit is given by

Velocity, $$\nu = \frac{1}{2\pi}\sqrt{\frac{1}{LC}}$$

$$C = \frac{1}{4\pi^2 \nu^2 L}$$

$$C = \frac{1}{4 \times (3.14)^2 \times (10^6)^2 \times 1}$$

$$C = 0.025 \times 10^{-12} \text{ Farad}$$

5. **Calculate the velocity of ultrasonic waves in a liquid in an acoustic grating experiment using the following data.**

Wavelength of light used = 600 nm

Frequency of ultrasonic waves = 100 MHz

Angle of diffraction for first order = 5°

Sol. We know that

Velocity $$v = \frac{2\, f\, n\lambda}{\sin\theta}$$

Given

$$f = 100 \text{ MHz} = 100 \times 10^6 \text{ Hz}$$

$$\lambda = 600 \text{ nm} = 600 \times 10^{-9} \text{ m}$$

$$\theta = 5^\circ \text{ for } n = 1$$

Hence

$$v = \frac{2\times100\times10^{6}\times600\times10^{-9}}{\sin 5^{o}}$$

$$v = \frac{2\times6\times10}{0.0872}$$

$$v = 1376 \text{ m/sec}$$

Conceptual Questions and Answers

1. **What are ultrasonic waves?**

Ans. Sound waves having frequency greater than 20 KHz are called ultrasonic waves.

2. **Explain magnetostriction effect?**

Ans. When a rod of ferromagnetic material like iron, cobalt and nickel is magentised longitudinally the rod undergoes a small change in length. This is called magnetostriction effect.

3. **What is piezoelectric effect?**

Ans. When pressure is applied to one pair of opposite faces of crystal like quartz, tourmaline, Rochelle salt etc, out with their faces perpendicular to its optic axis equal and opposite charges appear across its other faces. This is known as piezoelectric effect.

4. **What is inverse piezoelectric effect?**

Ans. If an alternating voltage is applied to one pair of opposite faces of the crystal, alternatively mechanical contraction and expansions are produced across the other opposite faces. This phenomenon is known as inverse piezoelectric effect.

5. **What is a SONAR?**

Ans. SONAR is a device which stands for Sound Navigation and Ranging. It is based on the echo sounding technique of ultrasound.

It is the acoustical technique used for finding the distance and direction of submarines, depth of sea, depth of rocks in the sea, etc.

6. **What is acoustic grating?**

Ans. When ultrasonic waves propagate in a liquid medium, the alternating compression and rarefactions vary the density of the medium. This change in density results in variation of refractive index of the liquid.

Under this condition, if a parallel beam of monochromatic light is passed through them at right angles to the waves, the liquid causes the light to undergo diffraction. Such a liquid column behaving like a grating is called acoustic grating.

Objective Questions

1. Ultrasonic waves are those waves which have
   (a) Frequencies less than 20 Hz
   (b) Frequencies between 20 Hz and 20 KHz
   (c) Frequencies above 20 KHz
   (d) Wavelength less than the wavelength of visible light
2. Which of the following cuts of quartz crystal is used in piezoelectric generator.
   (a) X-cut (b) Y-cut
   (c) Anyone (d) None of these
3. Which method is used for generation of ultrasonic waves
   (a) Magnetostriction generator (b) Piezoelectric generator
   (c) Galtons' whistle (d) All
4. Ultrasonic waves can be detected by
   (a) Galton's whistle (b) Piezoelectric oscillator
   (c) Piezoelectric detector (d) None of these
5. The wavelength of ultrasonic waves in air is of the order of
   (a) 1 cm (b) 5×10^{-8} cm
   (c) 5×10^{5} cm (d) 500 cm
6. Ultrasonic waves can be detected by
   (a) Hebb's method (b) Quincle's method
   (c) Kundt's method (d) A telephone
7. The wavelength of wave in a medium is related to frequency by
   (a) $v = f\lambda^2$ (b) $v = f/\lambda$
   (c) $v = f\lambda$ (d) $v = \lambda/f$

8. In piezoelectric effect the frequency of ultrasonic is related with Young's modulus

   (a) $f = 2t\sqrt{\rho/Y}$ (b) $f = \frac{1}{2t}\sqrt{\rho/Y}$

   (c) $f = \frac{1}{2t}\sqrt{Y/\rho}$ (d) $f = 2t\sqrt{Y/\rho}$

9. In magnetostriction method ultrasonic waves are produced by

   (a) Vibration of rod (b) Filament of valve

   (c) From LC circuit (d) Changing of magnetic field

10. Ultrasonic waves, infrasonic and audio travel in medium with speed V_u, V_i and V_a respectively then

    (a) $V_u < V_a < V_i$ (b) $V_u > V_a > V_i$

    (c) $V_a < V_u$ and $V_u = V_i$ (d) $V_a = V_u = V_i$

Theoretical Questions

1. Define piezoelectric effect? Draw a circuit diagram for the production of ultrasonic using a quartz crystal. Explain its working.
2. What is magnetostriction effect and show how magnetostriction oscillator produces ultrasonics waves?
3. Write short note on application of ultrasonic.
4. Explain the cavitation effect when an ultrasonic wave is passed through a liquid.
5. Write short note on uses of ultrasonics in echo sounding and flaw detection.
6. Explain the method of non-destructive testing.
7. Explain direct and inverse piezoelectric effect. Explain with neat diagram their uses in ultrasonics.
8. Write a short note on

   (a) Uses of ultrasonic wave as flaw detector

   (b) SONAR

   (c) Acoustic Grating

   (d) Cavitation

Numerical Problems

1. Calculate the depth of the sea if the time interval between the emitted signal and the echo received is 2 seconds in sonar studies. Take the velocity of sound in sea water as 1490 m/s.

   [1490 m/s]

2. Calculate the natural frequency of the thickness vibrations for Quartz plate of thickness 5.5×10^{-3} m, given that Young's modulus along X-axis is 8×10^{10} N/m^2 and density of crystal is 2.65×10^3 Kg/m^3.

   [500×10^3 Hz]

3. Calculate the frequency of the fundamental note emitted by piezoelectric crystal using the following data: vibrating length = 3 mm; Young's modulus = 8×10^{10} N/m^2 and density of crystal = 2.5 gm/cm^3.

   [0.943×10^6 Hz]

4. A quartz crystal in a ultrasonic interferometer produces stationary waves of frequency 1.5 MHz. If the distance between 6 consecutive nodes is 2.75 mm, find the velocity of the ultrasonic waves

   [1650 m/sec]

5. A quartz crystal of thickness 0.001 m is vibrating at resonance. Calculate the fundamental frequency of it. (Young's modulus of quartz = 7.9×10^{10} N/m^2, and its density = 2560 Kg/m^3)

   [2730 KHz]

Answers to Objective Questions

1. (c)	2. (a)	3. (d)	4. (c)	5. (a)	6. (c)
7. (c)	8. (c)	9. (a)	10. (d)		

CHAPTER 7

Electromagnetics

We can scarcely avoid the interference that light consists in the transverse undulations of the same medium which is the cause of electric and magnetic phenomena.

- James Cleark Maxwell

7.1 Introduction

In 1864 J.C. Maxwell put a mathematical theory that showed a close relationship between all electrical and magnetic phenomenon. To develop his theory, Maxwell used the following four concepts.

1. Electric field originates on positive charge and terminates on negative charge.
2. Magnetic field lines always form a close loops.
3. A varying magnetic field induces an e.m.f and hence an electric field.
4. Magnetic field are generated by moving charges or currents.

In fact entire theory of electromagnetic field is condensed in to four laws. These laws are namely, Gauss's law in electrostatics, Gauss's law in magnetism, Ampere's law and Faraday's law; govern the interaction of bodies which are magnetic or electrically charged or both.

A complete set of relation giving the connection between the charges at rest (electrostatics), charges in motion (current electricity), electric fields and magnetic fields (electromagnetism) were derived theoretically and summarized in four equations by Maxwell, called Maxwell's equations.

Maxwell's equation help us to explain why a compass needle points north, why light bends when it enters water, why thunderstorm occur, why we see aurora in the polar region, and many other natural phenomenon. These equations also form the basis for the operation of a large number of devices in use today e.g., electric motor, television, transmitters and receivers, microwave ovens computers radar etc.

7.2 Electromagnetic Waves

According to Maxwell's equation representing Faraday's law, a changing magnetic field produces an electric field; while according to equation representing Ampere-Maxwell's circuital law, a changing electric field produces a magnetic field. It means that a change in either field produces the other field. Maxwell worked out from his equations the variation in electric and magnetic fields would lead to a wave consisting of fluctuating electric and magnetic fields perpendicular to each other and also perpendicular to the direction of propagation of the wave. Such waves which can propagate in space without any material medium are called electromagnetic waves.

Fig. 7.1 shows a graphical representation of an electromagnetic wave in which the electric field vector $\bar{E}$ and the magnetic field vector $\bar{B}$ are vibrating along Y-axis and Z-direction respectively and the wave is propagation along X-direction. Both $\bar{E}$ and $\bar{B}$ vary with time and space and have the same frequency.

Maxwell predicted also that the electromagnetic waves would travel in free space (vacuum) with the speed of light. Thus he concluded that light itself is an electromagnetic wave which is transverse in nature, other examples of electromagnetic waves are radiowave, microwave, infrared rays, ultraviolet rays, X-rays and γ-rays.

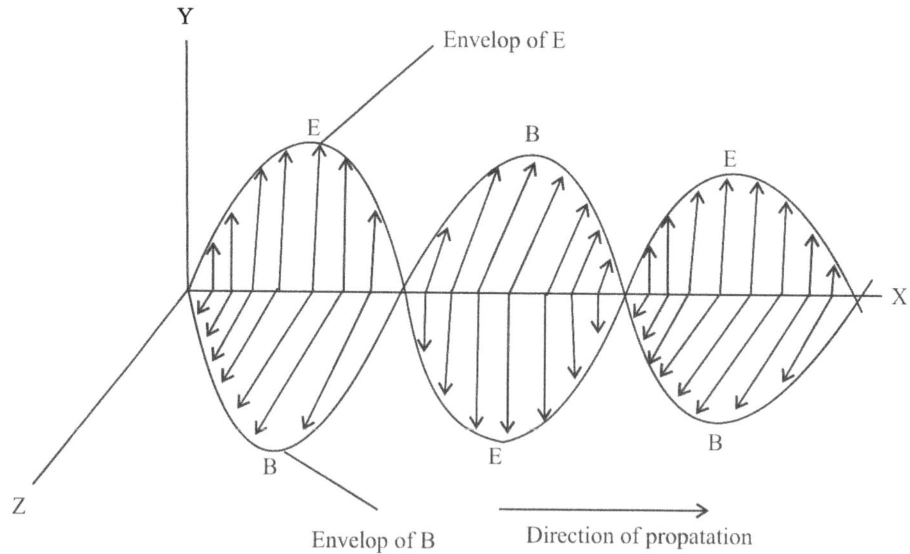

Fig. 7.1

7.3 Transverse Nature of Electromagnetic Waves

Let us consider a plane electromagnetic wave propagating along X-axis. The wave front will be a plane one parallel to YZ plane. For any position of the wavefront, the electric and magnetic fields $\bar{E}$ and $\bar{B}$, will be zero to the right of the wavefront, while to the left of it the values of $\bar{E}$ and $\bar{B}$ will depend on x and t. (but not upon y and z because we are considering a plane wave).

To establish transverse nature of electromagnetic waves, we shall show that $\bar{E}$ and $\bar{B}$ are perpendicular to the direction of wave propagation.

Let us consider an elementary parallelepiped ABCDEFGH with sides dx, dy and dz parallel to X, Y and Z axes respectively as shown in Fig. 7.2.

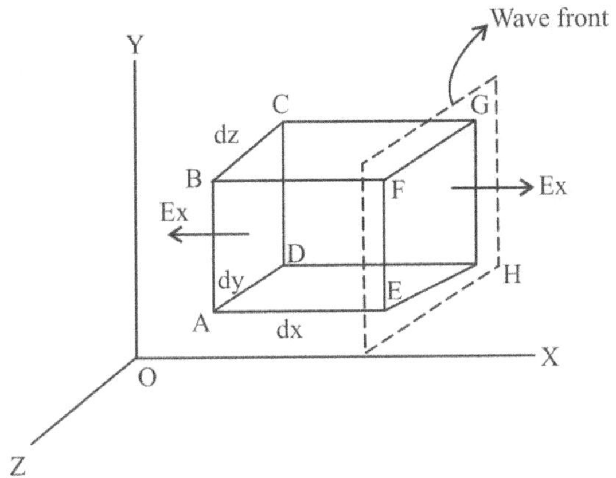

Fig. 7.2

Suppose the wavefront reaches the face EFGH at any time t. The field $\bar{E}$ and $\bar{B}$ are zero to the right of EFGH, while to the left of EFGH, their values vary with x-only. Since the parallelopiped does not enclose any charge, then according to Gauss's law for electricity, the total electric flux through its six faces is zero, that is

$$\oint \vec{E}.\vec{ds} = 0 \qquad .....(7.1)$$

$$\int_{ABCD} \vec{E}.\vec{ds} + \int_{EFGH} \vec{E}.\vec{ds} + \int_{BCGF} \vec{E}.\vec{ds} + \int_{ADHE} \vec{E}.\vec{ds} + \int_{ABFE} \vec{E}.\vec{ds} + \int_{DCGH} \vec{E}.\vec{ds} = 0 \qquad .....(7.2)$$

Since $\bar{E}$ does not vary with y, the electric flux through faces BCGF and ADHE (normal to Y-axis) are equal and opposite and hence the contributions of third and fourth integrals mutually cancel. Similarly the contributions of third and fourth integrals mutually cancel. Similarly the contribution of fifth and sixth integrals mutually cancel because $\bar{E}$ does not vary with Z also. Hence eq. 7.2 reduces to

$$\int_{ABCD} \vec{E}.\vec{ds} + \int_{EFGH} \vec{E}.\vec{ds} = 0 \quad .....(7.3)$$

If E_x and E_x' be the values of the x-components of $\bar{E}$ at the faces ABCD and EDGH respectively, than we have

$$\int_{ABCD} \vec{E}.\vec{ds} = -E_x(dy\,dz)$$

and

$$\int_{EFGH} \vec{E}.\vec{ds} = E_x'(dy\,dz)$$

where dy dz is the area of each face. The minus sign arises from the negative direction along x-axis) of the outward normal to the face ABCD.

Thus eq. 7.3 becomes

$$-E_x(dy\,dz) + E_x'(dydz) = 0$$

$$E_x' - E_x = 0 \qquad \text{since } dy\,dz \neq 0$$

Now there are two possibilities

1. E_x' is equal to E_x or
2. Both E_x' and E_x are separately zero

The possibility $E_x' = E_x$ means that the electric field is static. But a static field cannot propagate a wave of finite wavelength. Hence this possibility is ruled out and we conclude that

$$E_x = 0 \text{ and } E_x' = 0$$

It means that the electric field $\vec{E}$ has no component parallel to the direction of propagation. A similar argument holds for the magnetic field $\vec{B}$.

Hence we conclude that in an electromagnetic wave, both the electric and magnetic fields are perpendicular to the direction of propagation i.e., electromagnetic waves are transverse in nature.

7.4 Properties of Electromagnetic Waves

1. The electromagnetic waves are produced by accelerated charges.

2. These waves do not require any material medium for propagation.
3. These waves travel in free space with a speed given by $\frac{1}{\sqrt{\mu_o \in_o}}$, whose value equals the speed of light.
4. The direction of variation of electric and magnetic field are perpendicular to each other and also perpendicular to the direction of propagation. Thus electromagnetic waves are transverse in nature.
5. The variation in electric and magnetic field occur simultaneously and the field acquire their maximum value E_o and B_o at the same place and at the same time.
6. In free space the magnitude of electric and magnetic fields in electromagnetic wave are related by $\frac{E}{B} = C$.
7. The energy in electro magnetic wave is divided on the average, equally between electric and magnetic field.
8. In vacuum the average energy density is $\frac{1}{2} \in_o E^2$ and the average magnetic energy density is $\frac{B^2}{2\mu_o}$
9. The electric field vector is responsible for the optical effect of an electromagnetic wave.

7.5 Ampere's Law: Ampere Circuital Theorem

The theorem states that the line integral of the magnetic field around any closed path is equal to μ_o times the current enclosed by the path.

Consider a closed path of any shape around a current carrying conductor. Let d*l* be a small element in the closed path. Let B be inclined at an angle θ with the element d*l*. Then B cos θ is the tangential component of B. The line integral of magnetic field around the path is

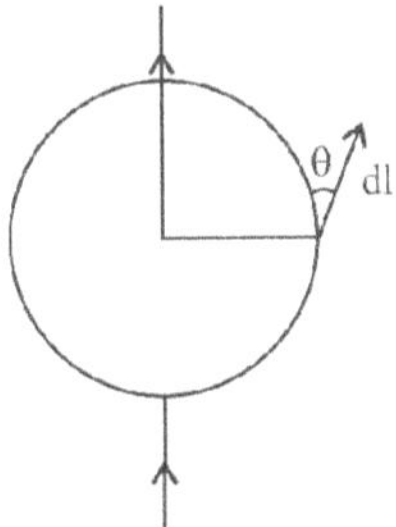

$$\oint B\, dl \cos\theta = \int \vec{B}.\vec{dl}$$

By Ampere circuital theorem

$$\oint \vec{B}.\vec{dl} = \mu_o I \qquad .....(7.4)$$

where I is the current enclosed by the closed loop and μ_o is the permeability of free space.

7.6 Applications of Amperes Circuital Law

7.6.1 To find B due to a Straight Conductor

Consider a closed path of any shape around a straight conductor carrying a current I. For simplicity, let us consider a circular path of radius r. At any point on this curve the value of B is constant in magnitude and is parallel to the element dl.

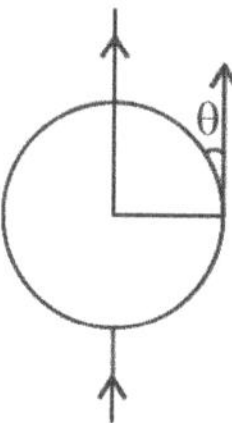

By Ampere circuital theorem

$$\oint \vec{B}.\vec{dl} = \mu_o I$$

$$B \oint dl = B \times 2\pi r$$

$$\therefore \qquad B = \frac{\mu_o I}{2\pi r} \qquad .....(7.5)$$

7.6.2 To Find B due to Toroid

A toroid is an endless solenoid. For this take a long solenoid and bent round in the form of a closed ring. Let a be the mean radius of the toroid and n be the number of turns per metre length of the toroid, as shown in Fig. 7.5.

Let the current flowing through the coil be I. Consider a closed circular path of radius 'a'. Along this path, by symmetry the magnetic field is uniform and tangential to the circular path.

By Ampere's circuital theorem

$$\oint \overline{B}.\overline{dl} = B \oint dl = B(2\pi a)$$

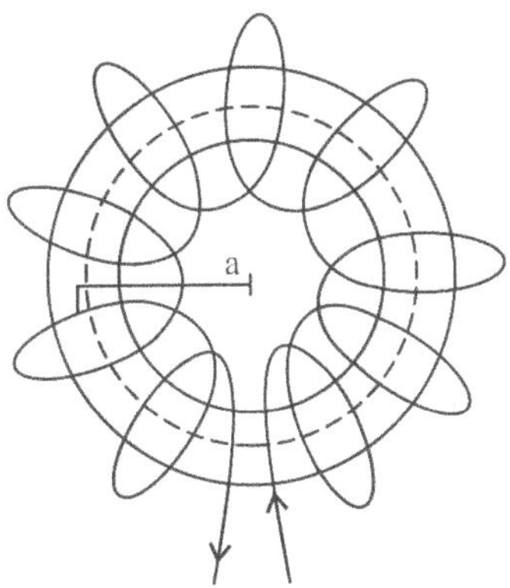

Fig. 7.5

The total current enclosed by the closed path

$$= 2\,\pi\,a\,n\,I$$

$$\therefore \qquad B(2\pi a) = \mu_o\,(2\,\pi\,a\,n\,I)$$

$$B = \mu_o\,n\,I \qquad \text{.....(7.6)}$$

7.6.3 Magnetic Field Due to Solenoid

Ampere's circuital theorem is given by

$$\oint \overline{B}.\overline{dl} = \mu_o\,I \qquad \text{.....(7.7)}$$

The figure for solenoid is shown in Fig. 7.6.

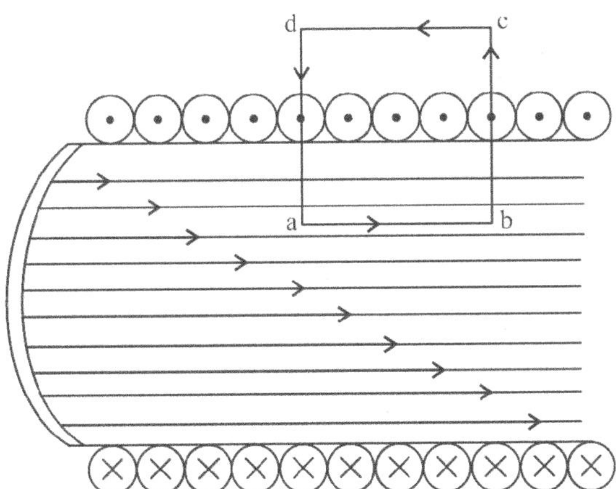

Fig. 7.6

Consider a rectangular path abcd in long solenoid as shown in Fig. 7.6.

$$\oint \bar{B}.\overline{dl} = \int_a^b \vec{B}.\vec{dl} + \int_b^c \vec{B}.\vec{dl} + \int_c^d \vec{B}.\vec{dl} + \int_d^a \vec{B}.\vec{dl}$$

$$\oint \bar{B}.\overline{dl} = \int B\, dl \cos\theta = B \times l \cos 0^\circ = Bl$$

In this case $dl = ab = l$, the angle between $\bar{B}$ and $d\bar{l}$ is $\theta = 0$

$$\therefore \qquad \int_b^c \overline{B}.\overline{dl} = \int_d^a \overline{B}.\overline{dl} = 0$$

Because $\bar{B}$ and $d\bar{l}$ are perpendicular; $\theta = 90^\circ$ and $\int_c^d \overline{B}.\overline{dl} = 0$ because the field outside the solenoid is zero.

$$\therefore \qquad \oint \bar{B}.\overline{dl} = Bl$$

Current passing through each turn is I. Let n be the number of turns per metre length of the solenoid and l be the length of the solenoid.

Then total current is nlI. Substituting this in equation (7.7)

We get

$$Bl = \mu_o\, n\, l\, I$$

$$B = \mu_o\, n\, I \qquad .....(7.8)$$

7.7 Poisson's and Laplace's Equation

7.7.1 Poisson's Equation

The intensity of electric field $\bar{E}$ and electric potential V at any point in an electric field are related by the equation

$$\vec{E} = -\text{grad } V \qquad .....(7.9)$$

$$\vec{E} = -\vec{\nabla} V$$

where

$$\vec{\nabla} = \hat{i}\frac{\partial}{\partial x} + \hat{j}\frac{\partial}{\partial y} + \hat{k}\frac{\partial}{\partial z}$$

We know that Gauss's law in differential form is given by

$$\text{div } \vec{E} = \rho/\varepsilon_o \qquad .....(7.10)$$

where ρ is the volume density of charge

Substituting the value of $\bar{E}$ from eq. 7.9 in eq. 7.10 we get

$$\text{div } (-\text{ grad } V) = \rho/\varepsilon_o$$

$$\text{div grad } V = -\rho/\varepsilon_o$$

$$\text{div grad } V = \vec{\nabla}.\vec{\nabla} V = \rho/\varepsilon_o \qquad .....(7.11)$$

Since

$$\vec{\nabla}.\vec{\nabla} = \left(\hat{i}\frac{\partial}{\partial x} + \hat{j}\frac{\partial}{\partial y} + \hat{k}\frac{\partial}{\partial z}\right) . \left(\hat{i}\,\partial/\partial x + \hat{j}\,\partial/\partial y + \hat{k}\,\partial/\partial z\right)$$

$$\therefore \quad \nabla^2 V = \frac{\partial^2}{\partial x^2} + \frac{\partial^2}{\partial y^2} + \frac{\partial^2}{\partial z^2} \qquad .....(7.12)$$

where ∇^2 is called the Laplacian operator. The eq. 7.11 can be written in term of Laplacian operator as

$$\boxed{\nabla^2 V = \rho/\varepsilon_o} \qquad ......(7.13)$$

Eq. 7.13 is called the Poisson's equation.

7.7.2 Laplace's Equation

For a charge free region, the volume density of charge (ρ) is zero. Hence, by substituting ρ = 0 in poission's equation

$$\nabla^2 V = \rho/\varepsilon_o$$

We have

$$\nabla^2 V = 0 \qquad .....(7.14)$$

This equation is known as Laplace's equation.

7.8 Basic Laws of Electricity and Magnetism

Maxwell's electromagnetic equations are based upon the following laws of electricity and magnetism.

7.8.1 The Gauss's Law of Electrostatics

The Gauss's law of electrostatics

$$\int_S \vec{E}.\vec{ds} = \frac{q}{\varepsilon_o}$$

$$\int_S \vec{E}.\vec{ds} = \frac{1}{\epsilon_o}\int_V \rho dv$$

and in terms of displacement current,

since $\vec{D} = \epsilon_o \vec{E}$

the above equation takes the form

$$\int_S \vec{D}.\vec{ds} = \int_V \rho dv$$

where $\int_V$ is the surface integral over the surface enclosing volume V.

From Gauss divergence theorem

$$\int_S \bar{D}.d\bar{s} = \int_V \text{div}\, D\, d\, v$$

Hence eq. (7.15) becomes

$$\int_V \text{div}\, D\, d\, v = \int_u \rho dv$$

or

$$\int_V (\text{div}\, D - \rho)\, dv = 0$$

or

$$\text{div}\, D - \rho = 0$$

$$\text{div}\, D = \rho \qquad \text{.....(7.16)}$$

7.8.2 The Gauss's Law in Magnetostatics

The magnetic flux density B is given by

$$\int_S \vec{B}.\vec{ds} = 0 \qquad \text{.....(7.17)}$$

From Gauss divergence theorem

$$\int_s \vec{B}.\vec{ds} = \int_v \text{div B dv}$$

Hence eq. 7.17 becomes

$$\int_v \text{div B dv} = 0$$

$$\text{div B} = 0 \qquad .....(7.18)$$

7.8.3 Faraday's Law of Electromagnetic Induction

$$\text{Curl } \vec{E} = -\frac{\partial \vec{B}}{\partial t}$$

$$\nabla \times \vec{E} = -\frac{\partial \vec{B}}{\partial t}$$

7.8.4 Ampere's Circuital Law

$$\oint_c \vec{H}.\vec{dl} = \int_s \vec{J}.\vec{ds} \qquad .....(7.19)$$

where H is magnetic field intensity and J is current density.

From stoke's theorem, we have

$$\int_s \text{Curl } \vec{H}.\text{d}\vec{s} = \int_s \text{J ds}$$

$$\text{Curl } \overline{H} = \overline{J} \qquad .....(7.20)$$

The fundamental laws governing electrical and magnetic phenomenon is given in Table 7.1.

Table 7.1 A list of the fundamental laws governing electrical and magnetic phenomenon.

S.No.	Law	What the law says	Mathematical state
1.	Gauss's law for the electric field	Charges gives rise to electric field; electric field lines begin and end on charges	$\int_s \overline{E}.d\overline{s} = \frac{q}{\varepsilon_o}$

Contd...

S.No.	Law	What the law says	Mathematical state
2.	Gauss's law for the magnetic field	Magnetic field lines close on themselves they do not begin or end at any point. This implies that an isolated magnetic pole does not exist	$\int_s \bar{B}.d\bar{s} = 0$ or $\nabla.\bar{B} = 0$
3.	Faraday's law of electromagnetic induction	Changing magnetic field gives rise to electric field	$\oint_c \bar{E}.d\bar{l} = -\frac{\partial\phi_B}{\partial t}$ or $\nabla\times\bar{E} = -\partial B/\partial t$
4.	Ampere's law (For steady current only)	Electric field give rise to magnetic field	$\oint_c \bar{B}.d\bar{l} = \mu_o I$ or $\nabla\times B = \mu_o J$ $\nabla\times\bar{H} = \bar{J}$

7.9 Equation of Continuity

The net amount of charge in an isolated system remains constant. The rate of flow of charge due to negatively charged electron constitute electric current i.e.,

$$I = -\frac{dq}{dt} = -\frac{d}{dt}\int_v \rho dv \quad .....(7.21)$$

The current I may also be expressed in terms of current density J as

$$I = \oint_s \vec{J}.\vec{ds}$$

$$\oint_s \vec{J}.\vec{ds} = -\frac{d}{dt}\int_v \rho dv$$

From Gauss divergence theorem, we know that

$$\oint_s \vec{J}.\vec{ds} = \int_v (\text{div } J)\, dv \quad ....(7.22)$$

Then

$$\int_v (\text{div } J)\, dv = -\int_v \left(\frac{\partial\rho}{\partial t}\right) dv$$

$$\int_v \left(\text{div } J + \frac{\partial\rho}{\partial t}\right) dv = 0$$

$$\text{div } J + \frac{\partial\rho}{\partial t} = 0 \quad .....(7.23)$$

This is called equation of continuity.

This equation also explain the fact that the electric charges can neither be created nor be destroyed,

For static fields

$$\frac{\partial \rho}{\partial t} = 0$$

$$\therefore \qquad \vec{\nabla}.\vec{J} = 0 \qquad \text{.....(7.24)}$$

This equation represents the continuity equation for static fields.

Note:

1. ***Gauss's divergence theorem:*** According the Gauss divergence theorem, the surface integral can be changed in to volume integral

$$\int_s \vec{A}.\vec{ds} = \int_v \text{div } \vec{A} \text{ dv} \qquad \text{.....(7.25)}$$

2. ***Stoke's theorem:*** According to stoke's theorem the line integral may be transform in to surface integral as

$$\int_s \text{Curl } \vec{A}.\vec{ds} = \int_l \vec{A}.\vec{dl} \qquad \text{.....(7.26)}$$

7.10 Maxwell's Displacement Current

A current in a conductor produces a magnetic field the qualitative relationship between the current and the magnetic field is given by Ampere's circuital law which states that the line integral of the magnetic field $\bar{B}$ around any closed loop is equal to μ_o times the net current I threading through the area enclosed by the loop.

Mathematically

$$\oint_c \bar{B}.d\bar{l} = \mu_o I \qquad \text{.....(7.27)}$$

where μ_o is the permeability constant and I is the conduction current through any surface bounded by a closed loop C around which the line integral is taken.

Maxwell in 1864, showed that the above equation was logically inconsistent when applied to charging or discharging of a capacitor. To understand it, let us consider a circuit containing a capacitor, a battery and a key as shown in Fig. 7.7. When the key is closed, a conduction current I flows through the connecting wires and the capacitor begins to charged. The current decreases as the charge on the capacitor grows. When the capacitor is fully charged to the voltage of the battery, the current vanishes. There is no current between the plates of the capacitor during charging.

Let us now consider a plane circular surface S_1 and a hemispherical surfaces S_2, both bounded by the same closed loop C, such that during charging the conduction current I threads through S_1 but not through S_2 (which lies in the region between the plates).

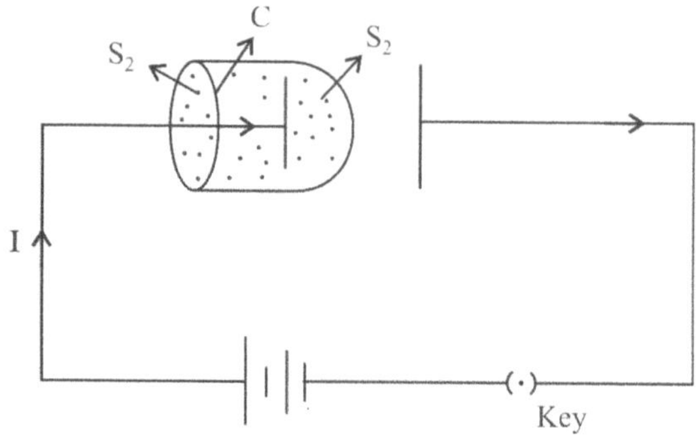

Fig. 7.7

Now, if the closed loop C is considered to bound the surface S_1, then by Ampere's circuital law, we have

$$\oint_c \bar{B}.d\bar{l} = \mu_o I \qquad .....(7.28)$$

Because the current I threads through S_1

If the closed loop C is considered to bound the surface S_2, then

$$\oint_c \bar{B}.d\bar{l} = 0 \qquad ......(7.29)$$

Because no current threads through S_2

Obviously eq. 7.28 and 7.29 both cannot be correct. Maxwell pointed out that this contradiction arises because of the assumed discontinuity in current (absence of current between the plates). He argued that during charging, a changing electric field exists in the region between the capacitor plate. The field is equivalent to a current which exists as long as the electric field is changing and produces the same magnetic effect as does the conduction current.

The current corresponding to the changing electric field is called the 'displacement current'. Thus during charging of capacitor, there is a conduction current I through the connecting wires and a displacement current I_d in the region between the plates of the capacitor, the current being thus continuous. The condition of continuity of current requires the displacement current I_d to be equal to $\epsilon_o \frac{d\phi_E}{dt}$, where ϵ_o is the permittivity constant and ϕ_E is the electric flux between the capacitor plates. Thus the Ampere's circuital law is modified as below

$$\oint \bar{B}.d\bar{l} = \mu_o (I + I_d) \quad .....(7.30)$$

$$\oint \bar{B}.d\bar{l} = \mu_o \left(I + \varepsilon_o \frac{d\phi_E}{dt} \right) \quad .....(7.31)$$

Note:

1. The displacement current in the region between the plates is precisely equal to the conduction current in the connecting wires. Thus the displacement and the conduction current together are continuous through the entire circuit including the capacitor although individually they are discontinuous.
2. There is however a difference in the magnetic effect produced by conduction and displacement currents. A conduction current in a thin wire produces a magnetic field at the surface of the wire much larger than the field produced by an equal displacement current at the edge of the capacitor plate. The reason is that the conduction current is confined to the thin wire but the displacement current is spread out over the entire surface area of the capacitor plates.

7.10.1 To Prove Conduction Current and Displacement Current are Equal

Let us now prove that the displacement current I_d in the region between the plates is equal to the conduction current in the connecting wires.

Let q be the charge on the capacitor plates at any instant. Then the instantaneous electric field between the plates is

$$E = \frac{q}{\epsilon_o A} \quad .....(7.32)$$

where A is the surface area of each plate.

From this we get

$$\frac{dE}{dt} = \frac{1}{\varepsilon_o A} \frac{dq}{dt} = \frac{1}{\epsilon_o A} I \quad .....(7.33)$$

where $I = \frac{dq}{dt}$ is the instantaneous current in the conducting wires. Thus

$$I = \epsilon_o A \frac{dE}{dt}$$

The displacement current I_d is given by

$$I_d = \epsilon_o \frac{d\phi_E}{dt}$$

But $$\phi_E = EA$$

$$\therefore \quad I_d = \epsilon_o A \frac{dE}{dt} \qquad .....(7.34)$$

$$\therefore \quad I_d = I \qquad .....(7.35)$$

which means that displacement current I_d is equal to the conduction current.

7.10.2 Characteristics of Displacement Current

1. Displacement current is a current in the sense that it produces a magnetic field. As it is not linked with the motion of charge, displacement current has no other properties of current. It has a finite value even in perfect vacuum, where there is no charge at all.

2. Displacement current serves the purpose to make the total current across the discontinuity in a conduction current.

3. Displacement current in good conductors is almost nil as compared to the conduction current at frequencies less than optical frequency $\approx 10^{15}$ Hz.

7.11 Maxwell's Electromagnetic Equations

7.11.1 Maxwell's Equation in Differential Form

1. $\nabla.\vec{D} = \rho$

2. $\nabla.\vec{B} = 0$

3. $\nabla.\vec{E} = -\frac{\partial \vec{B}}{\partial t}$

4. $\nabla.\vec{H} = \vec{J} + \frac{\partial \vec{D}}{\partial t}$

7.11.2 Maxwell's Equation in Integral Form

1. $\oint_S \vec{D}.d\vec{s} = \int_V \rho.dv$

2. $\oint_s \vec{B}.\vec{ds} = 0$

3. $\oint \vec{E}.\vec{dl} = -\int_s \frac{\partial \vec{B}}{\partial t}.\vec{ds}$

4. $\oint \vec{H}.\vec{dl} = \int_s \left[\vec{J} + \frac{\partial \vec{D}}{\partial t}\right].\vec{ds}$

7.11.3 Maxwell's Equation in Free Space

In free space $\rho = 0$ and $J = 0$ and hence the Maxwell's equations in differential form becomes

1. $\nabla.\vec{D} = 0$ or $\nabla.\vec{E} = 0$
2. $\nabla.\vec{B} = 0$ or $\nabla.\vec{H} = 0$
3. $\nabla \times \vec{E} = -\frac{\partial B}{\partial t} = -\mu_0 \frac{\partial \vec{H}}{\partial t}$
4. $\nabla \times \vec{H} = -\frac{\partial \vec{D}}{\partial t} = \epsilon_0 \frac{\partial \vec{E}}{\partial t}$

7.12 Physical Significance of Maxwell's Equations

1. ***Maxwell's first equation:*** Maxwell's first equation is given as

$$\nabla.\overline{D} = \rho \quad \text{(Differential form)} \quad .....(7.36)$$

$$\oint_s \overline{D}.\overline{ds} = \int_v \rho dv \quad \text{(Integral form)} \quad .....(7.37)$$

The total electric displacement through the surface enclosing a volume is equal to the total charge within the volume.

(a) It represents Gauss's law

(b) It is time independent or steady state equation

(c) The flux of the lines of electric force depends upon charge density

(d) Charge acts as a source or sink for the lines of electric force.

2. ***Maxwell's second equation:*** Maxwell's second equation is given as

$$\nabla \times \overline{B} = 0 \quad \text{(Differential form)} \quad .....(7.38)$$

$$\int_s \overline{B}.d\vec{s} = 0 \qquad \text{(Integral form)} \qquad .....(7.39)$$

The net magnetic flux emerging through any closed surface is zero.

(a) Isolated magnetic poles do not exist

(b) It states that equal number of lines of magnetic induction enter and leave a given volume

(c) There is no source or sink for lines of magnetic force

(d) It is a time independent equation

3. ***Maxwell's third equation:*** Maxwell's third equation is given as

$$\nabla \times \vec{E} = -\frac{\partial \vec{B}}{\partial t} \qquad \text{(Differential form)} \qquad .....(7.40)$$

$$\oint_l \vec{E}.d\vec{l} = -\frac{\partial}{\partial t}\int_s \vec{B}.d\vec{s} \qquad \text{(Integral form)} \qquad .....(7.41)$$

The electromotive force around a closed path is equal to the time derivative of the magnetic displacement through any surface bounded by the path

(a) It relates the electric field $\overline{E}$ and magnetic induction $\overline{B}$

(b) It is time dependent or time varying equation i.e., $\vec{E}$ is generated by time variation of $\overline{B}$

(c) It summarizes the Faraday's laws or Lenz's law of electromagnetic induction

(d) It relates the space variation of $\overline{E}$ with the time variation of $\overline{B}$ or $\overline{H}$

4. ***Maxwell's fourth equation:*** Maxwell's fourth equation is given as

$$\nabla \times \vec{H} = \vec{J} + \frac{\partial \vec{D}}{\partial t} \qquad \text{(Differential form)} \qquad .....(7.42)$$

$$\oint_l \vec{H}.d\vec{l} = \int_s \left[\vec{J} + \frac{\partial \vec{D}}{\partial t}\right] \qquad \text{(Integral form)} \qquad .....(7.43)$$

The magnetomotive force around a closed path is equal to the conduction current plus the displacement current through any surface bounded by the path

(a) It relates the magnetic field $\overline{B}$ with electric displacement vector $\overline{D}$ and the current $\vec{J}$

(b) It is also time dependent equation

(c) It represents Ampere's law

(d) It relates the space variation of $\overline{B}$ with the time variation of $\overline{D}$ or $\overline{E}$.

7.13 Derivation of Maxwell's Equations

7.13.1 Maxwell's First Equation or Gauss's Law in Electrostatics

According to this law the electric flux ϕ passing through any closed hypothetical surface in an electric field is $\frac{1}{\epsilon_o}$ times the total charge enclosed by the surface.

Thus

$$\phi = \oint_s \vec{E}.d\vec{s} = \frac{q}{\epsilon_0} \qquad \text{.....(7.44)}$$

where ϵ_0 is the permittivity of free space

Let the charge be distributed over a volume and let ρ be the volume charge density, then

$$\oint_s \vec{E}.d\vec{s} = \frac{1}{\epsilon_0}\oint_v \rho dv \qquad \text{.....(7.45)}$$

By using Gauss's divergence theorem

$$\oint_s \vec{E}.d\vec{s} = \oint_v \text{div } \vec{E}\ dv \qquad \text{.....(7.46)}$$

Hence

$$\oint_v \text{div } \vec{E}\ dv = \frac{1}{\epsilon_0}\oint_v \rho dv \qquad \text{.....(7.47)}$$

This is true for any arbitrary shape dv, so the integral must be equal

$$\therefore \qquad \text{div } \overline{E} = \rho/\epsilon_0 \qquad \text{.....(7.48)}$$

or

$$\nabla.\vec{E} = \frac{\rho}{\epsilon_0}$$

If $\vec{D}$ (electric displacement vector) = $\epsilon_0 \vec{E}$

Then

$$\nabla.\vec{D} = \rho \qquad \text{.....(7.49)}$$

7.13.2 Maxwell's Second Equation or Gauss's Law in Magnetism

According to this law the magnetic flux passing through a closed surface placed in a magnetic field is zero.

Thus

$$\oint_s \vec{B}.\vec{ds} = 0 \qquad .....(7.50)$$

From this equation it is clear that the existence of free pole is not possible which means that within volume bound by a closed surface only magnetic dipoles made up of equal and opposite poles may be possible, so that the net pole strength within the closed surface will be zero.

By using Gauss's divergence theorem

$$\oint_s \vec{B}.\vec{ds} = \oint_v \text{div}\,\overline{B}.dv \qquad .....(7.51)$$

From eq. 7.50 and 7.51 we have

$$\oint_v \text{div}\,\overline{B}.dv = 0$$

or

$$\text{div}\,\overline{B} = 0$$

$$\nabla.\overline{B} = 0 \qquad .....(7.52)$$

7.13.3 Maxwell's Third Equation or Faraday's Law of Electromagnetic Induction

According to Faraday, a time-varying magnetic field induces an e.m.f in a closed conducting loop. Hence induced e.m.f is equal to the rate of change of magnetic flux

$$\text{e.m.f} = -\frac{d\phi}{dt} \qquad ......(7.53)$$

Minus sign is an indication that the e.m.f is in such a direction as to produce a current whose flux, if added to the original flux, would reduced the magnitude of the e.m.f

The magnetic flux linked with the loop is given as

$$\phi = \oint_s \overline{B}.\overline{ds} \qquad .....(7.54)$$

The e.m.f voltage is related to the electric field $\overline{E}$ which acts tangential to the loop by the equation

$$\text{e.m.f} = \oint_l \overline{E}.\overline{dl} \qquad ......(7.55)$$

From eq. 7.53 and 7.55, we have

$$\oint_l \overline{E}.\overline{dl} = -\frac{d\phi}{dt} = -\frac{d}{dt}\int_s \vec{B}.d\vec{s} \qquad .....(7.56)$$

Eq. 7.56 is known as Faraday's law. This law states that the line integral of the electric field around any closed path equals the rate of change of magnetic flux through any surface area bounded by that path.

By using Stoke's theorem

$$\oint_l \bar{E}.d\bar{l} \oint_s (\nabla \times E).d\vec{s} \qquad .....(7.57)$$

$$\therefore \qquad \int_s \left(\nabla \times \vec{E}\right).d\vec{s} = -\frac{\partial}{\partial t} \int_s \vec{B}.d\vec{s} \qquad .....(7.58)$$

Since this statement must be true for any surface, the integrals may be equated

$$\nabla \times \vec{E} = -\frac{\partial \vec{B}}{\partial t} \qquad .....(7.59)$$

7.13.4 Maxwell's Fourth Equation or Modified Ampere's Law

Ampere established that the line integral of magnetic field taken about any closed path must equal the current enclosed by that path

$$\oint_l \bar{H}.d\bar{l} = I \qquad .....(7.60)$$

where H is the magnetic field strength

Suppose the current is distributed through space with a current density J, then the above equation must also be written as

$$\oint_l \vec{H}.d\vec{l} = \int_s \vec{J}.d\vec{s} \qquad .....(7.61)$$

Applying stoke's theorem, we have

$$\oint_l \vec{H}.d\vec{l} = \int_s \left(\nabla \times \vec{H}\right).d\vec{s} \qquad .....(7.62)$$

From 7.61 and 7.62 we get

$$\nabla \times \bar{H} = \vec{J} \qquad .....(7.63)$$

Ampere's law $\nabla \times \vec{H} = \vec{J}$ does not hold good for time-varying fields. Therefore Maxwells modified the above equation by introducing the concept of displacement current.

Eq. 7.63 in $\nabla \times \vec{H} = \vec{J}$

Taking divergence of this equation

$$\nabla . \left(\nabla \times \vec{H} \right) = \nabla . \vec{J} \quad .....(7.64)$$

But divergence of the curl of a vector is always zero.

$$\therefore \quad \nabla . \vec{J} \text{ should be zero but } \nabla . \vec{J} = -\frac{\partial \rho}{\partial t} \quad .....(7.65)$$

This means that the divergence of current density is zero. This is not true for time-varying fields.

From Gauss's law of electrostatics

$$\nabla . \vec{D} = \rho$$

$$\nabla . \vec{E} = \rho / \epsilon_0 \quad .....(7.66)$$

Since $\vec{D} = \epsilon_0 \vec{E}$ .....(7.67)

Differentiating we get

$$\nabla . \frac{\partial \vec{E}}{\partial t} = \frac{\partial \rho}{\partial t} . \frac{1}{\epsilon_0}$$

or

$$\epsilon_0 \nabla . \frac{\partial \vec{E}}{\partial t} = \frac{\partial \rho}{\partial t}$$

Adding $\nabla . \vec{J}$ on both sides we get

$$\nabla . \vec{J} + \nabla . \frac{\partial \left(\epsilon_0 \vec{E} \right)}{\partial t} = \nabla . \vec{J} + \frac{\partial \rho}{\partial t}$$

$$\nabla . \vec{J} + \frac{\nabla . \vec{D}}{\partial t} = \nabla . \vec{J} + \frac{\partial \rho}{\partial t} \quad .....(7.68)$$

According to general equation of continuity

$$\nabla . \vec{J} + \frac{\partial \rho}{\partial t} = 0 \quad .....(7.69)$$

Comparing eq. 7.68 and 7.69 we get

$$\nabla.\vec{J} + \frac{\nabla.\vec{D}}{\partial t} = 0$$

or $$\nabla.\left[\vec{J} + \frac{\partial \vec{D}}{\partial t}\right] = 0 \qquad .....(7.70)$$

$\left[\vec{J} + \frac{\partial \vec{D}}{\partial t}\right]$ is the total current density.

Maxwell modified the Ampere's law and the modified form of Ampere's law is

$$\nabla \times \vec{H} = \vec{J} + \frac{\partial \vec{D}}{\partial t} \qquad .....(7.71)$$

where $\vec{J}$ = conduction current density.

and $$\frac{\partial \vec{D}}{\partial t} = \vec{J}_d \text{ (displacement current density)} \qquad .....(7.72)$$

7.14 Wave Equation in Free Space

Let us consider the case of electromagnetic phenomena in free space or more generally, in a perfect dielectric containing no charges ($\rho = 0$) and no conduction current i.e., $J = 0$.

∴ The field equations become

$$\nabla.\vec{D} = 0 \qquad .....(7.73)$$

$$\nabla.\vec{B} = 0 \qquad .....(7.74)$$

$$\nabla \times \vec{E} = -\frac{\partial \vec{B}}{\partial t} \qquad .....(7.75)$$

$$\nabla \times \vec{H} = \frac{\partial \vec{D}}{\partial t} \qquad .....(7.76)$$

For free space $\mu = \mu_o$, $\in\ = \ \in_o$, $\sigma = 0$ and $\rho = 0$

$$\therefore \qquad \bar{D} = \in_o \bar{E} \qquad \text{and} \qquad \bar{B} = \mu_o \bar{H}$$

Taking Curl of eq. 7.75 on both sides, we have

$$\nabla \times (\nabla \times \bar{E}) = -\frac{\partial}{\partial t}(\nabla \times \bar{B})$$

$$\nabla \times (\nabla \times \bar{E}) = -\mu_o \nabla \times \frac{\partial \bar{H}}{\partial t} \qquad \text{.....(7.77)}$$

and

$$\nabla \times \bar{H} = \frac{\partial \bar{D}}{\partial t}$$

$$\nabla \times \bar{H} = \in_o \frac{\partial \bar{E}}{\partial t}$$

$$\nabla \times \frac{\partial \bar{H}}{\partial t} = \in_o \frac{\partial^2 \bar{E}}{\partial t^2} \qquad \text{.....(7.78)}$$

$$\therefore \qquad \nabla \times (\nabla \times \bar{E}) = -\mu_o \in_o \frac{\partial^2 \bar{E}}{\partial t^2}$$

since

$$\nabla \times (\nabla \times \bar{E}) = \nabla(\nabla . \bar{E}) - \nabla^2 \bar{E} \qquad \text{.....(7.79)}$$

and

$$\nabla . \bar{E} = \frac{1}{\in_o} \nabla . \bar{D} = 0 \qquad \text{.....(7.80)}$$

$$\therefore \qquad -\nabla^2 E = -\mu_o \in_o \frac{\partial^2 \bar{E}}{\partial t^2}$$

or

$$\nabla^2 \vec{E} = \mu_o \in_o \frac{\partial^2 \bar{E}}{\partial t^2} \qquad \text{.....(7.81)}$$

Similarly we can solve for the magnctic field

$$\nabla^2 \bar{H} = \mu_o \in_o \frac{\partial^2 \bar{H}}{\partial t^2} \qquad \text{.....(7.82)}$$

Eq. (7.81) and (7.82) represents wave equations

In general case, electromagnetic wave propagation involves electric and magnetic fields having more than one components, each dependent on all these co-ordinates, in addition to time.

The waves which serve as a building block in the study of these waves consisting of electric and magnetic fields that are perpendicular to each other and to the direction of propagation are uniform in planes perpendicular to the direction of propagation. These waves are called uniform plane waves.

Note:

1. $\nabla^2\bar{E} = \mu_o \in_o \dfrac{\partial^2 \bar{E}}{\partial t^2}$ can be expanded in cartesian coordinate to a set of three equation.

   (a) $\dfrac{\partial^2 E_x}{\partial x^2} + \dfrac{\partial^2 E_x}{\partial y^2} + \dfrac{\partial^2 E_x}{\partial z^2} = \mu_o \in_o \dfrac{\partial^2 E_x}{\partial t^2}$

   (b) $\dfrac{\partial^2 E_y}{\partial x^2} + \dfrac{\partial^2 E_y}{\partial y^2} + \dfrac{\partial^2 E_y}{\partial z^2} = \mu_o \in_o \dfrac{\partial^2 E_y}{\partial t^2}$

   (c) $\dfrac{\partial^2 E_z}{\partial x^2} + \dfrac{\partial^2 E_z}{\partial y^2} + \dfrac{\partial^2 E_z}{\partial z^2} = \mu_o \in_o \dfrac{\partial^2 E_z}{\partial t^2}$

7.15 Velocity of a Plane Electromagnetic Wave i.e., $C = \dfrac{1}{\sqrt{\mu_o \in_o}}$

In case of a plane electromagnetic wave the wave equations in free space may be calculated by Maxwell's equations.

$$\nabla.\bar{D} = 0 \quad \text{.....(7.83)}$$

$$\nabla.\bar{B} = 0 \quad \text{.....(7.84)}$$

$$\nabla \times \bar{E} = -\frac{\partial \bar{B}}{\partial t} \quad \text{.....(7.85)}$$

$$\nabla \times \bar{H} = \frac{\partial \bar{D}}{\partial t} \quad \text{.....(7.86)}$$

For free space

$$\mu = \mu_o, \in = \in_o, \sigma = 0 \text{ and } \rho = 0$$

$$\therefore \quad \bar{D} = \in_o \bar{E} \text{ and } \bar{B} = \mu_o \bar{H}$$

Taking curl of equation (7.85) on both sides, we have

$$\nabla\times\left(\nabla\times\vec{E}\right) = -\frac{\partial}{\partial t}\left(\nabla\times\vec{B}\right)$$

$$\nabla\times\left(\nabla\times\vec{E}\right) = -\mu_o\nabla\times\frac{\partial\vec{H}}{\partial t} \qquad \text{.....(7.87)}$$

Since $\vec{B} = \mu_o\,\vec{H}$

and $\nabla\times\vec{H} = \dfrac{\partial\vec{D}}{\partial t} = \epsilon_o\dfrac{\partial\vec{E}}{\partial t}$

$$\nabla\times\frac{\partial\vec{H}}{\partial t} = \epsilon_o\frac{\partial^2\vec{E}}{\partial t^2}$$

∴ eq. 7.87 becomes

$$\nabla\times\left(\nabla\times\bar{E}\right) = -\mu_o\,\epsilon_o\frac{\partial^2\bar{E}}{\partial t^2}$$

$$\nabla\left(\nabla.\bar{E} - \nabla\bar{E}\right) = -\mu_o\,\epsilon_o\frac{\partial^2\bar{E}}{\partial t^2}$$

$$\nabla\left(\nabla.\bar{E}\right) - \nabla^2 E = -\mu_o\,\epsilon_o\frac{\partial^2\bar{E}}{\partial t^2}$$

$$\therefore \quad \nabla^2\bar{E} = \mu_o\,\epsilon_o\frac{\partial^2 E}{\partial t^2} \qquad \text{.....(7.88)}$$

Since $\nabla.\bar{E} = \dfrac{1}{\epsilon_o}\nabla.\bar{D} = 0$

Similarly, for magnetic field we get

$$\nabla^2 H = \mu_o\,\epsilon_o\frac{\partial^2 H}{\partial t^2} \qquad \text{.....(7.89)}$$

Eq. 7.88 and 7.89 are the vector equations of same nature. This means that each of the components of $\bar{E}$ and $\bar{H}$ satisfies the scalar wave equation of the following kind

$$\nabla^2\upsilon = \mu_o\,\epsilon_o\frac{\partial^2\upsilon}{\partial t^2} \qquad \text{.....(7.90)}$$

where υ is a scalar quantity and it expresses any component of $\bar{E}$ and $\bar{H}$.

Eq. 7.8 is similar to general wave equation.

$$\nabla^2 \upsilon = \frac{1}{C^2}\frac{\partial^2 \upsilon}{\partial t^2} \qquad \text{.....(7.91)}$$

where C is the velocity of electromagnetic wave, comparing equation 7.90 and 7.91.

This means that the velocity of field vectors $\bar{E}$ and $\bar{H}$ in free space is C.

That is

$$C = \frac{1}{\sqrt{\mu_o \in_o}} \qquad \text{.....(7.92)}$$

7.15.1 Intrinsic Impedance

Let us consider a wave travelling in x direction; hence the electric field components are

$$E_x = 0,\ E_y = E_m \cos \beta\,(x - ct),\ E_z = 0$$

where E_m is the amplitude of the wave in y direction, β is phase constant.

Let $\beta C = \omega$ where $\omega = 2\pi f$ and f is frequency

$$E_x = 0$$

$$E_y = E_m \cos(\omega t - \beta x)$$

$$E_z = 0$$

as

$$\nabla \times \bar{E} = -\frac{\partial \bar{B}}{\partial t}$$

We can write this equation as

$$\frac{\partial \vec{B}}{\partial t} = -\begin{vmatrix} i & j & k \\ \frac{\partial}{\partial x} & \frac{\partial}{\partial y} & \frac{\partial}{\partial z} \\ \vec{E}_x & \vec{E}_y & \vec{E}_z \end{vmatrix}$$

$$\frac{\partial \vec{B}_x}{\partial t} = -\left(\frac{\partial \vec{E}_z}{\partial y} - \frac{\partial \vec{E}_y}{\partial z}\right)$$

$$\frac{\partial \vec{B}_y}{\partial t} = -\left(\frac{\partial \vec{E}_x}{\partial z} - \frac{\partial \vec{E}_z}{\partial x}\right)$$

$$\frac{\partial \vec{B}_z}{\partial t} = -\left(\frac{\partial \vec{E}_y}{\partial x} - \frac{\partial \vec{E}_x}{\partial y}\right)$$

as $\vec{E}_x = 0$; $\vec{E}_y = E_m \cos(\omega t - \beta x)$ and $\vec{E}_z = 0$

$$\frac{\partial \vec{B}_x}{\partial t} = 0; \qquad \frac{\partial \vec{B}_y}{\partial t} = 0$$

$$\frac{\partial \vec{B}_z}{\partial t} = -\frac{\partial \vec{E}_y}{\partial x} = -\ \beta E_m \sin(\omega t - \beta x) \qquad .....(7.98)$$

The components of magnetic field are found by integrating w.r.t time

$$\vec{B}_x = 0$$

$$\vec{B}_y = 0$$

$$\vec{B}_z = \frac{\beta}{\omega} E_m \cos(\omega t - \beta x)$$

$$\vec{B}_z = \frac{\beta}{\omega} \vec{E}_y$$

as $\frac{\beta}{\omega} = \frac{1}{C}$ where C = velocity of phase propagation

$$\vec{B}_z = \frac{1}{C} \vec{E}_y$$

$$\vec{E}_y = C\vec{B}_z$$

or

$$\frac{\vec{E}_y}{\vec{B}_z} = C \qquad .....(7.99)$$

as $$\vec{B}_z = \mu_0 H_z$$

$$\vec{E}_y = C\,\mu_o\,\vec{H}_z$$

$$C = \frac{1}{\sqrt{\mu_o \in_o}}$$

$$\therefore \quad \vec{E}_y = \frac{\mu_o}{\sqrt{\mu_o \in_o}}\vec{H}_z$$

or $$\frac{\vec{E}_y}{\vec{H}_z} = \sqrt{\frac{\mu_o}{\in_o}} \quad \text{.....(7.100)}$$

For free space

$$\mu_o = 4\pi \times 10^{-7}\ \text{H/m}$$

$$\in_o = 8.82 \times 10^{-12}\ \text{C}^2/\text{N-m}^2$$

$$\therefore \quad \sqrt{\frac{\mu_o}{\in_o}} = 120\pi \text{ or } 377$$

$$\frac{\vec{E}_y}{\vec{H}_z} = \sqrt{\frac{\mu_o}{\in_o}} = 120\pi \text{ or } 377 \quad \text{.....(7.101)}$$

It shows that the value of electric vector at any instant in the em wave is about 377 times the value of magnetic vector. It is because of this reason that while discussing the behaviour of light as em wave, we prefer the use of electric vector.

The quantity $\sqrt{\frac{\mu_o}{\in_o}}$ has the dimension of impedance and is known as intrinsic or characteristic impedance of free space, denoted by Z_o.

It is a constant quantity having value = 377 Ω

7.16 Electromagnetic Waves in Medium with Finite μ, ∈ and σ (i.e., conducting medium)

The Maxwell's equations are

$$\bar{\nabla} \times \bar{E} = -\frac{\partial \bar{B}}{\partial t} \quad \text{.....(7.102)}$$

$$\bar{\nabla} \times \bar{H} = \bar{J} + \frac{\partial \bar{D}}{\partial t} \quad \text{.....(7.103)}$$

$$\bar{\nabla}.\bar{D} = \rho \qquad .....(7.104)$$

$$\bar{\nabla}.\bar{B} = 0 \qquad .....(7.105)$$

Suppose $\sigma \neq 0$ i.e., there may be a conduction current in the medium but the charge density is zero everywhere. That is $\rho = 0$. This reduces the equation (7.104) to

$$\bar{\nabla}.\bar{D} = 0$$

or

$$\bar{\nabla}.(\in \bar{E}) = 0$$

$\Rightarrow$

$$\bar{\nabla}.\bar{E} = 0 \qquad .....(7.106)$$

Taking the curl of eq. 7.102 we get

$$\nabla \times (\bar{\nabla} \times \bar{E}) = \bar{\nabla} \times \left(-\frac{\partial B}{\partial t}\right) \qquad .....(7.107)$$

But curl E = grad div E – $\nabla^2 E$ .....(7.108)

$$\bar{\nabla} \times (\bar{\nabla} \times \bar{E}) = \nabla(\bar{\nabla}.E) - \nabla^2 E$$

But for $\rho = 0$; we have from eq. 7.106 $\nabla.E = 0$

$\therefore$

$$\bar{\nabla} \times (\bar{\nabla} \times \bar{E}) = -\nabla^2 E \qquad .....(7.109)$$

Also

$$\bar{\nabla} \times \left(-\frac{\partial \bar{B}}{\partial t}\right) = -\mu \frac{\partial}{\partial t}(\bar{\nabla} \times \bar{H}) \qquad .....(7.110)$$

Substituting the values in eq. 7.107, we get

$$-\nabla^2 E = -\mu \frac{\partial}{\partial t}(\bar{\nabla} \times \bar{H})$$

on substituting the value of $\bar{\nabla} \times \bar{H}$ from eq. 7.103

we have

$$\nabla^2 E = \mu \frac{\partial}{\partial t}\left[\bar{J} + \frac{\partial \bar{D}}{\partial t}\right]$$

Also $\bar{J} = \sigma \bar{E}$ and $\bar{D} = \in \bar{E}$ and

$$\nabla^2 \bar{E} = \mu \frac{\partial}{\partial t}\left[\sigma \bar{E} + \frac{\partial}{\partial t}(\in \bar{E})\right]$$

$$\nabla^2 \bar{E} = \mu\sigma \frac{\partial \bar{E}}{\partial t} + \mu \in \frac{\partial^2 \bar{E}}{\partial t^2}$$

or

$$\nabla^2 \bar{E} = \mu \in \frac{\partial^2 \bar{E}}{\partial t^2} + \mu\sigma \frac{\partial \bar{E}}{\partial t}$$

$$\nabla^2 \bar{E} - \mu \in \frac{\partial^2 \bar{E}}{\partial t^2} - \mu\sigma \frac{\partial \bar{E}}{\partial t} = 0 \qquad .....(7.111)$$

This is the general wave equation for the electric vector in an EM wave. In a similar way, by taking the curl of equation 7.103 we can obtain the general wave equation for the magnetic field vector as

$$\nabla^2 \vec{H} - \mu \in \frac{\partial^2 \vec{H}}{\partial t^2} - \mu\sigma \frac{\partial \vec{H}}{\partial t} = 0 \qquad .....(7.112)$$

7.17 Poynting Vector

The cross product of electric field vector $\bar{E}$ and the magnetic field vector $\vec{H}$ is called poynting vector. It is denoted by $\vec{S}$.

Thus $\vec{S} = \vec{E} \times \vec{H}$ it is named after

J.H poynting. The unit of S is Joule per second per square metre.

An electromagnetic wave has electric and magnetic field vectors oscillating perpendicular to each other, and the electric and magnetic disturbance travels in a direction perpendicular to both the electric and magnetic vectors. Suppose, a plane polarised em wave is propagating along the x-axis. Let the electric vector is directed along the y-axis, then the magnetic vector will be directed along the Z-axis.

Hence,

$$\vec{S} = \vec{E} \times \vec{H} = \hat{j}E_y + \hat{k}H_z$$

$$= \hat{j} \times \hat{k}\,(E_y H_z) = \hat{i}\,(E_y H_z) \qquad .....(7.113)$$

But E_y H_z measure the energy per second per unit area. Hence $\vec{S}$ represents the energy propagation along the x-axis in $J/S/m^2$. In other words, pointing vector gives the time rate of flow of electromagnetic waves energy per unit area of the medium.

7.17.1 Poynting Theorem

The rate at which energy in transmitted through unit area perpendicular to the direction of propagation of energy is called pointing vector and is represented by $\vec{P}$.

Two of the Maxwell's equations are given by

$$\nabla \times \vec{H} = \vec{J} + \frac{\partial \vec{D}}{\partial t} \quad .....(7.114)$$

$$\nabla \times \vec{E} = -\frac{\partial \vec{B}}{\partial t} \quad .....(7.115)$$

Taking dot product of both sides of eq. 7.114 and 7.115 with $\vec{E}$ and $\vec{H}$ respectively we get

$$\vec{E}.\left(\nabla \times \vec{H}\right) = \vec{E}.\vec{J} + \vec{E}.\frac{\partial \vec{D}}{\partial t}$$

$$\vec{E}.\left(\nabla \times \vec{H}\right) = \vec{E}.\vec{J} + \frac{\partial}{\partial t}\left[\frac{1}{2} \in E^2\right] \quad .....(7.116)$$

and

$$\vec{H}.\left(\nabla \times \vec{E}\right) = H\left[-\frac{\partial \vec{B}}{\partial t}\right] = -\frac{\partial}{\partial t}\left[\frac{1}{2}\mu H^2\right] \quad ......(7.117)$$

Subtracting eq. 7.116 from eq. 7.117 we get

$$\vec{E}.\left(\nabla \times \vec{H}\right) - \vec{H}.\left(\nabla \times \vec{E}\right) = \vec{E}.\vec{J} + \frac{\partial}{\partial t}\left[\frac{1}{2} \in E^2 + \frac{1}{2}\mu H^2\right] \quad .....(7.118)$$

But

$$\nabla.\left(\vec{E} \times \vec{H}\right) = \vec{H}.\left(\nabla \times \vec{E}\right) - \vec{E}.\left(\nabla \times \vec{H}\right)$$

Therefore eq. 7.118 becomes

$$-\nabla.\left(\vec{E} \times \vec{H}\right) = \vec{E}.\vec{J} + \frac{\partial}{\partial t}\left[\frac{1}{2}\mu H^2 + \frac{1}{2} \in E^2\right]$$

Let us now consider a volume V enclosed by a surface S. Integrating the above relation over the volume V we have

$$\frac{\partial}{\partial t}\int_V \left[\frac{\mu H^2}{2}+\frac{\in E^2}{2}\right]dv+\int_V \left(\vec{E}.\vec{J}\right)dv=-\int_V \nabla.\left(\bar{E}\times\bar{H}\right)dv$$

Using Gauss divergence theorem, we have

$$\int_V \nabla.\left(\vec{E}\times\vec{H}\right)dv=\oint_S \left(\vec{E}\times\vec{H}\right).ds$$

$$\therefore \qquad -\frac{\partial}{\partial t}\int_V \left[\frac{\mu H^2}{2}+\frac{\in E^2}{2}\right]dv-\int_V \left(\vec{E}.\vec{J}\right)dv=\oint_S \left(\vec{E}\times\vec{H}\right).\vec{ds} \qquad \text{.....(7.120)}$$

1. Hence $-\frac{\partial}{\partial t}\int_V \left[\frac{\mu H^2}{2}+\frac{\in E^2}{2}\right]dv$ represents the rate of decrease of energy stored in volume V due to electric and magnetic fields.

2. $-\int_V \left(\vec{E}.\vec{J}\right)dv$ represents the rate of at which electromagnetic energy is lost through Joule heating.

3. Hence $\oint_S \left(\vec{E}\times\vec{H}\right)\vec{ds}$ represents the rate of flow of energy over the surface S enclosing the volume V.

Therefore $\vec{E}\times\vec{H}$ gives the rate of flow of energy through unit area enclosing the volume V. This is denoted by $\vec{P}$, called poynting vector.

$$\vec{P}=\vec{E}\times\vec{H} \qquad \text{.....(7.121)}$$

The direction of $\vec{P}$ is perpendicular to both $\vec{E}$ and $\vec{H}$.

7.18 Skin Depth

We know that the amplitude of the Em wave decreases exponentially with a distance of penetration of the wave. Suppose the amplitude at a depth x is denoted by E_{ox}, then

$$E_{ox}=E_o e^{-Rx} \qquad \text{.....(7.122)}$$

where

$$R=\sqrt{\frac{\mu\sigma\omega}{2}} \qquad \text{.....(7.123)}$$

The decreases in the amplitude or the attenuation of the field vector is quantitatively expressed in terms of a quantity called skin depth. It is defined as the distance beyond the surface of the conductor inside it at which the amplitude of the field vector is reduced to $\frac{1}{e}$ of its value at the surface.

Let the skin depth be denoted by δ, then

$$E_{o\delta} = \frac{1}{e} E_{ox} \qquad \text{at } x = 0$$

$$E_o\, e^{R\delta} = \frac{1}{e} E_o e^{R.0} = \frac{E_o}{e}$$

$$\Rightarrow \qquad R\,\delta = 1$$

$$\delta = \frac{1}{R} \qquad .....(7.124)$$

$$\therefore \qquad \delta = \frac{1}{R} = \sqrt{\frac{2}{\mu\sigma\omega}} \qquad .....(7.125)$$

Hence $\qquad \delta \propto \frac{1}{\sqrt{\sigma}} \qquad$ and $\delta \propto \frac{1}{\sqrt{\omega}}$

It shows that the skin depth varies inversely as the square root of conductivity of the medium and frequency of EM wave, i.e., at higher frequencies, the skin depth is very-very small and EM waves almost fail to penetrate in to the conductor.

Note:

Skin depth is also given by

$$\text{Skin depth } (\delta) = \frac{\lambda_C}{2\pi}$$

where λ_C is the wavelength in the conductor

$$\therefore \qquad \delta = \frac{\lambda_C}{2 \times 3.14}$$

$$\delta = 0.16\,\lambda_C \qquad .....(7.126)$$

This shows that the skin depth is 16% of the value of wavelength in the medium.

7.19 Response of a Medium to the EM waves

The electric vector of an EM wave through a medium ($\sigma \neq 0$) is given by

$$\vec{E} = E_o\, e^{(i\omega t - \gamma x)} \qquad .....(7.127)$$

The current density in the medium is given by

$$\vec{J} = \sigma \vec{E} \quad \text{.....(7.128)}$$

The current density in the medium is given by

$$\frac{\partial \vec{D}}{\partial t} = \in \frac{\partial \vec{E}}{\partial t}$$

$$= i\omega\varepsilon\, E_o\, e^{(iwt-\sqrt{x})} \quad \text{.....(7.129)}$$

$$\frac{\partial \vec{D}}{\partial t} = i\omega \in \vec{E}$$

$$\therefore \quad \frac{\vec{J}}{\frac{\partial \vec{D}}{\partial t}} = \frac{\sigma \vec{E}}{i\omega \in \vec{E}} = \frac{\sigma}{i\omega \in} \quad \text{.....(7.130)}$$

Hence from eq. 7.130 following conclusion can be made

1. We find that the conduction current density and hence the current lags behind the displacement current by $\frac{\pi}{2}$.
2. For good conductor, the conduction current should be large as compared to the displacement current i.e., in such a media

   $$\sigma >> \omega \in$$

   (a) For dielectric on the other hand $\sigma << \omega \in$

   (b) If the material are good conductor then

   $$\sigma \cong \omega \in$$

   (c) If the material are dielectric then

   $$\delta \cong \frac{\omega \in}{100}$$

3. In between the limits $0.01 < \frac{\sigma}{\omega \in} < 100$, the medium is classified as semi conductor.
4. The insulator material in general behave as dielectric even at low frequencies because of the low values of frequencies of EM waves.
5. The response of a medium of the EM wave depends upon the frequency. That is, a medium with the given conductivity and permittivity may behave as a dielectric for higher frequencies and as a good conductor for lower frequencies.

Example

Copper behave as a conductor for EM wave for frequency of the order of 10^{16} Hz but as a dielectric for frequency of the order of 10^{20} Hz or more.

7.20 Electromagnetic Spectrum

The wavelength-ranges, frequency ranges, methods of production properties and uses of the various regions of electromagnetic spectrum are summarised below.

7.20.1 Gamma Rays

(a) Wavelength range: $1 \times 10^{-4} - 1 \times 10^{-10}$ m

(b) Frequency range: $3 \times 10^{22} - 3 \times 10^{18}$ Hz

(c) Production: By disintegration of atomic nuclei

(d) Properties: Chemical reaction on photographic plates, diffraction, highly penetrating and chargeless.

(e) Uses: Provide information about structure of atomic nuclei

7.20.2 X-Rays

(a) Wavelength range: $1 \times 10^{-11} - 3 \times 10^{-8}$ m

(b) Frequency range: $3 \times 10^{19} - 1 \times 10^{16}$ Hz

(c) Production: By sudden deacceleration of high speed electron

(d) Properties: All properties of gamma rays but less penetrating

(e) Uses: Reveal structures of inner atomic electron shell and crystal diagnosis.

7.20.3 Ultraviolet-radiation

(a) Wavelength range: $1 \times 10^{-8} - 4 \times 10^{-7}$ m

(b) Frequency range: $3 \times 10^{16} - 8 \times 10^{14}$ Hz

(c) Production: By sun, arc, vacuum spark and ionised gases

(d) Properties: All properties of gamma rays but less penetrating, produce photo-electric effect.

(e) Uses: In detection of invisible writing, forged documents, finger prints and to preserve food stuffs.

7.20.4 Visible Light

(a) Wavelength range: $4 \times 10^{-7} - 8 \times 10^{-7}$ m

(b) Frequency range: $8 \times 10^{14} - 4 \times 10^{14}$ Hz

(c) Production: Radiated by excited atoms in gases and incandescent bodies

(d) Properties: Reflection, refraction, interference, diffraction, polarisation, photoelectric effect, photographic action and sensation of sight.

(e) Uses: Reveal structure of molecules and arrangement of electrons in external shells of atoms.

7.20.5 Infrared Radiation

(a) Wavelength range: $8 \times 10^{-7} - 5 \times 10^{-3}$ m

(b) Frequency range: $4 \times 10^{14} - 6 \times 10^{10}$ Hz

(c) Production: From hot bodies

(d) Properties: Heating effect on thermopile and holometer reflection, diffraction, penetration through fog.

(e) Uses: In green houses to keep the plants warm and in warfare to look through haze.

7.21.6 X-Rays

(a) Wavelength range: $1 \times 10^{-3} - 3 \times 10^{-1}$ m

(b) Frequency range: $3 \times 10^{11} - 1 \times 10^{9}$ Hz

(c) Production: By oscillating currents in special vacuum tubes

(d) Properties: Reflection, polarisation

(e) Uses: In radar, long-distance wireless communication via satellites and in microwave ovens

7.20.7 Radiowaves

(a) Wavelength range: $1 \times 10^{-1} - 1 \times 10^{4}$ m

(b) Frequency range: $3 \times 10^{9} - 3 \times 10^{4}$ Hz

(c) Production: By oscillating electric circuits

(d) Properties: Reflection, diffraction

(e) Uses: In Radio and T.V. communication system

7.20.8 Long Waves

(a) Wavelength range: $5 \times 10^{6} - 6 \times 10^{-6}$ m

(b) Frequency range: 60-50 Hz

(c) Production: Weak radiation from a.c circuits

Solved Numerical Problems

1. A parallel plate capacitor with plate area A and plate separation d is charged by a constant current i. Consider a plane surface of area $A/3$ parallel to the plates and drawn symmetrically between the plates. Find the displacement current through this area.

Sol. Suppose the charge on the capacitor plate at any time t is Q. The electric field between the plates is

$$E = \frac{Q}{\epsilon_o A}$$

The electric flux through a plane surface of an area $A/3$ is

$$\phi_E = E \times \text{Area} = \frac{Q}{\epsilon_o A} \cdot \frac{A}{3} = \frac{Q}{3\epsilon_o}$$

The displacement current through the area considered is

$$i_d = \epsilon_o \frac{d\varphi_E}{dt} = \epsilon_o \left(\frac{1}{3\epsilon_o}\right)\frac{dQ}{dt} = \frac{i}{3}$$

2. Show that equation of continuity

$$\text{div } \vec{J} + \frac{\partial \rho}{\partial t} = 0 \text{ is contained in Maxwell's equations}$$

Sol. According to Maxwell's fourth equation

$$\text{Curl } \vec{H} = \vec{J} + \frac{\partial \vec{D}}{\partial t}$$

Taking divergence of both sides we get

$$\text{div}\left(\text{curl } \bar{H}\right) = \text{div}\left(\vec{J} + \frac{\partial \bar{D}}{\partial t}\right)$$

But

$$\text{div}\left(\text{curl } \vec{H}\right) = 0$$

$\therefore$ $$\text{div}\left(\vec{J}+\frac{\partial \bar{D}}{\partial t}\right) = 0$$

or $$\text{div}\ \vec{J} + \text{div}\left(\frac{\partial \bar{D}}{\partial t}\right) = 0$$

or $$\text{div}\ \vec{J} + \frac{\partial}{\partial t}\left(\text{div}\ \vec{D}\right) = 0$$

Since $$\text{div}\ \vec{D} = \rho$$

$\therefore$ $$\text{div}\ \vec{J} + \frac{\partial}{\partial t}(\rho) = 0$$

$$\text{div}\ \vec{J} + \frac{\partial \rho}{\partial t} = 0$$

3. Show that a plane electromagnetic wave equation can be represented in the form of

$$\left(\nabla^2 + K^2\right)\vec{E} = 0$$

where $\vec{K}$ is the propagation vector equal to $2\pi/\lambda$.

Sol. Differential equation of a plane electromagnetic wave propagating along X-axis is given by

$$\vec{E} = \vec{E}_o\ (\sin \omega t - kx)$$

$$\frac{\partial \vec{E}}{\partial t} = \omega \vec{E}_o\ \cos\ (\omega t - x)$$

Again differentiating w.r.t t we get

$$\frac{\partial^2 \vec{E}}{\partial t^2} = -\omega^2\ E_o\ \sin\left(\omega t - kx\right)$$

$$\frac{\partial^2 \vec{E}}{\partial t^2} = -\omega^2\ E_o\ \sin\left(\omega t - x\right) = -\omega^2 \vec{E}$$

$$\frac{\partial^2 \vec{E}}{\partial t^2} = -\left(\frac{2\pi C}{\lambda}\right)^2 \vec{E}$$

or $$\frac{\partial^2 \vec{E}}{\partial t^2} = -C^2\ K^2\ \vec{E} \qquad .....(2)$$

Since $$K = \frac{2\pi}{\lambda}$$

Substituting the value of $\frac{\partial^2 \vec{E}}{\partial t^2}$ in $\nabla^2\vec{E} = \mu_0\varepsilon_0 \frac{\partial^2 \bar{E}}{\partial t^2}$ we get

$$\nabla^2 \vec{E} = -\mu_o \in_o C^2 K^2 \vec{E}$$

Since $$C = \frac{1}{\sqrt{\mu_o \in_o}}$$

$$\nabla^2 \vec{E} = -K^2 \vec{E}$$

$$\left(\nabla^2 + K^2\right) \vec{E} = 0$$

4. Show that the ratio of the magnitude of electric ad magnetic field vectors in electromagnetic waves is equal to the speed of the light in free space.

Sol. The electric and magnetic field vectors for a plane electromagnetic wave propagating along X-axis are given by

$$\vec{E} = \vec{E}_o \sin\left(\omega t - Kx\right) \quad .....(1)$$

$$\vec{B} = \vec{B}_o \sin\left(\omega t - Kx\right) \quad .....(2)$$

where $$\vec{E} = E_y\hat{J} + E_z\hat{K}$$

and $$\vec{B} = B_y\hat{J} + B_z\hat{K}$$

we know that

$$E_z = -C\,B_y \quad .....(3)$$

and $$E_y = C\,B_z \quad .....(4)$$

Substituting these values in equation in

$$\vec{B} = B_y\hat{J} + B_z\hat{K}$$

$$\vec{B} = -\hat{J}\frac{E_z}{C} + \hat{K}\frac{E_y}{C}$$

Putting $$\hat{J} = \hat{K} \times \hat{I}$$

or $$-\hat{J} = \hat{I} \times \hat{K}$$

and $$\hat{K} = \hat{I} \times \hat{J}$$

$$\therefore \qquad \vec{B} = \frac{1}{C}\left[\left(\hat{I}\times\hat{K}\right)E_z + \left(\hat{I}\times\hat{J}\right)E_y\right]$$

$$\vec{B} = \frac{1}{C}\left[\hat{I}\times\left(E_z\hat{K} + E_y\hat{J}\right)\right]$$

$$\text{or} \qquad \vec{B} = \frac{1}{C}\left(\hat{I}\times\vec{E}\right)$$

$$\left|\vec{B}\right| = \frac{1}{C}\left(\hat{I}\times\vec{E}\right)$$

$$\text{or} \qquad \vec{B} = \frac{E}{C}$$

Since $\hat{i}$ and $\vec{E}$ are mutually perpendicular.

$$\therefore \qquad \frac{E}{B} = C$$

5. If the magnitude of $\vec{H}$ in a plane wave is 1 amp/meter, find the magnitude of $\vec{E}$ for plane wave in free space.

Sol. We know that

$$\frac{H_o}{E_o} = \sqrt{\frac{\varepsilon_o}{\mu_o}}$$

$$\text{or} \qquad E_o = \sqrt{\frac{\mu_o}{\epsilon_o}}H_o$$

Given,

$$H_o = 1\text{amp/meter}$$

$$\mu_o = 4\pi \times 10^{-7} \text{ wb/A-m}$$

$$\text{and} \qquad \epsilon_o = 8.85 \times 10^{-12} \text{ C}^2\text{/N-m}^2$$

$$\therefore \qquad E_o = 1\times\sqrt{\frac{4\pi\times10^{-7}}{8.85\times10^{-12}}}$$

$$E_o = 376.72 \text{ V/m}$$

6. If the earth receives 2 cal min^{-1} cm^{-2} solar energy, what are the amplitudes of electric and magnetic fields of radiation.

Sol. The energy transported by an electromagnetic wave per unit area per second during propagation is represented by

$$\vec{S} = \left(\vec{E} \times \vec{H}\right)$$

The energy flux per unit area per second is

$$\left|\vec{S}\right| = \left|\vec{E} \times \vec{H}\right| = EH \sin 90^\circ = EH$$

The energy flux per unit area per second at the earth surface

$$\left|\vec{S}\right| = 2 \text{ cal/min cm}^2$$

$$= \frac{2 \times 4.2 \times 10^4}{60} \text{ Joule m}^{-2} \text{ sec}^{-1}$$

$$\therefore \quad \left|\vec{S}\right| = EH = \frac{2 \times 4.2 \times 10^4}{60} = 1400 \quad .....(1)$$

But

$$\frac{E}{H} = \sqrt{\frac{\mu_o}{\epsilon_o}} = \sqrt{\frac{4\pi \times 10^{-7}}{8.85 \times 10^{-12}}} = 376.72 \text{ ohms} \quad .....(2)$$

Multiplying eqns 1 and 2 we get

$$EH \times (E/H) = 1400 \times 376.72$$

$$E^2 = 5.274 \times 10^5$$

$$E = 726.2 \text{ V/m}$$

Substituting this value of E in eq. 1, we get

$$(726.2)H = 1400$$

or

$$H = \frac{1400}{726.2}$$

$$H = 1.928 \text{ Amp-turn/m}$$

Therefore the amplitude of electric and magnetic fields of radiation are

$$E_o = E\sqrt{2} = 726.2\sqrt{2}$$

$$E_o = 1026.8 \text{ V/m}$$

and

$$H_o = H\sqrt{2} = 1.928\sqrt{2}$$

$$H_o = 2.726 \text{ Amp-turn/m}$$

7. Assuming that all the energy from a 1000 watt lamp is radiated uniformly, calculate the average values of the intensities of electric and magnetic fields of radiation at a distance of 2 m from the lamp.

Sol. If the total power ρ_o is radiated uniformly in all directions, than the power of energy flux per unit area per second at a distance r from the point source (or lamp) is

$$S_{av} = \frac{\rho_o}{4\pi r^2} = \frac{1000}{4\pi(2)^2} \text{ watt/m}^2$$

From the Poynting theorem

$$|\vec{S}| = |\vec{E} \times \vec{H}| = EH \sin 90^\circ$$

$\therefore$

$$EH = \frac{1000}{16\pi} \qquad \text{.....(1)}$$

But

$$\frac{E}{H} = \sqrt{\frac{\mu_o}{\epsilon_o}} = 376.72 \text{ ohm} \qquad \text{.....(2)}$$

Multiplying eqns. 1 and 2 we get

$$EH \times \frac{E}{H} = \frac{376.72 \times 1000}{16\pi}$$

$$E = \left(\frac{376.72 \times 1000}{16 \times 3.14}\right)^{1/2}$$

$$E = 86.59 \text{ V/m}$$

From eq. 1, we have

$$H = \frac{1000}{16 \times 3.14 \text{ E}}$$

$$H = \frac{1000}{16 \times 3.14 \times 86.59} = 0.23 \text{ Amp-turns/m}$$

8. The sunlight strikes the upper atmosphere of earth with energy flux 1.38 KW/m^2. What will be the peak values of electric and magnetic fields at the points.

Sol. Energy flux or pointing vector S is given by

$$\vec{S} = \frac{\vec{E} \times \vec{B}}{\mu_o}$$

or $$S = \frac{EB}{\mu_o} = \frac{E_{rms}^2}{\mu_o C}$$

Since $$E_{rms} = E_o/\sqrt{2}$$

∴ $$S = \frac{E_o^2}{2\mu_o C}$$

or $$E_o = \sqrt{2\mu_o CS}$$

where

$$\mu_o = 4\pi \times 10^{-7} \text{ H/m}$$

$$C = 3 \times 10^8 \text{ m/sec}$$

and

$$S = 1.38 \text{ KW/m}^2$$

$$= 13.80 \text{ W/m}^2$$

∴ $$E_o = \sqrt{2 \times 4 \times 3.14 \times 10^{-7} \times 3 \times 10^8 \times 1380}$$

$$E_o = 1.02 \text{ KV/m}$$

Peak value of magnetic field is given by

$$B_o = \frac{E_o}{C} = \frac{1.02 \times 10^3}{3 \times 10^8} = 3.4 \times 10^{-6} \text{ ωb/m}^2$$

9. A plane electromagnetic wave propagating along the x-direction has a wavelength 5.0 mm. The electric field is in the Y-direction and its maximum magnitude is 38 V/m. Write the equations of the electric and magnetic fields as a function of x and t.

Sol. The equations of electric and magnetic fields of a plane electromagnetic wave is given by

$$E = E_o \sin\left[\frac{2\pi}{\lambda}(ct - x)\right]$$

and $$B = B_o \sin\left[\frac{2\pi}{\lambda}(ct - x)\right]$$

Given

$$E_o = 38 \text{ V/m}$$

$$\lambda = 5.0 \text{ mm}$$

$$\lambda = 5 \times 10^{-3} \text{ m}$$

$$\therefore \qquad E = 38 \sin\left[\frac{2\pi}{5\times10^{-3}}(ct - x)\right]$$

or

$$E = 38 \sin\left[\frac{2\pi\times10^{3}}{5}(ct - x)\right]$$

The magnitude of the magnetic field is given by

$$B_o = \frac{E_o}{C} = \frac{38}{3\times10^{8}} = 1.27 \times 10^{-7} \text{ weber/m}^2$$

$$\therefore \qquad B = 1.27 \times 10^{-7} \sin\left[\frac{2\pi\times10^{3}}{5}(ct - x)\right]$$

10. The relative permittivity of distilled water is 81. Calculate the refractive index and velocity of light in it.

Sol. We have refractive index

$$n = \sqrt{\frac{\mu \in}{\mu_o \in_o}}$$

Given $\in = 81 \in_o$

and for distilled water $\mu = \mu_o$

$$\therefore \qquad n = \sqrt{81} = 9$$

and we know that

$$v = \frac{C}{n} = \frac{3\times10^{8}}{9} = 3.33 \times 10^{7} \text{ m/sec}$$

11. If in a time invariant field, the magnetic field associated with a plane electromagnetic wave has only z-components which is expressed as

$$H_z = 6x \cos\beta + 12\, y \sin\gamma$$

Then obtain an expression for current density $\vec{J}$

Sol. Maxwell's equation relating $\vec{H}$ and $\vec{J}$ is expressed as

$$\vec{\nabla}\times\vec{H} = \vec{J} + \frac{\partial\vec{D}}{\partial t} \qquad \text{.....(1)}$$

Since the field is time invariant therefore

$$\frac{\partial \vec{D}}{\partial t} = 0$$

∴ equation 1 is reduced to

$$\therefore \quad \vec{\nabla} \times \vec{H} = \vec{J}$$

$$\therefore \quad \vec{\nabla} \times \vec{H} = \begin{vmatrix} \hat{i} & \hat{j} & \hat{k} \\ \frac{\partial}{\partial x} & \frac{\partial}{\partial y} & \frac{\partial}{\partial z} \\ H_x & H_y & H_z \end{vmatrix}$$

Here $H_x = 0$; $H_y = 0$; and $H_z = 6x \cos\beta + 12y \sin\gamma$

$$\therefore \quad \vec{\nabla} \times \vec{H} = \begin{vmatrix} \hat{i} & \hat{j} & \hat{k} \\ \frac{\partial}{\partial x} & \frac{\partial}{\partial y} & \frac{\partial}{\partial z} \\ 0 & 0 & (6x \cos\beta + 12y \sin\gamma) \end{vmatrix}$$

On solving we get

$$\nabla \times \vec{H} = (12 \sin\gamma)\,\hat{i} - 6\cos\beta\,\hat{j}$$

$$\therefore \quad \vec{J} = (12 \sin\gamma)\,\hat{i} - (6 \cos\beta)\,\hat{j}$$

12. The electric field intensity of a uniform plane electromagnetic wave in air is 7.5 KV/m in the y-direction. The wave is propagating in the x-direction at a frequency of 2×10^9 rad/sec.

Determine (a) wavelength of electromagnetic wave (b) frequency (c) time period and (d) the amplitude and magnetic field intensity (Given $\mu_o = 4\pi \times 10^{-7}$ N/A^2 and

$$\epsilon_o = 8.854 \times 10^{-12}\ C^2 / N - m^2$$

Sol. The intensity of the electric field E is given by

$$E = E_y = E_o \cos(\omega t - \beta x)$$

$$= 7.5 \times 10^3 \cos(2 \times 10^9 t - \beta x)$$

(a) Wavelength of the electromagnetic wave

$$\lambda = \frac{C}{f} = \frac{3\times10^8 \text{ m/sec}}{\left(\omega/2\pi\right)} = \frac{3\times10^8\times2\times3.14}{2\times10^9}$$

$$\lambda = 0.942 \text{ m}$$

(b) Frequency $f = \frac{\omega}{2\pi} = \frac{2\times10^9}{2\times3.14} = 318.5\times10^6 \text{ H}_z$

$$= 318.5 \text{ MHz}$$

(c) Time period $T = \frac{1}{f} = \frac{1}{318.5\times10^6}$

$$T = \frac{1}{f} = 3.14\times10^{-9} \text{ sec}$$

We know that

$$\frac{E_o}{H_o} = \sqrt{\frac{\mu_o}{\epsilon_o}}$$

$$\frac{E_o}{H_o} = \sqrt{\frac{4\pi\times10^{-7}}{8.85\times10^{-12}}}$$

$$\frac{E_o}{H_o} = 376.72\ \Omega$$

$$\therefore \quad H_o = \frac{E_o}{376.72\ \Omega}$$

$$H_o = \frac{7.5\times10^3}{376.72\ \Omega}$$

$$H_o = 19.91 \text{ A/m}$$

or $H_z = 19.91\cos(2\times10^9\, t - \beta x)$

13. Show that in the electromagnetic wave the electrostatic energy density is equal to the magnetic energy density.

Sol. The electrostatic energy density at a point where electric field is $\vec{E}$, is given by

$$U_e = \frac{1}{2}\vec{D}.\vec{E} = \frac{1}{2}\in E^2$$

Also the magnetic energy density at a point where magnetic field is H, given by

$$U_m = \frac{1}{2}\vec{B}.\vec{H} = \frac{1}{2}\mu H^2$$

Since $$\frac{E}{H} = \sqrt{\frac{\mu}{E}}$$

$\therefore$ $$U_e = \frac{1}{2}\in\left(H\sqrt{\frac{\mu}{\in}}\right)^2$$

$$U_e = \frac{1}{2}\mu H^2 = U_m$$

$\therefore$ $U_e = U_m$ Hence proved

14. Show that the energy flux in a plane polarised EM waves in free space is the energy density times the velocity of the wave.

Sol. Let us consider that the plane polarised Em wave is propagating in x-direction. The electric and magnetic field vector may be written as

$$\vec{E} = \hat{j}E_y; \quad \vec{H} = \hat{k}H_z$$

$\therefore$ $$P = |P| = \left|\vec{E}\times\vec{H}\right| = E_y H_z$$

Also the total energy density in the EM wave is

$$U = \frac{1}{2}\in_o E_y^2 + \frac{1}{2}\mu_o H_z^2$$

But $$\frac{1}{2}\in_o E_y^2 = \frac{1}{2}\mu_o H_z^2$$

$\therefore$ $$U = \frac{1}{2}\in_o E_y^2 + \frac{1}{2}\in_o E_y^2 = \in_o E_y^2 \qquad .....(1)$$

Now

$$\frac{E_y}{H_z} = \sqrt{\frac{\mu_o}{\in_o}}$$

$$H_z = \sqrt{\frac{\mu_o}{\in_o}}E_y$$

$\therefore$ $$P = E_y\left[\sqrt{\frac{\mu_o}{\in_o}}E_y\right] = \frac{1}{\mu_o \in_o}\in_o E_y^2 + C\in_o E_y^2 \qquad .....(2)$$

Comparing eq. 1 and 2 we get

$$P = C u$$

$\therefore$ energy flux = wave velocity $\times$ energy density

15. Calculate the skin depth for a frequency of 10^{10} Hz for silver. Given $\sigma = 2 \times 10^7$ S/m and $\mu_o = 4\pi \times 10^{-7}$ H/m.

Sol. We know that skin depth is given by

$$\delta = \sqrt{\frac{2}{\mu\sigma\omega}} = \sqrt{\frac{2}{\mu\sigma 2\pi\nu}}$$

$$\delta = \sqrt{\frac{2}{4\pi\times10^{-7}\times2\times10^{7}\times2\times3.14\times10^{10}}}$$

$$\delta = 1.125\ \mu\text{m}$$

16. For silver $\mu = \mu_o$ and $\sigma = 3 \times 10^7$ mhos/m. Calculate skin depth at 10^8 Hz frequency.

Sol. We know that skin depth is given by

$$\delta = \sqrt{\frac{2}{\mu\sigma\omega}}$$

$$\delta = \sqrt{\frac{2}{2\pi\nu\sigma\mu_o}}$$

Given $\nu = 10^8$ Hz; $\sigma = 3 \times 10^7$ mhos/m and $\mu = \mu_o = 4\pi \times 10^{-7}$ N/A^2

$$\delta = \sqrt{\frac{1}{\pi\times10^{8}\times3\times10^{7}\times4\pi\times10^{-7}}}$$

$$\delta = \sqrt{\frac{1}{12\pi^2\times10^{8}}}$$

$$\delta = 0.09 \times 10^{-4}\ \text{m}$$

$$\delta = 9 \times 10^{-4}\ \text{cm}$$

17. Show that in a conductor, the magnitude of the electric vector reduces to about 10% at a distance of 0.733 λ_C, where λ_C is the length of EM wave in the conductor.

Sol. We know that

$$\lambda_C = 2\,\pi\,\delta$$

Also

$$E_{ox} = E_o\, e^{-Rx}$$

$$E_{ox} = E_o\, e^{-x/\delta}$$

$$E_{ox} = E_o\, e^{-2\pi x/\lambda_c}$$

$$\therefore \quad \frac{1}{100} = e^{-2\pi x/\lambda_C}$$

$$x = \frac{\lambda_C}{2\pi} \log_e (100)$$

$$x = \frac{\lambda_C}{2\pi} \times 2.303 \log_{10}^{100}$$

$$x = \frac{\lambda_C}{2\pi} \times 2.303 \times 2 = 0.7333\, \lambda_C$$

18. Determine the conduction current and displacement current densities in a material having conductivity of 10^{-3} mhos/m and relative permittivity $\in_r = 2.45$. The electric field in the material is given by

$$E = 4 \times 10^{-6} \sin (9 \times 10^9\, t) \text{ volt/m}$$

Sol. The conduction current density (J) is given by

$$J = \sigma E \qquad \text{.....(1)}$$

Here $\sigma = 10^{-3}$ mhos/m and

$$E = 4 \times 10^{-6} \sin (9 \times 10^9\, t) \text{ volt/m}$$

$$\therefore \quad J = 10^{-3} \times 4 \times 10^{-6} \sin (9 \times 10^9\, t)$$

or $$J = 4 \times 10^{-9} \sin (9 \times 10^9\, t) \text{ A/m}$$

Further, the displacement current density (J_d) is given by

$$J_d = \frac{\partial D}{\partial t} = \frac{\partial}{\partial t}(\in_o \in_r E)$$

or $$J_d = \in_o \in_r \frac{\partial E}{\partial t} \qquad \text{.....(2)}$$

Given $\in_o = 8.85 \times 10^{-12}$ $C^2/N\text{-}m^2$ and $\in_r = 2.45$

$$J_d = 8.85 \times 10^{-12} \times 2.45 \frac{\partial}{\partial t} [4 \times 10^{-6} \sin (9 \times 10^9\, t)]$$

$$J_d = 8.85 \times 10^{-12} \times 2.45 [4 \times 10^{-6} \times 9 \times 10^9] \cos (9 \times 10^9\, t)]$$

$$J_d = 7.84 \times 10^{-7} \cos (9 \times 10^9\, t)] \text{ A/m}^2$$

19. Show that for frequency $\leq 10^9$ Hz, a sample of silicon will act like a good conductor. For silicon, one may assume $\frac{\epsilon}{\epsilon_o} = 12$ and $\sigma = 2$ mhos/cm. Also calculate the penetration depth for this sample at frequency 10^6 Hz.

Sol. For a material to be conductor, the value of $\frac{\sigma}{\omega \epsilon}$ should be greater than 1

$\therefore$ Here $\sigma = 2$ mhos/cm $= 200$ mhos/m

$$\omega = 2\pi\nu = 2 \times 3.14 \times 10^9$$
$$\epsilon = 12\epsilon_o$$

and
$$\frac{1}{4\pi\epsilon_o} = 9 \times 10^9$$

$\therefore$
$$\frac{\sigma}{\omega\epsilon} = \frac{200}{2 \times \pi \times 10^9 \times 12\epsilon_o}$$

$$\frac{\sigma}{\omega\epsilon} = \frac{200 \times 2}{4\pi \times \epsilon_o \times 12 \times 10^9}$$

$$\frac{\sigma}{\omega\epsilon} = \frac{400 \times 9 \times 10^9}{12 \times 10^9}$$

$$\frac{\sigma}{\omega\epsilon} = 300$$

Since $\frac{\sigma}{\omega\epsilon}$ is greater than 1; therefore silicon is good conductor at frequency at $\leq 10^9$ Hz.

The skin depth or penetration depth is given by

$$\delta = \sqrt{\frac{2}{\omega\mu\sigma}}$$

Hence $\nu = 10^6$ Hz

$$\mu = \mu_o = 4\pi \times 10^{-7}$$

and $\sigma = 200$

$\therefore$
$$\delta = \sqrt{\frac{2}{2\pi \times 10^6 \times 4\pi \times 10^{-7} \times 200}}$$

$$\delta = 35.5 \text{ mm}$$

20. Ocean water can be assumed to be non-magnetic dielectric with $R = \left(\frac{\in}{\in_o}\right) = 80$ and $\sigma = 4.3$ mhos/m

(a) Calculate the frequency at which the penetration depth will be 10 cm

(b) Show that for frequencies less than 10^8 Hz, it can be considered as good conductor.

Sol. (a) The skin depth δ is given by

$$\delta = \sqrt{\frac{2}{\omega\sigma\mu}}$$

where $\omega = 2\pi\nu$

$\mu = \mu_o = 4\pi \times 10^{-7}$ N/A^2

$\sigma = 4.3$ mhos/m

and $\delta = 10$ cm $= 0.1$ m

$$0.1 = \sqrt{\frac{2}{2\pi\nu \times \mu_o \times \sigma}}$$

$$0.1 = \sqrt{\frac{2}{2 \times 3.15 \times \nu \times 4\pi \times 10^{-7} \times 4.3}}$$

On solving we get

$$\nu = 6 \times 10^6 \text{ Hz}$$

(b) For ocean water to be good conductor, the value of $\frac{\sigma}{\omega \in}$ should be greater than 1

Here $\sigma = 4.3$, $\omega = 2\pi\nu = 2 \times 3.14 \times 10^8$

and $\in = 80 \in_o$

Hence for frequency less than 10^8 Hz

$$\frac{\sigma}{\omega \in} > \frac{4.3}{2\pi \times 10^8 \times 80 \in_o}$$

$$\frac{\sigma}{\omega \in} > \frac{4.3 \times 2}{4\pi \in_o \times 10^8 \times 80}$$

$$\frac{\sigma}{\omega \in} > \frac{8.6 \times 9 \times 10^9}{80 \times 10^8}$$

$$\frac{\sigma}{\omega \in} > \frac{8.6 \times 9}{8}$$

$$\frac{\sigma}{\omega \in} > 9.67$$

Since the value of $\frac{\sigma}{\omega \in} >> 1$, therefore ocean water is good conductor at frequency less than 10^8 Hz.

Conceptual Questions

1. **Arrange in descending order of wavelength**

   γ-rays, x-rays, infrared rays, radio waves, microwaves.

Ans. Radio waves, microwave, infrared rays, X-rays, γ-rays.

2. **Arrange in ascending order of wavelength.**

   Microwaves, γ-rays, radio waves, ultraviolet light

Ans. γ-rays, ultraviolet light, micro waves, radio waves.

3. **Name the part of the electromagnetic spectrum which is used in green houses to keep plants warm.**

Ans. Infrared radiations

4. **Name the electromagnetic radiation used for viewing objects through haze and fog.**

Ans. Infrared radiation.

5. **What are the dimensions and value of** $\frac{1}{\sqrt{\mu_o \in_o}}$

Ans. $[LT^{-1}]$ and 3×10^8 m/sec

6. **What does a time-varying electric field yield.**

Ans. Magnetic field

7. **Induced electric fields due to changing magnetic fields are more readily observed than the induced magnetic fields due to changing electric fields, why?**

Ans. The induced electric field (i.e., induced emf) due to changing magnetic field can be easily enhanced by taking a (coil of large number of turns and moving it rapidly in a magnetic field. Hence it is readily observable. On the other hand, the induced magnetic field due to a changing electric field that is, due to a displacement current, is too small to be observed at moderate frequencies.

8. A closed loop of $\vec{B}$ is produced by a changing electric field. Are $\vec{E}$ and $d\vec{E}/dt$ necessary non-zero at all points on the loop and in the area enclosed by the loop.

Ans. No, the basic requirement for the production of a closed loop of $\vec{B}$ is that the total electric flux through the area enclosed by the loop should vary with time. The flux change may arise from any portion of the area. Elsewhere $\vec{E}$ and $d\vec{E}/dt$ may be zero.

9. Which of the Maxwell's equation signifies that electric filed lines are not closed curves?

Ans. $\oint \vec{E}.d\vec{s} = q / E_o$ (Gauss's law of electrostatics)

10. Which of the Maxwell's equation signifies that free magnetic poles do not exist?

Ans. $\oint \vec{B}.d\vec{s} = 0$

11. Light waves can travel in vacuum but sound waves require a material medium to travel. Why?

Ans. Light waves are electromagnetic waves which can travel in vacuum. But sound waves are mechanical waves which can travel only in a material medium.

12. Represent the direction of propagation of electromagnetic waves in terms of $\vec{E}$ and $\vec{B}$

Ans. $\vec{E} \times \vec{B}$

13. Which of the following pairs of components of $\vec{E}$ and $\vec{B}$ would generate a plane electromagnetic wave propagating along Z-direction?

(i) E_x, B_y **(ii) E_y, B_x**

(iii) E_x, B_x **(iv) E_y, B_y**

(v) E_x, B_y **(vi) E_z, B_x**

Ans. E_x, B_y and E_y, B_x

14. Magnetic filed lines can never start from a point nor end at a point, they form closed loops. But the field lines around a bar magnet do seem to start from the north Pole and end at the south pole. How will you reconcile the two statements?

Ans. The magnetic field lines of a bar magnet seem to start from the north pole and end at the south pole only outside the magnet. Inside the magnet the field line go from south pole to north pole, thus forming closed loops. The net flux of $\vec{B}$ over any fully enclosing north or south pole is zero.

15. If the earth had no atmosphere, would its average surface temperature higher or lower than what it is now.

Ans. Lower: The infrared radiation emitted by earth is retained by earth's atompshere due to green house effect and it keeps the earth warm. Or the earth had no atmosphere, there would be no green house effect and so its average temperature would be lower than it is now.

16. The ozone layer in the atmosphere is crucial for human survival on earth. Why?

Ans. The ozone layer is situated in the atmospheres at a height from 30 to 50 Km above the surface of the earth. It absorbs ultraviolet and other low-wavelength radiations coming from the sun and the stars which are otherwise very dangerous for living cells and plants. It thus protects us from the harmful effect of solar radiations.

17. Write down Maxwell's equation in Differential form?

Ans. The Maxwell's equation in Differential form are

1. $\nabla.\vec{D} = \rho$

2. $\nabla.\vec{B} = o$

3. $\nabla \times \vec{E} = \frac{-\partial \vec{B}}{\partial t}$

4. $\nabla \times \vec{H} = \vec{J} + \frac{\partial \vec{D}}{\partial t}$

18. Write down Maxwell's equation in integral form

Ans. The Maxwell's equation in integral form are

1. $\oint_s \vec{D}.d\vec{s} = \int_v \rho.dv$

2. $\oint \vec{B}.d\vec{s} = o$

3. $\oint \vec{E}.d\vec{l} = -\int_s \frac{\partial \vec{B}}{\partial t}.d\vec{s}$

4. $\oint \vec{H}.d\vec{l} = \int_s \left[\vec{J} + \frac{\partial \vec{D}}{\partial t}\right].d\vec{s}$

Objective Questions

1. The displacement current can be represented by

(a) $i_d = \epsilon_o \frac{d\phi_B}{dt}$ (b) $i_d = \epsilon_o \frac{d\phi_E}{dt}$

(c) $i_d = \mu_o \frac{d\phi_E}{dt}$ (d) $i_d = \mu_o \frac{d\phi_B}{dt}$

2. The meaning of div B = 0 is

(a) Flow if current is zero

(b) Curl B = 0

(c) There is no existence of an isolated magnetic pole

(d) B = 0

3. The velocity of light in free space is given by

(a) $C = \sqrt{\mu_o \epsilon_o}$ (b) $C = \sqrt{\frac{\mu_o}{\epsilon_o}}$

(c) $C = \sqrt{\frac{1}{\mu_o \epsilon_o}}$ (d) $C = \frac{\sqrt{\epsilon_o}}{\mu_o}$

4. The plane electromagnetic wave in free space is represented by

(a) $\nabla^2 E - \mu_o \epsilon_o \frac{\partial^2 B}{\partial t^2} = 0$

(b) $\nabla^2 E - \mu\sigma \frac{\partial E}{\partial t} + \mu t \frac{\partial^2 E}{\partial t^2} = 0$

(c) $\nabla^2 E - \mu_o \epsilon_o \frac{\partial^2 E}{\partial t^2} = 0$

(d) $\nabla^2 B - \mu_o \epsilon_o \frac{\partial^2 E}{\partial t^2} = 0$

5. The energy of electromagnetic wave in free space is given by

   (a) $\frac{E^2}{2\epsilon_o} + \frac{B^2}{2\mu_o}$ (b) $\frac{1}{2}\epsilon_o E^2 + \frac{1}{2}\mu_o B^2$

   (c) $\frac{E^2 + B^2}{2}$ (d) $\frac{1}{2}\epsilon_o E^2 + \frac{B^2}{2\mu_o}$

6. The displacement current arises due to
   (a) Positive charge only
   (b) Time varying electric field
   (c) Negative charge only
   (d) Both positive and negative charges

7. Which one of the following is not a part of Maxwell's equation.

   (a) $\nabla.E = 0$ (b) $\nabla \times E = -\frac{\partial B}{\partial t}$

   (c) $\nabla \times H = J + \frac{\partial D}{\partial t}$ (d) $\nabla . B = 0$

8. The amplitudes of electric and magnetic fields are related to each other by the relation.

   (a) $E_o B_o = C^2$ (b) $B_o = E_o C$

   (c) $E_o B_o = C$ (d) $E_o = B_o C$

9. The energy per unit time, per unit area perpendicular by the electromagnetic fields is expressed as

   (a) $\vec{S} = \vec{E} \times \vec{B}$ (b) $\vec{S} = \frac{1}{\epsilon_o}\left[\vec{E} \times \vec{B}\right]$

   (c) $\vec{S} = \mu_o\left[\vec{E} \times \vec{B}\right]$ (d) $\vec{S} = \frac{1}{\mu_o}\left[\vec{E} \times \vec{B}\right]$

10. "The work done on the charges by the electromagnetic force is equal to the decrease in energy stored in the field, less the energy which flowed out through the surface" is the statement of

    (a) Gauss's theorem (b) Stoke's theorem

    (c) Gauss's divergence theorem (d) Poynting theorem

Theoretical Questions

1. State and prove Ampere's circuital law?
2. Use Ampere's law to find the magnetic field induction $\vec{B}$ at a point with a current carrying

   (i) Long solenoid (ii) Toroid
3. Write the Maxwell's equations. Explain the physical significance of each equation?
4. Solve Maxwell's equations in free space?
5. Deduce the equation for the propagation of the plane EM wave in the free space?
6. Show that the electric, magnetic vectors are normal to each other and to the direction of the propagation of the wave?
7. Deduce Maxwell's equations for free space and prove that electromagnetic waves are transverse?
8. Prove that the velocity of plane electromagnetic wave in the vacuum is given by

$$C = \frac{1}{\sqrt{\mu_o \in_o}}$$

9. Deduce an expression for intrinsic impedance. Show that for free space its value is equal to 120π?
10. What is displacement current?
11. Explain the concept of Maxwell's displacement current and show how it led to modification of the Ampere's law?
12. State Ampere's law in differential and integral forms. Discuss the modification of Ampere's law made by Maxwell's taking displacement current in consideration. Explain the displacement current and its implication?
13. Write the Maxwell's equation in integral and differential forms. Explain the physical significance of each equation?
14. Deduce four Maxwell's equations in free space. Explain the concept of Maxwell's displacement current and show how it led to the modification of Ampere's law?
15. Show that if $\vec{B}$ is a magnetic field and its divergence $\vec{B}$=o every where, it means that an isolated magnetic pole is not possible.

16. Show that the faraday's law of electromagnetic induction can be expressed as $\nabla \times \vec{E} = \frac{-\partial \vec{B}}{\partial t}$

17. State and explain Poynting theorem for the flow of energy in electromagnetic waves?

18. Write the Maxwell's equations in integral as well as in differential forms and explain their physical significance. Show that the velocity of plane electromagnetic waves in the free space is given by $C = \frac{1}{\sqrt{\mu_o \in_o}}$

19. Prove that electromagnetic waves are transverse in nature?

20. Discuss the physical significance of Poynting theorem?

21. Deduce Poynting theorem for the flow of energy in an electromagnetic field?

22. What is skin depth; derive an expression for it?

Numerical Problems

1. The electric filed in an electro magnetic wave in given by E = E_o sin (wt – kx),where E_o = 1000 N/C Find the energy contained in a cylinder of cross section 10^{-3} m^2 and length 100 cm along the X-axis?

   $[4.425 \times 10^{-11}$ J]

2. A plane mono chromatic linearly polarized light wave is travelling eastward. The wave is polarised with E directed vertically up and down. Write expressions for E, H and B provided that E_o = 0.1V/m and frequency is 20 MH_z

   $$\left[E_2 = 0.1 \sin(4\pi \times 10^7 t - 0.419x);\ By = \frac{-E_z}{C} : H_Y = \frac{B_Y}{\mu_o}\right]$$

3. A plane electromagnetic wave propagating in the X-direction has a wavelength 7.0mm. The electric field is in the Y-direction and its maximum magnitude is 42V/m.Write suitable equations for the electric and magnetic fields as function of x and t.

   $$\left[E = 42 \sin\left\{\frac{2\pi}{7}(ct - x)\right\}, B = 1.4 \times 10^{-7} \sin\left\{\frac{2\pi}{7}(ct - x)\right\}\right]$$

4. Using Maxwell's equation,

$$\text{Curl } \vec{B} = \mu_o \left(\vec{J} + \frac{\partial \vec{D}}{\partial t} \right)$$

Prove that

$$\text{Div } \vec{D} = \rho$$

5. The maximum electric field in a plane electromagnetic wave is 10^2 N/C. The wave is going in the X-direction and the electric field is in the Y-direction. Find the maximum magnetic field in the wave and its direction

[$B_o = 3.33 \times 10^{-7}$ wb/m^2]

6. The electric field intensity of a plane electromagnetic wave in free space is expressed as

$$\vec{E} = E_o \cos w\left(t - \frac{z}{v} \right) \hat{J}$$

Find the expression for the components of magnetic field intensity $\vec{H}$ with the help of Maxwell's equation for free space

$$\begin{bmatrix} \vec{H}_Y = o : H_z = o \\ \vec{H} = \dfrac{-E_o}{\eta_o} \cos w(t - z/v)\hat{I} \end{bmatrix} \text{ where } \eta_0 = \sqrt{\frac{\mu_0}{\epsilon_0}}$$

7. The electric field of a uniform plane electromagnetic wave in free space is given by

$$\vec{E} = 45 \sin(6\pi \times 10^8 t - 2\pi x)\hat{J} + 15 \cos(6\pi \times 10^8 t - 2\pi x)\hat{K} \text{ volt/m}$$

Determine (a) Phase constant (b) angular frequency w (c) frequency f (d) intrinsic impedance x_o and (e) the corresponding magnetic field H

[$\beta = 2\pi$ rad/s; $w = 6\pi \times 10^8$ rad/s $f = 3 \times 10^8$ H_z;$x_o = 377\ \Omega$; $\vec{H} = 0.0398 \cos(6\pi \times 10^8 t - 2\pi x)\hat{J} + 0.119 \sin(6\pi \times 10^8 t - 2\pi x)\hat{K}$]

8. Find the skin depth δ at a frequency of 3.0×10^6 H_z in aluminium where $\sigma = 38.0 \times 10^6$ s/m and $\mu_r = 1$. Also find out the propagation constant and wave velocity.

[$\delta = 0.04716$ mm; $v = 888.51$ m/s; $\gamma = 29.986 \times 10^3 < 45° \text{ m}^{-1}$]

Answers to Objective Questions

1. (b)	2. (c)	3. (c)	4. (c)	5. (b)	6. (b)
7. (a)	8. (d)	9. (d)	10. (d)		

CHAPTER 8A

Superconductivity

The experiment left no doubt, as for as accuracy of measurement went, the resistance disappeared. At the same time, however, something unexpected occurred. The disappearance did not take place gradually but abruptly. From 1/500 the resistance at 4.2 K, it could be established that the resistance had become less than a thousand – millionth part of that at normal temperature. Thus the mercury at 4.2 K has entered a new state, which, owing to its particular electrical properties, can be called the state of superconductivity.

Heike Kamerlingh Onnes

8A.1 Introduction

The electrical resistivity of many metals and alloys drops suddenly to zero. When the specimen is cooled to a sufficiently low temperature, often a temperature in the liquid helium range. This phenomenon is called superconductivity, was observed first by Kamerlingh Onnes, in Leiden in 1911.

At a critical temperature T_C the specimen undergoes a phase transition from a state of normal electrical resistivity to a superconducting state as shown in Fig. 8A.1

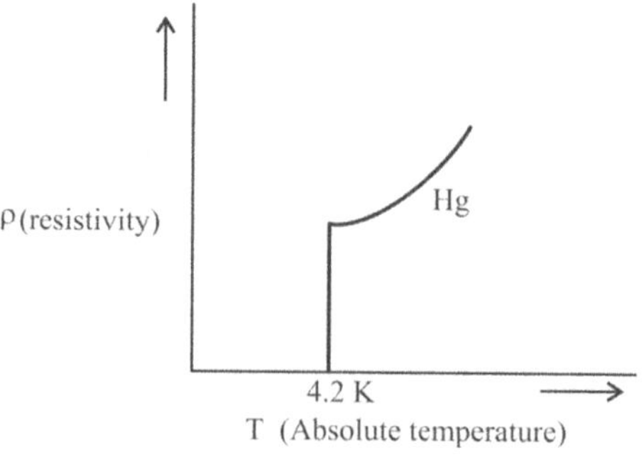

Fig. 8A.1

The temperature at which resisistivity of the specimen disappear is known as transition temperature or critical temperature (T_C). The value of T_C extends from 0.9 K to approximately 10 K (Nb). In the compounds of Nb, T_C is of the order of 14 K.

The most striking features of superconductors are;

1. A superconductor can behave as if it had no measurable DC electrical resistivity. Currents have been established in superconductors which, in the absence of any driving field.
2. A superconductor can behave as a perfect diamagnet. A sample in thermal equilibrium in an applied magnetic field, provided the field is not too strong, carries electrical surface currents. These currents give rise to an additional magnetic field that precisely cancels the applied magnetic field in the interior of the superconductor.
3. Estimates of the resistivity of superconducting phase derived from the experiment is of the order of 4×10^{-25} Ω–m while for metal it is of the order of 10^{-15} Ω–m. Comparison of the resistivity of a superconducting body to that of copper at room temperature reveals that superconducting body is at least 10^{17} times less resistive.
4. The sharpness of superconducting state transition depends on the state of purity of the sample. The sharp vertical discontinuity in Fig. 8A.2 below is indicative of that found for a single crystal of very pure element. The broad transition shown by broken line is typical of the transition shape observed for materials which are inhomogeneous. The temperature interval over which the transition from the normal and superconducting state has taken place may be of the order of 2×10^{-5} K.

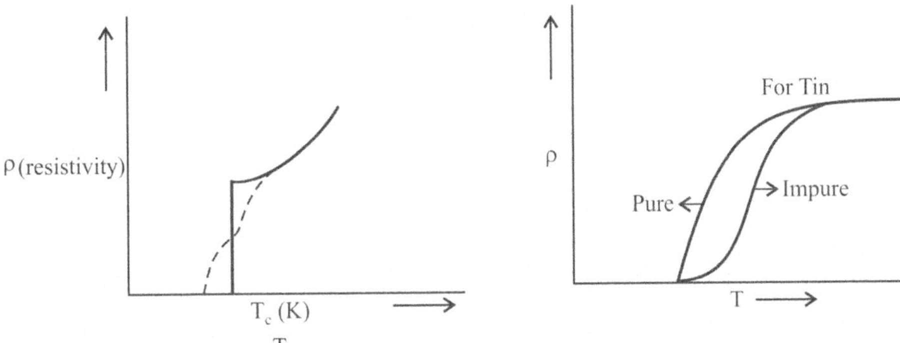

Fig. 8A.2

5. Zero resistance of a superconducting materials implies, transmission of current at any distance with no losses, the production of large magnetic fields or storage of energy. But these applications could not be realised because superconductors reversed to normal conductor at a relatively low current density $\vec{J}_C$ or relatively low magnetic field called the critical field $\vec{B}_C$. Thus the three parameter T_C, B_C and J_C becomes very important in the practical applications of superconductors.

   Fig. 8A.3 below shows the boundary in temperature-current density and magnetic field space separating normal and superconducting states. Boundary position for all the three parameters depends on the material. For the three parameters, values lying between origin and boundary corresponds to super conducting state, outside the boundary, conductors is normal.

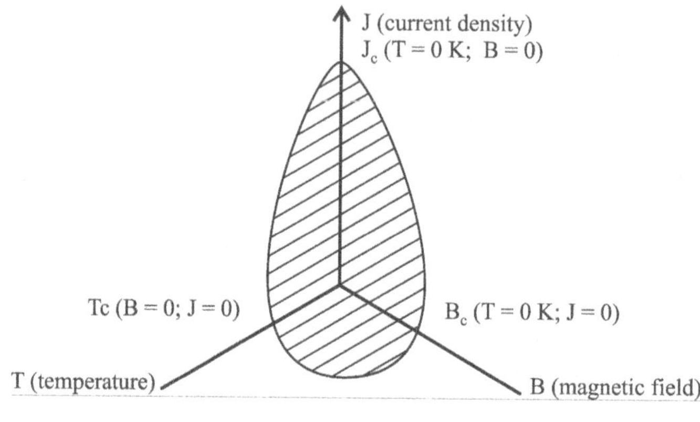

Fig. 8A.3

6. A superconductor usually behaves as if there were a gap in energy of width 2Δ centered about the fermi energy, in the set of allowed one-electron levels. Thus an electron of energy E can be accommodated by or extracted from a superconductor only if $E - E_F$ or $E_F - E$ exceeds Δ. The energy gap Δ increased in size as the temperature drops, leveling off to a maximum value $\Delta(0)$ at very low temperature.

8A.2 Critical Temperature (T_C)

The temperature at which electrical resistance is zero is called the critical temperature T_C and varies with the individual material.

For practical purposes, critical temperature are achieved by cooling materials with either liquid helium or liquid nitrogen.

The transition to the superconducting state is a sharp one in bulk specimens. Above a critical temperature T_C the properties of the metal are completely normal, below T_C superconducting properties are displayed, the most dramatic of which is the absence of any measurable D.C electrical resistance. Measured critical temperature range from a few millidegree Kelvin up of a little over 20 K. The corresponding thermal energy K_B T_C varies from about 10^{-7} e V up to a few thousands of an electron volt.

8A.3 General Properties of Superconductors

Following are the general properties of superconductors;

1. The transition temperature is different for different substances
2. Superconductivity is found to occur in metallic elements in which the number of valence electron (Z) lies in between 2 and 8
3. Transition metals having odd number of valence electrons are favourable to exhibit superconductivity while metals having even number of valence electrons are infavourable.
4. Ferromagentic and antiferromagnetic materials are not superconductors.
5. Materials for which $Z\rho>10^{10}$, where Z is the number of valence electron per m^3 and ρ is the resistivity in ohm-m at room temperature, shows superconductivity.
6. In addition to the drop in electrical resistivity to zero when cooled to transition temperature the following changes also occur;
   (i) The magnetic flux lines are rejected out of the superconductor. This property is known as Meissner effect.
   (ii) There is a discontinuous change in specific heat
   (iii) There are small changes in thermal conductivity and the volume of the material.

8A.4 The Meissner Effect

When a weak magnetic field is applied to a superconducting specimen at a temperature below transition temperature T_C, the magnetic flux lines are expelled as shown in Fig. 8A.4. The specimen acts as an ideal diamagnet. This effect is called Meissner effect. This effect is reversible i.e., when the temperature is raised from below T_C, at $T = T_C$, the flux lines suddenly start penetrating and specimen return back to the normal state.

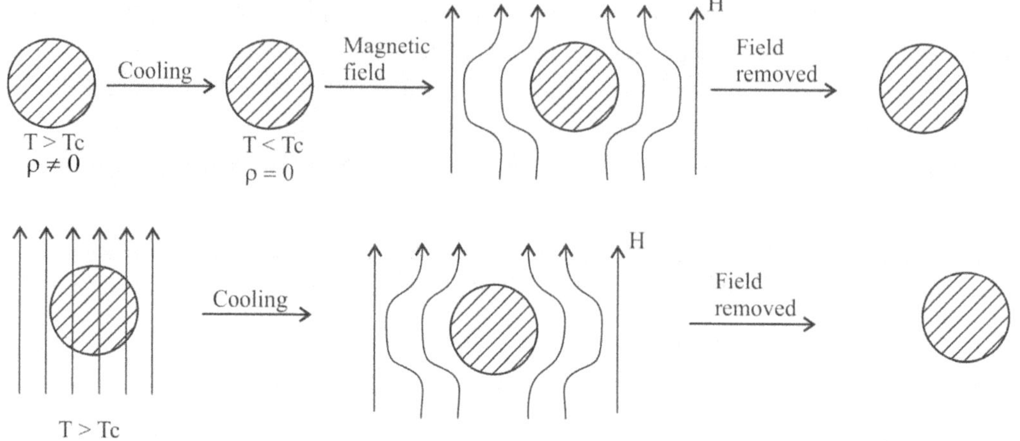

Fig. 8A.4

Under this condition, the magnetic induction inside the specimen is given by

$$B = \mu_0 (H + M) \quad .....(8A.1)$$

where H is the external applied magnetic field and M is the magnetisation produced inside the specimen.

When the specimen is in superconducting state, then according to Meissner effect inside the bulk superconductor

$$B = 0$$

Hence

$$\mu_0 (H + M) = 0$$

or

$$M = -H$$

Thus the material is perfectly diamagnetic.

Magnetic susceptibility can be expressed as

$$\chi = \frac{M}{H} = -1$$

Which is true for a perfect diamagnet.

Note: It is interesting to note that the perfect diamagnetic behaviour of a superconductor can not be explained singly considering its zero resistivity, such a perfect conductor would behave differently under different condition as shown in Fig. 8A.5.

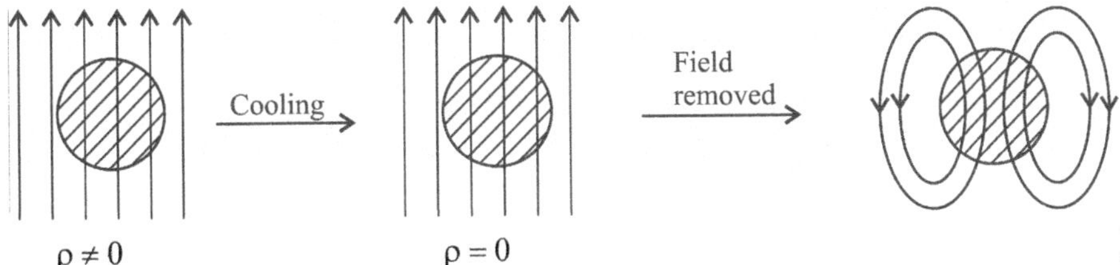

Fig. 8A.5

Since the resitivity ρ is zero for a perfect conductor, the application of ohm's law E = ρ J, indicates that no electric field can exist inside the perfect conductors.

Using the Maxwell's equation

$$\nabla \times \vec{E} = -\frac{\partial \vec{B}}{\partial t}$$

under super conducting condition since E is zero

or $$\frac{\partial \vec{B}}{\partial t} = 0$$

or B = constant

This means that the magnetic flux passing through the specimen should not change on cooling to the transition temperature. The Meissner effect contradicts this result.

Thus we conclude that the behaviour of a superconductor is different from that of a perfect conductor and the superconducting state may be considered as a characteristic thermodynamic phase of a substance in which the substance can not sustain steady electric and magnetic fields. Hence the two mutually independent properties defining the superconducting state are the zero resistivity and perfect magnetism i.e.,

$$E = 0 \text{ and } B = 0 \qquad \text{.....(8A.2)}$$

8A.5 Type I and Type II Superconductors

Superconductors have been classified as type I and type II depending upon their behaviour in an external magnetic field i.e., how strictly they follow the Meissner effect.

8A.5.1 Type I or Soft Superconductors

The superconductors which strictly follow the Meissner effect are called type I superconductors. The typical magnetic behaviour of lead, a type I superconductor is shown in Fig. 8A.6.

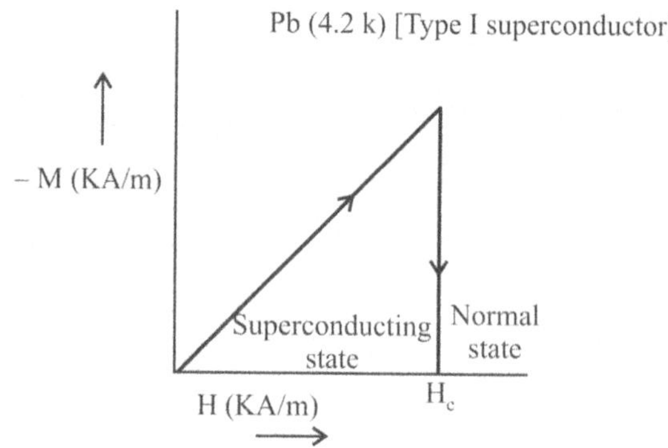

Fig. 8A.6

1. These superconductors exhibit perfect diamagnetism below a critical field H_C which for most of the cases is of the order of 0.1 Tesla.
2. As the magnetic field is increased beyond H_C, the field penetrates the material completely and the later abruptly reverts to its normal resistive state.
3. These materials give away their superconductivity at lower field strength and are referred to as the soft superconductors.
4. These materials have very limited technical application owing to the very low values of H_C.

8A.5.2 Type II or Hard Superconductors

These superconductors do not follow the Meissner effect strictly i.e., the magnetic field does not penetrate these materials abruptly at the critical field. The typical magnetization curve for Pb – Bi alloy shown in Fig. 8A.7 illustrates the magnetic behaviour of such a superconductor.

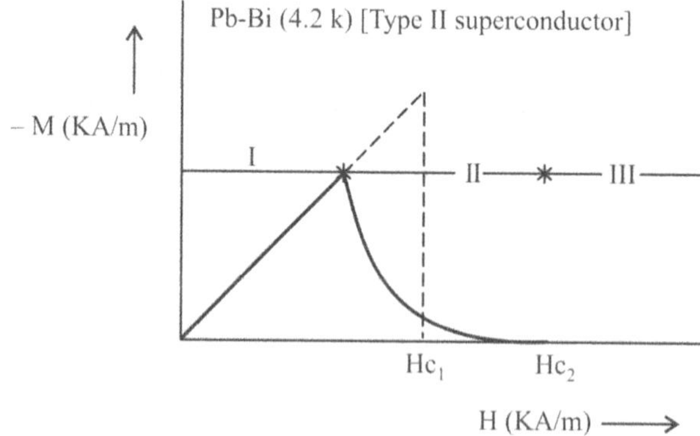

Fig. 8A.7 (Magnetization curve of a lead-Bismuth alloy at 4.2 k)

I → Superconducting state

II → Vortex or mixed or intermediate state

III → Normal state

1. It follows from the curve that for field less than H_{C_1} the material exhibits perfect diamagnetism and no flux penetration takes place. Thus for $H < H_{C_1}$; the material exists in the superconducting state.
2. As the field exceeds H_{C_1} the flux begins to penetrate the specimen and for $H = H_{C_2}$, the complete penetration occurs and the material becomes a normal conductor.
3. In the region between the fields H_{C_1} and H_{C_2} the diamagnetic behaviour of the material vanishes gradually and the flux density B inside the specimen remains non-zero i.e., the Meissner effect is not strictly followed. The specimen in this region is said to be existing in the vortex or intermediate state.
4. The type II superconductors are also called the hard superconductors because relatively large fields are needed to bring them back to the normal state.
5. A large magnetic hysteresis can be induced in these materials by appropriate mechanical treatment. Hence these materials can be used to manufacture superconducting wire which can be used to produce high magnetic fields of the order of 10 tesla.

8A.6 Comparison between Type I and Type II Superconductors

A comparison between type I and type II superconductors is given in Table 8A.1

Table 8A.1

	Type I superconductor		Type II superconductor
1.	These superconductors are called as soft superconductors	1.	These superconductors are called as hard superconductors
2.	Only one critical field exists for these superconductors	2.	Two critical fields H_{C_1} (lower critical field) and H_{C_2} (upper critical field) exists for these superconductors
3.	The critical field value is very low	3.	The critical field value is very high
4.	These superconductors exhibit perfect and complete Meissner effect	4.	These do not exhibit a perfect and complete Meissner effect
5.	These materials have limited technical applications because of very low field strength Examples: Pb, Hg, Zn etc	5.	These materials have wider technological applications because of very high field strength Example: Nb_3Ge, Nb_3si, $YBa_2 Cu_3O_7$
6.	The coherence length is larger than the penetration depth	6.	The coherence length is smaller than penetration depth

Persistent Current

Fig. 8A.8 displays the resitivity of a superconductor metal versus temperature as the critical temperature T_C is crossed. Above T_C the resistivity has the form characteristic of a normal metal.

$$\rho(T) = \rho_0 + \beta T \qquad .....(8A.3)$$

The constant term arising from impurity and defect scattering and the term in T arising from phonon scattering.

Below T_C these mechanism lose the power to degrade an electric current and the resistivity drops abruptly to zero. Current can flow in a superconductor, with no discernible dissipation of energy.

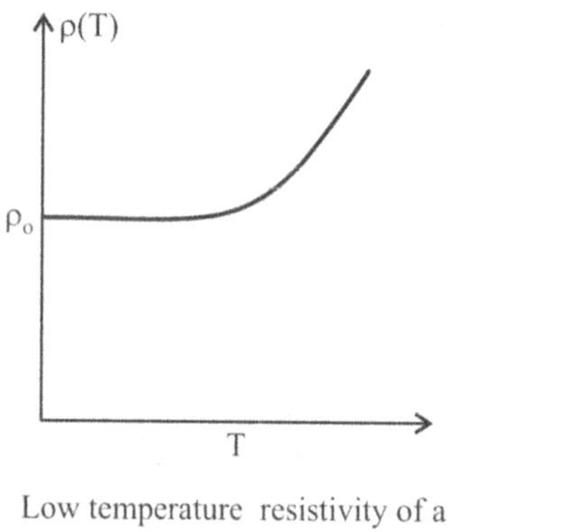

Low temperature resistivity of a normal metal $\rho(T) = \rho_0 + \beta T$ containing no magnetic impurity

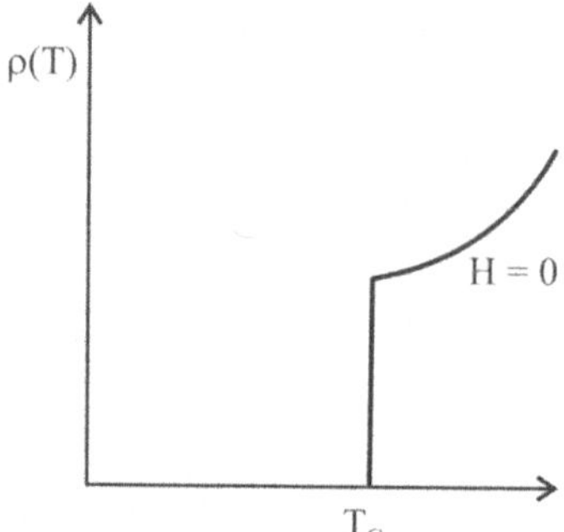

Low temperature resistivity of a superconductor, in zero magnetic field, containing non magnetic impurities. At $T = T_c$, ρ drops abruptly to zero

Fig. 8A.8

It is observed that current induced in a superconducting ring continues to flow without decay for several hours. In a typical case current of hundred amperes continues to flow in superconducting ring without decay for over a year. Such a current is known as persistent current.

There are however some limitations;

1. Superconductivity is destroyed by application of a sufficiently large magnetic field.
2. If the current exceeds a "critical current" the superconducting state will be destroyed (Silsbee effect). The size of the critical current, which can be as large as 100 amp in a 1mm wire, depends on the nature and geometry of the specimen, and is related to whether the magnetic field produced by the current exceeds the critical field at the surface of the superconductor.

3. A superconductor well below its transition temperature will also respond without dissipation to an A.C electric field provided that the frequency is not too large. The change from dissipation to normal response occurs at a frequency ω of the order of $\frac{\Delta}{\hbar}$ where Δ is the energy gap.

Critical Current (or Silsbee's Rule)

The minimum value of current that can be passed in a sample of superconductor without destroying its superconductivity is called critical current I_C. If a current of I ampere is flowing through a wire of radius r of a soft superconductor, then the surface magnetic field H_S associated with current is

$$H_S = \frac{I}{2\pi r} \quad .....(8A.4)$$

If H_S exceeds H_C, that is, $H_S > H_C$, the superconducting material becomes normal. In addition to this surface or internal magnetic field, if external magnetic field H is also present, then to sustain the superconducting state the total magnetic field strength should be less than or equal to the critical magnetic field. Thus, we have

$$H_C = H_S + 2H$$

or

$$H_S = H_C - 2H \quad .....(8A.5)$$

From eq. 8A.4

$$H_S = \frac{I_C}{2\pi r}$$

∴

$$I_C = 2\pi r\,(H_C - 2H) \quad .....(8A.6)$$

This is called Silsbee's rule. It is evident from equation 8A.6 that the critical current I_C decreases with the increase of applied field H until it reaches zero at $H = \frac{H_C}{2}$. If the applied field is zero, then

$$I_C = 2\pi r H_C$$

In other words, a bulk superconductor behaves in an applied external magnetic field H_{ext}, as if inside the sample H = 0

8A.7 Effect of Magnetic Field

Superconducting state of a metal mainly depends on temperature and strength of the magnetic field in which the metal is placed. Superconductivity disappear if the temperature of the specimen is raised above T_C or a strong magnetic field is applied.

At temperature below T_C, in the absence of any magnetic field, the material is in superconducting state. When the strength of the magnetic field applied reaches a critical value H_C the superconductivity disappear.

At $\qquad T = T_C, H_C = 0$

At temperature below T_C, H_C increases. The dependence of the critical field upon the temperature is given by

$$H_C(T) = H_C(0)\left[1-\left(\frac{T}{T_C}\right)^2\right] \qquad .....(8A.7)$$

where $H_C(0)$ is the critical field at 0k, $H_C(0)$ and T_C are constant and characteristic of the material.

Magnetic field temperature dependence of type I and type II superconductors are shown in Fig. 8A.9.

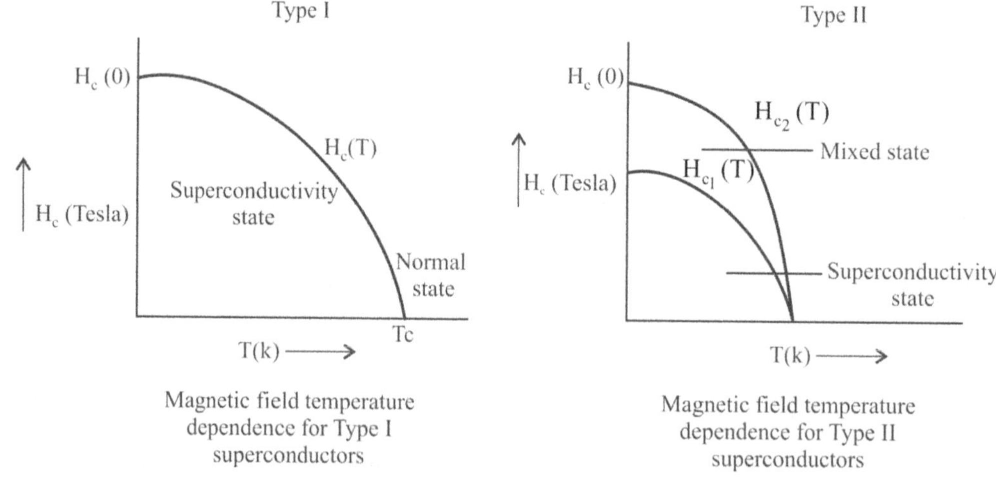

Magnetic field temperature dependence for Type I superconductors

Magnetic field temperature dependence for Type II superconductors

Fig. 8A.9

8A.8 Flux Quantisation

In 1950, Ginzburg and Landau introduced a complex wave function Ψ to describe the superconducting state. The density of the superconducting electron is

$$n_S \propto |\Psi|^2 \quad .....(8A.8)$$

The complex wave function Ψ is considered as an order parameter to describe the superconducting state. For a given temperature the "order parameter" Ψ is a function of position in the material. It is not a constant and vanishes above T_C.

The wave function Ψ can be written in terms of the magnitude and phase as

$$\Psi = |\Psi| \exp(i\,\phi) \quad .....(8A.9)$$

The magnetic flux ϕ enclosed by a superconducting ring is given by

$$\phi = \frac{nhc}{2e} = n\phi_0 \quad .....(8A.10)$$

where $\phi_0 = \frac{hc}{2e} = 2.07 \times 10^{-17}$ weber and is known as fluxoid or flux quantum. Thus the magnetic flux enclosed by the superconducting ring is quantised. This equation holds good when the ring is under superconducting condition only. This concept is called flux quantisation.

The flux through the ring is sum of the flux due to the external source and the flux due to the super current flowing through the ring i.e.,

$$\phi = \phi_{ext} - \phi_{SC} \quad .....(8A.11)$$

Thus flux is quantised.

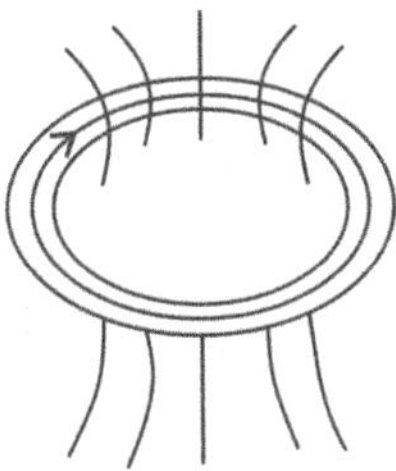

Fig. 8A.10

8A.9 London Equation

London proposed addition of two more equations to Maxwell's equation to explain Meissner effect. According to the two fluid model given by them, A superconductor is supposed to be composed of two distinct types of electrons i.e., normal electron and super electron. While normal electron behave in a usual manner, while the super electron behave in a much different way, they experience no scattering.

At temperature above T_C i.e., $T > T_C$ all the electrons are normal.

At temperature below T_C i.e., $T < T_C$, with decrease in temperature a large proportion of electron becomes super electrons and at T = 0k, all are super electrons.

The London equation is given by

$$\nabla \times \vec{J}_S = -\frac{n_S e^2}{m}\vec{B} \qquad .....(8A.12)$$

where J_S is the current density of super electron

n_S is the number of super electron

e is the charge on super electron

and m is its mass

8A.9.1 London's Penetration Depth

According to London equation, the magnetic flux does not suddenly drops to zero, at the surface of type I superconductors but decreases exponentially.

The depth up to which the flux can penetrate can be calculated as follows

We know that

$$\nabla \times B = \mu_0 J_S \qquad .....(8A.13)$$

Taking curl of this equation, we get

$$\nabla \times (\nabla \times B) = \mu_0 (\nabla \times J_S) \qquad .....(8A.14)$$

We know that

$$\nabla \times (\nabla \times B) = \nabla (\nabla . B) - \nabla^2 B$$

Since $(\nabla . B) = 0$ for a superconductor, Equation 8A.14 becomes

$$-\nabla^2 B = \mu_0 (\nabla \times J_S) \qquad .....(8A.15)$$

From equation

$$\nabla \times J_S = -\frac{n_S e^2}{m} B$$

We get

$$\nabla^2 B = \frac{\mu_0 n_S e^2}{m} B$$

$$\nabla^2 B = \frac{B}{\lambda^2} \qquad \text{.....(8A.16)}$$

where
$$\lambda = \left(\frac{m}{\mu_0 n_s e^2}\right)^{\frac{1}{2}} \qquad \text{.....(8A.17)}$$

is called the London penetrating depth

The solution of the differential equation 8A.16 is given by

$$H = H(0) \exp\left(-\frac{x}{\lambda}\right) \qquad \text{.....(8A.18)}$$

The Fig. 8A.11 shows the variation of flux density H with wavelength λ.

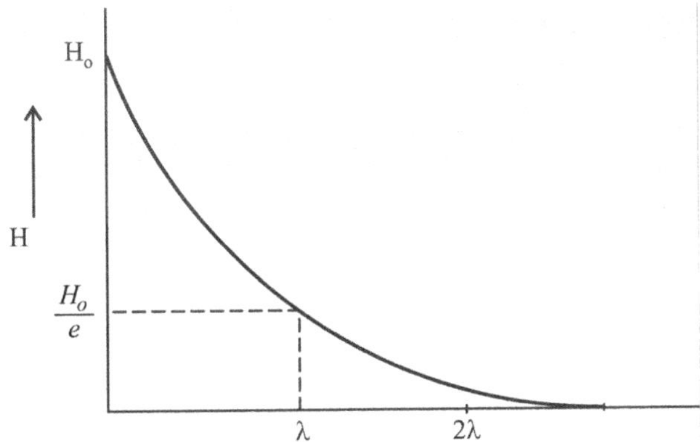

Fig. 8A.11.

Let
$$x = \lambda$$

∴
$$\frac{H}{H(0)} = e^{-1}$$

$$H = \frac{H(0)}{e} \qquad \text{.....(8A.19)}$$

Hence the penetration depth λ can be defined as the depth from the surface at which the magnetic flux density falls to $\frac{1}{e}$ of its initial value at the surface. Since it decreases exponentially, the flux inside the bulk of superconductor is zero and hence is in agreement with the Meissner effect.

The penetration depth is found to depend on temperature. Its dependence is given by the relation

$$\lambda(T) = \lambda(0)\left[1 - \frac{T^4}{T_C^4}\right]^{\frac{-1}{2}} \qquad .....(8A.20)$$

where $\lambda(0)$ is the penetration depth at T = 0 K.

According to equation 8A.20,

1. λ increases with the increase of temperature T
2. At $T = T_C$; λ becomes infinite.

At $T = T_C$, the substance changes from superconducting state to normal state and hence the field can penetrate to the whole specimen i.e., the specimen has an infinite depth of penetration.

8A.10 Josephson Effect

Let us consider a thin insulating layer sandwitched between two superconductors as shown in Fig. 8A.12. In addition to normal tunnelling of single electron, the super electron cooper pair also tunnel through the insulating layer from one superconductor to another without dissociation, even at zero potential difference across the junction. Their wave function on both the sides are highly correlated. This is known as Josephson effect.

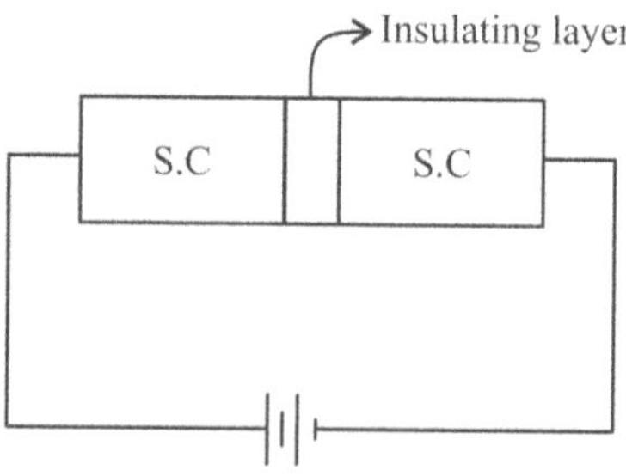

Fig. 8A.12

8A.10.1 D.C Josephson Effect

According to Josephson, when tunnelling occurs through the insulator, it introduces a phase difference $\Delta\phi$ between the two parts of the wave function on opposite side of the junction.

The tunnelling current is given by

$$I = I_0 \sin(\phi_0) \qquad .....(8A.21)$$

where I_0 is the maximum current that flows through the junction without any potential difference across the junction. I_0 depends on the thickness of the junction and the temperature; with no applied voltage a d.c. current flows across the junction. The magnitude of the current varies between I_0 and $-I_0$ according to the value of the phase difference.

$$\phi_0 = \phi_2 - \phi_1 \quad .....(8A.22)$$

This is called D.C Josephson effect.

8A.10.2 A.C Josephson Effect

Let us assume that a static potential V_0 is applied across the junction. This results in additional phase difference introduced by the cooper pair during tunneling across the junction. This additional phase difference $\Delta\phi$ at any time t can be calculated using quantum mechanics

$$\Delta\phi = \frac{Et}{\hbar} \quad .....(8A.23)$$

where E is the total energy of the system.

In this case $E = (2e) V_0$.

where 2e is the charge on cooper electron

Hence

$$\Delta\phi = \frac{2eV_0t}{\hbar} \quad .....(8A.24)$$

The tunneling current can be written as

$$I = I_0 \sin(\phi_0 + \Delta\phi)$$

$$I = I_0 \sin\left(\phi_0 + \frac{2eV_0t}{\hbar}\right) \quad .....(8A.25)$$

This is the form of

$$I = I_0 \sin(\phi_0 + \omega t) \quad .....(8A.26)$$

where $$\omega = \frac{2eV_0}{\hbar} \quad .....(8A.27)$$

This represents an alternating current with angular frequency ω. This is the A.C Josephson effect. When an electron pair crosses the junction a photon of energy $\hbar\omega = 2eV_0$ is emitted or absorbed. The current-voltage characteristic of a Josephson junction is shown in Fig. 8A.13.

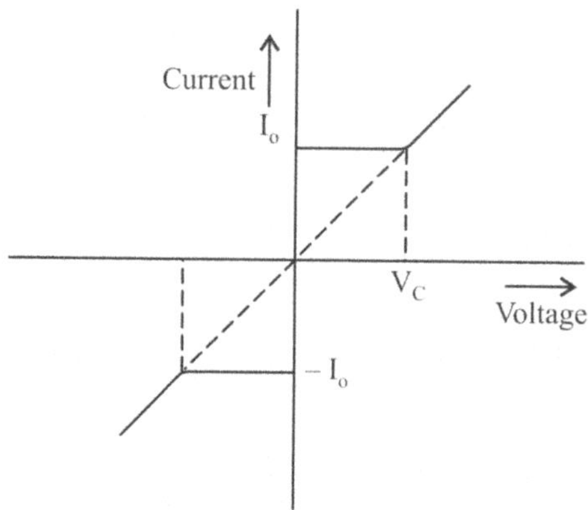

Fig. 8A.13 (Current voltage characteristic of a Josephson junction)

Results;

1. When $V_0 = 0$, there is a constant flow of d.c current i_c through the junction. This current is called superconducting current and the effect is called d.c Josephson effect.
2. So long $V_0 < V_C$, a constant d.c current i_0 flows
3. When $V_0 > V_C$, the junction has a finite resistance and current oscillates with a frequency

$$\omega = \frac{2eV_o}{\hbar}$$

This effect is called a.c Josephson effect.

8A.11 B.C.S Theory

B.C.S theory of superconductivity was put forward by Bardeen, Cooper and Schrieffer in 1957 and hence named as B.C.S theory. It can be explained by electron-electron interaction via lattice deformation.

1. Let us consider an electron passing through the lattice of positive ion. The electron is attracted by the neighbouring positive ion core as shown in Fig. 8A.14 and gets screened by them.

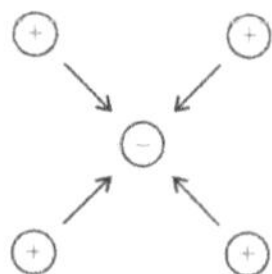

Fig. 8A.14

2. The screening greatly reduces the effective charge of this electron.
3. Due to the attraction between the electron and the ion core, the lattice gets deformed, lattice wave propagates, the quantum of energy of lattice wave is known as phonons.
4. Now if another electron passes by the side of the assembly of the said electron and the iron core, it gets attracted towards the assembly.
5. The second electron interacts with the first electron via lattice deformation. This interaction is said to be due to the exchange of a virtual phonon q, between the two electrons.

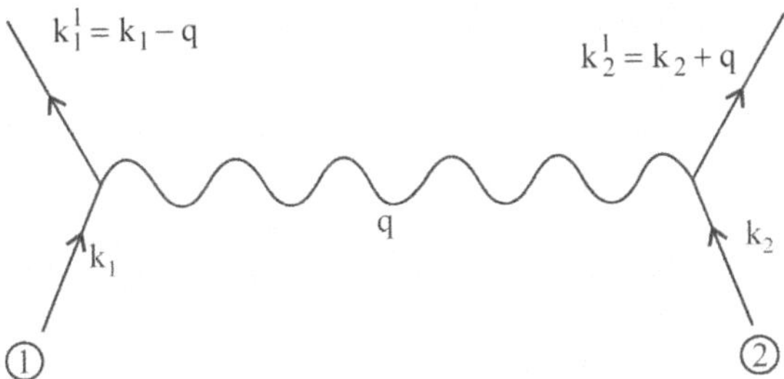

Fig. 8A.15

6. The interaction process can be written in terms of the wave vector as

$$K_1 - q = K_1^1 \qquad \text{and} \qquad K_2 + q = K_2^1$$

This gives

$$K_1 + K_2 = K_1^1 + K_2^1$$

i.e., the net wave vector of the pair is conserved

7. Only those electrons form cooper pairs which have equal and opposite momenta and spins

(a) momentum $\hbar \vec{K}_1 = -\hbar \vec{K}_2$

$$\vec{K}_1 = \vec{K}_2$$

where $\vec{K}$ is the wave vector of the electron

(b) $S_1 = -S_2$

(c) Cooper pairs are situated within the energy range $K_B\theta_D$ on the fermi surface, where K_B = Boltzmann constant and θ_D is the Debye temperature.

8A.11.1 Accomplishment of BCS Theory

1. The attractive interaction between electrons (cooper pairs) can lead to the ground state, separated from the excited states by an energy gap (E_g) of the order of 10^{-5} ev. Most of the thermal and electromagnetic properties are the consequences of the energy gap.
2. The electron phonon interaction leads to the energy gap of the observed magnitude. This is indirect interaction.
3. First electron distorts the lattice (phonon is created) and second electron obsorbed the phonon. The electron phonon interaction is attractive. The penetration depth and coherence length emerge as the natural consequence of BCS theory.
4. Critical temperature of superconductor is governed by the electron density of states at the fermi surface

$$T_C = 1.14\ \theta_D \exp\left[\frac{-1}{U\,D\,(E_F)}\right]$$

Where θ_D is the Debye temperature and $D(E_F)$ is the density of states.

5. The flux quantisation through superconducting ring in term of the pair charge 2e is the natural consequence of the BCS theory.

8A.12 Thermal Properties

8A.12.1 Entropy

Entropy is a measure of the disorder of a system. In normal metals with the decrease of temperature, entropy decreases linearly as shown in Fig. 8A.15.

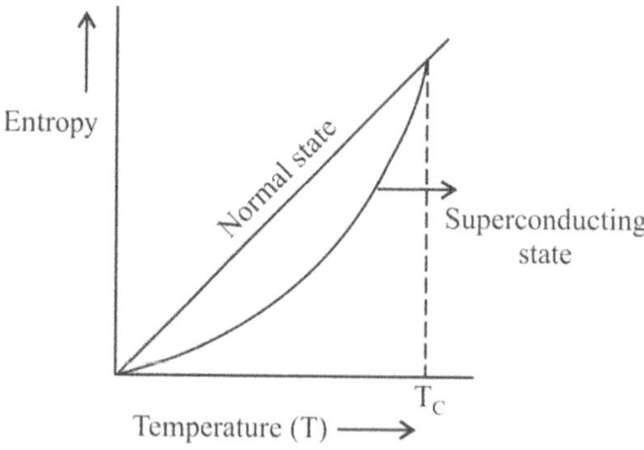

Fig. 8A.15

In superconducting metals like Aluminium, the entropy decreases linearly up to critical temperature T_C.

Cooling below this temperature the entropy decreases markedly. This means that the superconducting state is more ordered than the normal state. The decrease in entropy in superconducting state means that the electron thermally excited are more ordered in the superconducting state. It has been estimated that in type I superconductors there is a spatial order which extends over a distance of the order of 10^{-6} m. This range is called Coherence length.

8A.12.2 Specific Heat

The variation of specific heat with temperature for gallium in the superconducting states are shown in Fig. 8A.16.

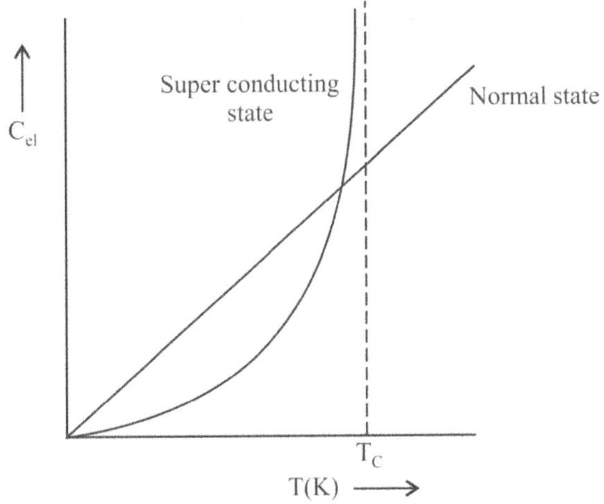

Fig. 8A.16

The later state can be obtained from the former by applying a magnetic field H greater the H_C, such a small field has no effect on the specific heat of the normal metal. For the normal metal, the specific heat obeys the following relationship

$$C_n = \gamma T + \beta T^3 \qquad \text{.....(8A.28)}$$

The first term on the right hand side of equation 8A.28 represents the electronic contribution to the specific heat, while the second one represents the contribution of lattice vibrations at low temperature. Since the later contribution remains unaffected in the superconducting state. Also, unlike in the normal state the electronic specific heat C_{el}, in the superconducting state does not show linear variation with temperature, rather it varies exponentially

$$C_{es} \propto \exp\left[\frac{-\Delta}{K_B T}\right] \quad .....(8A.29)$$

This type of dependence suggests that it requires a finite energy to excite an individual electron in a superconductor. Where as no such energy is required in a normal metal. This implies the existence of an energy gap E_g in a superconductor. The quantity Δ is related to E_g and is often called the energy gap parameter.

8A.12.3 Energy Gap

The temperature dependence of electronic specific heat indicates the existence of an energy gap E_g. in a superconductor. This energy gap is of entirely different nature compared with the energy gap in insulator. The gap is tied to the lattice in an insulator, where as it is tied to the fermi gas in a superconductor.

The gap separates the lowest excited state in a superconductor from the ground state as shown in Fig. 8A.17 and is related to Δ as

$$E_g = 2\Delta \quad .....(8A.30)$$

Therefore for Gallium

$$E_g = 2 \times 1.4\ K_B T_C$$

$$E_g = 2.8\ K_B T_C \sim 10^{-4}\ \text{ev}$$

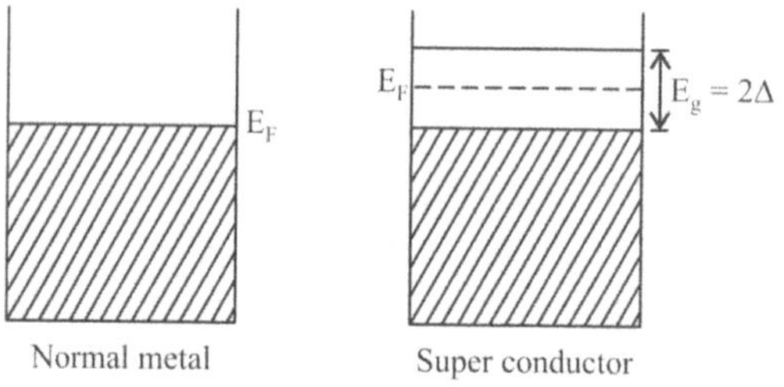

Fig. 8A.17

1. The electron present in the excited state behave as normal electron and create resistance where as those present below it behave as superconducting electron.
2. The energy gap varies with temperature. It is maximum at 0 k and decreases, continuously to zero as the temperature is increased to the critical temperature as shown in Fig. 8A.18.

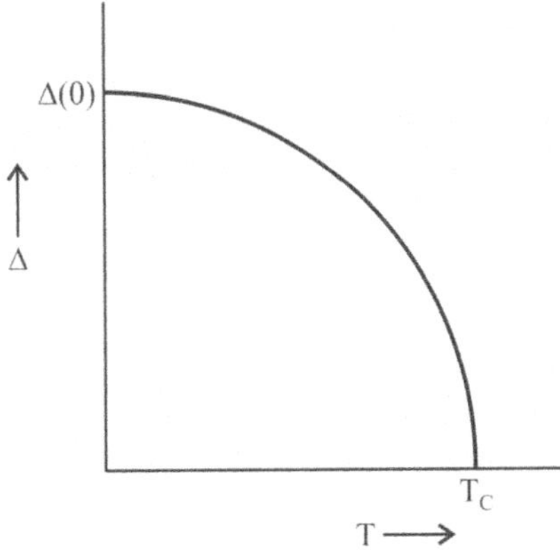

Fig. 8A.18

3. Thus at 0 k, there are no electrons above the gap and at $T = T_C$, all the superconducting electrons becomes normal electrons.
4. The variation of energy gap Δ with temperature is given by a mathematical expression

$$\Delta(T) = 1.74\,\Delta(0)\left[1-\frac{T}{T_C}\right]^{\frac{1}{2}} \qquad \text{.....(8A.31)}$$

Experimentally it is found that

$$\frac{2\Delta(0)}{K_B T_C} = 3.5 \text{ for most of the metals.}$$

8A.12.4 Isotope Effect

The transition temperature of a superconductor varies with its isotopic mass

$$T_C \propto M^{-1/2}$$

$$T_C\, M^{1/2} = \text{constant} \qquad \text{.....(8A.32)}$$

Thus larger the isotopic mass, lower is the transition temperature. For example, the transition temperature of mercury changes from 4.185 K to 4.146 K when its isotopic mass is changes from 199.5 to 203.4 amu

Now it is known that the mean square amplitude of atomic or lattice vibration at low temperature is proportional to $M^{-1/2}$ and the Debye temperature θ_D, of the phonon spectrum is related to M as

$$\theta_D\, M^{1/2} = \text{constant} \qquad \text{.....(8A.33)}$$

From 8A.32 and 8A.33 we obtain

$$\frac{T_C}{\theta_D} = \text{constant} \qquad \text{.....(8A.34)}$$

or in general, it can be written as

$$T_C \propto \theta_D \propto M^{-1/2} \qquad \text{.....(8A.35)}$$

The equation 8A.34 and 8A.35 indicates that the lattice vibrations are likely to be involved in causing superconductivity i.e., the electron–phonon interactions might be playing an important role for the occurrence of superconductivity.

8A.13 Characteristic of Superconductors in Superconducting States

We shall study the characteristics under the following two aspects:

1. Characteristics which do not change in superconducting transition
2. Characteristics which change in superconducting transition

8A.13.1 Properties which remain unchanged during superconducting Transitions

1. The appearance of similar x-ray diffraction pattern in both normal and superconducting states of a metal shows that no change in crystal lattice occurs during the transition from superconducting state to normal state.
2. The appearance of no appreciable change in the intensity distribution in superconducting transition inferred that the change in the electronic structure of metal in superconducting state is very minute.
3. The photo electric properties of a metal do not change in the superconducting transition from normal state.
4. The reflectivity of the superconductor in the visible and infrared region is same as that of the metals in normal superconducting states.
5. The elastic properties in both normal and superconducting states of a material are same.

8A.13.2 Properties which change in superconducting Transitions

1. The magnetic flux does not penetrate into the specimen of the superconductor. However, when $H > H_C$, the superconductor becomes a normal conductor.
2. When the current in the superconductor increases beyond a critical value, that is when $I > I_C$ superconducting material turns in to normal material.

3. All the thermoelectric effects disappear in the superconducting state.
4. The specific heat of the superconductor shows a jump at critical temperature. Since the superconductivity, mainly affects electron, it may be assumed that the lattice vibrational part remains unaffected. It increases as the temperature goes below the critical value, that is $T < T_C$.
5. In all superconductors, the entropy decreases significantly on cooling below the critical temperature.
6. The thermal conductivity of superconductor undergo a continuous change between the superconducting and normal phases. It is usually lower in superconducting phase.
7. All the properties that depend on the motion of the conduction band electron also changes as metal becomes superconductive, such as the thermo electric and magnetic resonance effects.
8. The superconductivity is lower for pure metals but higher for certain alloys.
9. The superconducting properties of metals can be changed by varying temperature, magnetic field, magnetic stress, size, impurities, atomic structure, isotopic mass, frequency of excitation of applied electric field etc.

8A.14 SQUID

A superconducting quantum interference device (SQUID) is a mechanism used to measure extremely weak signals, such as small changes in the human body's, electromagnetic energy field, using a device called a Josephson junction, a SQUID can detect a change of energy as much as 100 billion times weaker than the electromagnetic energy that moves a compass needle.

It consists of a ring of a superconducting material having two side arms A and B which act as an entrance and exit for the super current respectively. The insulating layer P and Q may be in general, have different thickness and let the currents through these layers be I_1 and I_2 respectively.

Both I_1 and I_2 vary periodically with the negative field, the periodicity of I_1 being greater than that of I_2. The variation of I_2 is an interference effect of the two junctions while that of I_1 is a diffraction effect that arises from the finite dimension of each junction. Since the current is sensitive to very small change in the magnetic field, the SQUID can be used as a very sensitive galvanometer (Fig. 8A.19).

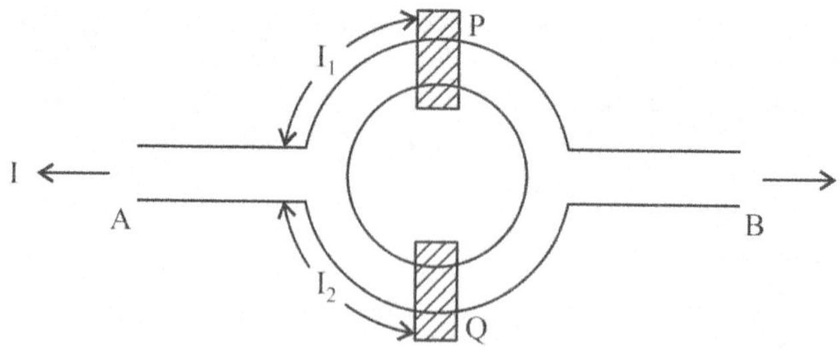

Fig. 8A.19

8A.15 Types of SQUIDS

8A.15.1 Radio Frequency (RF) SQUIDS

It is made up of one Josphson junction which is mounted on a superconducting ring. An oscillating current is applied to an external circuit, whose voltage changes as an effect of the interaction between it and the ring. The magnetic flux is then measured.

8A.15.2 Direct Current (DC) SQUIDS

It consists of two Josephson junctions employed in parallel so that electrons tunnelling through the junctions demonstrates quantum interference, dependent upon the strength of magnetic field within a loop. It is much more sensitive. It demonstrates resistance in response to even tiny variations in a magnetic field, which is the capacity that enables detection of such minute changes.

8A.15.3 High Temperature SQUIDS

These are more recently developed and are made of a substance called YBCO (chemical formula $YBa_2Cu_3\ O_{7-x}$) and are cooled by liquid nitrogen, which is cheaper and more easily handled than liquid helium. They are less sensitive than conventional "low temperature" SQUIDS but many applications do not require the extreme sensitively of the low temperature SQUID.

8A.16 Maglev (Magnetic Levitation)

Magnetic levitation or maglev is the process by which an object is suspended above another object with no other support but magnetic fields.

We know that, a diamagnetic substance repels a magnetic field. Thus, the perfect diamagnetic property of super conductors make them suitable for achieving friction less motion in motors and bearing.

The phenomenon of magnetic levitation is based on Meissner effect.

8A.16.1 How to Achieve Magnetic Levitation

The magnetic levitation is brought about by enormous repulsion between two highly powerful magnetic fields. If a small magnet is brought near a superconductor it will be repelled. This repulsion takes place due to the induced currents in the superconductor which is being generated by the magnetic field of the magnet. Because of zero resistance property of the superconductor this current persists and thus, the field due to this induced current repels the field due to the magnet. As a result, the magnet floats freely (i.e., levitated) above the superconductor. Thus, the levitation of the magnet or maglev demonstrates two critical properties of superconductors.

(i) Zero resistance and (ii) Meissner effect

8A.16.2 Applications

Magnetically levitated vehicles are called maglev vehicles. The utility of such levitation in vehicles is that in the absence of contact between the moving and stationary systems the friction is eliminated with such an arrangement great speeds could be achieved with very low energy consumption e.g. Maglev train.

8A.17 High Temperature Superconductors

Among the superconductors having transition temperature greater than 10 K in general are called high temperature superconductors. The high temperature superconductors belongs to five chemical system having general formula

1. $Ba\ Pb_{1-x} Bi_x O_3$
2. $La_{2-x} M_x Cu O_{4-x}$
3. $Ba_2 M Cu_3 O_{7-x}$
4. $Ba_2 La_{1+x} Cu_3 O_8$
5. $Bi_2 Ca Sr_2 Cu_2 O_8$

If the cell contains, one atom of rare earth, two atoms of Ba, three atoms of Cu and 7 atoms of oxygen, then such compounds are known as 123 superconductors.

8A.18 Applications of Superconductors

1. Superconductors are used for producing very strong magnetic field of the order of 20-50 tesla.
2. Magnetic energy can be stored in a large super conductor.
3. Josephson effect is used to generate microwave with frequency $\omega = \frac{2eV_0}{\hbar}$.

3. Josephson effect is used to generate microwave with frequency $\omega = \frac{2eV_0}{\hbar}$.
4. AC Josephson effect is used to measure very low temperature.
5. A Josephson junction is used for switching of signals from one circuit to another. The switching time is of the order of 1 pico second and hence very useful in high speed computers.
6. Superconductors are also being used in medical industry as SQUIDS which means superconducting quantum interference devices.

Solved Numerical Problems

1. A superconducting tin has a critical temperature of 3.7 K in zero magnetic field and a critical field of 0.0306 T at 0 K. find the critical field at 2 K.

Sol. The critical field H_c is given as

$$H_c = H_c(0)\left[1 - \left(\frac{T}{T_c}\right)^2\right]$$

Here $H_c(0) = 0.0306$ T

$T_c = 3.7$ K

and $T = 2$ K

$$\therefore \quad H_c = 0.0306\left[1 - \left(\frac{2}{3.7}\right)^2\right]$$

$$H_c = 0.0306\ [\ 1 - 0.292\ ]$$

$$H_c = 0.0306 \times 0.708$$

$$H_c = 0.0216 \text{ Tesla}$$

2. The transition temperature for Pb is 7.2 K. However at 5 K it loses the superconducting property subjected to a magnetic field of 3.3×10^4 A/m. Find the maximum value of H which will allow the metal to retain. its super conductivity at 0 K.

Sol The critical magnetic field varies with temperature as

$$H_c = H_c(0)\left[1 - \left(\frac{T}{T_c}\right)^2\right]$$

Here $T_c = 7.2$ K, $T = 5$ K and $H_c(T) = 3.3 \times 10^4$ A/m

$$H_C(o) = H_C(T)\left[1-\left(\frac{T}{T_c}\right)^2\right]^{-1}$$

$$H_c(0) = \frac{3.3\times10^4}{\left(1-\frac{5}{7.2}\right)^2}$$

$$H_c(0) = \frac{3.3\times10^4\times51.28}{26.28}$$

$$H_c(0) = 6.37 \times 10^4 \text{ A/m}$$

3. At what temperature is $H_c(T) = 0.1\, H_c(0)$ for lead Pb, having $T_c = 7.2$ K ?

Sol.
$$H_c(T) = H_c(0)\left[1-\left(\frac{T}{T_c}\right)^2\right]$$

In the given problem,

$$H_c(T) = 0.1\, H_c(0) \text{ and } T_c = 7.2 \text{ K}$$

$$0.1\, H_c(0) = H_c(0)\left[1-\left(\frac{T}{7.2}\right)^2\right]$$

$$0.1 = 1-\left(\frac{T}{7.2}\right)^2$$

or
$$T^2 = 0.90 \times 7.2 \times 7.2$$

$$T = 6.83 \text{ K}$$

4. Find the critical field in Pb at $T = 4.2$ K, Here $H_c(0) = 0.0803$ Wb/m^2 for Pb?

Sol. The variation of critical field with temperature is given by

$$H_c(T) = H_c(0)\left[1-\left(\frac{T}{T_c}\right)^2\right]$$

we know that

$$T_c \text{ for Pb} = 7.2 \text{ K}$$

$$H_c(T) = 0.0803\left[1-\left(\frac{4.2}{7.2}\right)^2\right]$$

$$H_c\ (T) = 0.0803\ [1-(0.58)^2]$$

$$H_c\ (T) = 0.0803\ (1-0.34)$$

$$H_c\ (T) = 0.0803 \times 0.66$$

$$H_c\ (T) = 0.0529 \text{ Tesla}$$

5. A long thin superconducting wire of a metal produces a magnetic field 105×10^3 amp/m on its surface due to the current through it at a certain temperature T. The critical field of the metal is 150×10^3 amp/m at T = 0 K. The critical temperature T_c of the metal is 9.3 K. What is the value of T?

Sol. Variation of magnetic field with temperature is

$$H_c(T) = H_c\ (0)\left[1-\left(\frac{T}{T_c}\right)^2\right]$$

Given $H_c\ (T) = 105 \times 10^3$ amp/m

$T = T$

$H_c\ (0) = 150 \times 10^3$ amp/m

$T_c = 9.2$ K

from the equation we have

$$\frac{H_c(T)}{H_c(0)} = 1-\left(\frac{T}{T_c}\right)^2$$

or

$$\left(\frac{T}{T_c}\right)^2 = 1 - \frac{H_c(T)}{H_c(0)}$$

$$T = T_c\left[1-\frac{H_c(T)}{H_c(0)}\right]^{1/2}$$

$$T = 9.2\left[1-\frac{105\times10^3}{150\times10^3}\right]^{1/2}$$

$$T = 9.2\left[\frac{45}{150}\right]^{1/2}$$

$$T = 9.2\sqrt{\frac{3}{10}}$$

or $\quad T = 5.03\ K$

6. The critical field for niobium is 1×10^5 A/m at 8 K and 2×10^5 A/m at 0 K. Calculate the transition temperature of the element.

Sol. $\quad H_c(T) = H_c(0)\left[1-\left(\frac{T}{T_c}\right)^2\right]$

$$\frac{H_c(T)}{H_c(0)} = 1-\left(\frac{T}{T_c}\right)^2$$

or $\quad T_c = \dfrac{T}{\left[1-\dfrac{H_c(T)}{H_c(0)}\right]^{1/2}}$

Here $\quad H_c(T) = 1\times 10^5$ A/m

$H_c(0) = 2\times 10^5$ A/m

and $\quad T = 8\ K$

$$T_c = \frac{8}{\left[1-\dfrac{1\times 10^5}{2\times 10^5}\right]^{1/2}} = \frac{8}{\sqrt{2}}$$

$$T_c = 5.656\ K$$

7. The transition temperature for lead is 7.26 K. The maximum critical field for the material is 8×10^5 A/m. Lead has to be used as a superconductor subject to a magnetic field of 4×10^4 A/m. What precaution will have to be taken?

Sol. Variation of magnetic field in superconductor may be express as

$$H_c(T) = H_c\ (0)\left[1 - \left(\frac{T}{T_c}\right)^2\right]$$

$$\left(\frac{T}{T_c}\right)^2 = \left[1 - \frac{H_c(T)}{H_c(0)}\right]$$

or $$T = T_c\left[1 - \frac{H_c(T)}{H_c(0)}\right]^{1/2}$$

Here $T_c = 7.26$ K

$H_c\ (T) = 4 \times 10^4$ A/m and
$H_c\ (0) = 8 \times 10^5$ A/m

$$T = 7.261\left[1 - \frac{4 \times 10^4}{8 \times 10^5}\right]^{1/2}$$

$$T = 7.26 \times \left(\frac{19}{20}\right)^{1/2}$$

$$T = 7.08 \text{ K}$$

The temperature of the metal should be below 7.08 K.

8. How much current can a Pb wire, 1.0 mm in diameter carry in its super conducting state at 4.2 K? Given B_c (4.2) = 0.0548 T.

Sol. Surface magnetic field associated with current I is

$$H_c = \frac{I_c}{2\pi.r}$$

$$I_c = 2\pi r\, H_c$$

$$I_c = \frac{2\pi r\ B_c\ (T)}{\mu_o}$$

$$I_c = \frac{\pi\, D\, B_c\ (T)}{\mu_o}$$

Here $D = 2r = 1.0\text{ mm} = 10^{-3}\text{ m}$ and $\mu_o = 4\pi \times 10^{-7}$

$$I_c = \frac{\pi \times 10^{-3} \times 0.0548}{4\pi \times 10^{-7}}$$

$$I_c = \frac{548}{4} = 137$$

$$I_c = 137 \text{ amp.}$$

9. A lead wire has a critical magnetic field of 6.5×10^3 A/m at 0K. The critical temperature is 7.18 K. At what temperature the critical field would drop to 4.5×10^3 A/m The diameter of wire is 2 mm. What is the critical current during at that temperature.?

Sol. Variation of magnetic field with temperature in a superconductor is given as

$$H_c(T) = H_c(0)\left[1 - \left(\frac{T}{T_c}\right)^2\right]$$

or
$$\frac{H_c(T)}{H_c(0)} = 1 - \left(\frac{T}{T_c}\right)^2$$

$$T = T_c = \left[1 - \frac{H_c(T)}{H_c(0)}\right]^{1/2}$$

here $H_c(0) = 6.5 \times 10^3$ A/m; $H_C(T) = 4.5 \times 10$; and $T_C = 7.18$ K

$$T = 7.18\left[1 - \frac{4.5 \times 10^3}{6.5 \times 10^3}\right]$$

$$T = 7.18\left[1 - \frac{0.9}{1.3}\right]^{1/2}$$

or
$$T = 7.18 \times \left[\frac{0.4}{1.3}\right]^{1/2} = 3.98 \text{ K}$$

the critical current,

$$I_c = 2\pi r H_c$$

Here
$$r = \frac{D}{2} = 1 \text{ mm} = 10^{-3}\text{m}$$

Critical current density

$$J_c = \frac{I_c}{\pi r^2}$$

$$J_c = \frac{2\pi r\, H_c\,(T)}{\pi r^2}$$

$$J_c = \frac{2H_c}{r} = \frac{2\times 4.5\times 10^{+3}}{10^{-3}}$$

$J_c = 9.0 \times 10^6$ A/m^2.

10. For a specimen of V_3 Ga, the critical fields are respectively 0.176 T and 0.528 T for 14 K and 13 K. Calculate the transition temperature and critical fields at 0 K and 4.2 K

Sol. We know that

$$H_c\,(T) = H_c\,(0)\left[1-\left(\frac{T}{T_c}\right)^2\right] \qquad (1)$$

Given At T = !4 K, H_c (T) = 0.176 T and at T = 13 K, H_c (T) = 0.528 T

Therefore $$0.176 = H_c\,(0)\left[1-\left(\frac{14}{T_c}\right)^2\right] \qquad (2)$$

and $$0.528 = H_c\,(0)\left[1-\left(\frac{13}{T_c}\right)^2\right] \qquad (3)$$

Dividing equation (3) by equation (2), we get

$$\frac{1-(13/T_c)^2}{1-(14/T_c)^2} = \frac{0.528}{0.176} = 3$$

$$1-\frac{169}{T_c^2} = 3-\frac{588}{T_c^2}$$

or $$\frac{419}{T_c^2} = 2$$

or $$T_c = \left(\frac{419}{2}\right)^{1/2} = 14.5 \text{ K}$$

Substituting this value of T_c in equation (2), we get

$$0.176 = H_c\ (0)\left[1-\left(\frac{14}{14.5}\right)^2\right]$$

$$0.176 = H_c\ (0)\left[1-(0.9655)^2\right]$$

$$H_c\ (0) = \frac{0.176}{1-0.932} = 2.588 \text{ Tesla}$$

for T = 4.2 K

$$H_c\ (T) = 2.588\left[1-\left(\frac{4.2}{14.5}\right)^2\right]$$

$$H_c\ (T) = 2.588\ [1-0.0839]$$

$$H_c\ (T) = 2.588 \times 0.9161 = 2.37 \text{ Tesla}$$

11. The penetration depth λ of Hg at 3.5 K is about 750 A°. Find penetration depth at 0 K. Given T_c for Hg = 4.153 K.

Sol. The variation of penetration depth with temperature is given by

$$\lambda(T) = \lambda(0)\left[1-\left(\frac{T}{T_c}\right)^4\right]^{-1/2}$$

Here λ (T) = 750 A° and T = 3.5 K and T_c for Hg = 4.153 K

$$\lambda(0) = \lambda(T)\left[1-\left(\frac{T}{T_c}\right)^4\right]^{1/2}$$

$$\lambda(0) = 750\left[1-\left(\frac{3.5}{4.153}\right)^4\right]^{1/2}$$

$$\lambda\ (0) = 750\ [\ 1 - 0.504\ ]^{1/2}$$

or $$\lambda\ (0) = 528.7 \text{ A}^\circ$$

12. For a superconductors the number of super electron is 10^{28} m^{-3} and $T_c = 3$ K, find the penetration depth at 0 K and 1 K

Sol. The penetration depth at T = 0K is given as

$$\lambda_o = \left(\frac{m}{\mu_o\, n_s\, e^2}\right)^{1/2}$$

Here $m = 9.1 \times 10^{-31}$ Kg

$\mu_o = 12.56 \times 10^{-7}$

$e = 1.6 \times 10^{-19}$ C and $n_s = 10^{28}/m^3$

$$\lambda_o = \left(\frac{9.1\times10^{-31}}{12.56\times10^{-7}\times10^{28}\times\left(1.6\times10^{-19}\right)^2}\right)^{1/2}$$

$$\lambda_o = 0.53 \times 10^{-7}\ \text{m} = 530\ \text{A}^o$$

we known that

$$\lambda\,(T) = \lambda_o\left[1-\left(\frac{T}{T_c}\right)^4\right]^{-1/2}$$

here $$\frac{T}{T_c} = \frac{1}{3}$$

or $$\frac{T^4}{T_c^{\,4}} = \frac{1}{81}$$

$$\lambda(T) = 530\left[1-\frac{1}{81}\right]^{-1/2}$$

$$\lambda(T) = 530\left[\frac{80}{81}\right]^{-1/2}$$

or $$\lambda(T) = \frac{530\times 9}{8.95} = 533\ \text{A}^o$$

thus λ (at 1K) = 533 A°

λ (at 1K) = 533 A°

13. Estimate the energy gap Eg for Nb (T_c = 9.25 K) at T = 0 K and find the minimum photon wavelength and frequency needed to break the cooper pair.

Sol. The energy gap (2Δ or E_g) at T = 0 K is given by

$$E_g \sim 3.53\, K_B\, T_C$$

$$E_g = 3.53 \times (1.38 \times 10^{-23}) \times 9.25$$

$$E_g = 4.5 \times 10^{-22}\text{ J}$$

$$E_g = 2.82 \text{ mev}$$
$$E_g \sim 2.82 \text{ mev}$$

further $$\lambda_{min} = \frac{hc}{Eg} = \frac{6.63 \times 10^{-34}\ (JS) \times 3 \times 1}{4.5 \times 10^{-22}}$$

$$\lambda_{min} = 4.42 \times 10^{-4}\text{ m}$$

This wavelength lie in the far infrared region of the electromagnetic radiations corresponding frequency

$$\nu = \frac{E_g}{h} = \frac{4.5\times10^{-22}}{6.63\times10^{-34}}$$

$$\nu = \frac{E_g}{h} = \frac{4.5}{6.63} \times 10^{11}\ S^{-1}$$

$$\nu = 6.79 \times 10^{11}\ S^{-1}$$

14. For Hg, the transition temperature for an average atomic mass 200. 59 u is 4.153 K, when for the isotope of mass 204 u, T_c = 4.118 K. Calculate the isotope effect coefficient α.

Sol. The isotope effect is represented as

$$T_c\, M^{\alpha} = \text{constant} = C$$

Taking log an both sides, we get

$$\log T_c + \alpha \log M = \log C$$

Differentiating, we get

$$\frac{dT_c}{T_c} + \alpha\frac{dM}{M} = 0$$

$$\alpha\frac{dM}{M} = -\frac{dT}{T_c}$$

or $$\alpha = -\frac{M}{dM}\frac{dT_c}{T_c}$$

here $dM = 204 - 200.59$

$dM = 3.41$ u

$dT_c = 4.118 - 4.153$ K

$dT_c = 0.035$ K

$$\alpha = -\left(\frac{200.59}{3.41}\right).\left(\frac{-0.035}{4.153}\right)$$

$$\alpha = \frac{7.02}{14.16} = 0.496 \quad \alpha = 0.496$$

Conceptual Questions

1. **Define superconductivity.**

Ans. Superconductivity is a phenomenon in which certain metals, alloys and ceramics conduct electricity without resistance when it is cooled below a certain temperature called the critical temperature T_c.

2. **What is the effect of impurity and stress on superconductor?**

Ans. Addition of impurities to a superconductor results in the lowering of T_c value. In superconductors, the increase in stress result in increase of the T_c value.

3. **Show graphically how H_c varies with respect to temperature?**

Ans. The material is said to be in the superconducting state within the curve and is in the normal state outside the curve.

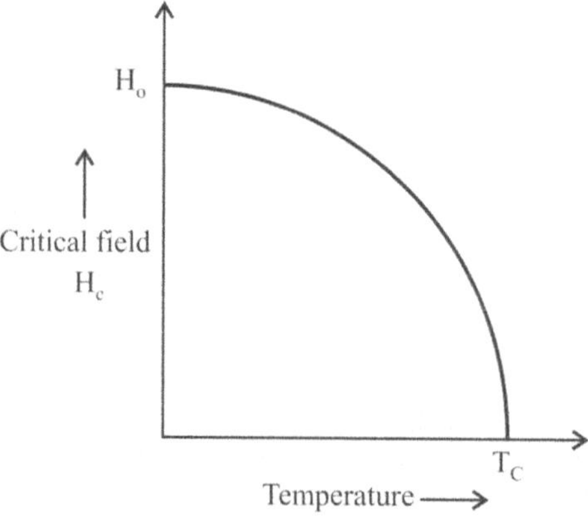

4. Explain persistent current?

Ans. When current is made to flow through a superconducting ring, it was found to flow steadily even after a year within the ring.

This steady flow of current (having no change in its value) in a superconducting ring without any potential driving it is called persistent current.

5. What is Meissner effect?

Ans. The complete expulsion of all magnetic field by a superconducting material is called Meissner effect.

6. What are the three important factors required to define a superconducting state?

Ans. The three important factors required to define a superconducting state are:

(i) The critical temperature T_C.

(ii) The critical current density J_C and

(iii) The critical magnetic field H_C

7. Compare type I and type II superconductors?

Ans.

Type I superconductor	Type II superconductor
1. These superconductors are called as soft super conductors.	1. These superconductors are called as hard superconductors.
2. Only one critical field exists for these superconductors.	2. Two critical fields H_{c1} (lower critical field) and H_{c2} (upper critical field) exist for these superconductors.
3. The critical field value is very low.	3. The critical field value is very very high.
4. These superconductors exhibit perfect and complete Meissner effect.	4. These do not exhibit a perfect and complete Meissner effect.

8. What is magnetic levitation?

Ans. Magnetic levitated vehicles are called Maglev vehicles. With the phenomenon of levitation the friction is completely eliminated. As a result, with such an arrangement great speeds could be achieved with very low energy consumption.

9. Explain Josephson effects?

Ans. Josephson effects are of two types?

(i) d.c Josephson effect and

(ii) a.c. Josephson effect.

(i) d.c. Josephson effect

The Tunnelling of electrons (copper pairs) across the insulator in a Josephson junction result in a net current which flows even in the absence of a potential difference. This is called d.c. Josephson effect.

(ii) a.c. Josephson effect

When a potential difference V is applied between the two sides of a Josephson junction, there is an oscillation of the tunnelling current with angular frequency $\omega = \frac{2eV}{\hbar}$. This is called the a.c Josephson effect.

10. What is a SQUID

Ans. SQUID is an abbreariation of superconducting Quantum interference Device. These are sensitive magnetometers, in which two Josephson junctions connected in parallel are used. These devices can be used to detect even small magnetic field changes.

11. What is cooper pair?

Ans. Cooper pair is a quasi-particle formed by the mutual attraction of two electrons with oppsote spins and opposite Linear momenta below a critical temperature T_c. It has a charge of – 2e and a mass 2 m_e but no net spin. The electrons are held together by the induced distortions and vibrations of the lattice of positive metal ions with which the electrons interact.

12. What is physical mechanism that generates the superconducting electrons?

Ans. Superconducting electrons exist in bound pairs (cooper paris). The pairing is because of an effective force between electrons of opposite spin and momentum that is produced as follows. One electron interacts with an atom, displacing it slightly, and the second electron is attracted to the displaced atom resulting in an effective attraction between electrons.

Objective Questions

1. A temperature at which the normal material turns in to a superconductor is called

   (a) Debye temperature (b) Neel temperature

   (c) Critical temperature (d) Curie temperature

2. The superconducting state of a superconductor is influenced by

   (a) Temperature (b) Magnetic field

   (c) Current (d) all of the three

3. A super conducting material exhibits

   (a) Zero conductivity and complete diamagnetism

   (b) Zero resistivity and complete paramagnetism

   (c) Infinite conductivity and complete paramagnetism

   (d) Zero resistivity and complete diamagnetism

4. The transition temperature of mercury is

   (a) 1 K (b) 1.14 K

   (c) 4.12 K (d) 9.22 K

5. A superconducting material when placed in a magnetic field will

   (a) Attract all the magnetic lines of forces passing through it

   (b) Repel all the magnetic lines of forces passing through it

   (c) Remain uninfluenced by the magnetic field

   (d) Attract magnetic field toward its centre

6. Critical magnetic field

   (a) does not depend on temperature

   (b) increase if temperature decreases

   (c) increases if temperature increases

   (d) does depend on temperature

7. The susceptibility of superconductor is

   (a) Positive and small (b) Negative and small

   (c) Positive and unity (d) Negative and unity

8. A superconducting material on being subjected to the critical field changes to
   (a) critical conductivity
   (b) remain uninfluenced
   (c) normal state
   (d) superconducting which is independent of temperatue
9. Superconductivity results due to
   (a) crystal structure having no atomic vibrations at 0K
   (b) crystal structure having infinite atomic vibrations at 0K
   (c) all electrons having Fermi energy at 0K
   (d) all electrons interact in the superconducting state.
10. The current in a superconductor
   (a) are confined to the surface of the superconductor
   (b) are confined to the middle portion of a superconductor
   (c) gives rise to a potential difference
   (d) can exist in quantised form
11. Soft superconductor observe
   (a) Meissner effect (b) Silsbee's rule
   (c) both (a) and (b) (d) None of the above
12. Hard superconductor observe
   (a) Breakdown of Silsbee's rule
   (b) incomplete Meissner effect
   (c) high critical field and transition temperature
   (d) all of these
13. Mark the combination of a material which is a type II superconductor
   (a) Cd, Sn (b) Al, Pb
   (c) Hg, Cr (d) Nb, V
14. The property that does not exhibit an appreciable change in superconducting state as compared to the normal state is
   (a) specific heat (b) thermal conductivity
   (c) entropy (d) volume

15. In superconducting state
    (a) entropy increases and thermal conductivity decreases
    (b) entropy and thermal conductivity both decreases
    (c) entropy and thermal conductivity both increases
    (d) entropy decreases and thermal conductivity increases

16. Transition temperature T_C and critical field H_C for a superconductor are related as

    (a) $H_C = H_0\left[1-\left(\frac{T}{T_C}\right)^2\right]$ (b) $T_C = T_0\left[1-\left(\frac{H_0}{H_C}\right)^2\right]$

    (c) $H_C = H_0\left[1-\left(\frac{T_C}{T}\right)^2\right]$ (d) $T_C = T_0\left[1-\left(\frac{H_C}{H_0}\right)^2\right]$

17. The variation of penetration depth with temperature is represented by the relation

    (a) $\lambda(T) = \lambda(0)\left[1-\left(\frac{T}{T_C}\right)^4\right]^{-\frac{1}{2}}$ (b) $\lambda(T) = \lambda(0)\left[1-\left(\frac{T}{T_C}\right)^4\right]$

    (c) $\lambda(T) = \lambda(0)\left[1-\left(\frac{T_C}{T}\right)^4\right]^{-1}$ (d) $\lambda(T) = \lambda(0)\left[1-\left(\frac{T_C}{T}\right)^4\right]$

18. The width of the energy gap of a superconductor at 0K is nearly
    (a) $K_B T_C$ (b) $3.5\ K_B T_C$
    (c) $3.5\ K_B T$ (d) $300\ K_B T_C$

19. The energy gap in a superconductor is maximum when the temperature is equal to
    (a) T_C (b) 0K
    (c) room temp (d) None of these

20. As the isotopic mass of H_g decreases
    (a) The critical temperature decreases slightly
    (b) The critical temperature increases slightly
    (c) The critical temperature remain unchanged
    (d) The critical temperature decreases drastically

21. The isotope effect coefficient is

(a) Zero (b) unity

(c) 0.5 (d) None of these

22. The magnetic flux Quanta, called fluxon produced in a superconductor has the value

(a) h/2e (b) 2h/e

(c) h/e (d) 2e/h

23. The total change on a cooper pair is

(a) Zero (b) + 2e

(c) – 2e (d) varies with T_C

24. The energy required to break a cooper pair is

(a) equal to the fermi energy of the electron

(b) equal to the vibrational energy of the atoms in superconductor

(c) equal to the energy gap in superconductor

(d) None of the above

25. A cooper pair is a system of two electron bound by exchange of

(a) photon between them (b) Proton between them

(c) phonon between them (d) neutron between them

26. The binding energy of a cooper pair is of the order of

(a) 10^{-12} ev (b) 10^{-4} ev

(c) 10^{-8} ev (d) 1 ev

27. Cooper pair are formed

(a) at high temperatures as the thermal energy is sufficient to form the cooper pair

(b) at very low temperature below T_c as the thermal energy is not sufficienct to disrupt the binding

(c) at a temperature > T_c as the thermal energy is sufficient to form cooper pair

(d) None of the above

28. A Magentic field can desotry superconductivity
    (a) By breaking the electron pair
    (b) By making both the electron spins point in the same direction
    (c) by either (a) or (b)
    (d) None of the these

29. The favourable condition for superconducting state of a matter is
    (a) a weak electron- phonon interaction
    (b) a strong electron-phonon interaction
    (c) a strong phonon- phonon interaction
    (d) a weak phonon-phonon interaction

30. BCS theory is valid for
    (a) low temperature superconductor
    (b) high temperature superconductor
    (c) both low and high temperature super conductor
    (d) Neither low nor high temperature superconductor

Theoretical Questions

1. What is superconductivity?
2. Explain few important properties of superconductors.
3. Explain meissner effect.
4. Define transition temperature.
5. Define critical magnetic field.
6. What do you mean by critical current density?
7. Explain the silsbee effect.
8. What are type I and type II superconductors
9. Distinguish between type I and type II superconductors.
10. How does the entropy changes with temperature in superconductors
11. How does the specific heat changes with temperature in superconductors
12. Explain the effect of isotope on superconductors.

13. Discuss the occurrence of energy gap in superconductors
14. Explain london penetration depth
15. What do you mean by flux quantization in superconductors?
16. What is Josephson effect, explain a.c and d.c Josephson effect?
17. Give the salient features of BCS theory of superconductivity.
18. How cooper pairs are formed in superconductors?
19. What is high temperature superconductivity?
20. Explain 1-2-3 Superconductors.

Numerical Problems

1. For superconducting specimen, the critical field are respectively 0.18 Wb/m^2 and 0.53 Wb/m^2 for 14 K and 13 K. Find the superconducting transition temperature and the critical field at 0 K?

   [14.47 K ; 2.73 Wb/m^2]

2. Calculate the critical current which can flow through a long thin superconducting wire of aluminium of diameter 10 $^{-3}$m. The critical magnetic field for aluminium is 7.9×10^3 amp/m.

   [24.81]

3. For lead, $T_c = 7.2$ K, Calculate the ratio of penetration depth at 3 K to that at 0 K.

   [1.015]

4. The penetration depth of mercury at 3.5 K is about 750 A°. Estimate the penetration depth at 0 K.

   [519 A°]

5. A voltage of 5.9 μ V is applied across a Josephson junction. What is the frequency of the radiation emitted by the junction.

   [$\nu = 2.851 \times 10^9$ Hz]

Answers to Objective Questions

1. (c)	2. (d)	3. (d)	4. (c)	5. (b)	6. (b)
7. (d)	8. (c)	9. (a)	10. (a)	11. (c)	12. (d)
13. (d)	14. (d)	15. (b)	16. (a)	17. (a)	18. (b)
19. (b)	20. (b)	21. (c)	22. (a)	23. (c)	24. (c)
25. (c)	26. (c)	27. (b)	28. (c)	29. (b)	30. (a)

CHAPTER 8B

Science and Technology of Nanomaterials

"Coal and diamonds, sand and computer chips, cancer and healthy tissue; throughout history, variations in the arrangement of atoms have distinguished the cheap from the cherished, the diseased from the healthy. Arranged one way, atoms make up soil, air and water arranged another, they make up ripe strawberries. Arranged one way, they make up frames and fresh air, arranged another, they make up ash and smoke".

Eric Drexler

8B.1 Introduction

The scientific community generally attributes the first acknowledgement of the importance of the nanoscale range to the brilliant Nobel Laureate Physicist Richard Feynnman in his famous 1959 Lecture, "There's Plenty of Room at the Bottom" in which he first proposed that the properties of materials and devices at the nanometer range would present future opportunities.

He said

"I want to build a billion tiny factories, models of each other, which are manufacturing simultaneously.... The principles of physics as far as I can see, do not speak against the possibility of manoeuvring things atom by atom. It is not an attempt to violate any laws; it is something in principle that can be done, but in practice, it has not been done because we are too big".

- *Richard Feynnman*

The term nonotechnology reached public awareness in 1986 with the publication of "Engines of creation, the coming era of nanotechnology" by Eric Drexler.

8B.2 Nanoscience

Nanoscience deals with the study of phenomenon at a very small scale of 10^{-9}m (1 nanometer) to 10^{-7} (100 nanometer) where properties of matter differ significantly from those at larger scales. The scale ranging from 1 nanometer (nm) to 100 nanometer is known as nanoscale and is difficult to visualize as human eye can see down to about 20 μm only.

8B.3 Visualization of Nanoscale

We know that with naked eyes, one can see only up to 20 μm. Even with the help of optical microscope, we cannot go beyond 1 μm. There are some tools by which we can visualize Nanoscale, some of them are;

1. SEM (Scanning electron microscope)
2. STM (Scanning tunnelling microscope)
3. AFM (Atomic force microscope)

8B.3.1 SEM (Scanning Electron Microscope)

The Scanning Electron Microscope or SEM is an incredible tool for seeing the unseen worlds of microspace.

Conventional light microscope use a series of glass lenses to bend light waves and create a magnified image. The SEM creates the magnified images by using electrons instead of light waves (Fig. 8B.1). The SEM shows very detailed 3-dimensional images at much higher magnification than is possible with a light microscope.

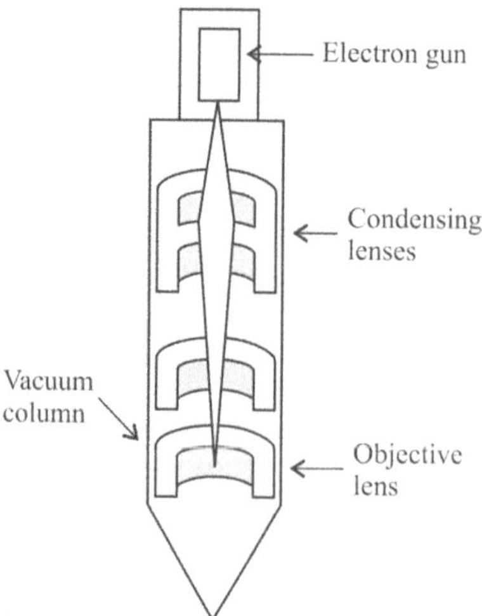

Fig. 8B.1 Scanning electron microscope.

8B.3.2 STM (Scanning Tunneling Microscope)

The Scanning Tunneling Microsocpe (STM) is considered as a type of electron microscope. The Scanning Tunneling microscope (STM) is a non-optical microscope that scans an electrical probe over a surface to be imaged to detect a weak electric current flowing between the tip and the surface (Fig. 8B.2).

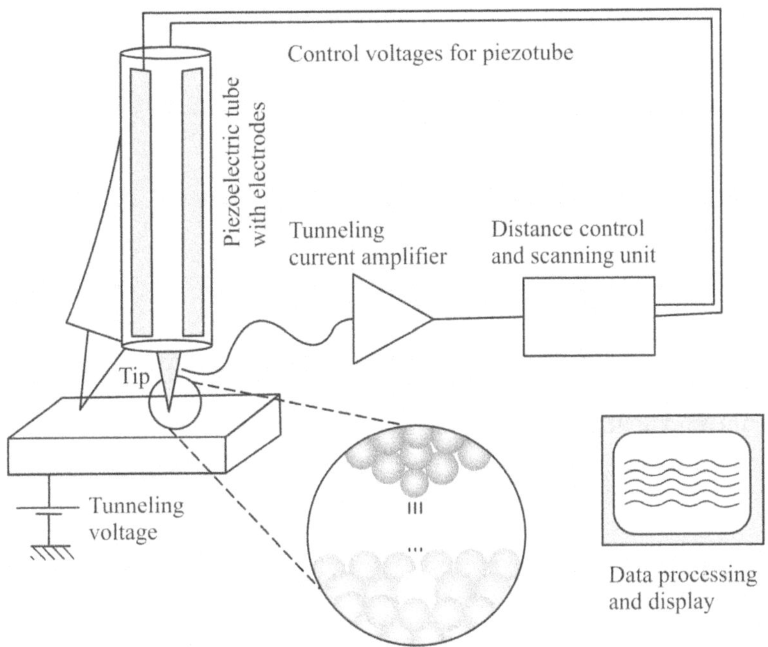

Fig. 8B.2 Schematic view of STM.

STM allows scientist to visualise region of high electron density and hence infer the position of individual atoms and molecules on the surface of a lattice.

The STM can obtain images of conductive surfaces at an atomic scale 2×10^{-10} m or 0.2 nanometer, and also can be used to manipulate individual atoms, trigger chemical reactions, or reversibly produce ions by removing or adding individual electrons from atoms or molecules

8B.3.3 AFM (Atomic Force Microscope)

The atomic force microscope (AFM) is a very high resolution type of scanning probe microscope, with demonstrated resolution of fraction of an A.U, more than 1000 times better than the optical diffraction limit (Fig. 8B.3).

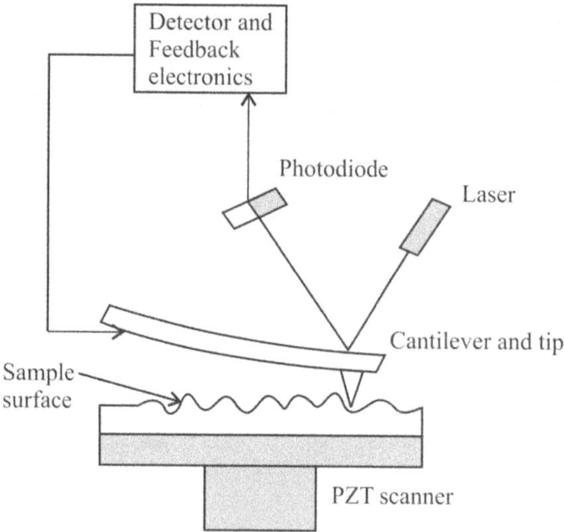

Fig. 8B.3 Atomic force microscope.

It uses a tiny tip that moves in response to the electromagnetic forces between the atoms of the surface and the tip. As the tip moves up and down, the motion is recorded and an electronic image of the atomic surface is formed.

8B.4 Advantages of AFM over SEM

The AFM has several advantages over the scanning electron microscope (SEM).

1. Unlike the electron microscope which provides a two dimensional projection or a two-dimensional image of a sample, an AFM provides a true three-dimensional surface profile.
2. Samples viewed by AFM do not require any special treatment [such as metal (carbon coatings)] that would irreversibly change or damage the sample.
3. While an electron microscope needs an expensive vacuum environment for proper operation, most AFM modes can work properly well in ambient air or even in a liquid environment. This make possible to study biological macromolecules and even living organism.

8B.5 Tools and Techniques

Two main approaches are used in nano technology.

8B.5.1 Bottom-Up Technique

Bottom-up technique is that technique where materials and devices are built up atom by atom. These processes are achieved by wet chemical synthesis route and by depositing the materials in the form of thin films in controlled environment.

8B.5.2 Top-down Technique

The top to down technique consists of cutting away material until you have a completed component or product of desired size. It can be achieved by milling the material using high energy ball mill.

Newer techniques such as Dual polarisation interferometry are enabling scientist to measure quantitatively the molecular interaction that take place at the nanoscale.

8B.6 Nanomaterials

The materials with average particle size less than 100 nm are known as nanomaterials.

Nanomaterials can be nanoscale in one dimension. e.g. Surface films.

Nanomaterials can be nanoscale in two dimension. e.g. Strands or fibers.

Nanomaterials can be nanoscale in three dimension. e.g. Particles.

Common types of nanomaterials include nanotubes, dendrimers, quantum dots (1-10 nm) and buckyballs.

The reduction in particle size increases the surface to volume ratio of nanoparticles than their bulk counter part, which can lead to greater chemical reactivity and affect their strength. Apart from increased surface to volume ratio, nano materials also show reduction in their coordination number as shown in Fig. 8B.4.

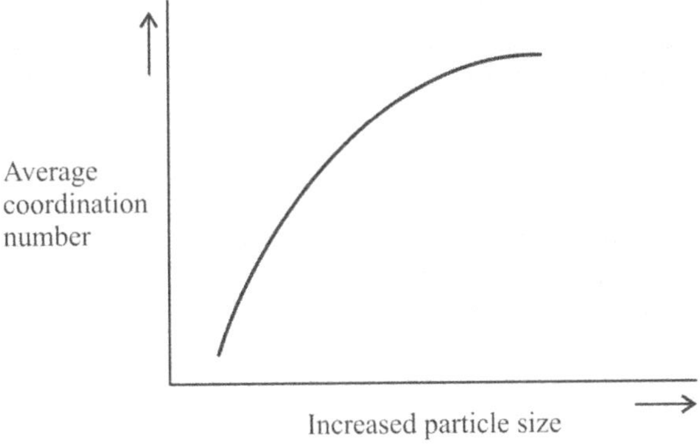

Fig. 8B.4 Variation of coordination number with changing particle size.

8B.6.1 Increase in Surface Area to Volume Ratio

Nanomaterials have a relatively larger surface area when compared to the same volume (or mass) of the material produced in a larger form. Let us consider a sphere of radius "R".

Then its surface area is given by $4\pi R^2$

Its volume is given by $\frac{4}{3}\pi R^3$

Therefore surface to volume ratio = $\frac{3}{R}$

Thus when the radius of the sphere decreases, its surface area to volume ratio increases. It makes materials more chemically reactive. As growth and catalytic chemical reactions occur at surfaces, this means that a given mass of material in nano particle state will be much more reactive than the same mass of material made up of larger particles.

8B.6.2 Quantum Confinement Effect

When atoms are isolated then their energy levels are discrete. When very large number of atoms are closely packed to form a solid, the energy levels split and form bands. Nanomaterials represents intermediate stage. When the dimension of wells are of the order of de-Broglie wavelength of electrons or mean free path of electrons i.e. within few tens of nanometer, the energy levels of electrons changes. This effect is called quantum confinement.

When the material is in sufficiently small size typically 10 nm or less, organisation of energy levels into which electrons can climb or fall change. This phenomenon results from electrons and holes being squeezed into a dimension that approaches a critical quantum measurement, called the 'excitation Bohr radius'. These can affect the optical, electrical and magnetic behaviour of materials, particularly as the structure or particle size approaches the smaller end of the nanoscale.

8B.7 Classification of Nanomaterials

The classification of the nanomaterials has been done on the basis of following seven main material categories:

1. Carbon based nanomaterials
2. Nanomaterials of metals and alloys
3. Nanopolymers
4. Nano-glasses
5. Nano-ceramics
6. Nano-composites
7. Biological nanomaterials

8B.8 Properties of Nano materials

The properties of nano materials are as follows:

1. Nanocrystalline materials are exceptionally strong
2. They are hard
3. They are ductile at high temperature
4. They are wear-resistant
5. They are erosion-resistant
6. They are corrosion-resistant
7. They are chemically very active.

8B.9 Carbon-Nanotubes

Carbon nanotubes (CNT) are sheets of graphite rolled up to make a tube i.e. carbon macro-molecules in cylindrical form. The nanotube dimensions are variable and can be as small as 0.4 nm in diameter.

8B.10 How to Produce Nanotubes

The carbon nanotubes (CNTs) are produced by using the following methods:

1. High pressure carbon monooxide deposition (HiPCo)
2. Chemical vapour deposition (CVD)
3. Vaccum sputtering method
4. Radio frequency (RF) plasma method
5. Ball-milling method
6. Sol-Gels method

8B.10.1 High Pressure Carbon Monooxide Deposition (HiPCo)

This method involves a heated chamber through which carbon monoxide and small clusters of iron atoms flow. When carbon-monoxide molecules land on the iron clusters, the iron acts as a catalyst. This helps a carbon monoxide molecule to breakup into a carbon atom and an oxygen atom. The carbon atom bonds with other carbon atom and a nanotube lattice is formed. The oxygen atom combines with another carbon monoxide molecule to form carbon dioxide gas. The gas then floats off into the air.

8B.10.2 Chemical Vapour Deposition (CVD)

In this method, a substrate is prepared with a layer of metal catalyst particles such as nickel, cobalt or iron. The substrate is heated to approximately 700 °C in a chamber. To initiate the growth of nanotubes, two gases are blown into the chamber.

One is a process gas such as ammonia, nitrogen or hydrogen. The other is a carbon containing gas such as methane. We know that methane contains one carbon atom and four hydrogen atoms. The high temperature in the chamber breaks the bonds between the carbon atoms and the hydrogen atoms in the methane molecules. This results in carbon atoms with no hydrogen atoms attached. These carbon atoms attach to the catalyst particles where they bond to other carbon atoms. This results in the formation of a nanotubes.

8B.10.3 Vaccum Sputtering Method

In this method, the target material is bombarded with a beam of inert gas ions and sputtered atoms are collected on a temperature controlled substrate. This method is utilized to fabricate multi layered nanocomposite materials. Sputtering controls the layer thickness to fractions of nano meter and can be used to make metallic or insulating sample.

8B.10.4 Radio Frequency (RF) Plasma Method

A thermal plasma can also deliver the energy necessary to cause evaporation of small micrometer size particles. The thermal plasma temperatures are in the order of 10,000 K. So that solid powder easily evaporates. A RF plasma method have been used to synthesize different nanoparticles such as oxides, carbides and nitrates of titanium and silicon.

8B.10.5 Ball-Milling Method

In this method the grain size of a powdered sample is reduced to a nanoscale by mechanical deformation created by ball-milling process. This process of nanoparticles synthesis is applicable to pure body centred cubic (bcc) structural metals, for example Zr, Hf and immiscible systems like FeAl, magnetic and catalytic nanoparticles are commonly grown by this method.

8B.10.6 Sol-Gels Method

In solutions molecules of nanometer size are dispersed and move around randomly and hence the solutions are clear. In colloids the molecules of size ranging from 20 μm to 100 μm are suspended in a solvent. When mixed with a liquid, colloids look cloudy or even milky, a colloid that is suspended in a liquid is called a sol. A suspension that keeps its shape is called a gel. Thus sol-gels are suspensions of colloids in liquids that keep their shape. Sol-gel formation occurs in different stages.

(a) Hydrolysis
(b) Condensation and polymerization of monomers to form particles
(c) Agglomeration of particles, this is followed by the formation of networks that extends throughout the liquid medium and forms a gel.

The rate of hydrolysis and condensation reactions are governed by various factors such as pH, temperature, H_2O/Si molar ratio, nature and concentration of catalyst and process of drying. Under proper conditions spherical nanoparticles are produced. Nanoparticles prepared by sol-gel method are $Si(OR)_3$ where R is alkyl groups of various types.

8B.11 Types of Nanotubes

Nanotubes can be of two types:

1. Single-walled Carbon Nanotubes (SWNT)
2. Multi-walled Carbon Nanotubes (MWNT)

8B.11.1 Single Walled Carbon Nanotubes (SWNT)

A SWNT can be thought of one atom thick layer of graphite called graphene wrapped in to a seamless cylinder (Fig. 8B.5). They are an important variety of carbon nanotubles because they exhibit electric properties which are not shared by MWNT.

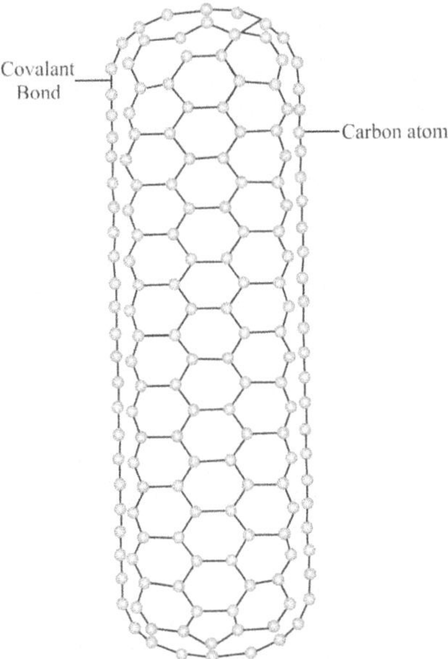

Fig. 8B.5 Carbon nanotube.

8B.11.2 Multiwalled Carbon nonotubes (MWNT)

It consists of multiple concentric nanotube cylinders i.e they consist of multiple layers of graphite rolled in on themselves to form a tube shape.

8B.12 Structure of Carbon Nanotubes

The bonding in carbon nanotubes is sp^2, with each atom joined to three neighbours similar to those in graphite structure. The tubes can therefore be considered as rolled up graphite sheets. The structure of nontubes can be specified by a chiral vector (n, m), which defined how the graphite sheet is rolled up (Fig. 8B.6) , where n and m are integers of the vector equation

$$R = n\,a_1 + m\,a_2$$

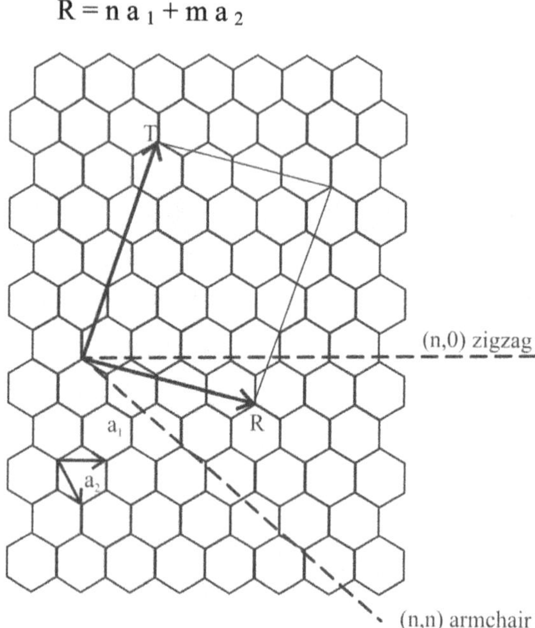

Fig. 8B.6 The (n,m) nanotube can be thought of as a vector $R = na_1 + ma_2$ in an infinite graphite sheet that describes how to "roll up" the sheet to make the nanotube. T represents the tube axis, and a_1 and a_2 are the unit vectors of graphite

1. The value of n and m determines the chirality, or twist of the nanotubes. The chirality affects the conductance, density, lattice structure and other properties.
2. A single walled C N T is considered metallic if the value of (n – m) is divisible by 3, otherwise the nanotube is semi conducting.
3. When the tubes are formed with random values of n and m, we would expect that two-third of nanotubes would be semi conducting, while the rest would be metallic.

4. The diameter of carbon nanotube can be determined by using the relationship

$$d = \left(n^2 + m^2 + n\,m\right)^{1/2} \times 0.0783 \text{ nm}$$

8B.12.1 Forms of Nanotubes

Depending upon the values of n and m, the carbon nanotubes have been divided into following categories.

(i) If m = 0 ; the nanotubes are called 'Zig Zag'

(ii) If n = 0 ; the nanotubes are called 'Armchair'

(iii) For any other combitation of n and m, nanotubes are called chiral.

Armchair, Zig Zag, and chiral form of carbon nanotubes are shown in Fig. 8B.7.

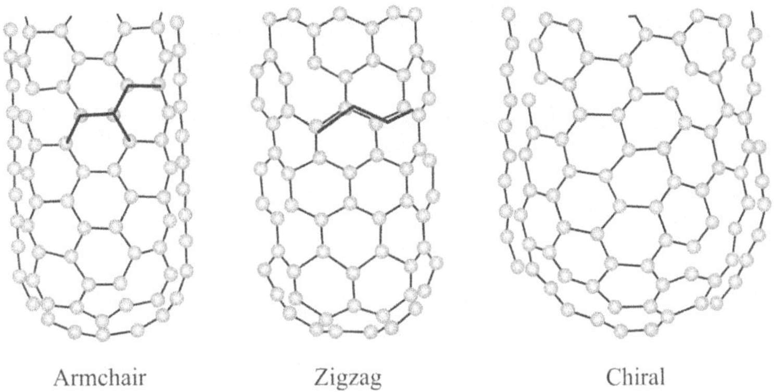

Fig. 8B.7 Armchair, Zig Zag and chiral Nanotubes.

8B.13 Properties of Nonotubes

Some of the useful and unique properties of carbon Nanotubes are as follows.

1. *Electrical conductivity:* Carbon nanotubes are metallic or semi conductors depending upon the diameter and how they are rolled Metallic nanotubes can have an electrical current density more than 1000 times greater than metals such as copper and silver.

2. *Strength:* Carbon nanotubes are the strongest materials on earth in terms of tensile strength. Note that tensile strength is a measure of the amount of force an object can withstand without tearing apart. The tensile strength of carbon nanotubes is approximately 100 times greater than that of steel of the same diameter.

3. *Stiffness:* Nanotubes are not only strong but also elastic. It takes a lot of force to bend a naotubes, but it returns to its original shape when the force is removed. Young's modulus for carbon nanotubes is about 5 times higher than that for steel.

4. *Thermal conductivity:* All nanotubes are very good thermal conductors along the tube. This property is known as "ballistic condition". Nanotubes have a thermal conductivity of more than 10 times that of silver. Nanotubes are also good insulator laterally to the tube axis.
5. Carbon nanotubes are light weight with a density of about one quarter to that of steel.
6. Carbon nantubes are a little bit sticky and forms a nonpolar molecules.

8B.14 Uses of Nanotubes

Nano tubes can be used in many applications as under:

1. Due to its unusual current conduction mechanism, wires made from nanotubes can conduct huge amount of current with less power wastage.
2. With nanotubes and nanowires, we can produce transistors and many devices about a nanometer wide. This can be used to reduce the size of the device and wires as the complexity of computer chips increases.
3. Nanotubes can produce materials with toughness unmatched by natural and man-made materials.Due to its great mechanical properties, nontubes can be used to produce every day items like clothes, sports gears, bullet proof Jackets and space suits.
4. Nano scale electric motors have also been developed using nanotubes.
5. In medical applications, the carbon nanotubes can be used as a vessel for transporting drugs in to the body. It is especially being used for treatment for cancer in destroying cancer cells.

8B.15 Bucky Balls

Bucky ball are very strong cage-like structure based on its interlocked shape. It also has the unique set of properties based on its structure. It was named Buckminster fullerene or in short "Bucky ball" after the famous American architect R.Buckminster fuller who had already designed domes in this structures.

The usual structure for this molecule is made of 60 carbon atoms, represented by C_{60} arranged in a soccer ball-like shape that is less than a nanometer in diameter. The C_{60} buckball is also known as fullerenes, is a truncated isosahedron, which is made up of 12 pentagons, and 20 hexagons of these no two pentagons share a vertex (F.g. 8B.8).

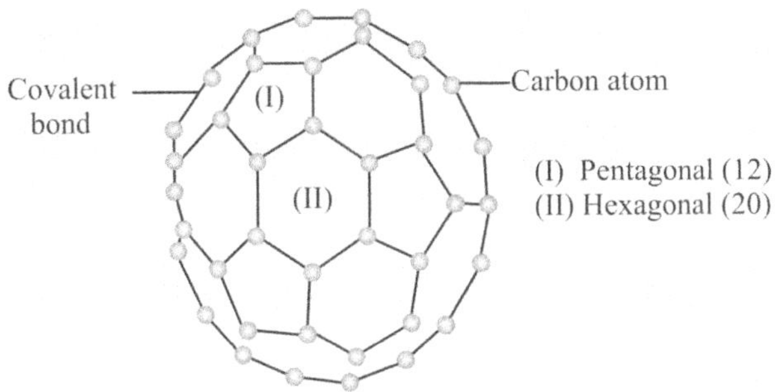

Fig. 8B.8 Bucky balls

8B.15.1 Creation of Bucky balls

Bucky balls can be created by vaporizing carbon placed between two carbon electrodes. When an electric arc is generated between two carbon electrodes placed very near to each other in a reaction chamber filled with low pressure of helium or neon, buky balls are generated along with carbon shoot. They are then separated by using solvent such as Benzene.

8B.15.2 Properties of Bucky balls

1. Bucky balls can be used in various applications because of their chemistry and unusual hollow, cage-like structure.
2. Because of spherical shape, bucky balls have extremely stable configuration which is resilient to impact and deformation i.e. if it is thrown against some object it will bounce back very much similar to soccer ball.
3. Bucky balls are also extremely stable and can withstand very high temperatures and pressures.
4. Bucky balls do not bond to one another. They do however stick together via vander waals forces.
5. The carbon atoms of bucky balls can react with atoms and molecules, leaving the stable, spherical structure still intact.
6. By doping bucky balls, they can be electrically insulating, conducting, semi conducting or even superconducting.

8B.15.3 Uses of Bucky balls

1. When a bucky ball is doped by inserting the right amount of potassium or cesium into empty spaces within the crystal, it becomes a super conductor. It is the best organic super conductor known to us.
2. Bucky balls are being used to develop stronger polymers.

3. Bucky balls are used for production of diamonds and carbides as cutting tools or hardening agent.
4. Anti ageing and anti-wrinkle creams are also being developed using bucky balls
5. The Scanning Tunnelling Microscope (STM) is one of the foremost tools in microscope which uses needle of bucky balls.
6. Attaching metals on the surface of fullerenes offers the possibility for bucky balls to become catalysts.
7. The C_{60} molecule can absorb one hydrogen atom for each carbon atom. The bucky ball structure is also not disturbed. This property suggests that bucky ball can be a better storage medium for hydrogen fuel than metal hydrides.
8. Bucky balls can deliver drug directly to the infected regions of the body. Bucky balls have the ability to act as antioxidants counter acting free radicals in the human body.

8B.16 Nanotechnology

The design, characterization, production and application of structures, devices and systems that take advantage of the unique properties at the nano scale by manipulating shape and size is known as Nanotechnology.

8B.17 Application of Nanotechnology

1. Rare earth Nanoparticles and rare earth oxide nanopowders are finding application in uses as varied enhanced fiber optic amplification (EDFA) to the removal of phosphate in the blood of patients with Hyper phosphotemia.

   Iron Nanoparticles, Iron oxide Nanopowder, cobalt Nanoparticles and several other elemental nanoparticles and alloys form a group of 'Magnetic Nanoparticles' with promising applications in medical treatment of cancer, magnetic storage and magnetic resonanace imaging (MRI).
2. Carbon nanotubes can behave like a conducting metallic or semiconductor depending on their structure, which is useful for nanoscale electronic devices.
3. Platinum Nanoparticles are being used to develop small proton exchange membrane fuel cells (PEM).
4. Ultra high purity silicon Nanoparticles are being used in new forms of solar energy cells.
5. Lithium Nanoparticles, lithium titatanate Nanopartcles and tantalum Nanoparticles will be found in next generation lithium ion batteries.
6. Nano composites are also seeing commercial use. Plastic nano composites are used for strong, lighter and rust proof car components.

Solved Numerical Problems

1. Carbon nanotubes of certain diameter are formed which have chiral vector (3, 5). Find the diameter of Carbon nanotube.

Sol.

Given n = 3

m = 5

Formula used

$$d = \left(n^2 + m^2 + nm\right)^{1/2} \times 0.0783 \text{ nm}$$

$$d = \left(3^2 + 5^2 + 3 \times 5\right)^{1/2} \times 0.0783 \text{ nm}$$

$$d = \left(9 + 25 + 15\right)^{1/2} \times 0.0783 \text{ nm}$$

$$d = \left(49\right)^{1/2} \times 0.0783 \text{ nm}$$

$$d = 0.5481 \text{ nm}$$

2. Find the ratio of diameters of carbon nanotubes having chiral vectors respectively (2, 4) and (4, 6).

Sol.

For first nanotube we have

n = 2

m = 4

∴ The diameter d_1 for the first carbon nanotube is

$$d_1 = \left(n^2 + m^2 + nm\right)^{1/2} \times 0.0783 \text{ nm}$$

$$d_1 = \left(4 + 16 + 2 \times 4\right)^{1/2} \times 0.0783 \text{ nm}$$

$$d_1 = \left(28\right)^{1/2} \times 0.0783 \text{ nm} \quad .....(1)$$

For the second carbon nanotube we have

n = 4

m = 6

Therefore the diameter of the second carbon nanotube is

$$d_2 = ((n^2 + m^2 + nm)^{1/2} \times 0.0783 \text{ nm}$$

$$d_2 = (16 + 36 + 24)^{1/2} \times 0.0783 \text{ nm}$$

$$d_2 = (76)^{1/2} \times 0.0783 \text{ nm} \qquad .....(2)$$

$$(1), (2), \Rightarrow \frac{d_1}{d_2} = \frac{(28)^{1/2}}{(76)^{1/2}} \times \frac{0.0783\text{nm}}{0.0783\text{nm}}$$

$$\therefore \frac{d_1}{d_2} = \sqrt{\frac{28}{76}}$$

$$\frac{d_1}{d_2} = \sqrt{\frac{1}{2.714}}$$

$$\frac{d_1}{d_2} = \frac{1}{\sqrt{2.714}}$$

Conceptual Questions

1. **What is Nanoscience?**

Ans: Nanoscience deals with the study of phenomena at a very small scale 10^{-9} m to 10^{-7} m where properties of matter differ significantly from those at larger scales.

2. **What is the range of the 'Nanoscale'?**

Ans. Nanoscale ranges from 1 to 100 nanometers (nm) in at least one dimension.

3. **What are nanomaterials?**

Ans. The materials with average prarticle size less than 100 nm are known as nanomaterials. Nanomateirals can be nanoscale in one dimesion e.g. surface films, Nanomaterials can be nanoscale in two dimentsion e.g. strands or fibres, Nanomaterails can be nanoscale in three dimesion e.g. particles. They can exist in single, fused, aggregated or agglomerated forms with spherical, tubular, and irregular shapes. Common types of nanomaterials include nanotubes, dendrimers, Quantum dots and buckyballs.

4. **Why are properties of nanoscale objects sometimes different than those of the same matericals at the bulk scale?**

Ans. Nanoscale materials behave very differently to the same material on a bulk scale.

There are two main reasons for this.

1. Nanomaterials have a very large surface compared to their volume.
2. On the nanoscale, Quantum effects begins to take over.

5. What are Nanotechnologies?

Ans. Nanosceinces and nanotechnologies are new approaches to research and development that aim to control the fundamental structure and behaviour of matter at the level of atoms and molecules. They offer the posiblity of furthering our understanding of phenonmena at the atomic level, and make possible the development of materials and devices with novel properties, functions and performance.

6. Write down the Advantage of AFM over SEM?

Ans. The AFM has several advantages over the scanning electron microscope (SEM)

1. Unlike the electron microscope which provides a two dimensional projection or a two dimestional image of a sample, the AFM provides a true three-dimetniosnal surface profile.
2. Samples viewed by AFM do not require any special treatments (such as metal/carbon coating) that would irreversibly change or damage the sample.
3. While an electron microscope needs an expensive vacuum environment for proper operation, most AFM modes can work perfectly well in ambient air or even a liquid environment. This makes it possible to study biological macromolecules and even living organism.

7. Write down the properties of Nanomaterials.

Ans. The properties of Nano materials are as follows.

1. Nanocrystalline materials are exceptionally storng
2. They are hard
3. The are ductile at high temperatures
4. They are wear-resistant
5. They are erosion-resistant
6. They are corrosion-resistant
7. They are chemically very active.

8. What are Nanowires?

Ans. A nanowire is a wire of dimensions of the order of a nanometer (10^{-9} meters). Nanowires can be defined as structures that have a lateral size constrained to tens of nanometers or less and an unconstrained longitudinal size.

At these scales, Quantum mechanical effects are important, hence such wires are also known as "Quantum wires".

9. Write down the types of Nanowires.

Ans. Types of Nanowires are:

1. Metallic Nanowires (e.g Ni, Pt, Au)
2. Semiconducting nanowires (e.g Si, Inp, GaN, etc)
3. Insulating nanowires (e.g SiO_2, TiO_2)
4. Molecular nanowires are composed of repeating molecular units either organic (e.g DNA) or inorganic (c.g $Mo_6 S_{8-x} I_x$, $Li_2 Mo_6 Se_6$)

10. What is Quantum dot?

Ans. When two nanowires acting as photon wave guides cross each other the junction acts as a Quantum dot.

Objective Questions

1. Properties of materials remarkably change as their size approach

   (a) Microscale (b) Verniner scale
   (c) main scale (d) nanoscale

2. Nano particles is defined as a small object that behave as a whole unit and has a size between

   (a) 1 and 100 cm (b) 1 and 100 nm
   (c) 1 and 100 μm (d) 1 and 100 mm

3. In nanotechnology, CVD (chemical vapour deposition) is a technique for the

   (a) identification of nanoprticles
   (b) synthesis of carbon nanotubes
   (c) determination of the size of nanoparticles
   (d) characterization of nanoparticles

4. Bucky ball is a spheroidal fullerene, a large molecule consisting of an empty case of
   (a) six hundered carbon atoms (b) sixty carbon atoms
   (c) six carbon atoms (d) None of the above
5. Armchair, Zig-Zag and chiral are the names of different types of
   (a) Buckyballs (b) Single walled nanotubes
   (c) Double walled nanotubes (d) multiwalled nanotubes
6. Magnetic nanoparticles are a class of nano particles which can be manipulated under the influence of
   (a) Magnetic field (b) Electrostatic field
   (c) Electric field (d) None of the above
7. C_{60} is actually a truncated isosahedron consisting of
   (a) 12 pentagons and 20 hexagons (b) 6 pentagons and 20 hexagons
   (c) 12 pentagons and 6 hexagons (d) None of the above

Theoretical Questions

1. Define nanoscience and nanotechonology.
2. How do Nanomaterials differ from their bulk counterpart.
3. Write a short note on carbon nanotubes and carbon buckyballs.
4. Give at least two procedures for synthesis of carbon nanotubes.
5. What are the three different types of carbon nanotubes.
6. Explain the various properties of nanotubes.
7. Give some uses of nanotubes.
8. What is bucky ball? How can we create a bucky ball and what are the uses of Bucky balls.
9. How are single-walled carbon nanotubes (SWNT) is different from multi-walled carbon nanotubes (MWNT).
10. What does chiral vector signify? On the basis of chiral vector, list various types of carbon nanotubes.
11. What are the various applications of nanotechonolgoy.

Numerical Problems

1. A armchair carbon nanotube (CNT) have diameter 1.35 nm. Calculate the Chiral vector for armchair CNT.

[10, 10]

2. Carbon nanotubes of certain diameter are formed which have chiral vector (6,3). Find the diameter of carbon nanotubes?

[0.621 nm]

Answers to Objective Questions

1. (d)	2. (b)	3. (b)	4. (b)	5. (b)	6. (a)
7. (a)					

APPENDIX

Fibre Optics

"I have heard a ray of light laugh and sing". We may talk by light to any visible distance without any conducting wire.

Alexander Graham Bell (1880)

Introduction

In a point to point communication link using electrical signal we use a conducting wire or a cable and the signal current is guided through the conductor cable wire. If we wish to use optical signals, then we need to have some cable or wire-type thing along which the light could travel in the desired or guided direction.

Light cannot pass through metal but can pass through a transparent glass, polymer or dielectric. Therefore optical transmission line called optical fibre, is made from any of these materials. Now questions arises why the optical fibres are made up of glass. The three important properties of glass which make it of great importance are:

1. There is a wide range of accessible temperatures where its viscosity is variable and can be well controlled unlike most materials, like water and metals which remains liquid until they are cooled down to their freezing temperatures and suddenly become solid. Glass on the other hand does not solidify at a discrete freezing temperature but gradually becomes stiffer and stiffer and eventually becoming hard. In the transition region it can be easily drawn into a thin fibre.
2. It is highly transparent. Today in most commercially available silica fibres, 96% of the power gets transmitted after propagating through 1 km of optical fibre.
3. The third most remarkable property is the intrinsic strength of glass. Its strength is about 2000.00 lb/in^2. So that a glass fibre of the type used in the telephone network and having a diameter (125 μm) of twice the thickness of a human hair can support a load of 40 lb.

Optical fibres are used in optical communication system. These systems are used in digital communications. The light pulses which are modulated according to the signal to be transmitted are guided along an optical fibre. The main advantage of optical fibre communication is its speed. The optical pulse repetition frequencies will be in the range of 1 MBPS (Mega bytes per second) to 1000 MBPS, small size, good overall compatibility, relative simplicity is the essential features of communication system.

The present day optical fibres have losses about 4.2 dB/km. By refining and purifying glasses used for fibre drawing, these losses can be reduced to 1 dB/km.

Classification of Optical Fibres

(a) Classification of optical fibres depending on material used in manufacturing of optical fibre cables.

1. Glass fibres
2. Plastic clad silica or p.c.s fibres
3. Plastic fibres

(b) Classification of optical fibres depending upon the requirement and usage.

1. ***Active fibre:*** The active fibre is one that emits light as well as guides part of it

2. ***Passive fibre:*** The passive fibre is one that guides light incident on it from an external source. The optical fibre communication system makes use of passive fibres.

3. ***Luminescent fibre:*** Luminescent fibre is one that emit luminescent radiation when excited by X-ray, ultraviolet or high energy particles.

4. ***Lasing fibre:*** Lasing fibre is one in which light amplification by stimulated emission of radiation takes place. These are small diameter fibres capable of providing higher pumping efficiencies and better mode selection.

5. **Multiple fibre:** Multiple fibre is one kind of fibre configuration that consists of multitude of smaller diameter fibres. These fibres have made possible the availability of very small fibres that are capable of high resolution.

6. ***Conical fibre:*** Conical fibre is one that is used for light condensing and aligned assemblies for magnification and demagnification purposes. In this type, conical condensing property is used.

Optical Fibres

A typical optical fibre as shown in Fig. (a) consists of:

(a) ***Core:*** Core of glass, silica or plastic with approximate diameter of 10 to100 μm with refractive index n_1.

(b) ***Cladding:*** The core is surrounded by a glass or a plastic cladding with refractive index n_2 such that $n_2 < n_1$. The difference $\Delta n = n_2 - n_1$ is typically very small of the order of 10^{-3}.

The refractive index of cladding can change either abruptly from n_1 to n_2 (a step-index fibre) as shown in Fig. (b) or gradually i.e., graded index fibre Fig. (c).

(c) ***Buffer:*** For providing safety and strength, a buffer plastic coating or housing encapsulates the core-cladding of the fibres.

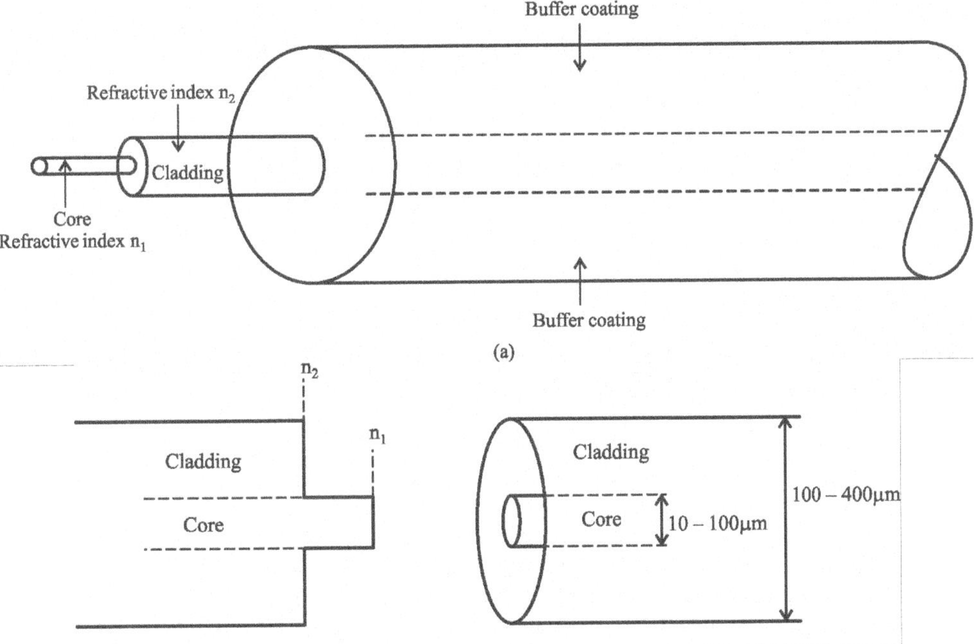

(b) Abruptly changing refractive indices at the core cladding boundary "a step-index fibre"

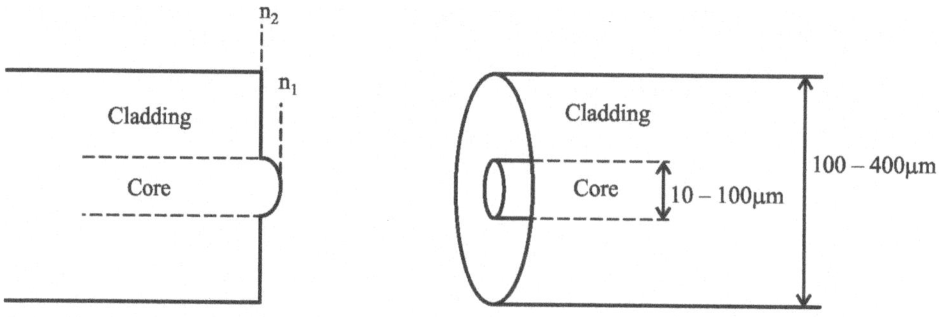

(c) Gradually changing refractive indices from core to cladding "graded index fibres"

Principle of Optical Fibres

The light launched inside the core through its one end propagates to the other end due to total internal reflection at the core and cladding interface. This is the principle of optical fibre. Total internal reflection at the fibre wall can occur only if the following two conditions are met:

1. The refractive index of the core material n_1 must be slightly higher than that of the cladding n_2 surrounding it.
2. At the core-cladding interface the angle of incidence ϕ_1 between the ray and the normal to the interface must be greater than critical angle, defined as

$$\sin \phi_c = \frac{n_2}{n_1} \qquad \text{.....(A.1)}$$

These conditions are illustrated as below.

The refractive index n is defined as the ratio of the speed of light in vacuum (C) to the speed of light in the medium (v), i.e.,

$$n = \frac{C}{v} \qquad \text{.....(A.2)}$$

The phenomenon of refraction of light at the interface between two transparent media of uniform indices of refraction is governed by Snells's law. Consider a ray of light passing from a medium of refractive index n_1 into a medium of refractive index n_2. Assume that $n_1 > n_2$ and that the angles of incidence and refraction with respect to the normal to the interface are respectively, ϕ_1, and ϕ_2 (Fig. (a)).

Then according to Snell's law

$$n_1 \sin \phi_1 = n_2 \sin \phi_2 \qquad \text{.....(A.3)}$$

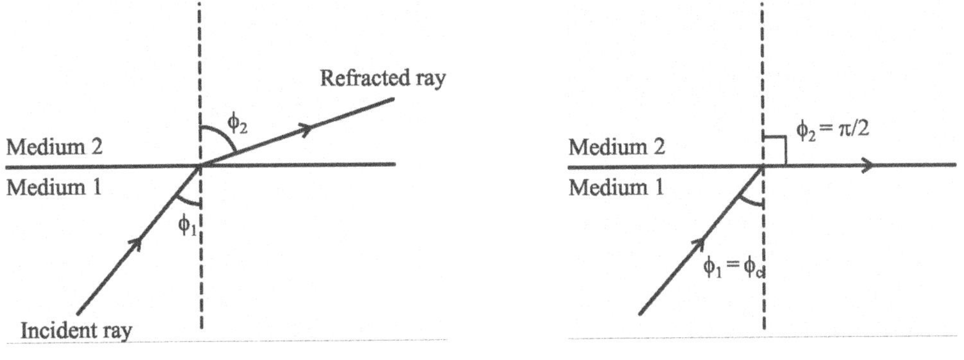

(a) Refraction of a ray of light

(b) Critical ray incident at $\phi_1 = \phi_c$ and refracted at $\phi_2 = \pi/2$

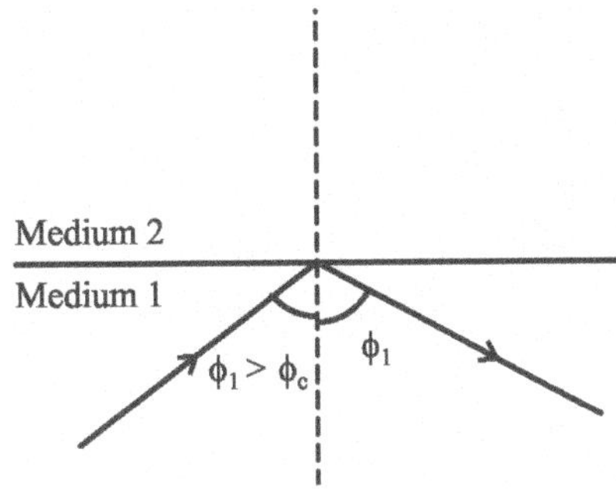

(c) Total internal reflection ($\phi_1 > \phi_c$)

As $n_1 > n_2$ if we increase the angle of incidence ϕ_1, the angle of refraction ϕ_2 will go on increasing until a critical situation is reached, when for a certain value of $\phi_1 = \phi_c$, ϕ_2 becomes equal to $\pi/2$ and the refracted ray passes along the interface (Fig.(b)). This angle $\phi_1 = \phi_c$ is called the critical angle. If we substitute the values of $\phi_1 = \phi_c$ and $\phi_2 = \pi/2$ in eq. 3 we get

$$n_1 \sin \phi_c = n_2 \sin \pi/2 = n_2$$

$$\sin \phi_c = \frac{n_2}{n_1} \qquad \text{.....(A.4)}$$

If the angle of incidence ϕ_1 is further increased beyond ϕ_c, the ray is no longer refracted but is reflected back in to the same medium (Fig.(c)). This is ideally expected. In practice however there is always some tunnelling of optical energy through this interface. The wave carrying away this energy is called the evanescent wave.

Acceptance Angle and Acceptance Cone

Not all the rays falling on the optical fibre will be transmitted through the fibre. Only those rays with sufficiently shallow grazing angle $\phi > \phi_c$ at the core-cladding interface are transmitted through the fibre.

Assume that such a fibre is placed in a medium (normally air) of refractive index n_a ($n_1 > n_2 > n_a$). If a ray of light enters from the flat end of the fibre at some angle α, from the medium in to the core, it will bend towards the normal, making an angle of refraction θ. This ray then strikes the core-cladding interface at an angle of incidence ϕ. If $\phi > \phi_c$, the ray will undergo total internal reflection. It will suffer multiple reflections at the core-cladding interface and emerge out of the fibre at the other end.

The propagation of rays by total internal reflection requires ϕ to be greater than ϕ_c and hence θ to be less than

$$\phi_m = \frac{\pi}{2} - \phi_c$$

Thus the angle of incidence α should be less than a certain angle α_m. The maximum value of α_m will correspond to the limit value of $\theta_m = \frac{\pi}{2} - \phi_c$

Applying Snell's law at the core-medium (air) interface, we get for $\alpha = \alpha_m$ and corresponding $\theta = \theta_m$

$$n_a \sin \alpha_m = n_1 \sin \theta_m = n_1 \cos \phi_c$$

for the incident ray at the core-cladding interface, we have

$$\sin \phi_c = \frac{n_2}{n_1}$$

Therefore

$$\cos \phi_c = \left[1 - \sin^2 \phi_c\right]^{1/2}$$

$$= \left[1 - \frac{n_2^2}{n_1^2}\right]^{1/2}$$

$$= \frac{\left(n_1^2 - n_2^2\right)^{1/2}}{n_1}$$

Thus

$$n_a \sin \alpha_m = n_1 \left[\frac{\left(n_1^2 - n_2^2\right)^{1/2}}{n_1}\right]$$

$$n_a \sin \alpha_m = \left(n_1^2 - n_2^2\right)^{1/2}$$

$$\sin \alpha_m = \frac{\left(n_1^2 - n_2^2\right)^{1/2}}{n_a}$$

$$\alpha_m = \sin^{-1}\left[\frac{\sqrt{n_1^2 - n_2^2}}{n_a}\right] \qquad \text{.....(A.5)}$$

The light collected and propagated by the fibre will thus depend on the value of α_m, which is fixed for a given optical fibre (n_1 and n_2 being constant). This limiting angle α_m is called the angle of acceptance or the acceptance cone half angle of the fibre. This means that all the ray incident within a cone of half-angle α_m will be collected and propagated by the fibre (Fig.(a).

Rotating the acceptance angle about the fibre axis is shown in Fig. (a), describes the acceptance cone of the fibre. Light launched at the fibre end within this acceptance cone will be accepted and propagated to the other end of the fibre by total internal reflection. Larger acceptance angles makes launching easier.

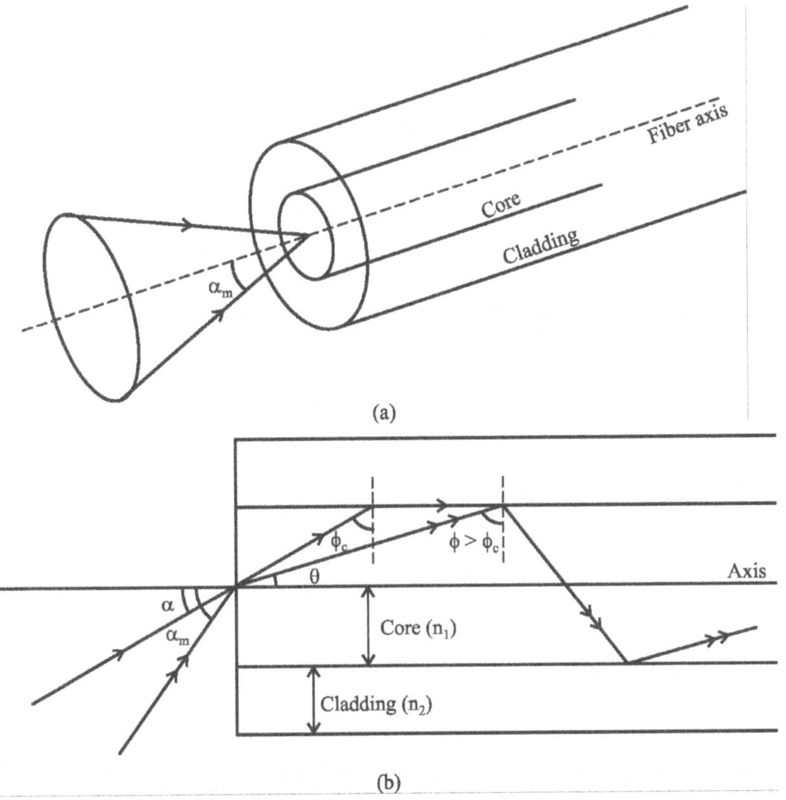

Numerical Aperture

Light collecting capacity of the fibre is expressed in terms of acceptance angle using the term "Numerical Aperture". It is given by

$$NA = n_a \sin \alpha_m = \left(n_1^2 - n_2^2\right)^{1/2} \quad .....(A.6)$$

NA can also be expressed in terms of the relative refractive index difference Δ, of the fibre, which is defined as follows.

$$\Delta = \frac{n_1^2 - n_2^2}{2 n_1} \quad .....(A.7)$$

Hence
$$NA = \left(n_1^2 - n_2^2\right)^{1/2}$$
$$NA = n_1 \sqrt{2\Delta} \quad .....(A.8)$$

The variation of NA with acceptance angle is shown in Fig. below

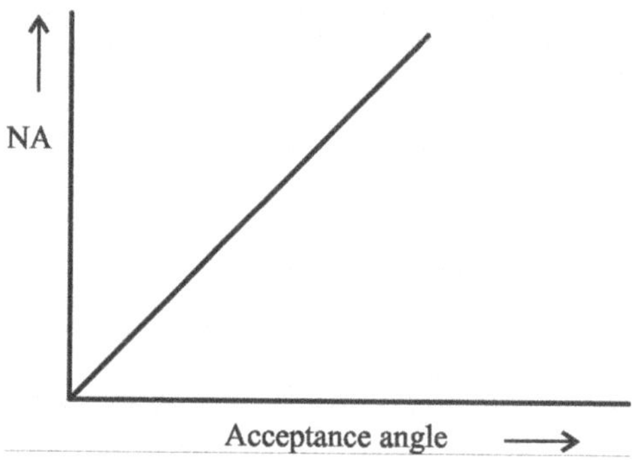

The numerical aperture is effectively dependent on the refractive indices of the core and cladding material and is not a function of the fibre dimensions.

Fig. below shows the variation of NA as a function of the ratio $\frac{n_1}{n_2}$. As the ratio $\frac{n_1}{n_2}$ increases, NA also increases.

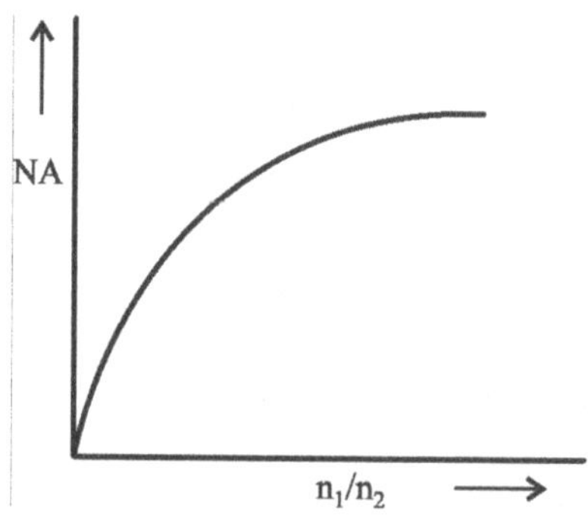

Normalised Frequency and Cut-off Wavelength of a Fibre or V-Parameter

The normalised frequency or cut-off parameter or V number or fibre characteristic term is given by

$$V = \frac{2\pi a}{\lambda}\left(n_1^2 - n_2^2\right)^{1/2} \qquad \text{.....(A.9)}$$

where a is the core radius

λ is the wavelength of light used

n_1 is the refractive index of the core

n_2 is the refractive index of the cladding

We know that

$$NA = \left(n_1^2 - n_2^2\right)^{1/2}$$

$$\Delta = \frac{n_1 - n_2}{n_1} = 1 - \frac{n_2}{n_1}$$

$$\frac{n_2}{n_1} = 1 - \Delta$$

Putting this we get

$$NA = \left\{n_1^2\left(1 - \frac{n_2^2}{n_1^2}\right)\right\}^{1/2}$$

$$= n_1\left[1 - (1-\Delta)^2\right]^{1/2}$$

$$= n_1\left[2\Delta - \Delta^2\right]^{1/2}$$

Δ^2 is very small, hence can be neglected

$$NA = n_1\,(2\Delta)^{1/2} \qquad \text{.....(A.10)}$$

Eq. (9) modifies to

$$V = \frac{2\pi a}{\lambda}.n_1\sqrt{2\Delta} \qquad \text{.....(A.11)}$$

From this expression it is clear that normalized frequency depends only on the fibre characteristic and the wavelength of light being propagated, and this normalised frequency determines the number of modes propagated along the wave guide.

To achieve single mode transmission in an idealised step index fibre, V must be less then 2.405. The wavelength at which the fibre becomes single mode is known as the cut-off wavelength, determined λ_c. Value of cut-off wavelength is given by

$$\lambda_c = \frac{2\pi a}{2.405}\left(n_1^2 - n_2^2\right)^{1/2}$$

$$\lambda_c = \frac{2\pi a\, n_1}{2.405}\sqrt{2\Delta} \qquad \text{.....(A.12)}$$

The modes and their corresponding cut-off V number is given below.

	Mode	Cut-off V number
1.	HE_{11}	0
2.	TE_{01}, TM_{01}	2.405
3.	HE_{21}	2.42
4.	HE_{12}, EH_{11}	3.83
5.	HE_{31}	3.86

Comparison of Step Index and Graded Index Fibres

1. Step index fibre has a bandwidth of about 50 MHz. Band width is the difference between maximum and minimum frequencies of electromagnetic wave that can propagate through fibre. Graded index fibres can have large (infinite) bandwidth.
2. Graded index fibres will have lower attenuation than step index fibres.
3. Step index fibres can be single mode fibres whereas graded index fibres are always multimode fibres.
4. The attenuation in graded index fibre is of the order of 5 $^{dB}/_{km}$ and the numerical aperature is 0.16 – 0.2, whereas the step index fibres of the same physical size shows a loss of 12 $^{dB}/_{km}$ and numerical aperture of 0.2 – 0.35.

Properties of different types of Optical Fibres

The Properties of the different types of optical fibres is given in Table below.

Sl.no	Factors	Single mode step index fibre	Multimode step index fibre	Multimode graded index fibre
1.	Core diameter	5-10 mm	50-100 mm	30-100 mm
2.	Cladding diameter	125 mm	125-500 mm	100-150 mm
3.	Buffer jacket	250-1000 mm	250-1000 mm	250-1000 mm
4.	Numerical aperture	0.08-0.15	0.16-0.5	0.2-3
5.	Attenuation	2-5 $^{dB}/_{km}$	2.6-5 $^{dB}/_{km}$	2-100 $^{dB}/_{km}$
6.	Band width	> 500 MHz	6-50 MHz	300-3000 MHz

Comparison of different types of Optical Fibres

The comparison of different types of fibres is shown in Table below.

Sl.no	Factors	Single mode step index fibre	Multimode step index fibre	Multimode graded index fibre
1.	Source	Laser	Laser	Laser or light emitting diode
2.	Cost	Cheap	Costly	Cheapest
3.	Coupling	Very difficult	Difficult	Easy
4.	Application	Under water cables	Date lines	Telephone lines
5.	Losses	Least	Intermediate	Maximum

Attenuation in Optical Fibres

The loss in optical fibre is defined differently depending on the type of loss described and the design of the fibre. Attenuation loss is generally measured in terms of the decibel (dB) which is a logarithmic unit. The decibel loss of optical power in a fibre is calculated through the formula

$$\text{Loss of optical power} = -10 \log \left(\frac{P_{out}}{P_{in}} \right) \text{ dB} \qquad .....(A.13)$$

where P_{out} is the power emerging out of the fibre

P_{in} is the power launched in to the fibre

Most of the manufacturers characterize attenuation loss by the number of decibels loss per kilometre of fibre. This value can be calculated as

$$\alpha = \text{loss}/\text{km} = \frac{-10}{L} \log \left(\frac{P_{out}}{P_{in}} \right) \text{ dB}/\text{km} \qquad .....(A.14)$$

where P_{out} is the power coming out of the fibre

P_{in} is the power launched in to the fibre

L is the length of the fibre tested

The following are the loss mechanism due to which the intensity of light decreases while propagating through fibre.

Absorption Losses

Three different mechanism contribute to absorption losses in glass fibres. These are ultraviolet absorption, infrared absorption and ion resonance absorption. In pure fused silica, absorption of ultraviolet radiation around 0.14 μm results in ionization of valence electrons into conduction band. Thus there is loss of light due to ionization. Also during the fabrication of fibre, to change the refractive index of the glass to any desired value, GeO_2 is doped. This causes shift in the UV absorption band towards larger wavelength region. Absorption of infrared photon by atoms within the glass molecule results in increase of random mechanical vibrations and hence heating. This infrared absorption also exhibits a main spectral peak corresponding to silicon at 8 μm, with minor peaks at 3.2, 3.8 and 4.4 μm. The peaks are broad and tail off into the visible part of the spectrum. OH^- ions are present in the material due to trapping of minute quantities of water molecules during manufacturing. The ions absorb energy at peaks of 0.95, 1.25 and 1.39 μm with the main peak at 1.39 μm, as shown in Fig. below

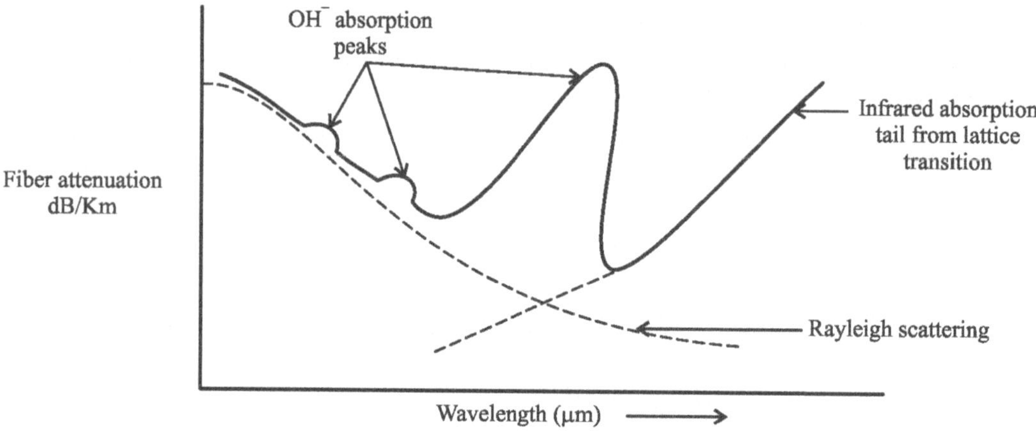

Fig. Absorption loss effect infused silica glass fibres

The presence of other impurities such as iron, copper and chromium may also create unacceptable losses within the unusable portion of the spectrum and hence extra care has to be taken while purification of silicon to avoid them.

Bending Losses

The distortion of the fibre from the ideal straight-line configuration may also result in fibre looses. Let us consider a wave front that travels perpendicular to the direction of propagation. In order to maintain this, the part of the mode which is on the outside of the bend has to travel faster than that on the inside.

Each mode extends an infinite distance into the cladding though the intensity falls exponentially. Since the refractive index of the cladding is less than that of the core, the part of the mode travelling in the cladding will attempt to travel faster. As per Einstein's theory of relativity, since part of the mode cannot travel faster, the energy associated with this particular part of the mode is lost by radiation. The loss can be represented by an absorption coefficient (α_β)

$$\alpha_\beta = C \exp\left[\frac{-R}{R_C}\right]$$

where C is constant, R is the radius of curvature of the fibre bend and R_C is given by $R_C = \frac{a}{(NA)^2}$, where a is the radius of the fibre. Since bends with radii of curvature is of the order of magnitude of the fibre which gives rise to heavy losses, it has to be avoided.

Scattering Losses

The glass material used for the optical fibres will have many microscopic inhomogeneties. They are because of the defects and changes in the structure and results in light scattering.

(a) ***Rayleigh scattering:*** It is because of small fluctuation in refractive index and changes in the density (Fig. 4A.15). These inhomogenities are fundamental and intrinsic to the glass used for fibre preparation and cannot be completely avoided.

(b) ***Mie scattering:*** This is due to the irregularities in the core-clad interface and refractive indices, diameter fluctuation, strain, bubble etc. These defects normally depend on the material, design and method and conditions of manufacturing of the optical fibres. These defects can be reduced considerably by proper control of different processes during manufacturing of the fibres.

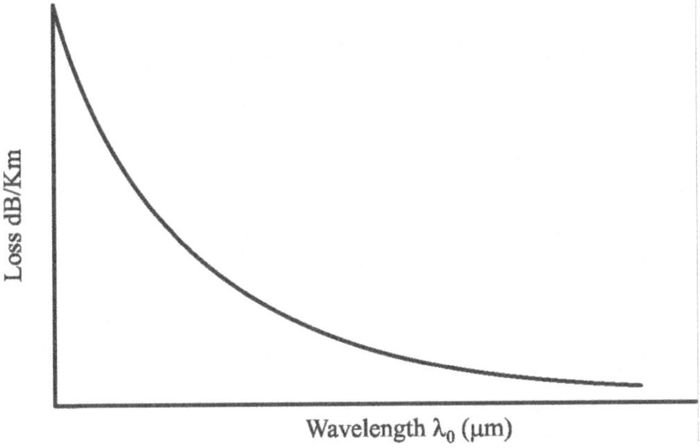

Fig. Rayleigh scattering losses in silica fibres

Dispersion in Optical Fibres

The distortion or broadening of optical signal propagating along the fibre is called the dispersion. In a digital communication system, the information to be transmitted is coded in the form of pulses. These pulses are transmitted through the optical fibre and then decoded at the receiver end. As the time required by the light to be propagated along the optical fibre of given length depends on the angle of incidence which may vary from 0 to θ_m; therefore the light pulses entering at different angles take different times; consequently the pulses broaden at the output end.

If broadening is larger, it becomes quite difficult to decode the information. This pulse broadening due to light travelling along different paths is called the pulse dispersion.

Smaller is the pulse dispersion, greater is the information carrying capacity of the fibre.

Types of Pulse Distortion

1. ***Wave guide dispersion:*** Because of wave guide structure, the light rays in the fibre follow different paths and take different times during travel of these paths. This dispersion is called waveguide dispersion.
2. ***Material dispersion or spectral dispersion:*** The refractive index of fibre core depends on wavelength (or frequency) of light, so different frequency components of input pulse travel with different velocities and hence pulse broadens. This is called the material dispersion or spectral dispersion. These two factors are present in each individual mode.
3. ***Inter modal dispersion:*** In multimode fibres, the broadening of signal is caused by the spread in group velocities (i.e., velocities with which energy of various frequencies propagates along the axis of the fibre). This dispersion is called inter mode dispersion.

   If the pulse broadening times due to waveguide dispersion, material dispersion and inter mode dispersion are $\Delta\tau_g$, $\Delta\tau_m$ and $\Delta\tau_N$ respectively, then the total dispersion is approximately given by

   $$\Delta\tau = \Delta\tau_g + \Delta\tau_m + \Delta\tau_N \qquad .....(15)$$

   In multimode fibres, the inter mode dispersion generally predominates and their relative values are such that

   $$\Delta\tau_N >> \Delta\tau_m > \Delta\tau_N \qquad .....(16)$$

   In single mode fibre, the inter mode dispersion $\Delta\tau_N$ is absent, so single mode fibres have much less dispersion and hence they can handle more information.

Solved Numerical Problems

1. Work out the critical angle for light at the boundary between the core and the cladding of a step index fibre, if the refractive index of the core is 1.48 and of the cladding is 1.45.

Sol. Here $n_1 = 1.48$

$n_2 = 1.45$

Since $\sin \phi_c = \dfrac{n_2}{n_1}$

$$\sin \phi_c = \frac{1.45}{1.48}$$

$$\sin \phi_c = 0.9797$$

$$\phi_c = 78.4^\circ$$

2. Calculate the critical angle and the angle of acceptance if the refractive indices of the glass core and the cladding are 1.7 and 1.5 respectively.

Sol. Here $n_1 = 1.7$ and $n_2 = 1.5$

The critical angle for glass core and cladding as the two media is given by

$$\sin \phi_c = \frac{n_2}{n_1} = \frac{1.5}{1.7} = 0.8824$$

or $\phi_c = 61.9^\circ$

The acceptance angle for the optical fibre is given by

$$\sin \theta_m = \sqrt{n_1^2 - n_2^2}$$

$$\sin \theta_m = \sqrt{(1.7)^2 - (1.5)^2}$$

$$\sin \theta_m = 0.8$$

$$\theta_m = 53.1^\circ$$

3. Calculate the numerical aperture and hence the acceptance angle for an optical fiber given that refractive indices of the core and the cladding are 1.45 and 1.40 respectively.

Sol. The numerical aperture is given by

$$NA = \sqrt{n_1^2 - n_2^2}$$

Given that

$n_1 = 1.45$ and $n_2 = 1.40$

$$\therefore \quad NA = \sqrt{(1.45)^2 - (1.40)^2}$$

$$NA = \sqrt{0.1425}$$

$$NA = 0.3775$$

Acceptance angle θ_m is given by

$$\sin \theta_m = \sqrt{n_1^2 - n_2^2}$$

$$\theta_m = \sin^{-1} \sqrt{n_1^2 - n_2^2}$$

$$\theta_m = \sin^{-1} (NA)$$

$$\theta_m = \sin^{-1} (0.3775)$$

$$\theta_m = \sin 22.18^o$$

4. A glass clad fiber is made with core glass of refractive index 1.5 and the cladding is doped to give a fractional index difference of 0.0005. Find

(a) The cladding index

(b) The critical internal reflection angle

(c) The external critical acceptance angle

(d) The numerical aperture

Sol. Given that $n_1 = 1.5$ and $\Delta = 0.0005$

(a) Let n_2 be the refractive index of cladding

$$\Delta = \frac{n_1 - n_2}{n_1}$$

$$0.005 = \frac{1.5 - n_2}{1.5}$$

$$\therefore \quad n_2 = 1.5 - 1.5 \times 0.0005$$

$$n_2 = 1.49925$$

(b) Let ϕ_c be the critical internal reflection angle

$$\sin \phi_c = \frac{n_2}{n_1}$$

$$\phi_c = \sin^{-1} \left[\frac{n_2}{n_1}\right] = \sin^{-1} \left[\frac{1.49925}{1.5}\right]$$

$$= \sin^{-1} (0.9995)$$

$$= 88.2^o$$

(c) Let θ_m be the external critical acceptance angle

$$\sin\theta_m = \frac{\sqrt{n_1^2 - n_2^2}}{n_a} \qquad \text{where } n_a = 1$$

$$\sin\theta_m = \sqrt{n_1^2 - n_2^2}$$

$$\theta_m = \sin^{-1}\sqrt{n_1^2 - n_2^2}$$

$$= \sin^{-1}\left[\left\{(1.5)^2 - (1.4995)^2\right\}\right]^{1/2}$$

$$= \sin^{-1}(0.0474)$$

$$= 2.72^\circ$$

(d) Numerical aperture

$$(NA) = n_1\sqrt{2\Delta}$$

$$= n_1\sqrt{2\times 0.0005}$$

$$= 1.5\sqrt{0.0010}$$

$$= 0.0474$$

5. An optical fiber has a NA of 0.20 and a cladding refractive index of 1.59 g. Determine the acceptance angle for the fiber in water which has a refractive index of 1.33.

Sol. Numerical aperture (NA)

$$= \frac{\sqrt{n_1^2 - n_2^2}}{n_a}$$

where $n_a = 1$ for air

$$0.20 = \sqrt{n_1^2 - n_2^2}$$

$$n_1 = \sqrt{(0.20)^2 + (1.59)^2}$$

$$n_1 = 1.6025$$

In water $\qquad n_w = 1.33$

$$NA = \frac{\sqrt{n_1^2 - n_2^2}}{n_w}$$

$$NA = \frac{\sqrt{(1.6025)^2 - (1.59)^2}}{1.33}$$

$$NA = 0.15$$

∴ acceptance angle

$$\sin \theta_m = NA$$

$$\theta_m = \sin^{-1}(NA)$$

$$\theta_m = \sin^{-1}(0.15)$$

$$= 8.6^o$$

6. Calculate a fractional difference between core and cladding refractive indices for a step index fibre having core and cladding refractive indices 1.55 and 1.50 respectively.

Sol. The fractional difference Δ is given by

$$\Delta = \frac{n_1 - n_2}{n_1}$$

where n_1 and n_2 are the refractive indices of core and cladding materials respectively.

Here $n_1 = 1.55$ and $n_2 = 1.50$

$$\therefore \quad \Delta = \frac{1.55 - 1.50}{1.55}$$

$$\Delta = 0.032$$

7. If the fractional difference between the core and the cladding refractive indices of a fibre is 0.0135 and numerical aperture NA is 0.2425, calculate the refractive indices of the core and cladding materials.

Sol. We know that

$$NA = n_1 \sqrt{2\Delta}$$

and

$$\Delta = \frac{n_1 - n_2}{n_1}$$

where n_1 and n_2 are the refractive indices of core and cladding materials respectively.

Here $NA = 0.2425$

and $\Delta = 0.0135$

$$n_1 = \frac{NA}{\sqrt{2\Delta}}$$

$$n_1 = \frac{0.2425}{\sqrt{2 \times 0.0135}}$$

$$n_1 = \frac{0.2425}{0.1643}$$

$$n_1 = 1.476$$

$$\Delta = 0.0135 = \frac{n_1 - n_2}{n_1}$$

$$0.0135 = \frac{1.476 - n_2}{1.476}$$

$$1.476 - n_2 = 0.0135 \times 1.476$$

$$n_2 = 1.476 - 0.02$$

$$n_2 = 1.456$$

8. The core diameter of multi mode step index is 60 μm. The difference in refractive indices is 0.013. The core refractive index is 1.46. Determine the number of guided modes when the operating wavelength is 0.75 μm.

Sol. The V-number is given by

$$V = \frac{2\pi a}{\lambda}\sqrt{n_1^2 - n_2^2}$$

$$V = \frac{2 \times 3.14 \times 30\ \mu m}{0.75\ \mu m}\sqrt{(1.46)^2 - (1.447)^2}$$

$$V = \frac{188.4 \times 0.0378}{0.75}$$

$$V = 9.493$$

The number of guided modes

$$N = \frac{V^2}{2} = \frac{90.118}{2} = 45.06$$

9. A single mode fiber is made with a core diameter of 10 μm and is coupled to a laser diode that produces 1.3 μm light. Its core glass has a refractive index of 1.55.

Find

(a) The maximum value required for the normalised index difference

(b) The refractive index required for cladding glass

(c) The fibre acceptance angle

Sol. (a) We know that

$$V = \frac{2\pi a}{\lambda}(NA) \qquad \text{Given } 2a = 10\ \mu m; \qquad \lambda = 1.3\ \mu m$$

$$(NA)_{max} = \frac{V_{max}\lambda}{2\pi a}$$

$$= \frac{2.405 \times 1.3\ \mu m}{3.14 \times 10\ \mu m}$$

$$(NA)_{max} = 0.0995$$

Normalised index difference

$$\Delta = \frac{1}{2}\left(\frac{NA}{n_1}\right)^2$$

$$\Delta = \frac{1}{2}\left(\frac{0.0995}{1.55}\right)^2$$

$$\Delta = 0.00206$$

(b) The cladding index required is

$$n_2 = n_1 (1 - \Delta)$$

$$n_2 = 1.55 (1 - 0.00206)$$

$$n_2 = 1.547$$

(c) The maximum acceptance angle for the fiber is

$$\sin \theta_m = (NA)_{max}$$

$$\theta_m = \sin^{-1} (NA)_{max}$$

$$\theta_m = \sin^{-1} (0.0995)$$

$$\theta_m = 5.710^{\circ}$$

10. A step index fibre has core refractive index 1.466, cladding refractive index 1.46. If the operating wavelength of the rays is 0.85 μm, calculate the cut-off parameter and the number of modes which the fibre will support. The diameter of the core = 50 μm.

Sol. We know that, cut-off parameter or cut-off number is

$$V = \frac{2\pi a}{\lambda}\sqrt{n_1^2 - n_2^2}$$

where a is the radius of the core, n_1 is the refractive index of the core, n_2 is the refractive index of cladding and λ operating wavelength.

Here $$a = \frac{50}{2} = 25\ \mu m$$

$$\lambda = 0.85\ \mu m$$

$$n_1 = 1.466$$

$$n_2 = 1.46$$

$$V = \frac{2 \times 3.14 \times 25}{0.85}\sqrt{(1.466)^2 - (1.46)^2}$$

$$V = 184.70\ \sqrt{2.149 - 2.131}$$

$$V = 184.70 \times 0.134$$

$$V = 24.75$$

11. The core diameter of a multimode fibre is 70 μm and the refractive index difference is of 1.5%. It operates at a wavelength of 0.85 μm. If the refractive index of the core is 1.46, calculate

(a) The refractive index of the cladding

(b) The normalised frequency V-number of the fibre

(c) The total number of guided modes in the fibre

Sol. Here $$a = \frac{70}{2} = 35\ \mu m$$

$$n_1 = 1.46$$

and $$\Delta = \frac{n_1 - n_2}{n_1} = 1.5\%$$

(a) $$\Delta = \frac{n_1 - n_2}{n_1} = \frac{1.5}{100}$$

$$n_1 - n_2 = 0.015\ n_1$$

$$n_2 = (1 - 0.015)\ n_1$$

$$n_2 = 0.985 \times 1.46$$

$$n_2 = 1.438$$

(b) Normalised frequency V-number

$$V = \frac{2\pi a}{\lambda}\sqrt{n_1^2 - n_2^2}$$

$$V = \frac{2 \times 3.15 \times 35}{0.85} \sqrt{(1.46)^2 - (1.438)^2}$$

$$V = 258.589 \sqrt{2.132 - 2.068}$$

or $$V = 258.589 \times 0.253$$

$$V = 65.42$$

(c) The total number of guided mode in the step index fibre is

$$N = \frac{V^2}{2} = \frac{(65.42)^2}{2}$$

$$N = 2139 \text{ modes}$$

12. A step index fibre has core refractive index 1.466, cladding refractive index 1.46. compute the maximum radius allowed for a fibre, if it supports only one mode at a wavelength 1300 nm.

Sol. The fibre will support only one mode if its cut-off parameter V is less than 2.405.

i.e., $$V < 2.405$$

$$\frac{2\pi a}{\lambda} \sqrt{n_1^2 - n_2^2} < 2.405$$

$$a < \frac{2.405\,\lambda}{2\pi\sqrt{n_1^2 - n_2^2}}$$

Here $\lambda = 1300 \text{ nm} = 1300 \times 10^{-9} \text{ m}$

$$= 1300 \times 10^{-3}\ \mu\text{m}$$

$n_1 = 1.466$; $n_2 = 1.46$

$\therefore$ $$a < \frac{2.405 \times 1300 \times 10^{-3}\ \mu\text{m}}{2 \times 3.14 \sqrt{(1.466)^2 - (1.46)^2}}$$

$$a < \frac{3.1265\ \mu\text{m}}{6.28\sqrt{2.149 - 2.131}}$$

$$a < \frac{3.1265\ \mu\text{m}}{6.25 \times 0.1341}$$

$$a < 3.17\ \mu\text{m}$$

The maximum radius allowed for a fibre

$$= 3.71\ \mu\text{m}$$

13. Calculate the number of modes at 820 nm in a graded index fibre having a parabolic profile a 25 μm core radius, n_1 = 1.48 and n_2 = 1.46. How does this compare to a step index fibre?

Sol. For a graded index fibre, the number of modes is given by

$$N = \frac{\alpha}{\alpha+2}\left(\frac{2\pi a}{\lambda}\right)^2 n_1^2 \Delta$$

where $$\Delta = \frac{n_1 - n_2}{n_1}$$

For a parabolic profile

$\alpha = 2$ and Here $a = 25\ \mu m$

$\lambda = 830\ nm = 0.82\ \mu m$

$n_1 = 1.48$ and $n_2 = 1.46$

$$\therefore \quad N = \frac{2}{4}\left[\frac{2\pi \times 25}{0.82}\right]^2 \times (1.48)^2 \left[\frac{1.48 - 1.46}{1.48}\right]$$

$$N = 543$$

For step index fibre, the number of modes are given by

$$N = \left(\frac{2\pi a}{\lambda}\right)^2 \frac{(n_1^2 - n_2^2)}{2}$$

$$N = \left(\frac{2\pi \times 25}{0.82}\right)^2 \times \left[\frac{(1.48)^2 - (1.46)^2}{2}\right]$$

$$N = 1079$$

This means that we can propagate in step index fibre as many as twice the number of modes as in graded index fibre.

14. A signal of 10 mw is injected in to a fibre. The outcoming signal from the other end is 40 mw. What is the attenuation loss in dB.

Sol. We know that

$$dB = -10 \log_{10}\left(\frac{P_{out}}{P_{in}}\right)$$

$$dB = -10 \log_{10}\left(\frac{40}{100}\right)$$

$$dB = -10 \log_{10}\left[\frac{2}{5}\right]$$

$$dB = -10 \left[\log_{10}{}^{2} - \log_{10}{}^{5}\right]$$

$$dB = -10\,[0.3010 - 0.6990]$$

$$dB = 3.98 \text{ dB}$$

15. An optical fibre of length 150 m has input power of 10 μw and output power of 9 μw. Compute loss in dB/km.

Sol.

$$\text{Loss in dB} = -10 \log_{10}\left(\frac{P_{out}}{P_{in}}\right)$$

$$dB = -10 \log_{10}\left(\frac{9}{10}\right)$$

$$dB = -10 \left[\log_{10}{}^{9} - \log_{10}{}^{10}\right]$$

$$dB = -10\,[0.9542 - 1] = 0.458$$

$$dB = 0.458 \text{ dB}$$

0.458 dB is the loss for 150 m length

∴ loss for 1 km is

$$= 0.458 \times \frac{1}{0.15}$$

$$= 3.053 \text{ dB/km}$$

16. A 100 km fibre is used in communication system. The fibre has a 3.0 dB/km loss. What will be the output power when the input power fed at the input of the fibre is 500 μw. Comments

Sol.

Loss per km

$$dB = -\frac{10}{L} \log_{10}\left(\frac{P_{out}}{P_{in}}\right)$$

Total loss in dB

$$= 10 \log_{10}\left(\frac{P_{out}}{P_{in}}\right)$$

$$= 100 \times (-3.0)$$

$$= -300 \text{ dB}$$

$$\log_{10}\left(\frac{P_{out}}{P_{in}}\right) = -30$$

or

$$10^{-30} = \frac{P_{out}}{P_{in}}$$

$$P_{out} = 10^{-30} \times P_{in}$$

$$P_{out} = P_{in} \times 10^{-30}$$

$$P_{out} = 10^{-30} \times 500\ \mu w$$

$$P_{out} = 5.0 \times 10^{-28}\ \mu w$$

This is very insignificant output power.

17. Attenuation can be expressed in decibels (dB) according to the equation $-10\log_{10}\left(\frac{I_{out}}{I_{in}}\right)$. Work out the attenuation in dB per km for a fibre in which the intensity falls by 50% over a distance of 50 km.

Sol. The attenuation in dB is given by

$$dB = -\,10\log_{10}\frac{I_{out}}{I_{in}}$$

Therefore, when the light falls by 50% i.e.,

when $\quad I_{out} = 0.5\ I_{in}$

$$\text{Attenuation} = -10\log_{10}\left[\frac{0.5\,I_n}{I_n}\right]$$

$$= -10\log_{10}[0.5]$$

$$= -\,10 \times (-\,0.3010)$$

$$= +\,3.01\text{ dB}$$

$$\therefore\ \text{attenuation/km} = \frac{+3.01\,\text{dB}}{50}$$

$$= +\,0.06\text{ dB/km}$$

18. The intensity of a light pulse travelling along a fibre decreases exponentially with distance according to the equation.

$$I = I_o\,e^{-\alpha x},$$

where I_o is the intensity at $x = 0$ and α is the attenuation constant. Show that the intensity reduces by 50% over a distance equal to $\frac{\log_e{}^2}{\alpha}$

Sol. The intensity of light pulse over a distance x,

$$I = I_o\,e^{-\alpha x} \qquad .....(1)$$

Suppose that the intensity of light pulse reduces by 50% over a distance n'. Thus

when $x = x'$, $I = \frac{50}{100} \times I_o$

$$I = 0.5\ I_o$$

$\therefore$ $0.5\ I_o = I_o\ e^{\alpha x'}$

or $e^{\alpha x'} = \frac{1}{0.5} = 2$

or $\alpha x' = \log_e{}^2$

or $x' = \frac{\log_e{}^2}{\alpha}$

19. The material dispersion in an optical fibre is defined by $\left|\frac{d^2n}{d\lambda^2}\right|$ is $4.0 \times 10^{-2}\ \mu m^{-2}$. Estimate the pulse broadening per km due to material dispersion within the fibre when it is illuminated with an LED source with a peak wavelength of 0.9 μm and a rms spectral width of 45 nm.

Sol. Given that

$$\left|\frac{d^2n}{d\lambda^2}\right| = 4.0 \times 10^{-2}\ \mu m^{-2}$$

$\lambda = 0.9\ \mu m$ and $\sigma_\lambda = 45$ nm

The pulse broadening due to material dispersion within the fibre is

$$\sigma_{mat} = D_{mat}(\lambda)\ L\ \sigma_\lambda$$

$$\sigma_{mat} = \frac{L}{C}\left|\lambda \frac{d^2n}{d\lambda^2}\right| \sigma_\lambda$$

$$\sigma_{mat} = \frac{1}{3 \times 10^8}\left|(900\ nm) \times \frac{4.0 \times 10^{-2}}{10^{-6}}\right| 45\ nm$$

$$\sigma_{mat} = 5.4\ ns/km$$

Objective Questions

1. The optical fibre is a guiding medium in which the information carrying
   (a) Light wave could be transmitted
   (b) Sound waves could be transmitted
   (c) Both (a) and (b)
   (d) None of these
2. The optical fibre consisting of a central core is cladded by material of
   (a) Slightly higher refractive index
   (b) Slightly lower refractive index
   (c) Equal refractive index
   (d) Very high refractive index
3. The phenomenon which takes place inside the optical fibre is
   (a) Reflection (b) Total internal reflection
   (c) Refraction (d) Transmission
4. The phenomenon of guidance by multiple total internal reflections in optical fibre was demonstrated by
   (a) Alexander Graham Bell (b) John Tyndall
   (c) Huygens (d) None of these
5. If n_1 and n_2 are the refractive indices of the core and cladding respectively. The values of n_1 and n_2 are
   (a) $n_1 \approx 1.48, n_2 \approx 1.46$ (b) $n_1 \approx 1.38, n_2 \approx 1.36$
   (b) $n_1 \approx 1.46, n_2 \approx 1.48$ (d) $n_1 \approx 1.36, n_2 \approx 1.38$
6. Numerical aperture signifies
   (a) Light reflecting power (b) Radius of curvature
   (c) Light gathering powers (d) None of these
7. The numerical aperture for optical fibre is
   (a) 14° (b) 13°
   (c) 12° (d) 11°
8. The simplest type of optical fibre in which the refractive index of the core remains constant is known as
   (a) Graded index fibre (b) Parabolic index fibre
   (c) Step index fibre (d) Simply optical fibre

9. The information carrying capacity of the fibre will be grater if
   (a) Pulse dispersion is small
   (b) Pulse dispersion is high
   (c) Pulse dispersion is neither too high nor too low
   (d) None of these

10. The pulse dispersion in step index fibre is given as
    (a) $\Delta\tau = \frac{n_1 L}{C}\left[\frac{n_2}{n_1} - 1\right]$ (b) $\Delta\tau = \frac{n_2 L}{C}\left[\frac{n_1}{n_2} - 1\right]$
    (c) $\Delta\tau = \frac{n_1 n_2 L}{C}$ (d) $\Delta\tau = \frac{n_1 L}{C}\left[\frac{n_1}{n_2} - 1\right]$
    where n_1, n_2 are refractive indices of core and cladding respectively and L is the distance travelled by a ray.

11. The ray path in parabolic index fibre is
    (a) Straight line (b) Trajectory
    (c) Sinusoidal (d) Circular

12. For the ray to be guided through a step index fibre, the following condition is satisfied $n_2 < \beta < n_1$. The value of β is
    (a) $\beta = n_1 \cos\theta_1$ (b) $\beta = n_1 \sin\theta_1$
    (c) $\beta = n_2 \cos\theta_2$ (d) $\beta = n_2 \sin\theta_2$

13. The pulse dispersion in the parabolic index fibre is
    (a) $\frac{n_1 L}{2C}\Delta^2$ (b) $\frac{n_2 L}{2c}\Delta$
    (c) $\frac{n_2 L}{2C}\Delta^2$ (d) None of these

    where $\Delta = \frac{n_1 - n_2}{n_2}$,

    n_1, n_2 are the refractive indices of the core and cladding respectively.

14. The pulse dispersion formula for the material dispersion is given by
    (a) $\Delta\tau = -\frac{\lambda_o L}{C}\frac{d^2 n}{d\lambda_o^2}\Delta\lambda_o$ (b) $\Delta\tau = -\frac{L}{C}\frac{d^2 n}{d\lambda_o^2}\Delta\lambda_o$
    (c) $\Delta\tau = -\lambda_o L\frac{d^2 n}{d\lambda_o^2}\Delta\lambda_o$ (d) $\Delta\tau = -\frac{\lambda_o L}{C}\Delta\lambda_o$
    where $\Delta\lambda_o$ is the spectral width, L distance travelled by the fibre, n is the varying refractive index.

15. The group velocity in material dispersion is given by

(a) $V_g = \frac{C}{n(\omega)}$ (b) $V_g = \frac{C}{\omega \frac{dn}{d\omega}}$

(c) $V_g = \frac{C}{n(\omega) + \omega \frac{dn}{d\omega}}$ (d) None of these

where ω is the angular frequency. N is the varying refractive index.

16. The importance of material dispersion is in

(a) Step index fibre (b) Graded index fibre

(c) Parabolic index fibre (d) Both a and b

17. In order to increase the information carrying capacity it is necessary to

(a) Increase the material dispersion

(b) Reduce the material dispersion

(c) First increase and then reduce the material dispersion

(d) None of these

18. For a 1 km length of the optical fibre, the corresponding bit rate would be about

(a) 400 M bits/second (b) 500 M bits/second

(c) 600 M bits/second (d) 700 M bits/second

19. Preferable source of light used in fiber optic communication is

(a) LED (b) Sodium lamp

(c) Cadmium red light (d) Laser

20. Angle of acceptance is maximum for a fibre if

(a) The critical angle is minimum (b) The critical angle is negative

(c) The critical angle is maximum (d) The critical angle is zero

21. Optical waveguides guides light waves by

(a) Total internal reflection of waves (b) Interference of waves

(c) Simple refraction of waves (d) Causing diffraction

22. Meridinal rays propagate along

(a) The axis of the fibre (b) Along cladding of the fibre

(c) Along the walls of the fibre (d) Both (a) and (c)

23. Optical fibres are made of
   (a) Metallic conductor
   (b) Magnetic oxides
   (c) Dielectric materials
   (d) Plastic doped with metallic impurities

24. High information carrying capability of optical fibres is measured with their
   (a) Low losses
   (b) High band width
   (c) Low costs
   (d) High efficiency

25. The total acceptance is
   (a) Half the minimum of acceptance angle
   (b) Half the maximum of acceptance angle
   (c) Twice the maximum of acceptance angle
   (d) Twice the minimum of acceptance angle

Answers to Objective Questions

1. (a)	2. (b)	3. (b)	4. (b)	5. (a)	6. (c)
7. (a)	8. (c)	9. (a)	10. (d)	11. (c)	12. (a)
13. (c)	14. (a)	15. (c)	16. (d)	17. (b)	18. (a)
19. (d)	20. (a)	21. (a)	22. (c)	23. (c)	24. (b)
25. (c)					

www.ingramcontent.com/pod-product-compliance
Lightning Source LLC
Chambersburg PA
CBHW080157170726
48282CB00005B/54

* 9 7 8 9 3 8 5 4 3 3 1 6 0 *